Prokaryotes and Evolution

Jean-Claude Bertrand • Philippe Normand
Bernard Ollivier • Télesphore Sime-Ngando
Editors

Prokaryotes and Evolution

Springer

Editors
Jean-Claude Bertrand
Unité Mixte de Service, UMS 3470,
OSU Pythéas
Aix Marseille Université
Marseille Cedex, France

Philippe Normand
Laboratoire d'Ecologie Microbienne,
UMR 5557
Université Claude Bernard Lyon 1
Villeurbanne, France

Bernard Ollivier
Aix Marseille Université, Université de
Toulon, CNRS, IRD, MIO UM 110
Marseille, France

Télesphore Sime-Ngando
Laboratoire "Microorganismes: Génome
et Environnement" (LMGE), CNRS
UMR 6023
Université Clermont Auvergne
Clermont-Ferrand, France

ISBN 978-3-319-99782-7 ISBN 978-3-319-99784-1 (eBook)
https://doi.org/10.1007/978-3-319-99784-1

Library of Congress Control Number: 2018962555

This Springer imprint is published by the registered company Springer Nature Switzerland AG
The registered company address is: Gewerbestrasse 11, 6330 Cham, Switzerland

Foreword: The Importance of Prokaryotic Evolution in the Study of Biology

Even though evolution is one of the major organizing theories in biology, it has had relatively little impact on the study of prokaryotes. Presumably, the absence of easily studied morphological features, a detailed fossil record, and readily identifiable populations created a methodological barrier that was difficult to penetrate prior to the development of modern molecular and chemical techniques. Prokaryotes also appear to be remote from the everyday. Invisible to casual observation, only specialists fully appreciate their rich evolutionary history. The result is a general lack of interest among many biologists. In fact, some biology texts and natural history museums present life's evolution on a logarithmic time scale where the last million years appear longer than the first three billion years. In this way, evolution on the early earth, a time dominated by prokaryotes, has been selectively minimized.

Nevertheless, the prokaryotes have a lot to tell biology. The early earth before the evolution of eukaryotes is the time when the most salient features of life evolved. As the first ancestors of the earliest life forms, the prokaryotes provide one of the few windows into these events. This is the time when the basic cellular design of life evolved. The capacity to energize cellular membranes with ion gradients and to use this energy to form high-energy chemical bonds, motility, and transport all evolved. The major catabolic and anabolic pathways of central metabolism, such as glycolysis and gluconeogenesis and the tricarboxylic acid cycle, evolved. The major biosynthetic pathways for the building blocks of the cell, including amino acids, nucleotides, lipids, and coenzymes, all evolved. Likewise, the basic mechanisms of information processing, including replication, transcription, and translation systems, evolved. Lastly, early prokaryotic life transformed the planet, forming the first biogeochemical cycles that maintained the early biosphere.

Most of the major prokaryotic phyla also evolved in the early earth, likely predating the origin of the eukaryotes. The modern prokaryotes likely descended from about 100 lineages that were present on earth 2 billion years ago. While some lineages appear to be represented by only a few small clades, many have undergone remarkable diversification. A major implication of this sequence of events is that eukaryotes evolved in a world fully colonized by prokaryotes and within an ecosystem comprised largely of prokaryotes. Thus, modern eukaryotes were shaped not only by the chemistry of the environment in which they evolved but also by the biology of prokaryotes.

Not only does the evolution of prokaryotes span a much longer time than the eukaryotes, the population sizes are many orders of magnitude larger. For instance, the population of humans now approaches 10^{10} individuals and that of an abundant group of insects, termites, is about 10^{17} individuals. In contrast, estimates of the number of a relatively minor gut bacterium *Escherichia coli* are about 10^{20} cells and of some of the abundant marine phototrophs are 10^{27} cells. This difference in scale regarding the population size implies that prokaryotic evolution is fundamentally different from that of eukaryotes. For instance, mutations which are rare in populations of eukaryotes are common in populations of prokaryotes. As a consequence, the opportunities for diversification are correspondingly larger.

Due to differences in scale, methodology, and familiarity, the central questions of prokaryotic evolution are fundamentally different from those of eukaryotes. This volume addresses many of them. In contrast to the major groups of eukaryotes, which are well known even outside of biology, most biologists are ignorant about many of the basic features of prokaryotic life. This question of familiarity is addressed by Normand and Caumette, who review the phylogeny, classification, and properties of prokaryotes in Chap. 2. The naming and classification of prokaryotes are fundamentally different from that used in eukaryotes. In addition, Normand and Caumette provide an overview of the properties of the 36 prokaryotic phyla with cultured representatives. The metabolic diversity is illustrated. For instance, phototrophy and lithotrophy are shown to be widely distributed. Likewise, the physical conditions able to support life span temperatures from below the freezing point to above the boiling point of water, salinities from freshwater to saturated salt solutions, and pHs from strongly acidic to strongly basic. Multicellular lifestyles and the variety of resting forms are described. There is no "typical" prokaryote, and these microorganisms possess varied and complex lifestyles.

The relationship of the prokaryotes to eukaryotes is another central question. In Chap. 1, Bertrand et al. discuss the alternative interpretations of recent phylogenomic analyses. The prokaryotes are known to comprise two phylogenetic domains, the Bacteria and Archaea. Are the Eukarya a third domain or a fundamentally different type of organism? Bertrand et al. show that the prokaryotes are a fundamentally different type of organism that are united by a large number of shared characteristics, including small size, genome structure, absence of meiosis, coupling of transcription and translation in the cytoplasm, formation of a characteristic cytoskeleton, functionally specialized cytoplasmic membranes, absence of phagocytosis to assimilate nutrients, and simple patterns of cellular differentiation. They further point out that recognition of the differences in the structure of the prokaryotic and eukaryotic cells enriches rather than contradicts the phylogenetic analysis.

Given the profound differences between prokaryotes and eukaryotes, is their evolution also fundamentally different? In Chap. 4, Bertrand et al. discuss this question in detail. First, they examine the relevance to prokaryotes of the major theories of evolution, from natural selection to neutral evolution. Of particular interest are features that might explain the success of prokaryotes and their ability to dominate the biosphere for 3.5 billion years. Particularly important seem to be their small size, efficient reproduction, short generation times, large populations, capacity for horizontal gene transfer between even distantly related lineages, formation of dormant or resting cells, and ability to colonize an enormous variety of biomes. Moreover, the study of prokaryotes has greatly enriched the theories of evolution from Darwin to the modern day.

The first cellular organisms of earth were probably prokaryotes, and the study of the origin of life is fundamental to the evolution of prokaryotes. Given the antiquity of the events and the enormous changes that have occurred on earth in the last 3.8 billion years, it is remarkable that anything can be deduced about the origin of life. In Chap. 3, Ollivier et al. discuss the state of our knowledge. Two lines of inquiry dominate. In the first approach, the properties of modern organisms inform the possibilities for ancient life. In this regard, mechanisms of lithotrophy are especially informative. Given the likely chemistry of ancient earth, what would the bioenergetics of early organisms look like? What types of enzyme catalysts might be present? In the second approach, ancient fossils and sediments are examined for evidence about ancient life. While stromatolites provide strong evidence for prokaryotic life 3 billion years ago, the evidence becomes increasingly ambiguous prior to that time.

Eukaryotes almost certainly evolved in a world dominated by prokaryotes, but the same is true of the modern prokaryotic lineages. Sime-Ngando et al. explore the consequences of this insight in Chap. 5. Most modern prokaryotes are members of communities comprised of other prokaryotes but also viruses, plants, animals, fungi, and other eukaryotes. The communities form ecosystems whose functions sustain the lives of its members. These interactions with these communities are ancient and highly complex, formed at multiple scales of biological organization, from gene to organism to ecosystem. They shape ecosystem functioning through major evolutionary forcing processes such as Red Queen dynamics, inter-actomics, molecular dialogue, host manipulation, coevolution, effects on food webs, and biogeochemical cycles. Studies on the interactions within these communities continue to provide novel insights into the evolution of these organisms.

There is also no reason to assume that prokaryotic evolution has not continued in the modern world. The enormous growth in populations of humans and domestic animals, large-scale conversion of forests and prairies to farms and pastures, and massive introduction of chemicals into the biosphere have all created novel habitats for prokaryotes. One challenge facing prokaryotic biology is to identify and understand how the prokaryotes have and will respond to these changes. As described by Wielgoss et al. in Chap. 6, one approach to this problem is experimental evolution or evolution in the laboratory. This allows detailed examination of the mechanisms of mutation, selection, and evolution. Alternative approaches examine the evolution of prokaryotes in response to well-documented modern events. For instance, the release of antibiotics in the last 80 years constitutes a "natural" experiment and makes it possible to chart the prevalence and nature of resistance in natural populations. Likewise, vascular plants colonized the continents about 400 million years ago, leading to the formation of soil and the rhizosphere habitats. The adaptation of nodulating bacteria, such as *Rhizobium*, to these new environments elucidates the capacity of prokaryotes to respond to these enormous changes. Laboratory studies can then complement observations of natural populations to examine the mechanism of this process in detail.

Lastly, life on earth is dominated by prokaryotes. If an alien came to earth from another planet and said, "Take me to your leader," it would not expect to go to Paris, Moscow, Tokyo, Beijing, or Washington. It would want to go to the soils, oceans, and sediments, all habitats densely populated by prokaryotes. Knowledge of the evolution of the prokaryotes is key to our understanding of all aspects of modern life, from its molecular biology and biochemistry to its biogeochemistry.

Department of Microbiology
University of Georgia
Athens, GA, USA

William B. Whitman

Contents

Chapter 1
Prokaryote/Eukaryote Dichotomy and *Bacteria/Archaea/Eukarya* Domains: Two Inseparable Concepts

Jean-Claude Bertrand, Pierre Caumette, Philippe Normand, Bernard Ollivier, and Télesphore Sime-Ngando

Abstract The various schemes proposed to classify microorganisms in the living world have long been subject of heated debates. The classical dichotomic distinction between *Prokaryotae* (cells without nucleus) and *Eukaryotae* (cells with nucleus) functional and phenotypic categories was deeply changed by rRNA gene-based analysis that divided the living world into three phylogenetic domains: the *Bacteria*, the *Archaea* (originally *Archaebacteria*), and the *Eukarya*. In this chapter, we review the terms of this debate between the prokaryotic/eukaryotic functional and phenotypic dichotomy and the 16S/18S phylogenetic dichotomy that separates prokaryotes into two distinct domains. The specific characteristics that emphasize the organizational and functional complexity of prokaryotes and justify maintaining this terminology are discussed. We conclude that the organizational and functional concept of a prokaryotes/eukaryotes dichotomy can be easily supplemented by the phylogenetic concept *Bacteria/Archaea/Eukarya*. The two concepts are not irrecon-

J.-C. Bertrand (✉)
Unité Mixte de Service, UMS 3470, OSU Pythéas, Aix Marseille Université, Marseille Cedex, France
e-mail: jean-claude.bertrand@univ-amu.fr

P. Caumette
Institut des Sciences Analytiques et de Physico-Chimie pour l'Environnement et les Matériaux, UMR 5254, CNRS, Université de Pau et des Pays de l'Adour, Pau, France

P. Normand
Laboratoire d'Ecologie Microbienne, UMR 5557, Université Claude Bernard Lyon 1, Villeurbanne, France

B. Ollivier
Aix Marseille Université, Université de Toulon, CNRS, IRD, MIO UM 110, Marseille, France

T. Sime-Ngando
Laboratoire "Microorganismes: Génome et Environnement" (LMGE), CNRS UMR 6023, Université Clermont Auvergne, Clermont Ferrand, France

J.-C. Bertrand et al. (eds.), *Prokaryotes and Evolution*,
https://doi.org/10.1007/978-3-319-99784-1_1

cilable but complementary, resulting in a consensual proposal that integrates both phenotypic and genotypic criteria.

Keywords Anammoxosome · *Bacteria*/*Archaea*/*Eukarya* domains · Horizontal gene transfers · Multicellularity and differentiation · Organelles in prokaryotes · Planctomycetes · Prokaryote/eukaryote dichotomy · Prokaryotic cytoskeleton · Prokaryotic membranes · Transcription-translation coupling

1.1 The Debate

The need for a classification of the living world is very ancient, and it has been felt to categorize microscopic organisms in the great living world as soon as they were discovered by van Leeuwenhoek. Various schemes have been proposed, and these have been the subject of heated debates.

Chatton was a zoologist who was expert on protozoa. He first proposed a classification of the living world into two cell types: the prokaryotes (cells without nucleus) and the eukaryotes (cells with nucleus). According to Chatton's classification, the prokaryotes comprised *Cyanophyceae*, *Bacteriaceae*, and *Spirochaetaceae* (Chatton 1925).

Stanier and van Niel (1962) adopted this classification, and after studying the "bacteria and blue-green algae," they proposed "the concept of a bacterium." They described the essential differences between the prokaryotic and eukaryotic cellular organization as follows. In contrast to the eukaryotic cell, the prokaryotic cell cytoplasm lacks components that provide specific functions (especially respiration and photosynthesis). As an example, within prokaryotes, respiratory and photosynthetic systems are not contained into a membrane-bound organelle. Moreover, the electron transfer chains which are located in the cytoplasmic membrane of bacteria are "functionally analogous to the internal membrane system of the mitochondrion," and in photosynthetic bacteria (blue-green algae or *Cyanobacteria* and purple bacteria or *Proteobacteria*), the sites of photosynthetic energy conversion are located in lamellar structures or spherical vesicles. In addition, the bacterial nucleus is not separated from the surrounding cytoplasm by a nuclear membrane. Bacteria contain a single chromosome which is normally circular except for *Streptomyces* where it is linear. The same authors pointed out that if the prokaryotic gene transfer allows the exchange of information between cells, this process cannot be considered equivalent to the sexual process that exists in eukaryotic organisms. Motility also differs. While there are flagella, axial filaments, and gliding movement in prokaryotes, eukaryotes possess a kinetic apparatus which contains invariably 11 fibrils.

Finally, a detailed study of the chemical composition of the cell wall of prokaryotes was given. The strength of the wall is provided by a mucopeptide which synthesis is inhibited in the presence of penicillin resulting under isotonic conditions in the formation of spherical cell structures called spheroplasts. For Stanier and van Niel, "it thus seems probable that the existence of the mucopeptide in the wall constitutes a supplementary specific character for the definition of the prokaryotic cell."

The criteria proposed by these authors for defining the eukaryotic cell were not questioned; however, it turned out that this definition was incomplete. For instance, it is known now that the presence of a mucopetide in the cell wall of prokaryotes is "exclusive," and not "universal" as stated by these authors.

In 1977, Woese and Fox (Woese and Fox 1977) deeply changed our vision of the living world. A molecular analysis based on the 16S rRNA gene sequence, permitted to divide the living world into three domains: the *Bacteria*, the *Archaea* (originally *Archaebacteria*), and the *Eukarya* (cf. Chap. 2 "Phylogeny and Biodiversity of Prokaryotes"). However, it is noteworthy that Woese's subdivision did not touch on the criteria of the presence or absence of a nucleus in cells. Woese's works provided evidence, for the first time, of a new evolutionary view of the living word. "Prokaryotes" were shown to be polyphyletic and also that *Archaea* shared many genes with the *Eukarya*. However, if *Eukarya* contained archaeal genes (mostly informational genes), they also contained bacterial genes (mostly operational genes).

In addition, the *Archaea-Eukarya* relationships concern only a small part of the eukaryotic genome. However, *Eukarya* only possess genetic informations allowing sexual reproduction and generating multicellular organisms, recognized by Darwin as the "most beautiful and most wonderful" forms of the living world.

According to their molecular studies, Woese and Fox rejected a division of the living world in prokaryotic and eukaryotic as reported in their original article: "Dividing the living world into Prokaryotae and Eukaryotae has served, if anything, to obscure the problem of what extant groupings represent the various primeval branches from the common line of descent. The reason is that Eukaryote/Prokaryote is not primarily a phylogenetic distinction, although it is generally treated so." Woese and his colleagues were right to claim that prokaryote/eukaryote dichotomy was not a phylogenic dichotomy, which in evolutionary terms, it was meaningless. Nevertheless, we have to keep in mind that *the concept of a bacterium*, as proposed by Stanier and van Niel, has been established earlier based on morphological and functional features to describe a biological organization, with no aim to provide a reliable phylogenetic classification of the living world. It should be also emphasized that these authors did not intend, at that time, to present prokaryotes as a monophyletic group and even that phylogeny was beyond the reach of the current experimental methods given the lack of reliable markers.

Taking into account all these phenotypic and phylogenetic considerations as reported by Stanier and van Niel, and Woese and Fox, respectively, Pace (2006) wrote one article entitled "Time for a change" in which he claimed that the distinction between prokaryotes and eukaryotes was "obsolete", "based on morphological subjectivities" and that "no one can define what is a Prokaryote, only what is not." He concluded that maintaining such a model leads to teach "wrong idea."

That same year, Martin and Koonin promptly reacted to these criticisms, by defending the concept of prokaryotes and enunciating the positive characteristics that define them. Dolan and Margulis (2007) immediately provided arguments in favor of the validation of the use of the term "prokaryotic." De Duve (De Duve 2007) argued in the same manner in an essay on the origin of eukaryotes, defending the term prokaryote.

Later on, Whitman (2009) established "The modern concept of the Prokaryotes," admitting the existence of the prokaryote/eukaryote dichotomy by writing "the prokaryotic-eukaryotic dichotomy was founded upon the recognition of two very different types of cellular organisms and not the phylogenetic relationships between them." The author highlighted the main phenotypic and physiological features related to prokaryotes that could explain their evolutionary success. They include morphology, cell size, metabolism, physiological diversity, cell division, linkage of transcription-translation with transcriptional regulation coupled to metabolism, multifunctionality of the cytoplasmic membrane in which are imbedded the transporters involved in energy conservation (respiration, photosynthesis), specific nutrient uptake (extracellular digestion of macromolecules), etc. A major difference between *Archaea* and *Bacteria* was their lipid composition and cell wall organization that he considered to be significant traits to maintain the *Bacteria/Archaea* dichotomy (Bertrand et al. 2015). However there is evidence that besides ester lipids, some members of diverse phyla within the domain *Bacteria* (e.g., *Thermotogae*, *Acidobacteria*) contain minor amounts of ether lipids in their membrane (DeRosa et al. 1988; Sinninghe et al. 2007; Sinninghe et al. 2014). Because of the presence of both these lipids in the order *Thermotogales* (e.g., *Thermotoga* spp.), a deep branching lineage within the phylogenetic tree, it was suggested that their production developed early during microbial evolution (Sinninghe et al. 2007).

Nevertheless, maintaining the prokaryote/eukaryote dichotomy was immediately rejected by Pace (2009b) in an article entitled "Rebuttal: the modern concept of the Prokaryote" where he took up the arguments first developed in 2006 (Pace 2006). He emphasized the fact that the criteria used to describe prokaryotes were arbitrary and that the concept of prokaryote "is discredited by the evolutionary data" (Pace 2009b). He further concluded that the concept of the prokaryote whatever it is, either "traditional" or "modern", was "misleading" and had to "disappear from our textbooks and language" (Pace 2009a).

In the same article (Pace 2009a), Pace rather recommended the term "microbe," giving to the latter the following terminology: "which captures the microbial quality of size and includes, as well, the poorly acknowledged microbial Eukaryotes." This proposal leads to one major remark, notably the difficulty to gather "microbes" in size since we may have microorganisms pertaining to the three domains with similar sizes but exhibiting fundamental differences in cellular organization, ecological role, metabolic versatility, and a different evolutionary history (Massana and Logares 2013).

With regard to what is mentioned above, it seems that the reconciliation of those who defend the term *prokaryotic* with those who do not is impossible. However, the discoveries made in microbiology during the last decades remain in agreement with the phylogenetic *Bacteria/Archaea* dichotomy as proposed by Woese and colleagues, together with the early definition by Stanier and Van Niel of prokaryote/eukaryote dichotomy based on structure and function.

It was established a long time afterward that the 16S-/18S-derived phylogeny established a phylogenetic dichotomy between prokaryotes and eukaryotes that matched the former phenotypic and biochemical differences.

1.2 Current Characterization of Prokaryotes

The many characteristics that emphasize the organizational and functional complexity of prokaryotes and justify maintaining this terminology will now be discussed. Concerning the dichotomy *Bacteria/Archaea*, cf. Chap. 2 "Phylogeny and Biodiversity of Prokaryotes."

1.2.1 Size of Cells

Typically, the size of a eukaryotic cell is considerably larger than that of a prokaryotic cell: >10μm and <5μm, respectively. However, some prokaryotic cells may have larger size, such as *Thiomargarita namibiensis* (size, 750 μm; cell volume 200,000 μm^3), *Epulopiscium fishelsoni* (600 micrometers by 80), *Oscillatoria* (7 μm in diameter), *Prochloron* (8 to 30 μm), etc.

In contrast, some *Bacteria* and *Archaea*, called "nanobacteria," whose cells are less than 0.2 μm in diameter and a volume of less than 0.1 μm^3, were isolated from different environments, for instance, *Candidatus* Pelagibacter ubique, with a cell volume of about 0.01 μm^3 (Rappe et al. 2002); in the same way, Candidatus Actinomarina minuta has an average cell ca. 0.013 μm^3 (Ghai et al. 2013). Nanobacteria kept their size independently of the growth conditions, even in nutrient-rich medium. Small size increases S/V ratio and gives several advantages compared to all other living organisms: nutrient absorption and diffusion of molecules in the cell that are more effective, protection against predators, and occupation of microenvironments.

If the size of eukaryotic cells is most commonly greater than prokaryotic cell, some eukaryotes (picoeukaryotes) are as small as prokaryotes (Vaulot et al. 2008); for example, the mean cell size of *Ostreococcus tauri* is 0.97 ± 0.28 in length and 0.70 ± 0.17 in width (Courties et al. 1994; Derelle et al. 2006).

1.2.2 Genome

In contrast to prokaryotic cells, eukaryotic cells contain multiple linear chromosomes associated with histones; absent in *Bacteria*, homologous proteins to histones of *Eukarya* are present in *Archaea* (Pereira and Reeve 1998). While prokaryotes typically contain a single chromosome together with several megaplasmids (Orlandini et al. 2014), some of them were shown to contain two chromosomes (e.g., *Rhodobacter sphaeroides*) and even more as reported for some aerobic and anaerobic members within the *Euryarchaea* (Soppa 2014).

The bacterial chromosome is generally circular, sometimes linear (*Borrelia burgdorferi*, *Streptomyces coelicolor*). Chromosome is always circular in *Archaea*.

In addition to the chromosome(s), *Bacteria* and *Archaea* contain extrachromosomal genetic elements known as plasmids. The occurrence of plasmids is very rare in *Eukarya* (Bapteste et al. 2009).

1.2.3 Sexuality

Cell division in prokaryotes is usually performed by binary fission; in eukaryotes, cell division involves mitosis. In the overwhelming majority of cases, reproduction is sexual (meiosis) in eukaryotes and always asexual in prokaryotes. Two other basic differences exist between eukaryotes and prokaryotes where introns are absent and a large fraction of genes are clustered in operons.

Concerning replication, all *Bacteria* duplicate their chromosome bidirectionally from a single replication origin (*ori*C). By contrast in the *Archaea* and *Eukarya*, chromosome replication is initiated at multiple sites. For instance, *Sulfolobus acidocaldarius* and *Sulfolobus solfataricus*, belonging to the Crenarchaeota phylum within the *Archaea*, both contain three origins of replication in their single chromosome (Lundgren et al. 2004).

Besides the vertical transmission of DNA (from parents to offsprings), organisms can exchange genes by horizontal gene transfer (HGT or LDT). This horizontal gene transfer between free-living organisms (endosymbiotic gene transfer is excluded) is particularly important in prokaryotes (Brochier-Armanet and Moreira 2015). A noteworthy fact, in prokaryotes, is the ability to transfer genetic information by HGT between (1) unrelated species, (2) organisms belonging to two prokaryotes domains (e.g., reverse gyrase), and (3) dead cell and living cell (transformation). HGT is a phenomenon certainly marginal in eukaryote (Bapteste et al. 2009).

1.2.4 Transcription-Translation Coupling in Prokaryotes

In eukaryotes, DNA replication and transcription take place in the nucleus, while translation takes place in the cytoplasm. Thus, the translation of the genetic message is topologically separated from their transcription. In contrast, prokaryotic transcription and translation are not spatially separated and are tightly linked. However, in very rare cases, these two processes can be decoupled. For example, Lewis et al. (2000) demonstrated that in *Bacillus subtilis*, the RNA polymerase is concentrated within the nucleoid and that the ribosomes are distributed on the cell periphery. In the same way, Nevo-Dinur et al. (2011) observed migration of some mRNA from nucleoid to other sites of the cells where they are translated in *E. coli*. Spatial transcription-translation-segregation was also demonstrated in *Gemmata obscuriglobus* which has an extensive endomembrane network. By using immunofluorescence and immunoelectron microscopy, the authors have shown that a substantial proportion of active protein synthesis (using anti-EF-Tu antibodies) take place in peripheral ribosomes, in regions distant from nucleoid (Gottshall et al. 2014). In spite of these exceptions, coupling transcription-translation remains the norm among prokaryotes.

1.2.5 Prokaryotes Possess a Cytoskeleton

For a long time, prokaryotes were considered organisms without a cytoskeleton, a feature used to distinguish prokaryotes from eukaryotes. However, during the last few decades, it has been demonstrated that, in addition to a rigid cell wall ("exoskeleton"), prokaryotes, with the exception of *Mollicutes*, possess structures homologous to those which constitute the cytoskeleton of eukaryotes (e.g., actin, tubulin, and intermediate filament proteins). It was a characteristic unknown by Stanier. These prokaryote counterparts exhibit considerable diversity at the structural and functional level. In addition to actin, tubulin, and intermediate filament proteins, prokaryotes may have specific cytoskeletal proteins, with no eukaryotic homologs. Prokaryotic cytoskeletal proteins are involved in cell division, DNA segregation, cell morphogenesis, and motility (cf. Graumann 2007; Cabeen and Jacobs-Wagner 2010; Ingerson-Mahar and Gitai 2012; Lin and Thanbichler 2013; Ozyamak et al. 2013; Cho 2015; Duggin et al. 2015). The occurrence of cytoskeleton was a prokaryotic "invention" and has been, therefore, an early event in the history of life on Earth. Here below, the main prokaryotic cytoskeletal elements will be described.

1.2.5.1 Prokaryotic Tubulin Homologs

The most studied is FtsZ protein – present in most *Bacteria* and several *Archaea* – which plays a central role in cell division (*Lord of the ring* (Cabeen and Jacobs-Wagner 2010)). FtsZ is also present in mitochondria and chloroplasts, thus providing further evidence of the evolutionary relationship between these organelles and prokaryotes. During cell division, FtsZ proteins polymerize to form a ring at the middle of the cell (the Z-ring) and attach other division proteins to form the divisome (Fig. 1.1a). The divisome is a set of proteins that assemble into the region where the septum takes shapes and directs cell division processes. In *E. coli*, ten proteins are recruited to form the ring (FtsZ, FtsA, ZipA, FtsK, FtsQ, FtsL, FtsB, FtsW, FtsI, FtsN). The FtsZ ring constricts during the division process (Fig. 1.1b), and the result leads to the formation of separate newborn daughter cells (Fig. 1.1c). FtsA and ZipA anchor the ring into the cytoplasmic membrane. Another bacterial tubulin, TubZ, is implicated in plasmid segregation (*Clostridium* and *Bacillus* species). BtubA and BtubB proteins (probably a result of HGT), present in *Prosthecobacter* species, have very similar structures to tubulin with function remaining unknown so far. In *Archaea*, CetZ proteins (CetZ1 and CetZ2) control cell shape.

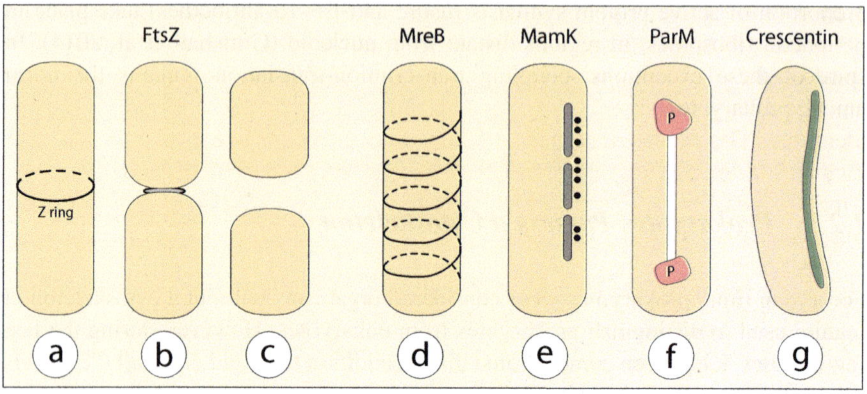

Fig. 1.1 Scheme and localization of some prokaryotic cytoskeletal elements.
FtsZ, forming a ring in the middle of the cell (**a**, **b**, **c**); MreB, forming a helical filament around inside the cell, below the cytoplasmic membrane (**d**); MamK, ensuring organization of magnetosomes (are indicated in black points) into linear chain (**e**); ParM, mediating plasmid (P) segregation (**f**); Crescentin (as indicated in green), impacting the inner curvature of the cell (**g**). (Modified and redrawn from Ozyamak et al. 2013; Graumann 2007; Cho 2015; Margolin 2005)

1.2.5.2 Prokaryotic Actin Homologs

The best-studied are MreB, ParM, AlfA, MamK, FtsA, and Alp7A. In rod-shaped bacteria, MreB proteins (*masters of morphogenesis* (Cabeen and Jacobs-Wagner 2010)) have as main function to control the cell shape and activity of cell wall biosynthetic enzymes. As a result, depletion of these proteins induces the formation of rounded cells (coccus shape). MreB is missing in cocci. This protein forms helical filaments that encircle the cell beneath the cytoplasmic membrane (Fig. 1.1d). MreB is involved in other functions (e.g., elongation of polar stalks in *Caulobacter crescentus*, motility in *Myxococcus xanthus*, normal sporulation in *Streptomyces coelicolor*). In *Magnetospirillum magneticum*, a protein called MamK (*Making a compass needle* (Cabeen and Jacobs-Wagner 2010)) is required for the linear arrangement of magnetosomes (Fig. 1.1e). The other proteins, ParM, AlfA, and Alp7A, are involved in plasmid partitioning (Fig. 1.1f). FtsA interacts with FtsZ for Z-ring formation.

1.2.5.3 Bacterial Intermediate Filament-Like Proteins

In *Caulobacter crescentus*, crescentin (CreS) is a protein which ensures, with MreB, the curved shape of the cell (Fig. 1.1g). In the absence of crescentin, the cell looks like a rod. Another bifunctional protein, the metabolic enzyme CtpS (CTP synthase enzyme), is also involved in the cell shape, independently of its enzymatic activity (Ingerson-Mahar and Gitai 2012). Other intermediate filament-like proteins were identified. They include Scc and CfpA in spirochetes, AglZ in *Myxococcus xanthus*, and in *Streptomyces coelicolor*.

1.2.5.4 Prokaryote-Specific Cytoskeletal Proteins

In addition to the "canonical" cytoskeleton proteins – tubulin, actin, and intermediate filaments – various cytoskeleton proteins, specific to prokaryotes, have been identified. The main components appeared during the evolution of prokaryotes including Walker A cytoskeletal ATPases (WACAs), bactofilins, ESCRT system, CtpS, CCRPs, DivIVA (targeting of proteins to septa or cell poles), PopZ (targeting of proteins to cell poles), etc.

WACAs, found in *Bacteria* and *Archaea*, include ParA which mediates plasmid and chromosome segregation (Cho 2015) and Min system (Ingerson-Mahar and Gitai 2012). Min system (MinC, MinD, MinE proteins) is an important inhibitor system, essential for the precise position of the FtsZ ring, to make sure that the FtsZ ring and divisome complex form only at the cell center and not at the cell poles. Bactofilins are widespread proteins in *Bacteria* and are involved in cell shape (*Caulobacter crescentus*, *Myxococcus xanthus*) (Ingerson-Mahar and Gitai 2012). Bactofilin is responsible for maintaining the helical cell shape of *Helicobacter pylori*. In some Crenarchaeota (e.g., *Sulfolobus*), which lack FtsZ homologs, the archaeal ESCRT system may mediate cell division (Samson et al. 2008) (Ingerson-Mahar and Gitai 2012). CCRPs are proteins involved in cell shape; four CCRPs have been identified in *Helicobacter pylori* (Ccrp58, Ccrp59, Ccrp1143, Ccrp1142) (Lin and Thanbichler 2013). In *Spiroplasma melliferum*, three major putative structural proteins have been identified; these have been called Fib (the major component), MreB, and, an elongation factor, Tu (Trachtenberg et al. 2008).

1.2.6 Cytoplasmic Membrane

1.2.6.1 Membrane Lipids

Prokaryotic and eukaryotic membranes consist of lipid and protein bilayers, whereas in hyperthermophilic *Archaea*, the basic structure is a monolayer structure (Bertrand et al. 2015). One of the major differences between prokaryotic and eukaryotic membranes is that eukaryotic membranes contain sterols which are usually absent in prokaryotes, except in few cases (methanotrophic bacteria and the mycoplasma are major exceptions). Instead of sterols, many bacterial membranes contain hopanoids (pentacyclic triterpenoids) which play a role similar to that of sterols in eukaryotic cells. Membranes of *Bacteria* and *Eukarya* are made of acetogenic lipids, generally composed of fatty acids ester-linked to a *sn*-1,2 glycerol, whereas *Archaea* biosynthesize isoprenoid lipids, composed of isoprenoid alkyl chains ether-linked to a *sn*-2,3 glycerol. These chemical distinctions between the *Eukarya*/*Bacteria* on the one hand and the *Archaea* on the other hand are not without exceptions (except for the stereochemistry of the glycerol backbone). Indeed, some *Eukarya* and *Bacteria* possess non-isoprenoid ether-linked lipids which exhibit an intriguing combination of structural characteristics of lipids of *Bacteria*/*Eukarya* and *Archaea* (Grossi et al. 2015).

1.2.6.2 Function

Prokaryotic cytoplasmic membrane plays several functions including permeability barrier, nutrient capture by membrane transport systems, sites of respiration and photosynthesis, lipid synthesis, and synthesis of components of the cell wall.

Furthermore, membrane invaginations are present in many prokaryotes. In *anoxygenic phototrophic bacteria*, photosynthetic apparatus is linked to cytoplasmic membrane. In anoxygenic phototrophic purple bacteria, photosynthetic systems are located in characteristic invaginations of cytoplasmic membrane, whose shape varies depending on the species (shapes of vesicles, lamellae, tubules, etc.) (Fig. 1.2a–e). In anoxygenic phototrophic green bacteria, photosynthetic pigments are located in structures called chlorosomes which are attached to the inner surface of the cytoplasmic membrane; the reaction centers are inserted in the cytoplasmic membrane (Petersen et al. 2010) (Fig. 1.2f). The presence of invaginations of cytoplasmic membrane and chlorosomes allows an increase in photosynthetic efficiency (Murat et al. 2010).

Some aquatic bacteria, called **magnetotactic bacteria** (e.g., *Aquaspirillum magnetotacticum*, *Magnetospirillum magnetotacticum*), transform extracellular iron to crystals of magnetite and build real "intracellular magnetic compasses," called magnetosomes. Chains of magnetosomes are membrane-bound (Murat et al. 2010; Saier and Bogdanov 2013).

In **nitrifying prokaryotes** (oxidation of ammonium to nitrate), many species have internal membranes on which are located key enzymes of nitrification (ammonia monooxygenase and nitrite oxidoreductase) (van Gool 1972).

Methane-oxidizing bacteria form an intracytoplasmic membrane system when they are grown in the presence of methane and methanol (Hanson and Hanson 1996). Depending on the organization of their internal membrane, methanotrophs were classified into three categories: type I (e.g., *Methylomonas*, *Methylobacter*, *Methylomicrobium*, *Methylosphaera*, *Methylovulum*, *Methylomagnum*), type II (e.g., *Methylosinus*, *Methylocystis*), and type X (e.g., *Methylococcus capsulatus*). In type I, membranes form parallel layers, filling most of the space within the cell, and are more or less perpendicular to the cell wall (De Boer and Hazeu 1972). Type I methanotrophs possess methane monooxygenase (pMMO) and use the ribulose monophosphate cycle (RuMP). Within members of the type II, the intracytoplasmic membrane is arranged parallel to the cell wall (Best and Higgins 1981). They all possess pMMO and use the serine cycle. They have often a soluble methane monooxygenase (sMMO). Representatives of type X have stacked membranes, use RuMP cycle, have elements of serine cycle, and possess sMMO.

1.2.7 Organelles in Prokaryotes

Organelle designates a membrane-enclosed structure which contains components required for a specific cell function. In contrast to eukaryotes, typically, prokaryotes do not contain internal organelles. However, such structures exist in some

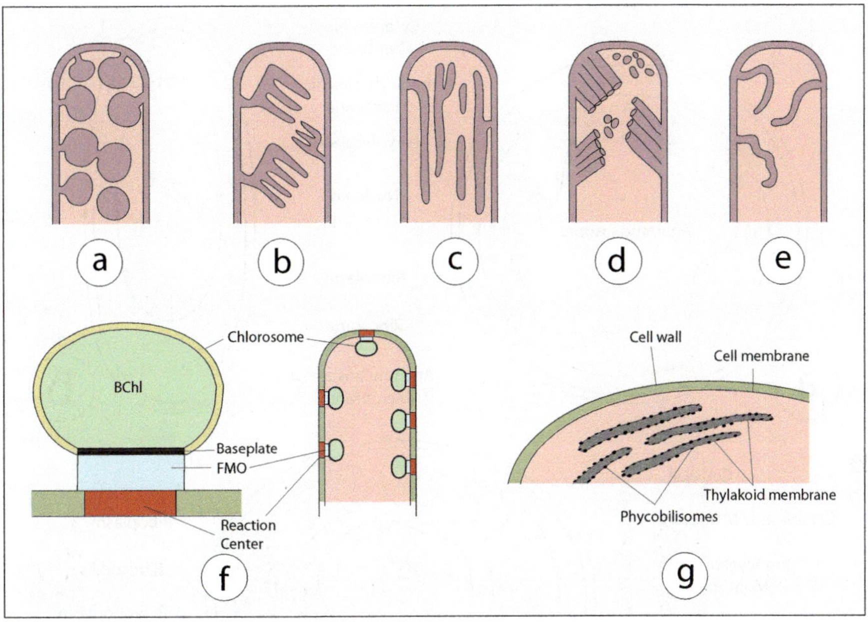

Fig. 1.2 Arrangements of cytoplasmic membranes invaginations in different genera and species of anoxygenic phototrophic purple bacteria: (**a**) vesicles (e.g., *Chromatium*, *Thiocystis*, *Lamprocystis*, *Thiocapsa roseopersicina*); (**b**) stacks (e.g., *Ectothiorhodospira*, *Rhodospirillum fulvum*); (**c**) membranes (e.g., *Rhodopseudomonas palustris*, *Rhodomicrobium vannielii*, etc.); (**d**) bundled tubes (e.g., *Thiocapsa pfennigii*); (**e**) tubes (e.g., *Rhodospirillum tenue*, *Rhodopseudomonas gelatinosa*)
Schematic model of the photosynthetic apparatus of anoxygenic green sulfur bacteria (**f**). The cholorosome contains bacteriochlorophyll c, d, or e (BChl), molecules surrounded by a protein-lipid monolayer. The energy is transmitted from bacteria chlorophyll through baseplate, through Fenna-Matthews-Olson (FMO) proteins, and finally to reaction center
Photosynthetic apparatus in cyanobacteria: thylakoids. Phycobilisomes, granular structures, high-harvesting complex, are attached to this thylakoid membrane (**g**). (Modified and redrawn from Liberton et al. 2006; Nevo et al. 2007; Murat et al. 2010; Petersen et al. 2010)

prokaryotes. For example, *Cyanobacteria* are characterized by an internal complexity. Photosynthesis and respiration take place in organelles, the thylakoids (except in *Gloeobacter violaceus* that does not have them). The thylakoids are stacked interconnected membranes (Nevo et al. 2007), which are most often arranged in parallel to the cytoplasmic membrane but without connection with it (Liberton et al. 2006; Murat et al. 2010) (Fig. 1.2g).

The **anammoxosome**, an organelle, is present in some anaerobic chemoautotrophic bacteria belonging to the family of *Planctomycetaceae* in the phylum of *Planctomycetes*: the anammox bacteria (van Niftrik and Jetten 2012) (Fig. 1.3a). These organisms have a characteristic cellular organization comprising a wall, a cytoplasmic membrane (outermost membrane), and an intracytoplasmic membrane. The intracytoplasmic membrane surrounds a compartment that contains the ribosomes

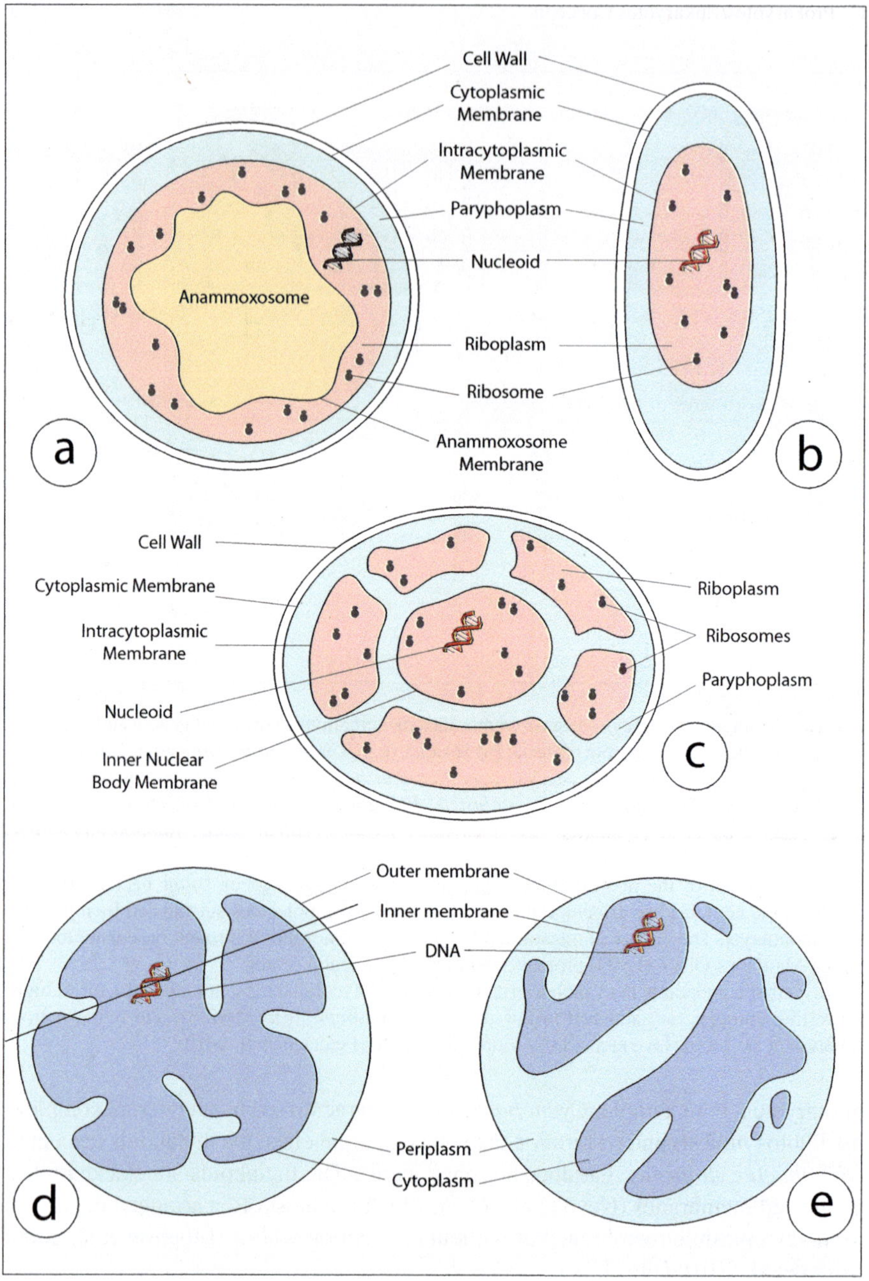

Fig. 1.3 Schematic representation of anammox bacteria (**a**)
Schematic representation of *Pirellula* (**b**)
Organization of *Gemmata obscuriglobus* proposed by Fuerst's team (**c**). Two-dimensional architecture
Organization of *Gemmata obscuriglobus* proposed by Devos's team. Two-dimensional architecture obtained from a three-dimensional reconstruction of whole cell (**d**). Viewed under favorable angle, DNA is membrane surrounded (**e**). (Modified and redrawn from van Niftrik and Jetten 2012; Devos 2014a; Sagulenko et al. 2014)

and nucleoid, the pirellulosome or riboplasm. The riboplasm itself contains a second membrane-bound compartment, the anammoxosome, which is surrounded by a membrane containing a particular lipid: the ladderane. In "*Candidatus* Kuenenia stuttgartiensis," a typical peptidoglycan was detected (van Teeseling et al. 2013). The anammoxosome is the site of energy production. These bacteria perform anaerobic oxidation of ammonium into dinitrogen with nitrite as electron acceptor and hydrazine (N_2H_4) and hydroxylamine (NH_2OH) as intermediates. The toxicity of these compounds is probably at the origin of the presence of particular fatty acids, ladderane – pentacycloanammoxic acid – in the membrane delimiting the anammoxosome to prevent the diffusion of these toxic compounds in the cytosol. The transfer of protons across the anammoxosome membrane and a proton motive force are established over the anammoxosome membrane, which generates ATP molecules (presence of ATP synthases anammoxosome membrane).

In most prokaryotes that use the Calvin cycle, protein-bounded organelles without lipid bilayer are present. They are called **carboxysomes** (polyhedral bodies) and contain RuBisCO, a key enzyme of the Calvin cycle involved in CO_2 fixation.

Other microcompartments were identified in *Bacteria*: (Eut) which is implicated in the utilization of ethanolamine and (*Pdu*) which is involved in 1,2 propanediol utilization (e.g., *Salmonella enterica*) (Kerfeld et al. 2010; Murat et al. 2010; Chowdhury et al. 2015).

Gas vacuoles do not have a metabolic function; instead, they control buoyancy of cells, positioning the cells at the optimum illumination in the water column (Murat et al. 2010) and at the top of the water table in the soil.

1.2.8 A Nuclear Membrane in Planctomycetes

Stanier et van Niel (1962) wrote, concerning the organization of prokaryotic cell, the "irrefutable evidence for the absence of a distinct nuclear membrane shows with particular clarity the very sharp separation between the nuclear and cytoplasmic regions." From that date, it was recognized that a fundamental difference between the organization of the eukaryotic and prokaryotic cell is the absence of a nuclear membrane in the latter. But, in the well-known text book, *Biology of Microorganism* (Madigan et al. 2015), it is written that in the bacterium *Gemmata*, "the nucleoid itself is surrounded by a 'nuclear envelope' consisting of a double membrane layer as occurs in the nuclear membrane of Eukaryotes." This description of a nuclear membrane in *Gemmata* – already announced by other authors (see below) – which challenges the dichotomy prokaryotes – eukaryotes, is however not universally accepted. Indeed, the existence of a membrane that circumscribes the nucleoid is still a matter of debate. While some researchers support this idea (Fuerst's team), others refute (Devos's team) the existence of a membrane surrounding the nucleoid.

Fuerst's team (2011) described and compared the cellular organization of two *Planctomycetes*, *Pirellula* and *Gemmata obscuriglobus* (Sagulenko et al. 2014). In *Pirellula*, the riboplasm is surrounded by a membrane called intracytoplasmic membrane. The riboplasm contains a naked nucleoid (Fig. 1.3b). In contrast, in *Gemmata obscuriglobus*, observed on sectioned cells, the paryphoplasm (cf. "anammoxosome") would contain several vesicles (riboplasm) surrounded by an intracytoplasmic membrane. Unlike *Pirellula*, the nucleoid would be surrounded by a membrane, single in some places and double in others. From their observations, the authors estimate (1) "that planctomycetes do represent an example of bacteria possessing a structural analogue of the nucleus of Eukaryotes" (Sagulenko et al. 2014) and (2) "planctomycetes cannot be referred to as typical Gram-negative bacteria," such as *Escherichia coli* (Sagulenko et al. 2014) (Fig. 1.3c).

Devos's team refutes Fuerst's proposal (Santarella-Mellwig et al. 2013; Devos 2014a). For these authors, the cellular organization of *Gemmata obscuriglobus* would be as follows: (1) an outer membrane; (2) an inner membrane, with a complex membrane system that would form wide invaginations inside the cytoplasm; and (3) a cell that would not nucleate. However, according to the angle of vision in 2D, the nucleoid seems surrounded by a membrane (Fig. 1.3d, e). Thus, cellular organization of *G. obscuriglobus* would be a variation of Gram-negative cell plan, complicated by the presence of an exceptional endomembrane system (Devos 2014b).

The identification of a peptidoglycan in planctomycetes by Jeske et al. (2015) and a lipopolysaccharide in *Gemmata obscuriglobus* (Mahat et al. 2016) is in favor of the proposal formulated by Devos.

1.2.9 Nutrient Capture: Endocytosis Within Prokaryotes?

Almost all eukaryotic cells feed by phagocytosis: they take up particulate material by engulfment which is digested in lysosomes. Prokaryotes utilize membrane transporters to assimilate nutrients dissolved in environment.

Endocytosis, which is a characteristic of eukaryotic cells, is absent in prokaryotes with a single exception. Indeed, an endocytosis-like process by *Planctomycetes* has been proposed by Lonhienne et al. (2010). They demonstrated the ability to take up proteins (green fluorescent protein) using vesicle formation (50–200 nm wide), as an energy-dependent process. Jeske et al. (2015) revealed the presence of peptidoglycan in *Planctomycetes* and considered that the vesicles formed during the endocytosis in these bacteria could not cross the barrier formed by the PG layer with a typical mesh size of 1.6–2.0 nm. For these authors, the endocytosis process in *Planctomycetes* must therefore be reconsidered.

1.2.10 Cellular Multicellularity and Differentiation

Multicellularity is very common in prokaryotes and can take different forms in nature (Claessen et al. 2014; Lyons and Kolter 2015): (1) association as linear or branched filaments, single or multiple layers (*Cyanobacteria*, *Actinomycetes*,

Chloroflexi, *Beggiatoa*, etc.); (2) aggregates (swarms, biofilms, mats, stromatolites), where grouped cells communicate, interact, and can have a coordinated behavior (e.g., *Myxobacteria*); and (3) magnetotactic multicellular prokaryotes (MMPs).

The multicellular prokaryotes have a unicellular stage in their life cycle, except magnetotactic multicellular prokaryotes (MMPs), which are always multicellular at all stages of their life. A multicellular organism divides into two identical multicellular organisms. If a cell is separated from a multicellular organism, it dies (Abreu et al. 2006; Zhang et al. 2014). Multicellularity may be transient (biofilms) or permanent (filamentous cyanobacteria, MMPs). Selective disadvantages of multicellularity (energetic coast) are largely overcome by advantages of this lifestyle (cf. Chap. 5 "Evolution of Living Beings Started with Prokaryotes and in Interaction with Prokaryotes"). The multicellularity in prokaryotes appeared very early in the evolution of life (cf. *Stromatolites* in Chap. 3: "Importance of Prokaryotes in the Functioning and Evolution of the Present and Past Geosphere and Biosphere") and allows to explore the origin of eukaryotic multicellularity.

Cell differentiation – specialization in different functions of cells derived from a single clone – exists in prokaryotes. For example, in cyanobacteria, two types of genetically identical cells provide two incompatible tasks: photosynthesis that generates oxygen and atmospheric nitrogen fixation, a reductive reaction that requires anaerobiosis. In the same way, the collective migration of cells over a solid surface, in *Bacillus subtilis*, involves two subpopulations that perform distinct tasks, one surfactant-producing and the other producing a matrix; together they form highly organized bundle, called "van Gogh bundles" (van Gestel et al. 2015). Many cells differentiate spores that can withstand episodes of desiccation or high temperature, for instance, the *Firmicutes* or the *Actinobacteria*.

Concerning biofilms, in natural environment, they are clonal, but they often involve several species.

If cell differentiation also exists in prokaryotes, it never reaches the complexity found in eukaryotes, where more than 200 different cell types are present in animals and where these cell types are arranged in tissues and organs in a defined plane.

1.2.11 Metabolic Pathways

The cellular organization of prokaryotes is poorly differentiated compared to eukaryotes, but functionally prokaryotes possess a much wider metabolism diversity which is not found in eukaryotes (e.g., chemolithotrophy, anoxygenic photosynthesis, etc.) (Table 1.1). Unlike eukaryotes, prokaryotes express all the functions that are necessary to the functioning of biogeochemical cycles. This metabolic plasticity of prokaryotes allows them to adapt to extreme environmental conditions incompatible with eukaryotic life (cf. Chap. 4 "Evolutionary Success of Prokaryotes"). Furthermore, by their metabolic activity, they have created environmental conditions in which eukaryotes could appear, grow, and diversify.

1.3 Conclusion

The "concept of bacterium" proposed by Stanier and van Niel, in 1964, was *a vital moment in the history of biology* (Sapp 2006). They were the first to describe in details the prokaryote/eukaryote dichotomy. Since then, the description of the prokaryotic cell has been enriched and completed. The cellular organization of the prokaryotes never compares to the complexity found in eukaryotes and "Only the eukaryotic cell appears, however, to have contained the potentialities for the development of highly differentiated multicellular biological systems" (Stanier and van Niel 1962). However, right now, it is clear that prokaryotes are not simple *unorganized bags of enzymes*; the bacterial cell is highly organized. In addition, some traits of eukaryotes are present to a (very) limited extent in prokaryotes: transcription-translation decoupling, the presence of a cytoskeleton, presence of endomembrane systems and organelles, a proteasome for protein recycling and endocytosis. The presence of eukaryote traits in prokaryotes – especially in *Planctomycetes* and in *Lokiarchaeota* – could represent intermediate steps of "the long road from prokaryotes to eukaryotes" and provide clues to explain the prokaryote/eukaryote transition that changed the history of life (Fuerst and Sagulenko 2011; Embley and Williams 2015; Spang et al. 2015). However, the origin of the eukaryotic cell is the subject of intense debates. The answer to the question of the origin of the eukaryotic cell remains pending.

Woese and Fox (1977) has revolutionized our vision of the living world by showing that within the prokaryotes, there are two different phylogenetic groups: *Bacteria* and *Archaea*. He concluded that the prokaryote terminology should be abandoned. As a result of this statement, a debate was opened between those in favor of maintaining the dichotomy prokaryote/eukaryote (Mayr 1998; Martin and Koonin 2006; Dolan and Margulis 2007; Whitman 2009) and those who consider this dichotomy obsolete (Woese and Fox 1977; Pace 2006). For the latter, prokaryotes are described in "negative terms" and "based on morphological subjectivities." But the morphological and functional criteria for the description of prokaryotes are neither subjective nor negative. Indeed, if prokaryotes do not have many traits present in eukaryotes, in contrast they possess many specific characteristics, lacking in eukaryotes, which have allowed prokaryotes to survive for 3.5 billion years and enabled the emergence and expansion of multicellular eukaryotes: very short doubling time, high mutation rates, plasticity of metabolism, genetic exchanges between cells (intensity of HTGs), etc. (Table 1.1). A eukaryotic organization would not have survived the harsh conditions that prokaryotes had to face during their evolutionary history, and even today, eukaryotes are unable to live in extreme conditions of life where many prokaryotes proliferate (cf. Chap. 4 "Success of Prokaryotes"). Furthermore, thanks to their extensive biochemical diversification, they colonized the entire planet, and, currently, they specifically fulfill many steps of the biogeochemical cycles, essential to sustain life on Earth. Using the term prokaryote, Woese (1994) claimed "Prokaryotes are the real chemists of this planet." In the absence of prokaryotes, biogeochemical cycles would stop function-

Table 1.1 Comparison between prokaryotic (*Bacteria* and *Archaea*) and eukaryotic cells

	Prokaryotes		Eukaryotes
	Bacteria	*Archaea*	*Eukarya*
Size of cells	<5 μm	<5 μm	>10 μm
Circular DNA	Present	Present	Absent
Number of chromosomes	1 (a)	1 (a)	Multiple linear chromosomes
Introns	Absent	Absent	Present
Operons	Present	Present	Rare
Ribosome size (in Svedberg units)	70 S	70 S	80 S
Cell division	Binary fission	Binary fission	Mitosis
Reproduction	Always asexual	Always asexual	Asexual and sexual (meiosis)
HGT	Important	Important	Marginal (b)
Transcription-translation	Coupled (a)	Coupled	Uncoupled
Cytoplasmic membrane lipids	Ester-linked	Ether-linked (a)	Ester-linked (a)
Membrane-enclosed organelles	**Present in some species**	Absent	Present (c)
Nuclear membrane	Absent (d)	Absent	Present
Metabolic pathways			
Chemolithotrophy	Present	Present	Absent
Anaerobic respiration	Present	Present	Generally absent
Oxygenic photosynthesis	Present	Absent	Present
Anoxygenic photosynthesis	Present	Absent	Absent
Methanogenesis	Absent	Present	Absent
Nitrogen fixation	Present	Present	Absent

(a) Exceptions; see text
(b) Except endosymbiosis gene transfer
(c) Exception, examples: mitochondrial groups (*Giardia intestinalis*, *Entamoeba histolytica*, *Trichomonas vaginalis*, *Trachipleistophora hominis*) which have secondary loss mitochondria (van der Giezen 2009)
(d) Ongoing debate; see text

ing, leading to the disappearance of all eukaryotes. Prokaryotes possess specific functions non-existing in eukaryotes and very important for the functioning of biogeochemical cycles, such as, anaerobic respirations, anoxygenic photosyntheses, most of fermentations, nitrogen fixation, specific sulfur metabolism, etc. (Table 1.1).

Organisms *are not mere assemblages of genes*, and the description of an organism in its entirety involves morphological, functional, and genetic traits. The term "prokaryote" – based on morphological and functional criteria – always has a specific meaning for many scientists. It is impossible to delete the term "prokaryote," which is still used abundantly in publications and textbooks (Rosenberg 2014). It should also be noted that the description of new species, genera, and phyla is based both on genetic-based methods and phenotypic characteristics (Tindall et al. 2010; Oren and Garrity 2014) (Tindall et al. 2010, Oren and Garrity 2014).

We suggest maintaining the distinction prokaryotes/eukaryotes (an organizational and functional concept) by supplementing it with *Bacteria*/*Archaea* dichotomy (a phylogenetic concept). The two concepts are not irreconcilable, but complementary. This consensual proposal, which coincides with that proposed by Christian De Duve (2007), integrates both phenotypic and genotypic criteria, *reconciling the equally valid demands of cell biology and phylogenies.*

References

Abreu F, Silva K, Martins J, Lins U (2006) Cell viability in magnetotactic multicellular prokaryotes. Int Microbiol 9:267–272

Bapteste E, O'Malley M, Beiko R, Ereshefsky M, Gogarten J, Franklin-Hall L, Lapointe F, Dupré J, Dagan T, Boucher Y et al (2009) Prokaryotic evolution and the tree of life are two different things. Biol Direct 4:34

Bertrand J-C, Brochier-Armanet C, Gouy M, Westall F (2015) For three billion years, microorganisms were the only inhabitants of the earth. In: Bertrand J, Caumette P, Lebaron P, Matheron R, Normand P, Sime-Ngando T (eds) Environmental microbiology: fundamentals and aplications. Springer, Dordrecht/Heidelberg/New York/London, pp 75–106

Best D, Higgins I (1981) Methane-oxidizing activity and membrane morphology in a methanol-grown obligate methanotroph, *Methylosinus trichosporium* OB3b. J Gen Microbiol 125:73–84

Brochier-Armanet C, Moreira D (2015) Horizontal gene transfer in microbial ecosystems. In: Bertrand J, Caumette P, Lebaron P, Matheron R, Normand P, Sime-Ngando T (eds) Environmental microbiology: fundamentals and applications. Springer, Dordrecht/Heidelberg/New York/London, pp 471–512

Cabeen M, Jacobs-Wagner C (2010) The bacterial cytoskeleton. Annu Rev Genet 44:365–392

Chatton E (1925) *Pansporella perplexa*. Réflexions sur la biologie et la phylogénie des protozoaires. Annales des Sciences Naturelles Zoologie et Biologie Animale 10-VII: 1–84

Cho H (2015) The role of cytoskeletal elements in shaping bacterial cells. J Microbiol Biotechnol 25:307–316

Chowdhury C, Chun S, Pang A, Sawaya M, Sinha S, Yeates T, Bobik T (2015) Selective molecular transport through the protein shell of a bacterial microcompartment organelle. Proc Natl Acad Sci U S A 112:2990–2995

Claessen D, Rozen D, Kuipers O, Søgaard-Anderson L, van Wezel G (2014) Bacterial solutions to multicellularity: a tale of biofilms, filaments and fruiting bodies. Nature Rev Microbiol 12:115–124

Courties C, Vaquer A, Troussellier M, Lautier J, Chrétiennot-Dinet M, Neveux J, Machado C, Claustre H (1994) Smallest eukaryotic organism. Nature 370:255

De Boer W, Hazeu W (1972) Observations on the fine structure of a methane-oxidizing bacterium. Antonie Van Leeuwenhoek 38:33–47

De Duve C (2007) The origin of eukaryotes: a reappraisal. Nature Rev 8:395–403

Derelle E, Ferraz C, Rombauts S, Rouzé P, Worden AZ, Robbens S, Partensky F, Degroeve S, Echeynié S, Cooke R et al (2006) Genome analysis of the smallest free-living eukaryote *Ostreococcus tauri* unveils many unique features. Proc Natl Acad Sci U S A 103:11647–11652

DeRosa M, Gambacorta A, Huber R, Lanzotti V, Nicolaus B, Stetter K, Trincone A (1988) A new 15,16-diméthyl-30-glyceryloxy-triacontanoic acid from lipids of *Thermotoga maritima*. J Soc Chem Commun 1988:1300–1301

Devos D (2014a) PVC bacteria: variation of, but not exception to, the gram-negative cell plan. Trends Microbiol 22:14–20

Devos D (2014b) Re-interpretation of the evidence for PVC cell plan supports a gram-negative origin. Antonie Van Leeuwenhoek 105:271–274

Dolan M, Margulis L (2007) Advances in biology reveal truth about prokaryotes. Nature 445:21

Duggin I, Aylett C, Walsh J, Michie K, Wang Q, Turnbull L, Dawson E, Harry E, Whitchurch C, Amos L et al (2015) CetZ tubulin-like proteins control archaeal cell shape. Nature 519:362–365

Embley T, Williams T (2015) Steps on the road to eukaryotes. Nature 521:169–170

Fuerst J, Sagulenko E (2011) Beyond the bacterium: planctomycetes challenge our concepts of microbial structure and function. Nature Rev Microbiol 9:403–413

Ghai R, Mizuno CM, Picazo A, Camacho A, Rodriguez-Valera F (2013) Metagenomics uncovers a new group of low GC and ultra-small marine Actinobacteria. Sci Rep 3:2471

Gottshall E, Seebart C, Gatlin J, Ward N (2014) Spatially segregated transcription and translation in cells of the endomembrane-containing bacterium *Gemmata obscuriglobus*. Proc Natl Acad Sci U S A 111:11067–11072

Graumann P (2007) Cytoskeletal elements in bacteria. Annu Rev Microbiol 61:589–618

Grossi V, Mollex D, Vinçon-Laugier A, Hakil F, Pacton M, Cravo-Laureau C (2015) Mono- and dialkyl glycerol ether lipids in anaerobic bacteria: biosynthetic insights from the mesophilic sulfate reducer *Desulfatibacillum alkenivorans* PF2803T. Appl Environ Microbiol 81:3157–3168

Hanson R, Hanson T (1996) Methanotrophic bacteria. Microbiol Rev 60:439–471

Ingerson-Mahar M, Gitai Z (2012) A growing family: the expanding universe of the bacterial cytoskeleton. FEMS Microbiol Rev 36:256–266

Jeske O, Schüler M, Schumann P, Schneider A, Boedeker C, Jogler M, Bollschweiler D, Rohde M, Mayer C, Engelhardt H et al (2015) Planctomycetes do possess a peptidoglycan cell wall. Nat Commun 6:7116

Kerfeld C, Heinhorst S, Cannon G (2010) Bacterial microcomparments. Ann Rev Microbiol 64:391–408

Lewis P, Thaker S, Errington J (2000) Compartmentalization of transcription and translation in *Bacillus subtilis*. EMBO J 19:710–718

Liberton M, Berg R, Heuser J, Roth R, Himadri B, Pakrasi H (2006) Ultrastructure of the membrane systems in the unicellular cyanobacterium *Synechocystis* sp. strain PCC 6803. Protoplasma 227:129–138

Lin L, Thanbichler M (2013) Nucleotide-independent cytoskeletal scaffolds in Bacteria. Cytoskeleton 70:409–423

Lonhienne T, Sagulenko E, Webb R, Lee K-C, Franke J, Devos D, Nouwens A, Caroll B, Fuerst J (2010) Endocytosis-like protein uptake in the bacterium *Gemmata obscuriglobus*. Proc Natl Acad Sci U S A 107:12883–12888

Lundgren M, Andersson A, Chen L, Nilsson P, Bernander R (2004) Three replication origins in *Sulfolobus* species: synchronous initiation of chromosome replication and asynchronous termination. Proc Natl Acad Sci U S A 101:7046–7051

Lyons N, Kolter R (2015) On the evolution of multicellularity. Curr Opinion Microbiol 24:21–28

Madigan M, Bender K, Buckley D, Sattley W, Stahl D (2015) Brock biology of microorganisms. Pearson, New York

Mahat R, Seebart C, Basile F, Ward N (2016) Global and targeted lipid analysis of *Gemmata obscuriglobus* reveals the presence of lipopolysaccharide, a signature of the classical gram-negative outer membrane. J Bacteriol 198:221–236

Margolin W (2005) FtsZ and the division of prokaryotic cells and organelles. Nat Rev Mol Cell Biol 6: 862–871

Martin W, Koonin E (2006) A positive definition of prokaryotes. Nature 442:868

Massana R, Logares R (2013) Eukaryotic versus prokaryotic marine picoplankton ecology. Environ Microbiol 15:1254–1261

Mayr E (1998) Two empires or three ? Proc Natl Acad Sci U S A 95:9720–9723

Murat D, Byrne M, Komeili A (2010) Cell biology of prokaryotic organelles. Cold Spring Harb Perspect Biol 2:a000422

Nevo R, Charuvi D, Shimoni E, Schwarz R, Kaplan A, Ohad I, Reich Z (2007) Thylakoid membrane perforations and connectivity enable intracellular traffic in cyanobacteria. EMBO J 26:1467–1473
Nevo-Dinur K, Nussbaum-Shochat A, Ben-Yehuda S, Amster-Choder O (2011) Translation-independent localization of mRNA in *E. coli*. Science 331:1081–1084
Oren A, Garrity G (2014) Then and now: a systematic review of the systematics of prokaryotes in the last 80 years. Antonie Van Leeuwenhoek 106:43–56
Orlandini V, Emiliani G, Fondi M, Maida I, Perrin E, Fani R (2014) Network analysis of plasmidomes: The Azospirillum brasilense Sp245 case. Int J Evol Biol:951035
Ozyamak E, Kollman J, Komeili A (2013) Bacterial actins and their diversity. Biochemistry 52:6928–6939
Pace N (2006) Time for change. Nature 44:289
Pace N (2009a) Problems with "Procaryote". J Bacteriol 191:2008–2010
Pace N (2009b) Rebuttal: the modern concept of the prokaryote. J Bacteriol 191:2006–2007
Pereira S, Reeve J (1998) Histones and nucleosomes in archaea and eukarya: a comparative analysis. Extremophiles 2:141–148
Petersen MØ, Linnanto J, Frigard N-U, Nielsen N, Miller M (2010) A model of the protein-pigment baseplate complex in chlorosomes of photosynthetic bacteria. Photosynth Res 104:233–243
Rappe MS, Connon S, Vergin K, Giovannoni S (2002) Cultivation of the ubiquitous SAR11 marine bacterioplankton clade. Nature 418:630–633
Rosenberg E (2014) The prokaryotes; Alphaproteobacteria and Betaproteobacteria. Springer, Berlin
Sagulenko E, Morgan G, Webb R, Yee B, Lee K-C, Fuerst J (2014) Structural studies of Planctomycete *Gemmata obscuriglobus* support cell compartmentalisation in a bacterium. PLoS One 9:3
Saier M, Bogdanov M (2013) Membranous organelles in bacteria. J Mol Microbiol Biotechnol 23:5–12
Samson R, Obita T, Freund S, Williams R, Bell S (2008) A role for the ESCRT system in cell division in archaea. Science 322:1710–1713
Santarella-Mellwig R, Pruggnaller S, Roos N, Mattaj I, Devos D (2013) Three-dimensional reconstruction of bacteria with a complex endomembrane system. PLoS Biol 11:e1001565
Sapp J (2006) Two faces of the prokaryote concept. Int Microbiol 9:163–172
Sinninghe DJ, Rijpstra W, Hopmans E, Shouten S, Balk M, Stams AJ (2007) Structural characterization of diabolic acid-based tetraester, tetraether and mixed ether/ester, membrane-spanning lipids of bacteria from the order *Thermotogales*. Arch Microbiol 188:629–641
Sinninghe DJ, Rijpstra W, Hopmans E, Foesel B, Wüst P, Overmann J, Tank M, Bryant D, Dunfield P, Houghton K et al (2014) Ether- and ester-bound iso-diabolic acid and other lipids in members of *Acidobacteria* subdivision 4. Appl Environ Microbiol 80:5207–5218
Soppa J (2014) Evolutionary advantages of polyploidy in halophilic archaea. Biochem Soc Trans 41:339–343
Spang A, Saw JH, Jorgensen SL, Zaremba-Niedzwiedzka K, Martijn J, Lind AE, van Eijk R, Schleper C, Guy L, Ettema TJ (2015) Complex archaea that bridge the gap between prokaryotes and eukaryotes. Nature 521:173–179
Stanier R, van Niel C (1962) The concept of a bacterium. Arch Mikrobiol 42:17–35
Tindall B, Rosselló-Móra R, Busse H-J, Ludwig W, Kämpfer P (2010) Notes on the characterization of prokaryote strains for taxonomic purposes. Int J Syst Evol Microbiol 60:249–266
Trachtenberg S, Dorward L, Speransky V, Jaffe H, Andrews S, Leapman R (2008) Structure of the cytoskeleton of *Spiroplasma melliferum* BC3 and its interactions with the cell membrane. J Mol Biol 378:778–789
van der Giezen M (2009) Hydrogenosomes and mitosomes: conservation and evolution of functions. J Eukaryot Microbiol 56: 221–231
van Gestel J, Vlamakis H, Kolter R (2015) From cell differentiation to cell collectives: *Bacillus subtilis* uses division of labor to migrate. PLoS Biol 13:e1002141

van Gool A (1972) Ultrastructure of *Nitrosomonas europaea* cells as revealed by freeze-etching. Archiv Mikrobiol 82:120–127

van Niftrik L, Jetten M (2012) Anaerobic ammonium-oxidizing bacteria: unique microorganisms with exceptional properties. Microbiol Mol Biol Rev 76: 585–596

van Teeseling MC, de Almeida NM, Klingl A, Speth D, Op den Camp H, Rachel R, Jetten M, van Niftrik L (2013) A new addition to the cell plan of anammox bacteria: "*Candidatus* Kuenenia stuttgartiensis" has a protein surface layer as the outermost layer of the cell. J Bacteriol 196:80–89

Vaulot D, Eikrem W, Viprey M, Moreau H (2008) The diversity of small eukaryotic phytoplankton (< or =3 microm) in marine ecosystems. FEMS Microbiol Rev 32:795–820

Whitman W (2009) The modern concept of the prokaryote. J Bacteriol 191:2000–2020

Woese C (1994) There must be a prokaryote somewhere: microbiology's search for itself. Microbiol Rev 58:1–9

Woese CR, Fox GE (1977) Phylogenetic structure of the prokaryotic domain: the primary kingdoms. Proc Natl Acad Sci U S A 74:5088–5090

Zhang R, Chen Y-R, Du H-J, Zhang W-Y, Pan H-M, Xiao T, Wu L-F (2014) Characterization and phylogenetic identification of a species of spherical multicellular magnetotactic prokaryotes that produces both magnetite and greigite crystals. Res Microbiol 165:481–489

Chapter 2
Phylogeny and Biodiversity of Prokaryotes

Philippe Normand and Pierre Caumette

Abstract Creating a hierarchical system for microbes began as soon as microbes were discovered and has been constantly evolving ever since. The larger categories, the techniques to define taxa, and the bibliographical conventions, all these are regularly changing. At the moment, there are 30 bacterial phyla and 6 archaeal phyla that are described. The impact of genomes on taxonomy and phylogeny is also discussed.

Keywords Evolution · Family · Genome · Genus · Kingdom · Locus · Order · Phylogeny · Species · Taxonomy

2.1 The Challenge of Naming Microbes

The description of the living world in terms of taxonomy has started along with philosophy and science around 2500 years ago. It was then realized, for plants and animals, that some individuals were more similar to one another than others, that closely related individuals could mate with related individuals of the opposite sex yielding fertile offsprings, and that some species were more similar to one another than others. The common habit of naming plants and animals also stems from the need for a common and stable referential system to link these species with a value in terms of nutrition, toolmaking, medicine, and religion. These efforts were first formalized in Linnaeus' work, *Systema Naturae* (1753), that defined species, genera, and other levels, based on morphological and sexual compatibility features that

---

P. Normand (✉)
Laboratoire d'Ecologie Microbienne, UMR 5557, Université Claude Bernard Lyon 1, Villeurbanne, France
e-mail: philippe.normand@univ-lyon1.fr

P. Caumette
Institut des Sciences Analytiques et de Physico-Chimie pour l'Environnement et les Matériaux, UMR 5254, CNRS, Université de Pau et des Pays de l'Adour, Pau, France

J.-C. Bertrand et al. (eds.), *Prokaryotes and Evolution*,
https://doi.org/10.1007/978-3-319-99784-1_2

were and still are evaluated for their pertinence and reliability. That vision was based on the premise that each species had been created as it was and only later was the concept of evolution grafted onto it. The concept of species was the one most precisely defined by many from the first proposed by Ray as "(having) features that perpetuate themselves in propagation from seed" (Ray 1686) to that of Mayr as "a group of actually or potentially interbreeding natural populations, which are reproductively isolated from other such groups" (Mayr 1963).

In the case of microbes, an effort was undertaken early on to classify them that eventually became a full-fledged part of taxonomy. The absence of real microbial sexuality proved a major drawback, but classification attempts were nevertheless made, using those features that were equivalent to those of superior organisms, such as shape, respiratory characteristics, staining reaction, or biotope. In the middle of the twentieth century, the reference book that attempted to sum up the taxonomic vision of microbiologists was the *Bergey's Manual of Determinative Bacteriology* (1923). This book, reedited and updated several times, defined around 20 categories of microbes using a dichotomous key with among other features photosynthesis, Gram stain, and respiration type. The main problem of this approach was the dichotomous nature of the scheme with a lot of emphasis on characters that may be absent from a given taxon, such as the Gram stain that may yield ambiguous results or based on a feature, such as the wall that is altogether absent in some lineages such as the *Mollicutes*, or such as photosynthesis that has been lost in many lineages.

A turning point in bacterial taxonomy was the proposal not to rely on a few key phenotypic elements but instead to define a number as large as possible of phenotypic features in order to minimize the weight of any single feature (Adanson 1763), an approach that would become feasible (Sokal and Sneath 1963) only when computers became available. Computers are necessary to treat a large number of features for a large number of strains and also on automated phenotypic determination possible with approaches such as the Biolog where carbon sources arrayed on microplates are monitored with a stain that depends on the ability of the microbial cells to respire the compound (Armon et al. 1990; Miller and Rhoden 1991). This approach based on the ability to use organic sources as electron donors is now widely used in microbial species description.

The successive editions of the Bergey's over 60 years provide a fascinating vision of the evolving outlook of microbiologists on bacterial taxonomy. Over a backdrop of constant increases in the number of taxa, the successive rearrangements provide a testimony to the frustrating search for a reliable basis on which to base bacterial taxonomy. The whole enterprise was discouraging, especially given the several instances of convergent evolution, but the process was maintained because of the universal need to have a coherent naming system for bacteria. The search for a universal golden standard was considered by some to have finally met with success when Woese and Fox (1977) proposed to use the ubiquitous 16S rRNA gene as molecular clock and consequently as a way to identify new isolates. The existing classifications of microbes are on the whole based on this gene that is ubiquitous, with a universal function, that of translation of mRNAs into proteins that is highly conserved yet has highly variable regions. The 16S ribosomal gene is not perfect because the ribosomal operon is sometimes plasmid-borne (Kunnimalaiyaan et al. 2001) or sometimes present in several divergent copies (Yap et al. 1999) thus raising some doubt as to its representativity and sometimes has a markedly divergent mutation rate in some lineages such as the mycoplasma (Rogers et al. 1985) thus introducing artifacts such as long branch attraction in phylogenetic reconstructions.

This system, with the Bacteriological Code that aims to formalize its rules, has its shortfalls that will be described in this chapter, yet it has provided microbiologists with a solid backbone that is still evolving today with the advent of ever cheaper and reliable sequencing technologies. It has also evolved to become more inclusive, for instance, it now includes higher ranks such as the phylum (Oren et al. 2015).

The evolution of prokaryotes has taken place over long temporal scales, and attempts have been made to anchor some datable geological events with the emergence of bacterial taxa. One of the best-known such events is the great oxidation event that took place 2.2–2.4 BY ago and was caused by oxygenic photosynthesis done by cyanobacteria (Schopf 2014). Other similar events are fossils of eukaryotic hosts in dated geological strata. Ochman and Wilson (1987) studied several such events and concluded that eubacteria, whatever their growth rate, had a constant substitution rate for 16S rRNA of about 1%/50 Myr, similar to the average rate for 18S rRNA in vertebrates and flowering plants. For instance, *Escherichia coli* that lives in the digestive tract of mammals and *Salmonella typhimurium* that does not probably diverged at the time of emergence of mammals (Ji et al. 2002) around 125 MYA, while their 16S rRNA gene have a distance of 1.8–2%, a date that would correspond to 90–100 MYA.

A major factor for changes in taxonomy is the basic principle that taxon structure must be coherent with phylogeny. Conversely, a paraphyletic taxon, for instance, is a group of organisms that have a recent common ancestor but that do not comprise some of its descendants. This was the case of genus *Rhizobium*, created to accommodate symbiotic nitrogen-fixing microbes that was found to contain some pathogenic *Agrobacterium* species, which lead to the creation of several coherent genera such as *Mesorhizobium*, *Sinorhizobium*, etc. to exclude those lineages that did not fit the phenotypic description and to accommodate the outlying strains.

2.2 Phylogenetic Classification and Biodiversity of Prokaryotes

The seven different taxonomical levels.

Species, genus, family, order, class, phylum, and kingdom are the main levels used in bacterial taxonomy nowadays, to which one must add the subspecies. There is also the recently described status of "*Candidatus*" (Murray and Stackebrandt 1995), to be used for bacteria not isolated in pure cultures but nevertheless characterized as distinct from known taxa based on isolated strains. Such characterization of unisolated strains, in the absence of a pure culture, can only be tentative, and definitive assessment of the type of metabolism will have to wait for a pure culture.

This category, "*Candidatus*", is not covered by the Rules of the Bacteriological Code. Consequently, a name included in the category "*Candidatus*" cannot be validly published, and it also cannot be designated as sp. nov. or gen. nov. The way to write such a name is, for example, "*Candidatus* Frankia datiscae" (Persson et al. 2015); it is a temporary designation since several such strains are investigated and end up being isolated as is the case of the *Frankia* symbionts of the plant *Datisca* (Gtari et al. 2015), and their status will thus have to be reexamined, and the "*Candidatus*" name should then be dropped.

– Species. The species is the only taxonomical level with an operational definition that has evolved over the years. Of course, bacteria do not have the sexuality of eukaryotes as discussed below. Species are commonly defined as a group of microorganisms that share a number of phenotypic and chemical features and also that share 70% of their DNA as measured in the DNA-DNA hybridization (DDH) assay used for the first time in the 1960s (Johnson and Ordal 1968) and formalized 20 years later (Wayne et al. 1987). The DNA of one organism is labeled, for instance, with a radioactive label, denatured, and mixed with the denatured unlabeled DNA with which it is to be compared. The mixture is incubated to allow DNA strands to associate and to form hybrid double-stranded DNA. Hybridized strands with a high degree of similarity will bind more firmly and thus require a higher temperature to separate. The assay will consist in measuring the proportion of DNA that resists melting.

  DNA-DNA hybridization has been used in countless studies for over 40 years as the technique of choice to define microbial species, but it is labor-intensive, requires all strains to be grown in pure cultures in the laboratory doing the assays, and requires large amounts of DNA for all the species to be tested.

  A number of approaches have been proposed as proxies for the DDH assay such as 16S RNA comparison (Stackebrandt and Goebel 1994), AFLP (Mougel et al. 2002), or MLSA (Glaeser and Kampfer 2015) and more recently whole-genome comparisons such as the average nucleotide identity (ANI) metric with a threshold at 95% (Goris et al. 2007). This last approach takes one genome sequence and cuts it into consecutive 1020 nt fragments, and these fragments are used to search against the whole genomic sequence of the other genome in the pair using the BLASTN algorithm (Altschul et al. 1997). The best BLASTN matches above 90% identity are then saved, and the lengths of the alignable regions for all such matches are summed, and that sum is divided by the total length of the genomic DNA of the query genome to provide a genome size-independent measurement of the percentage of the query's DNA that is conserved in the reference genome (Goris et al. 2007).

  Some species do not conform to the currently accepted working definition, the best-known case being that of *Shigella* strains that are very close to *Escherichia coli* (Zuo et al. 2013) but maintained in a different genus for public health reasons and mostly habit; although this positioning is frequently criticized on the ground, there is no difference between the two genera (Pettengill et al. 2015). Another species that has been the object of a controversy is *Rhizobium*, into which (Young et al. 2001) it was proposed to include all *Agrobacterium* species on the ground of phylogeny, proposal that was vigorously contested on the ground of habit of associating the pathogens with an old name (Farrand et al. 2003). The community debated the question for some time until a consensus was reached that proposed to maintain the two genera as distinct entities (Lindström and Young 2011).

Some other species share many phenotypic features yet are clearly very large based on genomic markers; they have thus been labeled species complexes, to indicate they are in the process of being split into better defined and more restrained entities. This is the case, for instance, of *Agrobacterium tumefaciens* (Lassalle et al. 2011), *Burkholderia cepacia* (Bloodworth et al. 2015), or *Bacillus subtilis* (Rooney et al. 2009).

There is one subspecific rank, that of the subspecies, that groups strains that share features. *Francisella tularensis* subsp. *tularensis*, for instance, is a virulent pathogen of humans causing tularemia, with fever and severe pneumonia, *F. tularensis* subsp. *holarctica* is a less virulent pathogen of humans, *F. tularensis* subsp. *mediasiatica* is moderately virulent to human, while *F. tularensis* subsp. *novicida* is not virulent on human (Sridhar et al. 2012). The concepts of "biovar" or "pathovar" are used in a similar manner to group strains within a species. In species *Rhizobium leguminosarum*, several strains can establish symbiosis with peas but not with clover, while others will have the ability to form nodules with clover and not with peas, an ability linked to a large plasmid. The former will be called *Rhizobium leguminosarum* bv. *viciae*, while the latter will be called *Rhizobium leguminosarum* bv. *trifolii*.

- Genus (pl. genera). A genus is a group of species that share a number of features. Some genera are very large, and others are much narrower depending on the history of the group, and there is no consensus on the desirable threshold to create a genus nor any of the larger categories, although this may change with the availability of genomes. Genera are also the best-known level since the genus name is given first in the full species name in the binomial system. There is no ending for genera.
- Family. A family is a group of genera that share a number of features. The ending for families is "-aceae" as in *Enterobacteriaceae*, the family of genus *Enterobacter*; however the type genus of the family is *Escherichia*.
- Order. An order is a group of families that share a number of features. The ending for orders is "-ales" as in *Enterobacteriales*, the order of family *Enterobacteriaceae*.
- Class. A class is a group of orders that share a number of features. There is no ending for classes such as *Gammaproteobacteria*, the class of order *Enterobacteriales*.
- Phylum. A phylum is a group of classes. There is no special ending for phyla, such as for *Proteobacteria*, the phylum of class *Gammaproteobacteria*. Taxonomic categories above the rank of class (phylum and above) were not covered by the Rules of the Bacteriological Code (Lapage et al. 1992); however this has changed recently, and these categories can now be validly published (Oren et al. 2015).
- Kingdom. A kingdom is a group of phyla. There are two kingdoms that interest us, the *Archaea* and the *Bacteria*. The names "empire" and "domain" are sometimes used for such groups.

- Recognized taxa.
  Microbial taxa at all levels are constantly being modified to accommodate new isolates, resolve polyphyletic taxa, and take into account new approaches that reveal proximities. The official reference journal for publishing such new taxa is the *International Journal of Systematic and Evolutionary Microbiology* or IJSEM published by the SGM (called *International Journal of Systematic Bacteriology* or IJSB and published by the ASM until 1999). Other journals (such as typically *Systematic and Applied Microbiology*, *Antonie van Leeuwenhoek*, etc.) also publish descriptions of new taxa, but the IJSEM as reference journal publishes a regular update in the form of a "Validation of publication of new names and new combinations previously effectively published outside the IJSB/IJSEM." Several individuals and institutions try to keep up with the flow and to organize it in a comprehensive and coherent form, the best known being the *Bergey's Manual of Systematic Bacteriology*, the most recent edition of which was published between 2001 and 2012 (Garrity et al. 2001; Brenner et al. 2005; Vos et al. 2009; Krieg et al. 2010; Whitman et al. 2012). Two web sites make a most remarkable effort at keeping up with published taxa, one is the http://www.bacterio.net maintained by Dr. Jean P. Euzéby from the École Nationale Vétérinaire de Toulouse, France, and the other is the NCBI (http://www.ncbi.nlm.nih.gov/Taxonomy). A mention can also be made of the microbewiki site (https://microbewiki.kenyon.edu/index.php/MicrobeWiki) that has photos and schematics published elsewhere, but it is quite lacunar. The present chapter will use the overall structure of the *Bergey's Manual of Systematic Bacteriology*, 2nd edition, which was published as a set of five volumes over the period 2001–2012 with a few updates.

2.2.1 Description of the Archaeal Phyla[1]

Archaea were given the status of domain by Woese in 1987 on the basis of 16S rRNA-based phylogeny and the fact these microorganisms had distinct biochemical features such as their ether-linked membrane lipids (Woese 1987). The archaeal groups he recognized as the equivalent of phyla were then the methanogens/halobacteria/*Thermoplasma*/*Thermococcus* group, on the one hand, and the extreme thermophiles group on the other hand. He later named these two phyla *Euryarchaeota* and *Crenarchaeota*, respectively (Woese et al. 1990). Later, other groups were recognized to accommodate new lineages such as the *Korarchaeota* (Barns et al. 1996), the *Nanoarchaeota* (Huber et al. 2002), or the *Thaumarchaeota* (Brochier-Armanet et al. 2008). Many more archaeal phyla have been proposed recently such as the "Aenigmarchaeota," the "Geoarchaeota," the "Nanohaloarchaeota," or the "Parvarchaeota," sometimes to accommodate deep-branching single-celled genomes (Rinke et al. 2013), but these have not yet been published in the IJSEM or in the Bergey's most recent edition and

[1] This number has been actualized based on the bacterio web site (http://www.bacterio.net)

will thus not be considered further. Recent proposals of two superphyla have also been made with "DPANN" and "TACK" to accommodate the *Diapherotrites*, the *Parvarchaeota*, the *Aenigmarchaeota*, the *Nanoarchaeota*, and the *Nanohaloarchaeota* (DPANN) (Rinke et al. 2013) and the *Thaumarchaeota*, the *Aigarchaeota*, the *Crenarchaeota*, and the *Korarchaeota* (TACK) (Guy and Ettema 2011), respectively; besides the *Euryarchaeota*, however, these superphyla have not gained official status yet and will not be considered further here.

2.2.1.1 Phylum-A1: *Crenarchaeota* (Woese et al. 1990)

The *Crenarchaeota* comprise a single class, *Thermoprotei*, and six orders (*Acidilobales*, *Cenarchaeales*, *Desulfurococcales*, *Fervidicoccales*, *Sulfolobales*, and *Thermoproteales*), each with one or two families. These microbes are coccoid-shaped hyperthermophiles which optimal growth temperature approaches that of boiling water, and they currently hold the temperature record at 113 °C (Blochl et al. 1997). Some are aerobic and others anaerobic, and they use hydrogen and sulfur as sources of energy as well as various electron acceptors, while *Thermoproteus* uses hydrogen as electron donor and sulfur as electron acceptor. Besides geothermal springs, they are found in acid mine drainages, soils, marine sponges, or lake sediments (Polonia et al. 2016). The name derives from ancient Greek κρήνη that means "fountain" and αρχαίος, ancient, as a reference to the thermal springs where many isolates have been found and their ancient characters.

2.2.1.2 Phylum-A2: *Euryarchaeota* (Woese et al. 1990)

The *Euryarchaeota* comprises eight classes, the *Methanomicrobia* (methanogens), the *Archaeoglobi*, the *Halobacteria* or *Halomebacteria* (halophiles), the *Methanobacteria* (methanogens), the *Methanococci* or *Methanothermea* (methanogens), the *Methanopyri* (methanogens), the *Thermococci* or *Protoarchaea* (thermophiles), and the *Thermoplasmata* (thermophiles), each with one to three orders. The phylum groups salt-loving microbes known as halophiles as well as methane producers (some of which are also halophilic), as well as and some extremely thermophilic aerobes (e.g., *Thermoplasma*) and anaerobes (e.g., *Pyrococcus*). The methane producers are found in oxygen-depleted organic matter-rich biotopes such as the gastrointestinal tract of animals, sludge, manure heaps, soils, marine sediments, landfills, and rice agriculture. The halophiles are mostly aerobic heterotrophs found in biotopes where the salt concentration is near saturation (NaCl 5 M or > 30% w/w) such as salt lakes, inland seas, and evaporating ponds of seawater such as salterns. The thermophiles, both aerobic and anaerobic, are found in hydrothermal springs. The name originates from ancient Greek ευρύς that means "wide" to underline the large range of ecological niches where they are found.

2.2.1.3 Phylum-A3: *Korarchaeota* (Barns et al. 1996)

The *Korarchaeota* is a small group of bacteria with a single class of uncultivated and relatively rare microorganisms present in hot springs. The *Korarchaeota* are thus hypothesized to have retained the physiology of the earliest cells that have appeared in thermal springs. Genomic approaches (Elkins et al. 2008) have permitted to study "*Candidatus* Korarchaeum cryptofilum," which exhibits an ultrathin filamentous morphology. The name refers to the position of the 16S genes at the base of the phylogenetic tree and originates from ancient Greek -κόρos meaning "young man" or - κόρη "young woman."

2.2.1.4 Phylum-A4: *Nanoarchaeota* (Huber et al. 2002)

The *Nanoarchaeota* with a single class/order are a very small group of uncultivated small cells living in close proximity to larger *Ignicoccus islandicus* cells, a *Korarchaeota* with which it lives in coculture in hot springs. Very few occurrences of this lineage have been described besides the *Nanoarchaeum*, which is most probably symbiotic but without knowledge of the physiology of the relationship. The name is from the Greek νάνος that means "dwarf" referring to the small size of the cells.

2.2.1.5 Phylum-A5: *Thaumarchaeota* (Brochier-Armanet et al. 2008)

The *Thaumarchaeota* with a single class/order are a small group of microbes that were described (Brochier-Armanet et al. 2008) as uncultivated mesophilic archaea. The name from the Greek θαύμα that means "wonder" was chosen to indicate the sense of awe felt when considering the ecology and physiology of these microbes that are relatively rare and recovered from soils, sea, or freshwater bodies and in particular contain the archaeal anaerobic ammonia oxidizers.

2.2.1.6 Phylum x: *Lokiarchaeota* (Spang et al. 2015)

The *Lokiarchaeota*, which are a candidate superphylum, called "Asgard" together with the *Thorarchaeota*, the *Odinarchaeota*, and the *Heimdallarchaeota* (Zaremba-Niedzwiedzka et al. 2017), found in deep marine sediments have many eukaryotic features such as several genes that were previously regarded as specific to eukaryotes (Spang et al. 2015). The four potential phyla would be among the archaeal deepest branches, closest to eukaryotic lineages and postulated to be close to the ancestor of eukaryotes. The name derives from Loki's Castle, a field of five active hydrothermal vents in the mid-Atlantic Ocean, located at 73 ° north on the Mid-Atlantic Ridge between Greenland and Norway at a depth of 2352 m where the first strains were detected.

2.2.2 Description of the Bacterial Phyla

Description of the 30 bacterial phyla in terms of physiological features and biota.

Phyla B1 to B9 are positioned close to the root of the phylogenetic tree based on 16S rRNA bacteria gene sequences in the 2012 edition of the Bergey's. These bacteria have extremophilic characters and are often isolated from extreme, in particular, thermal environments. However, this position is not found in most phylogenetic analyses based on protein markers (Lopez-Garcia and Moreira 2008). Many species belonging to the phyla B1 to B9 can live in environments considered extreme, either very hot or highly contaminated by metals or with high ionizing radiation levels. The phyla *Aquificae* and *Thermotogae* and genus *Thermus* contain the best-known thermophilic members among domain Bacteria. Some of the phyla described below may be grouped later like, for instance, *Chlamydiae*, *Planctomycetes*, *Verrucomicrobia*, and *Lentisphaerae* in the superphylum PVC (Lagkouvardos et al. 2014) or the *Actinobacteria*, *Deinococcus*, and *Cyanobacteria* proposed to form the Terrabacteria associated with an early colonization of land 2.5–3.2 BYA (Battistuzzi et al. 2004).

2.2.2.1 Phylum-B1: *Aquificae* (Gupta and Lali 2013)

This small phylum includes a single class (*Aquificae*) with a single order, *Aquificales*, in which there are three families, the *Aquificaceae* (Reysenbach 2001a), the *Desulfurobacteriaceae*, and the *Hydrogenothermaceae*. Among the genera described so far are *Aquifex*, *Calderobacterium*, *Hydrogenivirga*, *Hydrogenobacter*, *Hydrogenobaculum*, and *Thermocrinis*. These Gram-negative, microaerophilic autotrophic and hydrogenotrophic bacteria can grow at temperatures reaching 95 °C. Representatives of these genera are all hyperthermophilic bacteria isolated from marine and terrestrial hot springs and are either chemoorganotrophs or chemolithotrophs with hydrogen, sulfur, or thiosulfate to be used as electron donors. *Aquifex*, with individual cells of 2–6 μm in length, forms aggregates composed of 100 individual cells or less. These microorganisms occupy a basal position in the phylogenetic tree of *Bacteria* based on 16S rRNA. However, this phylogenetic position is controversial, and these microorganisms may actually be descendants of mesophilic microorganisms which are secondarily adapted to life at high temperature. Lateral gene transfer (LGT) appears to have played an important factor in the evolutionary history of the taxon, mainly from *Thermotogae* or *Epsilonproteobacteria* (Eveleigh et al. 2013). *Aquifex* means maker of water in Latin and refers to the fact that its respiratory metabolism yields water.

2.2.2.2 Phylum-B2: *Thermotogae* (Reysenbach 2001b)

It is represented by a single class and four orders (the *Kosmotogales*, the *Mesoaciditogales*, the *Petrotogales*, and the *Thermotogales*), each with one family (Bhandari and Gupta 2014). These are thermophilic or hyperthermophilic bacteria with the exception of *Mesotoga* spp. which are the only mesophilic representatives of this phylum (see (Ben Hania et al. 2011)). They are all Gram-negative and anaerobic and have an outer membrane larger than the bacterial body that appears to be wrapped around it (Reysenbach 2001b). One family (*Thermotogaceae*) includes several genera including *Fervidobacterium* and *Thermosipho* (*Fervidobacteriaceae*), *Oceanotoga*, *Pseudothermotoga*, *Thermotoga*, and *Thermopallium* (*Thermotogaceae*), isolated for the most part from deep and hot environments such as aquifers, oil reservoirs, or underground volcanic sources. These bacteria are fermentative with the exception of *Mesotoga* spp. which grow only in the presence of electron acceptors (i.e., elemental sulfur) and thus rather oxidize their substrates; some may use either thiosulfate or elemental sulfur as electron acceptors to be reduced to H_2S. The name derives from the ancient Greek θερμus "warm" and the Latin *toga*, gown.

2.2.2.3 Phylum-B3: *Thermodesulfobacteria* (Garrity and Holt 2001c)

The *Thermodesulfobacteria* is a phylum with a single order, the *Thermodesulfobacteriales*; a single family, the *Thermodesulfobacteriaceae*; and a single genus initially (Garrity and Holt 2001c), *Thermodesulfobacterium*. These thermophilic anaerobic bacteria have been isolated from thermal springs and deep-sea hydrothermal vents and are able to use sulfate as electron acceptor to metabolize lactate and pyruvate under thermophilic conditions (Bhatnagar et al. 2015) or to dismutate elementary sulfur. Later, other genera with closely related phenotypic features were added to the phylum such as *Caldimicrobium* (Miroshnichenko et al. 2009), *Thermodesulfatator* (Moussard et al. 2004), *Thermosulfurimonas* (Slobodkin et al. 2012), and *Geothermobacterium* shown to reduce Fe(III) (Kashefi et al. 2002). The name derives from the ancient Greek θερμus "warm," the Latin *sulfur*, sulfur, and ancient Greek βακτήριον for "rod."

2.2.2.4 Phylum-B4: *Deinococcus-Thermus* (Weisburg et al. 1989)

There are 1 class and 2 orders (*Deinococcales* and *Thermales*) having 57 species of genus *Deinococcus*, 1 species of genus *Deinobacterium*, 19 species of genus *Thermus*, and about 50 species in the other genera of the *Thermaceae*. The two genera that give their names to the phylum are among the best known in relation to extreme environments. One, *Deinococcus*, has been isolated from sludge, soil, and other biotopes and has been studied in details for its unusual ionizing radiation resistance through unusual mechanisms (Zahradka et al. 2006) to withstand levels of gamma radiation that kill all other soil microbes (Rainey et al. 2005). The other,

Thermus, isolated from thermal springs (Brock and Freeze 1969) is best known to have yielded the thermoresistant *Taq* polymerase that has made PCR a ubiquitous technique in biology (Brock 1997). The two have been grouped into a unique phylum (Weisburg et al. 1989) despite a high-GC bias that tends to group it with other thermophilic eubacteria, e.g., *Thermotoga maritima* and *Thermomicrobium roseum*. The two genera are aerobic chemoheterotrophs and stain Gram-positive, but it was also shown that *Deinococcus* was capable of anaerobically using as electron acceptor Fe(III)-nitrilotriacetic acid coupling it to the incomplete oxidation of lactate to CO_2 and acetate even though it was unable to link this process to growth (Fredrickson et al. 2000). The name derives from the ancient Greek δεινός (deinos) and κόκκος (kokkos) meaning terrible grains to refer to the unusual resistance to radiations, while the name of the second derives from the ancient Greek θερμus "warm."

2.2.2.5 Phylum-B5: *Chrysiogenetes* (Garrity and Holt 2001d)

The *Chrysiogenetes* comprise a single order, the *Chrysiogenales*; a single family, the *Chrysiogenaceae*; and one genus *Chrysiogenes* (Garrity and Holt 2001d) to which two others were subsequently added, *Desulfurispira* (Sorokin and Muyzer 2010) and *Desulfurispirillum* (Sorokin et al. 2007), for a grand total of four species. Species of these two last genera use small organic molecules as electron donors and minerals (metals, metalloids, elementary sulfur) as electron acceptors. The best-known species, *Chrysiogenes arsenatis*, was isolated from a gold mine wastewater containing arsenate that was found able to be used as electron acceptor with acetate as electron donor, yielding CO_2 and arsenite (Macy et al. 1996; Coil et al. 2013). Nitrate or nitrite can also be used as electron acceptors, producing NH_4. Other members of the phylum are anaerobic elemental sulfur- and nitrate-reducing bacteria isolated from a sulfide-removing bioreactor (Sorokin et al. 2007) or a soda lake (Sorokin and Muyzer 2010). The name derives from the ancient Greek χρυσός, gold, and γένος, type.

2.2.2.6 Phylum-B6: *Chloroflexia* (*Chloroflexi*) (Gupta et al. 2013)

The *Chloroflexia* comprise seven classes (*Anaerolineae*, *Ardenticatenia*, *Caldilineae*, *Chloroflexia*, *Ktedonobacteria*, *Thermoflexia*, and *Thermomicrobia*). The group was initially known as the green non-sulfur bacteria (Oyaizu et al. 1987), changed to *Chloroflexi* (Garrity and Holt 2001b) but was later corrected to *Chloroflexia*, with "ia" which is the usual ending for bacterial phyla (Gupta et al. 2013). The phylum has little homogeneity which contains anoxygenic photoautotrophic microbes, aerobic chemoheterotrophs, thermophilic organisms, as well as anaerobic organisms that obtain energy by reductive dehalogenation of organic chlorinated compounds such as tetrachloroethane and trichloroethane or through fermentation. The name derives from the ancient Greek χλωρός, light green, which refers to the light-harvesting capability, and the Latin *flexus*, bended, which refers to the filamentous form of the cells.

The *Thermomicrobia* were proposed as a deep-branching phylum of thermophilic non-photosynthetic bacteria recovered from hot springs or spring soils (Garrity and Holt 2001e), and it has been proposed to be moved to the *Chloroflexia* (Hugenholtz and Stackebrandt 2004; Gupta et al. 2013). The type species *Thermomicrobium roseum* is a red-pigmented motile aerobic CO-oxidizing Gram-negative rod (Wu et al. 2009).

2.2.2.7 Phylum-B7: *Thermomicrobia*

The *Thermomicrobia*, initially considered a distinct phylum, have now been moved to Phylum B6, the Chloroflexia (described above).

2.2.2.8 Phylum-B8: *Nitrospirae* (Garrity and Holt 2001f)

Phylum *Nitrospirae* is a small group of one class/one order/four families. It comprises deep-branching bacteria (Garrity and Holt 2001f) recovered from deep-sea vent niches, soils, lakes, and acid mine drainage. Initially characterized as a group of chemolithotrophs that played a major role besides genus *Nitrobacter* in nitrate oxidation (Hovanec et al. 1998), it was also found to be able to fix nitrogen under acidic conditions (Tyson et al. 2005). The name derives from the Latin *nitro*, saltpeter, which refers to the substrate metabolized, and *spira*, a coil, which refers to the spiraling shape of the cells.

2.2.2.9 Phylum-B9: *Deferribacteres* (Garrity and Holt 2001a)

Phylum *Deferribacteres* comprises a single order, the *Deferribacteres*; a single family, the *Deferribacteraceae*; and several genera, prominent among which are *Deferribacter* and *Geovibrio* (Garrity and Holt 2001a). These are straight or curved anaerobic rods isolated from marine sediments or oil tanks and are chemoorganotrophic using nitrate, iron, manganese, or cobalt as electron acceptors for anaerobic respirations. The name from the Latin *ferrum* refers to the rare ability to use iron as electron acceptor and from the ancient Greek βακτήριον for "rod."

2.2.2.10 Phylum-B10: *Synergistetes* (Jumas-Bilak et al. 2009)

Phylum *Synergistetes* comprises a single class (*Synergistia*), a single order (*Synergistales*), and a single family (*Synergistaceae*). It is a small group of rod-shaped Gram-negative, anaerobic, chemo- and fermentative organotrophic isolated from the rumen of animals, other animal niches, and anaerobic bioreactors (Jumas-Bilak et al. 2009). These bacteria respire anaerobically in the rumen of animals and the human intestinal flora and have been associated recently with periodontal disease (Yu et al. 2016) and with wastewater (Hamdi et al. 2015). The name derived from the English "synergist," a coworker, which presumably refers to metabolism of the rumen bacteria that degrade toxins and thus permit other microbes to digest plant material in the rumen.

2.2.2.11 Phylum-B11: *Cyanobacteria* (Woese et al. 1985)

If the ten first bacterial phyla have been discovered and described recently, phylum #11 is a major ecological contributor that has been known for a long time. Described in the first editions of the Bergey's as the first dichotomy, they are oxygenic photosynthetic at the root of the food web in many aquatic biotopes. There are thousands of references on cyanobacteria illustrating the ecological engineering role they play, fixing CO_2, generating O_2, and yielding food as phytoplankton. There are several phylogenetic treatments; the one in the Bergey's considers there are 14 lineages and several unrelated strains (Castenholz 2001), among which the best knowns are the heterocystous cluster, the *Prochlorococcus/Synechococcus* lineage, and the *Prochlorothrix* lineage. These Gram-negative bacteria perform oxygenic photosynthesis with photosystems I and II inserted in membrane systems (thylakoids) and are considered to be the ancestors of present-day chloroplasts. In eutrophic biotopes, they cause blooms visible from the air that can rot and cause mass killing of fish and land animals. They are also the origin of the stromatolites (Chap. 3). This phylum is one of the few where all members are photosynthetic, except for a recently described basal lineage of non-photosynthetic species called *Vampirovibrio chlorellavorus* (Soo et al. 2015). They are also one of the best examples of multicellular prokaryotes since they have specialized non-photosynthetic nitrogen-fixing cells called heterocysts that exchange with neighboring cell photosynthates for fixed nitrogen (Wolk 2000). They are considered responsible for the great oxidation event that 2.2–2.4 BY ago transformed the earth from an anaerobic to a predominantly aerobic world (Schopf 2014). Their name derives from the Greek κυανός (*kyanós*), blue, because of their chlorophyll pigments and from the ancient Greek βακτήριον for "rod."

2.2.2.12 Phylum-B12: *Chlorobi* (Garrity and Holt 2001g)

The *Chlorobi* is a small phylum with three classes (the *Chlorobia*, the *Dehalococcoidia*, and the *Ignavibacteria*), each with one order and one or two families (*Chlorobiales/Chlorobiaceae*, *Dehalococcoidales/Dehalococcoidaceae* and *Ignavibacteriales/Melioribacteraceae* and *Ignavibacteriaceae*) and initially five genera, among which the *Chlorobium*, considered to branch close to the *Bacteroidetes* and *Fibrobacteres* phyla (Gupta 2004). These bacteria are Gram-negative, anaerobic, obligately phototrophic (*Chlorobia*), or chemotrophic (*Dehalococcoidia* and *Ignavibacteria*). Some species use sulfide or thiosulfate as electron donor for CO_2 fixation, and some accumulate sulfur globules on the outside of cells (*Chlorobia*) *Dehaloccoides* are important microorganisms that intervene in dehalogenation of chlorinated compounds. Ammonia or dinitrogen are used as nitrogen source. They can establish a symbiosis with a *Betaproteobacterium*, exchanging motility, which is fundamental to phototrophic bacterium, for nitrogen and carbon sources (Cerqueda-Garcia et al. 2014). Initially described as "green sulfur bacteria," their name derives from the Greek "χλωρος" pale green and "βιος" life.

2.2.2.13 Phylum-B13: *Proteobacteria* (Woese et al. 1985)

The *Proteobacteria* are the most abundant group of bacteria, with six classes: the *Alphaproteobacteria*, the *Betaproteobacteria*, the *Gammaproteobacteria*, the *Deltaproteobacteria*, the *Epsilonproteobacteria*, and the *Zetaproteobacteria.* They comprise 16 (alpha), 11 (beta), 20 (gamma), 9 (delta), 2 (epsilon), and 1 order (zeta) and share only a few features besides their Gram-negative wall (Woese 1987). It has been hypothesized they derive from a photosynthetic ancestor, hence their first name "the purple bacteria and their relatives," a characteristic that would have been lost in most branches. They constitute one of the most successful bacterial lineages with plant pathogens, animal pathogens, saprophytes, etc., with a negative Gram stain and an outer membrane outside of the cytoplasmic membrane separated by the periplasmic space. These bacteria perform oxidative phosphorylation and are considered to be the ancestors of present-day mitochondria. The name derives from the ancient Greek Πρωτέας, the name of the god of the seas Proteus who was capable of assuming different shapes to refer to the many shapes and physiological abilities of the different lineages. *Proteobacteria* comprise *Agrobacterium* (Chap. 4), *Rhizobium* (Chap. 5), etc.

2.2.2.14 Phylum-B14: *Firmicutes* (Woese et al. 1985)

The *Firmicutes* are one of the most abundant groups of bacteria, comprising two major classes (the *Bacilli* and the *Clostridia*) and five minor classes (*Erysipelotrichia*, *Limnochordia*, *Negativicutes*, *Thermolithobacteria*, and the *Tissierellia*) with 26 families and 223 genera altogether (Schleifer 2001). Their main features shared by a large proportion of strains are a Gram-positive cell wall, a low genomic G + C, and the production of endospores. The *Bacilli* are aerobic with a few anaerobes and comprise the orders *Bacillales* and *Lactobacillales*; they are found in all biotopes and comprise a few well-known pathogens such as *B. anthracis*. *Clostridia* are generally anaerobic; comprise the orders *Clostridiales*, *Halanaerobiales*, and *Thermoanaerobacterales*; and comprise a few well-known pathogens such as *C. tetani* but are on the whole considered ubiquitous. The *Erysipelotrichia* comprise only one order with one family, the *Erysipelotrichaceae*, are anaerobic Gram-positive rods without endospore, and are aerobic to facultatively anaerobic. The *Mollicutes* that were treated as part of the *Firmicutes* in previous editions (2001) are now considered as sufficiently different on phenotypical ground to warrant a placement separate from that of the *Firmicutes*. The name derives from the Latin *firmus*, strong, and *cutis*, skin, referring to the Gram-positive strong cell wall. *Firmicutes* comprise *Bacillus* (Chap. 5), etc.

2.2.2.15 Phylum-B15: *Tenericutes (Mollicutes)* (Murray 1984)

The *Tenericutes* comprise one class, the *Mollicutes*, with five orders (*Acholeplasmatales*, *Anaeroplasmatales*, *Entomoplasmatales*, *Haloplasmatales*, and *Mycoplasmatales*), each with one or two families, each with one (or two) genera, and many *Candidatus*. They are an elusive group of bacteria formerly grouped on the basis of their 16S rRNA together with the *Firmicutes* and previously designated *Mollicutes* (Edward and Freundt 1967). Their main feature is the lack of a cell wall that only permits them to survive in isotonic media such as animal or plant tissues. The *Mollicutes* are also known to have rapid mutation rates (Woese et al. 1984) and an alternative genetic code where UGA codes for trp (and not "stop" as in other bacterial lineages) (Yamao et al. 1985) and CGG codes for termination (and not for arg) (Oba et al. 1991). They are widespread pathogens of plants and animals (Browning and Citti 2014), a common contaminant of cell cultures (Siqueira et al. 2017) and also free-living as seen recently from enrichment of methane seeps (Skennerton et al. 2016). Their former name *Mollicutes* originates from the Latin *mollis*, soft, and *cutis*, skin, referring to their wall-less envelope. The new name *Tenericutes* also originates from the Latin *tener* meaning "tender" and "young," referring to the same envelope feature.

2.2.2.16 Phylum-B16: *Actinobacteria* (Stackebrandt et al. 1997)

The *Actinobacteria* are a large group of mainly soil bacteria. They comprise six classes: the *Acidimicrobiia* (with 1 order the *Acidimicrobiales* with 3 families the *Acidimicrobiaceae*, the *Iamiaceae*, and the *Microthrixaceae*), the *Actinobacteria* (with 18 orders the *Acidothermales*, the *Actinomycetales*, the *Actinopolysporales*, the *Bifidobacteriales*, the *Catenulisporales*, the *Corynebacteriales*, the *Frankiales*, the *Geodermatophilales*, the *Glycomycetales*, the *Jiangellales*, the *Kineosporiales*, the *Micrococcales*, the *Micromonosporales*, the *Nakamurellales*, the *Propionibacteriales*, the *Pseudonocardiales*, the *Streptomycetales*, and the *Streptosporangiales* with too many families to be listed here), the *Coriobacteriia* (with 2 orders the *Coriobacteriales* with 2 families the *Atopobiaceae* and the *Coriobacteriaceae* and the *Eggerthellales* with 1 family the *Eggerthellaceae*), the *Nitriliruptoria* (with 4 orders each with a family the *Egibacterales/Egibacteraceae*, the *Egicoccales/Egicoccaceae*, the *Euzebyales/Euzebyaceae*, and the *Nitriliruptorales/Nitriliruptoraceae*), the *Rubrobacteria* (with 2 orders each with a family the *Gaiellales/Gaiellaceae*, the *Rubrobacterales/Rubrobacteraceae*), and the *Thermoleophilia* (with 2 orders the *Solirubrobacterales* with 4 families the *Conexibacteraceae*, the *Parviterribacteraceae*, the *Patulibacteraceae*, and the *Thermoleophilales* with 1 family the *Thermoleophilaceae*). They have been recovered mostly from soils but also from freshwaters and brackish waters, and they include the most deadly human and animal pathogens, the tuberculosis agent *Mycobacterium tuberculosis* among many. *Actinobacteria* have several unique features such as the production of sporangiospores carrying exospores and the presence of a proteasome to recycle damaged or misfolded proteins, and although not unique they also have a Gram-positive wall, a high G + C% genome, and hyphae well adapted to a lacunar biotope such as the soil and synthesize a large array of secondary metabolites well adapted to oligotrophic biotopes that favors slow-growing cells. Although generally described as large-celled, high-GC large genome like, for instance, *Streptomyces* with 4 μM long (Szeszak et al. 1973) and 73% GC, 10 Mb genome cells (Bentley et al. 2002), metagenomics has revealed the existence of small (average diameter of 0.292 μm lowest described for any planktonic prokaryote), low GC (33%) small genome (<1 Mb) cells associated with picocyanobacteria and named Candidatus Pelagibacter ubique *Pelagibacter ubique* (Ghai et al. 2013). The presence of a proteasome has been considered by some (Cavalier-Smith 2002) as an argument for the actinobacteria being ancestors to the eukaryotes, but this proposal has not been accepted by most. The anaerobic *Atopobium* and the *Coriobacteriia* have been considered ancestral based on 16S phylogeny (Embley and Stackebrandt 1994) or on molecular signatures (Gupta and Bao 2012). The positioning of anaerobic lineages at the root of the actinobacteria is coherent with an anaerobic early earth that changed to an aerobic one 2.2–2.4 BY ago (Schopf 2014). The initial name derives from the Greek ακτίνα (ray) and μύκητες (fungi) to illustrate the aspect of colonies with radiating hyphae mistakenly thought at first to be fungi, now replaced to "bacteria" from the ancient Greek βακτήριον for "rod." *Actinobacteria* comprise the symbiont *Frankia* (Chap. 5), etc.

2.2.2.17 Phylum-B17: *Planctomycetes* (*Planctobacteria*) (Ward 2011)

The *Planctomycetes* are a small group of aquatic bacteria for which the name *Planctobacteria* has been proposed to eliminate any reference to eukaryotes (Cavalier-Smith 2002). They comprise two classes: the *Phycisphaerae* (with two orders the *Phycisphaerales* with one family the *Phycisphaeraceae* and the *Tepidisphaerales* with one family the *Tepidisphaeraceae*) and the *Planctomycetia* (with one order the *Planctomycetales* with three families: the *Gemmataceae*, the *Isosphaeraceae*, and the *Planctomycetaceae*). They are recovered from freshwater and brackish water bodies, on sea corals, from various soils, and from oxidation ditches where municipal sewage water is treated and where *Planctomyces* were found to proliferate to reach 13.5% of the microbes present (Chen et al. 2016). *Planctomycetes* are the only bacteria capable of oxidizing anaerobically ammonium (Strous et al. 1999), a metabolic transformation called anammox whereby nitrite supplies the oxygen to transform ammonium into dinitrogen, a process very useful for treatment of ammonium-rich wastewater, based on a specific enzyme, the hydrazine synthase (Dietl et al. 2015). The name derives from the Greek πλανκτον (to float) and μύκητες (fungi) to illustrate the fact large masses of cells thought to be eukaryotic were seen floating, now replaced by "bacteria" from the ancient Greek βακτήριον for "rod."

2.2.2.18 Phylum-B18: *Chlamydiae*

The phylum contains one class (*Chlamydiae*), one order (*Chlamydiales*), and four families (*Chlamydiaceae*, *Parachlamydiaceae*, *Simkaniaceae*, and *Waddliaceae*). It forms together with *Planctomycetes* and *Verrucomicrobia*, the super phylum PVC. It contains obligatory intracellular parasitic and pathogenic bacteria responsible for lung and urogenital infections. Chlamydia are also often found in amoebae where they are considered as symbionts (Horn 2008). Chlamydial infections are often asymptomatic and they are thus considered a highly successful pathogen (Mohamad et al. 2014). The name derives from the ancient Greek χλαμύδιον, small cloak, referring to their apparent ability to cover or cloak the nuclei of infected cells (Byrne 2003); however what was originally thought to be the cloak is in fact a host-derived cytoplasmic vesicle containing numerous individual organisms and is termed an inclusion.

2.2.2.19 Phylum-B19: *Spirochaetes* (Cavalier-Smith 2002)

The *Spirochaetes* comprise one class (*Spirochaetes*) and four orders: the *Brachyspirales* (with one family, the *Brachyspiraceae*), the *Brevinematales* (with one family, the *Brevinemataceae*), the *Leptospirales* (with one family, the *Leptospiraceae*), and the *Spirochaetales* (with two families the *Borreliaceae* and the *Spirochaetaceae*). They are a group that has one conspicuous morphological feature, a spiraling shape linked to their axial filaments that run along the length of the cells and cause a twisting motion. Most strains stain Gram-negative and are heterotrophic, free-living, and anaerobic. Their size is unusual with a length that ranges from 3 to 500 μM. They contain a few well-known pathogens such as *Treponema pallidum*, the agent of syphilis, yaws, and other diseases; *Borrelia burgdorferi*, the agent of Lyme disease or borreliosis; and *Leptospira* spp., the agents of leptospirosis. *Spirochaetes* have been hypothesized (Margulis 1981) to have been the source of the eukaryotic flagellum although this has been disputed on the ground that they carry no DNA unlike mitochondria and chloroplasts and that the sequence of the proteins have no homology. The name derives from the ancient Greek σπείρα, a coil or a spiral, and χαίτη "mane of hair."

2.2.2.20 Phylum-B20: *Fibrobacteres* (Garrity and Holt 2001h)

The *Fibrobacteres* comprise three classes: the *Chitinispirillia* (with one order the *Chitinispirillales* and one family, the *Chitinispirillaceae*), the *Chitinivibrionia* (with one order the *Chitinivibrionales* and one family, the *Chitinivibrionaceae*), and the *Fibrobacteria* (with two orders/each with one family, the *Fibrobacterales/*

Fibrobacteraceae and the *Fibromonadales/Fibromonadaceae*). This small phylum is related to the phyla *Bacteroidetes* and *Chlorobi* with which it shares molecular signatures and thus they may form a single phylum in the future (Gupta 2004). They constitute a group of bacteria recovered from the rumen and thus thought to contribute to the digestion of plant fibers. Their main feature is their ability to degrade under anaerobic conditions the cellulose present in mammal guts as well as in landfills, freshwater, or termites guts (Ransom-Jones et al. 2012). The name derives from the Latin *fibra*, fiber, and from the ancient Greek βακτήριον for "rod."

2.2.2.21 Phylum-B21: *Acidobacteria* (Thrash and Coates 2011)

The *Acidobacteria* comprise four classes: the *Acidobacteriia* (with one order the *Acidobacteriales* and one family, the *Acidobacteriaceae*), the *Blastocatellia* (with several taxa), the *Holophagae* (with two orders/each with one family, the *Acanthopleuribacterales/Acanthopleuribacteraceae* and the *Holophagales/Holophagaceae*), and the *Solibacteres* (with one order the *Solibacterales* and two families, the *Bryobacteraceae* and the *Solibacteraceae*). They constitute an elusive group of bacteria that have been often identified in molecular catalogs, especially those made on soils but with very few representatives in culture. However, since most strains can only be defined on the basis of a 16S sequence, the concept of divisions is also often used, and there would be 26 such divisions in the phylum (Barns et al. 2007). Their main feature is their acidophily although this is not a consistent characteristic. Most isolates are heterotrophic, aerobic, or microaerophilic, and some species (*Telmatobacter bradus*, *Acidobacterium capsulatum*) are facultative anaerobic bacteria (Kielak et al. 2016). The first recognized strain and species of the phylum *Acidobacteria* was *Acidobacterium capsulatum* obtained from an acid mine drainage in Japan (Kishimoto and Tano 1987). The name derives from the Latin *acidus*, sour, and from the ancient Greek βακτήριον for "rod."

2.2.2.22 Phylum-B22: *Bacteroidetes* (Woese et al. 1985)

The *Bacteroidetes* (Woese et al. 1985) are a group of bacteria, comprising six classes: the *Bacteroidia*, the *Chitinophagia*, the *Cytophagia*, the *Flavobacteriia*, the *Saprospiria*, and the *Sphingobacteriia*, with 25 families and more than 350 genera altogether (Schleifer 2001). They are obligatory anaerobes, Gram-negative devoid of endospores present abundantly in the feces of mammals. The *Bacteroidetes* are a poorly known phylum because in part of their strict anaerobic requirement and thus the difficulties in isolating them. They have been associated with the fight against obesity since a low *Bacteroidetes/Firmicutes* ratio has been found in obese mice (Ley 2010). Their name derives from the ancient Greek βακτήριον for "rod" and εἶδος, form or kind.

2.2.2.23 Phylum-B23: *Fusobacteria* (Garrity and Holt 2001h)

The *Fusobacteria* comprise one class: the *Fusobacteriia* (with one order the *Fusobacteriales*) and two families the *Fusobacteriaceae* and the *Leptotrichiaceae.* They are a small group of Gram-negative obligate anaerobic bacteria recovered from a variety of biotopes such as ulcers, caries (Dige et al. 2014), or sea surface oil slicks (Gutierrez et al. 2016). They are known to be involved in a wide spectrum of human infections ranging from tissue necrosis and septicemia to more recently described cases of intra-amniotic infection and preterm births (Gauthier et al. 2011). They have also been found to be normal gut microbes although they have been found more abundant in colorectal cancer patients (Suehiro et al. 2017). The name originates from the Latin *fus*, spindle, and refers to the shape of the cells that are tapered at both ends, although many species do not have that shape (Gharbia et al. 2010).

2.2.2.24 Phylum-B24: *Verrucomicrobia* (Hedlund et al. 1997)

The *Verrucomicrobia* are a small group of soil, seawater, and eukaryotes gut bacteria. They comprise four classes: the *Gemmatimonadetes* (with one order the *Methylacidiphilae* and one family the *Methylacidiphilaceae*), the *Opitutae* (with two orders each with a family, the *Opitutales*/*Opitutaceae* and the *Puniceicoccales*/*Puniceicoccaceae*), the *Spartobacteria* (with one order the *Chthoniobacterales* and one family the *Chthoniobacteraceae*), and the *Verrucomicrobiae* (with one order the *Verrucomicrobiales* and one family the *Verrucomicrobiaceae*). They have been recovered from freshwater, soils, and human feces. Uncultivated strains have been identified as symbionts of eukaryotes such as nematodes. One isolate, a moderately halophilic, obligately anaerobic, and saccharolytic bacterium that thrives in the suboxic transition zones of hypersaline microbial mats, was published very recently (Spring et al. 2016). The name originates from the Latin *verruca*, wart, because of the wart-like morphology of colonies.

2.2.2.25 Phylum-B25: *Gemmatimonadetes* (Zhang et al. 2003)

The *Gemmatimonadetes* comprise two classes: the *Gemmatimonadetes* (with one order the *Gemmatimonadales* and one family the *Gemmatimonadaceae*) and the *Longimicrobia* (with one order the *Longimicrobiales* and one family the *Longimicrobiaceae*). They are a very small group of soil bacteria identified through molecular cataloging as making up 2% of the soil community (DeBruyn et al. 2011) but that are in general resistant to cultivation. They have a large phylogenetic breadth (19% 16S rDNA sequence divergence). One representative, *Gemmatimonas aurantiaca*, is a Gram-negative, aerobic, rod-shaped polyphosphate-accumulating microorganism and the first isolate of the phylum (Zhang et al. 2003). The name originates from the Latin *gemma*, buds, and the Greek μονάσ, unit, and refers to the reproduction of cells by binary fission with buds, which suggests asymmetrical cell division.

2.2.2.26 Phylum-B26: *Lentisphaerae* (Cho et al. 2004)

The *Lentisphaerae* together with the three phyla *Planctomycetes*, *Verrucomicrobia*, and *Chlamydiae* form a superphylum (Lagkouvardos et al. 2014), a concept that has no official value but nevertheless indicates a recent common ancestor after which the four phyla have diverged to occupy different niches. They comprise two classes: the *Lentisphaeria* (with two orders, the *Lentisphaerales* with one family the *Lentisphaeraceae* and the *Victivallales* with one family the *Victivallaceae*) and the *Oligosphaeria* (with one order the *Oligosphaeriales* and one family the *Oligosphaeriaceae*). A few isolates have been obtained from seawater; they are Gram-negative, strictly aerobic, chemoheterotrophic, and facultatively oligotrophic sphere-shaped bacteria and synthesize copious amounts of exo-polymers (Cho et al. 2004). The name derives from the Latin *lentus*, sticky, and *sphaera*, sphere, and refers to the abundant gummy exo-polymers produced by the cells.

2.2.2.27 Phylum-B27: *Dictyoglomi* (Patel 2011)

The *Dictyoglomi* are a very small group of Gram-negative thermophilic bacteria recovered from volcanic hot spring environments or paper-pulp factory effluent. They comprise one class: the *Dictyoglomia* (with one order the *Dictyoglomales* and one family the *Dictyoglomaceae*). They are known to be thermophilic and anaerobic chemoorganotrophs (Saiki et al. 1985) and have been characterized as specialized carbohydrate fermentor due to their genomes having a wide range of glycosyl hydrolases among which active xylanases (Brumm et al. 2016). Their morphology is unusual with filaments, bundles, and spherical bodies. The name originates from ancient Greek δίκτυον, a net, and the Latin *glomus*, a sphere, to refer to two aspects of the morphology of the culture.

2.2.2.28 Phylum-B28: *Caldiserica* (Mori et al. 2009)

The *Caldiserica* are a very small phylum with one class, the *Caldisericia* (with one order the *Caldisericales* and one family the *Caldisericaceae*) of Gram-negative thermophilic bacteria initially recovered from hot spring environments (Mori et al. 2009) and found later in anaerobic sludge reactor (Aida et al. 2014) or in freshwater (Kadnikov et al. 2012). They are known to be thermophilic and obligately anaerobic chemoorganotrophs with multicellular filaments, a single polar flagellum, forming complex aggregates. The name originates from the Latin *caldus*, hot, and *sericum* silk presumably to refer to the morphological aspect of the filaments in culture.

2.2.2.29 Phylum-B29: *Elusimicrobia* (Geissinger et al. 2009)

The *Elusimicrobia* comprise one class: the *Elusimicrobia* (with one order the *Elusimicrobales* and one family the *Elusimicrobiaceae*). They are a very small group of Gram-negative mesophilic bacteria initially recovered from the gut of a beetle larva (Geissinger et al. 2009) found later in freshwater and soils (Wang et al. 2016) and in the gut of varied mammals (Wong et al. 2016) for which very few isolates exist. They are a typically rod-shaped ultramicrobacteria with a very small volume and a very small genome (1.64 Mb). The name originates from the Latin *elusus*, deceived, and the ancient Greek μικρό, small, and βιος, life, to refer to the elusive aspect of the cells that are hard to cultivate.

2.2.2.30 Phylum-B30: *Armatimonadetes* (Tamaki et al. 2011)

The *Armatimonadetes* comprise two classes: the *Armatimonadia* (with one order the *Armatimonadales* and two families) and the *Fimbriimonadia* (and one family). They are a very small group of aerobic, mesophilic chemoheterotrophic bacteria initially recovered from the rhizoplane of an aquatic plant, a reed (Tamaki et al. 2011). Cells are Gram-negative and ovoid to rod-shaped. They are a typically rod-shaped ultramicrobacteria with a very small volume and a very small genome (1.64 Mb). The name originates from the Latin *armati*, armored, and the ancient Greek μονάσ, unit, to refer to the thick aspect of the cell walls (Fig. 2.1).

2.3 Reflections on Genome-Based Phylogeny: The Classification of the Future?

Due in part to the seminal work of Woese (Fox et al. 1980), the 16S rRNA gene has become the golden standard of bacterial taxonomy, the first gene to be sequenced to identify an unknown isolate, the target of choice to amplify a complex community, and as a consequence probably the gene with the highest number of entries in data banks. However, this gene suffers from several drawbacks, such as the fact it is sometimes on a plasmid as in *Paracoccus* (Battermann et al. 2003), sometimes present in divergent copies as in *Thermomonospora* (Yap et al. 1999), and the most important problem being that it is a single marker, thus subject to errors. It has thus been found to be less reliable than another single marker, the longer 23S rRNA gene (Sen et al. 2014).

A basic tenet of biology is that any finding should be supported by two or more independent approaches, which is why several works have aimed at establishing phylogeny of various taxa using other markers. The conclusion of such works has been that on the whole these other markers did yield comparable topologies "with exceptions." For instance, the RecA-based phylogeny did not permit to cluster the *Firmicutes* (Eisen 1995). One reason for such differences may have been lateral transfers, and another may have been errors or functional constraints. The multi-locus sequence analysis (MLSA) by determining five housekeeping genes was an attempt at overcoming these problems (Maiden et al. 1998). This approach by using several housekeeping genes scattered on the genome that are considered less prone to lateral transfers is considered to yield more reliable topologies.

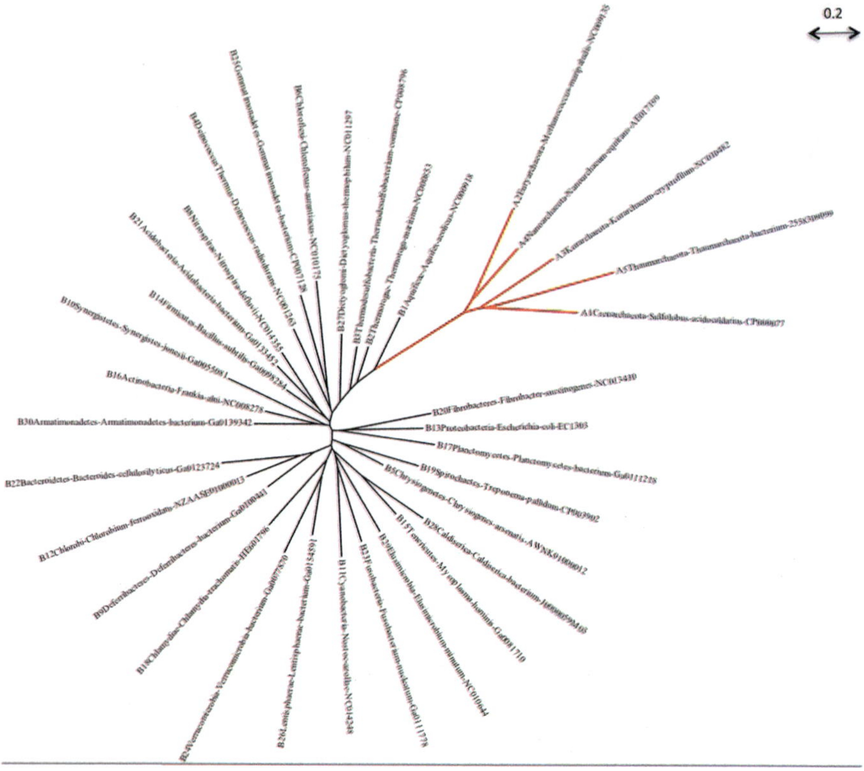

Fig. 2.1 23S rRNA-based phylogenetic tree. In red the archaeal lineages
A neighbor-joining (Saitou and Nei 1987) 23S rRNA-based phylogenetic tree of the bacterial and archaeal phyla described above recovered from the JGI site. In red are the archaeal phyla

However given the constantly sinking sequencing costs and the availability of complete genomes, the question has arisen of the use of whole genomes to compare strains. In recent years, several approaches have been proposed to compare whole genomes, the best known being the ANI where the DNA of genome A is cut into 1000 nt fragments, and these are compared with the ultrafast algorithm BlastN to genome B yielding a metric that has been compared to the DNA/DNA reference (Goris et al. 2007). This method is robust and can process large number of sequences (Fig. 2.2).

Several other approaches have been aimed to improve the robustness of the calculations to repeated sequences and to gaps in draft genomes (Meier-Kolthoff et al. 2013). These metrics carry different names such as Genome Blast Distance Phylogeny (GBDP), genome-genome distance (GGD), or in silico DNA-DNA hybridization (isDDH) (Colston et al. 2014), and in general they correspond to wet-laboratory-derived DDH. The ANI metrics appear to be the most cited (629 in April 2017), while the GBDP has been cited 296 times.

The Prunier approach (Abby et al. 2010) aims to provide an estimate of the robustness of a tree independent of the marker(s) used. The bootstrap developed by

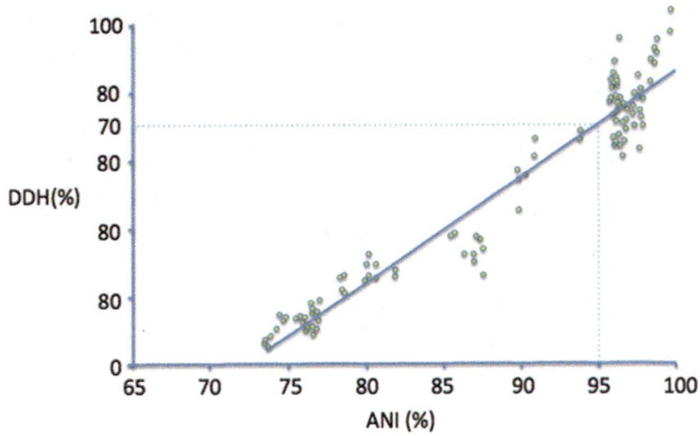

Fig. 2.2 Correspondence between DDH and in silico ANI results
Correspondence between the reference DDH method and the in silico ANI method (Goris et al. 2007). Derived from *Bacillus*, *Burkholderia*, *Escherichia*, *Pseudomonas*, *Shewanella*, and *Streptococcus*, this shows that a DDH of about 70% which is the threshold for species delineation corresponds to an ANI of 95%

Felsenstein (Felsenstein 1985) and widely used is based on a subsampling approach to in effect estimate if all positions in an alignment yield the same topology. In contrast, the Prunier approach uses two categories of proteins, one being the UU (universal unicopy) which is the most conserved ones and the UN (universal nonunicopy). The former are used to build a tree, and the latter are used to quantify the topology differences. Applied to *Actinobacteria* this approach has resulted in marked reassessments of the former 16S rRNA-based tree and emendation of several orders (Sen et al. 2014).

The rising availability of genomes and molecular markers also has made the naming of species based on a reduced number of strains. In the age of phenotypes, polyphasic taxonomy aimed at identifying groups of strains that shared phenotypic markers and designated them as species. A recent tendency is to focus on a single isolate, and provided it is distant enough in its 16S rRNA gene (<97%) or in its whole genome (<95%), then this isolate will be considered a representative of a species to which other strains may or not be aggregated later.

2.4 Discussion on the Definition of Species in Prokaryotes

The concept of species has been developed for plants and animal and is based on the success of sexual reproduction. It is often defined as a group of organisms in which two individuals can produce fertile offsprings by sexual reproduction (Mayr 1963). This broad definition has several exceptions such as those species that have no or infrequent sexuality like the yeasts or higher organisms that have bouts of parthenogenesis. Another problem is hybridization between related species.

Bacteria and Archaea can exchange genetic material, but these exchanges do not involve the whole genome, are infrequent, and can involve very distant organisms. The ability of microbes to exchange genetic material is also dependent on the mismatch system that can mutate, thereby promoting divergence to be reacquired by lateral transfer to create barriers to the emerging lineages (Vulic et al. 1999). Therefore the ability to exchange genes in the bacterial or archaeal domains cannot be the basis on which to define species but may contribute to the emergence of new physiological functions.

Nevertheless, the use of the notion of species in asexual bacteria and archaea is likely to continue, if only because of the need for a universal nomenclatural system.

2.5 Conclusion

Microbiology is a science that comprises today thousands of scientists that attempt to describe the microbial world in terms that need to be understood by colleagues speaking different languages in faraway locales and in different fields of science. This colossal effort applies to human, animal, and plant diseases, to industrial and agricultural processes, and to environmental biotope descriptions, and it requires a generally agreed-upon system to name the objects of the science, the bacterial taxa. Even if considered by some as bothersome and pedantic, bacterial taxonomy is and will continue to be necessary to microbiology.

Genomes in databases have increased in number at an almost exponential rate since the first bacterial genome publication in 1995 (Fleischmann et al. 1995) to reach the present (April 2017) tally of 96,352, of which 1077 are archaeal (http://www.ncbi.nlm.nih.gov/genome) with no sign of slowing. Already these numbers are exceeding those of taxa described in the Bergey's Manual, and this tendency should be reinforced with progress in sequencing techniques and the accompanying drop in prices. Data and database management may thus start to be the real limiting factor, especially if other "omics" such as transcriptomics, proteomics, metabolomics, or epigenetics are added. All these should provide an ever richer, more complex view of bacterial diversity in the different biotopes.

References

Abby SS, Tannier E, Gouy M, Daubin V (2010) Detecting lateral gene transfers by statistical reconciliation of phylogenetic 0020 forests. BMC Bioinformatics 11:324

Adanson M (1763) Familles des plantes. Vincent, Paris

Aida AA, Hatamoto M, Yamamoto M, Ono S, Nakamura A, Takahashi M, Yamaguchi T (2014) Molecular characterization of anaerobic sulfur-oxidizing microbial communities in up-flow anaerobic sludge blanket reactor treating municipal sewage. J Biosci Bioeng 118:540–545

Altschul SF, Madden TL, Schaffer AA, Zhang J, Zhang Z, Miller W, Lipman DJ (1997) Gapped BLAST and PSI-BLAST: a new generation of protein database search programs. Nucleic Acids Res 25:3389–3402

Armon R, Potasman I, Green M (1990) Biochemical fingerprints of *Legionella* spp. by the BIOLOG system: presumptive identification of clinical and environmental isolates. Lett Appl Microbiol 11:290–292

Barns SM, Delwiche CF, Palmer JD, Pace NR (1996) Perspectives on archaeal diversity, thermophily and monophyly from environmental rRNA sequences. Proc Natl Acad Sci U S A 93:9188–9193

Barns SM, Cain EC, Sommerville L, Kuske CR (2007) *Acidobacteria* phylum sequences in uranium-contaminated subsurface sediments greatly expand the known diversity within the phylum. Appl Environ Microbiol 73:3113–3116

Battermann A, Disse-Kromker C, Dreiseikelmann B (2003) A functional plasmid-borne rrn operon in soil isolates belonging to the genus *Paracoccus*. Microbiology 149:3587–3593

Battistuzzi FU, Feijao A, Hedges SB (2004) A genomic timescale of prokaryote evolution: insights into the origin of methanogenesis, phototrophy, and the colonization of land. BMC Evol Biol 4:44

Ben Hania W, Ghodbane R, Postec A, Brochier-Armanet C, Hamdi M, Fardeau ML, Ollivier B (2011) Cultivation of the first mesophilic representative ("mesotoga") within the order *Thermotogales*. Syst Appl Microbiol 34:581–585

Bentley SD, Chater KF, Cerdeno-Tarraga AM, Challis GL, Thomson NR, James KD, Harris DE, Quail MA, Kieser H, Harper D et al (2002) Complete genome sequence of the model actinomycete *Streptomyces coelicolor* A3(2). Nature 417:141–147

Bergey DH, Harrison FC, Breed RS, Hammer BW, Huntoon FM (1923) Bergey's manual of determinative bacteriology, 1st edn. The Williams and Wilkins Co., Baltimore

Bhandari V, Gupta RS (2014) Molecular signatures for the phylum (class) *Thermotogae* and a proposal for its division into three orders (*Thermotogales, Kosmotogales* ord. nov. and *Petrotogales* ord. nov.) containing four families (*Thermotogaceae, Fervidobacteriaceae* fam. nov., *Kosmotogaceae* fam. nov. and *Petrotogaceae* fam. nov.) and a new genus *Pseudothermotoga* gen. nov. with five new combinations. Antonie Van Leeuwenhoek 105:143–168

Bhatnagar S, Badger JH, Madupu R, Khouri HM, O'Connor EM, Robb FT, Ward NL, Eisen JA (2015) Genome sequence of a sulfate-reducing thermophilic Bacterium, *Thermodesulfobacterium commune* DSM 2178T (Phylum Thermodesulfobacteria). Genome Announc 3:e01490–e01414

Blochl E, Rachel R, Burggraf S, Hafenbradl D, Jannasch HW, Stetter KO (1997) *Pyrolobus fumarii,* gen. and sp. nov., represents a novel group of archaea, extending the upper temperature limit for life to 113 degrees C. Extremophiles 1:14–21

Bloodworth RA, Selin C, Lopez De Volder MA, Drevinek P, Galanternik L, Degrossi J, Cardona ST (2015) Draft genome sequences of *Burkholderia contaminans*, a *Burkholderia cepacia* complex species that is increasingly recovered from cystic fibrosis patients. Genome Announc 3:e00766–e00715

Brenner DJ, Krieg NR, Staley JT, Garrity GM (2005) Bergey's manual of systematic bacteriology. Springer-Verlag, New York

Brochier-Armanet C, Boussau B, Gribaldo S, Forterre P (2008) Mesophilic Crenarchaeota: proposal for a third archaeal phylum, the Thaumarchaeota. Nat Rev Microbiol 6:245–252

Brock T (1997) The value of basic research: discovery of *Thermus aquaticus* and other extreme thermophiles. Genetics 146:1207–1210

Brock TD, Freeze H (1969) *Thermus aquaticus* gen. n. and sp. n., a nonsporulating extreme thermophile. J Bacteriol 98:289–297

Browning GF, Citti C (2014) Mollicutes: molecular biology and pathogenesis. Caister Academic Press, Norfolk

Brumm PJ, Gowda K, Robb FT, Mead DA (2016) The complete genome sequence of hyperthermophile *Dictyoglomus turgidum* DSM 6724 reveals a specialized carbohydrate fermentor. Front Microbiol 7:1979

Byrne GI (2003) Chlamydia uncloaked. Proc Natl Acad Sci U S A 100:8040–8042

Castenholz RW (2001) Phylum BX, *Cyanobacteria*. In: Boone DR, Castenholz RW (eds) Bergey's manual of systematic bacteriology: the Archaea and the deeply branching and phototrophic bacteria, vol 1. Springer, New York, pp 473–599

Cavalier-Smith T (2002) The neomuran origin of archaebacteria, the negibacterial root of the universal tree and bacterial megaclassification. Int J Syst Evol Microbiol 52:7–76

Cerqueda-Garcia D, Martinez-Castilla LP, Falcon LI, Delaye L (2014) Metabolic analysis of *Chlorobium chlorochromatii* CaD3 reveals clues of the symbiosis in '*Chlorochromatium aggregatum*'. ISME J 8:991–998

Chen Y, Ye L, Zhao F, Xiao L, Cheng S, Zhang XX (2016) Bacterial community shift during the startup of a full-scale oxidation ditch treating sewage. J Microbiol Biotechnol https://doi.org/10.4014/jmb.1607.07007

Cho J, Vergin K, Morris R, Giovannoni S (2004) *Lentisphaera araneosa* gen. nov., sp. nov, a transparent exopolymer producing marine bacterium, and the description of a novel bacterial phylum, *Lentisphaerae*. Environ Microbiol 6:611–621

Coil DA, Lo JR, Chen R, Ward N, Robb FT, Eisen JA (2013) Draft genome sequence of the arsenate-respiring bacterium *Chrysiogenes arsenatis* strain DSM 11915. Genome Announc 1:e00953–e00913

Colston SM, Fullmer MS, Beka L, Lamy B, Gogarten JP, Graf J (2014) Bioinformatic genome comparisons for taxonomic and phylogenetic assignments using *Aeromonas* as a test case. MBio 5:e02136

DeBruyn JM, Nixon LT, Fawaz MN, Johnson AM, Radosevich M (2011) Global biogeography and quantitative seasonal dynamics of *Gemmatimonadetes* in soil. Appl Environ Microbiol 77:6295–6300

Dietl A, Ferousi C, Maalcke WJ, Menzel A, de Vries S, Keltjens JT, Jetten MS, Kartal B, Barends TR (2015) The inner workings of the hydrazine synthase multiprotein complex. Nature 527:394–397

Dige I, Gronkjaer L, Nyvad B (2014) Molecular studies of the structural ecology of natural occlusal caries. Caries Res 48:451–460

Edward DG, Freundt FA (1967) Proposal for *Mollicutes* as name of the class established for the order *Mycoplasmatales*. Int J Syst Bacteriol 17:267–268

Eisen JA (1995) The RecA protein as a model molecule for molecular systematic studies of bacteria: comparison of trees of RecAs and 16S rRNAs from the same species. J Mol Evol 41:1105–1123

Elkins JG, Podar M, Graham DE, Makarova KS, Wolf Y, Randau L, Hedlund BP, Brochier-Armanet C, Kunin V, Anderson I et al (2008) A korarchaeal genome reveals insights into the evolution of the Archaea. Proc Natl Acad Sci U S A 105:8102–8107

Embley TM, Stackebrandt E (1994) The molecular phylogeny and systematics of the actinomycetes. Annu Rev Microbiol 48:257–289

Eveleigh RJ, Meehan CJ, Archibald JM, Beiko RG (2013) Being *Aquifex aeolicus*: untangling a hyperthermophile's checkered past. Genome Biol Evol 5:2478–2497

Farrand SK, Van Berkum PB, Oger P (2003) *Agrobacterium* is a definable genus of the family *Rhizobiaceae*. Int J Syst Evol Microbiol 53:1681–1687

Felsenstein J (1985) Confidence limits on phylogenies: an approach using the bootstrap. Evolution 39:783–791

Fleischmann RD, Adams MD, White O, Clayton RA, Kirkness EF, Kerlavage AR, Bult CJ, Tomb JF, Dougherty BA, Merrick JM et al (1995) Whole-genome random sequencing and assembly of *Haemophilus influenzae* Rd. Science 269:496–512

Fox GE, Stackebrandt E, Hespell RB, Gibson J, Maniloff J, Dyer TA, Wolfe RS, Balch WE, Tanner RS, Magrum LJ et al (1980) The phylogeny of prokaryotes. Science 209:457–463

Fredrickson JK, Kostandarithes HM, Li SW, Plymale AE, Daly MJ (2000) Reduction of Fe(III), Cr(VI), U(VI), and Tc(VII) by *Deinococcus radiodurans* R1. Appl Environ Microbiol 66:2006–2011

Garrity G, Holt J (2001a) Phylum BIX, *Deferribacteres phy. nov*. In: Boone DR, Castenholz RW (eds) Bergey's manual of systematic bacteriology: the Archaea and the deeply branching and phototrophic bacteria, vol 1. Springer, New York, pp 465–471

Garrity G, Holt J (2001b) Phylum BVI, *Chloroflexi phy. nov*. In: Boon DR, Castenholz RW (eds) Bergey's manual of systematic bacteriology: the Archaea and the deeply branching and phototrophic bacteria, vol 1. Springer, New York, pp 427–446

Garrity GM, Holt JG (2001c) Phylum BIII, *Thermodesulfobacteria phy. nov*. In: Boone DR, Castenholz RW (eds) Bergey's manual of systematic bacteriology: the Archaea and the deeply branching and phototrophic bacteria, vol 1. Springer, New York, pp 389–393

Garrity GM, Holt JG (2001d) Phylum BV. *Chrysiogenetes phy. nov.* In: Boone DR, Castenholz RW (eds) Bergey's manual of systematic bacteriology, (The Archaea and the deeply branching and phototrophic Bacteria), vol 1. Springer-Verlag, New York, pp 421–425

Garrity GM, Holt JG (2001e) Phylum BVII. *Thermomicrobia phy. nov*. In: Boone DR, Castenholz RW (eds) Bergey's manual of systematic bacteriology: the Archaea and the deeply branching and phototrophic Bacteria, vol 1. Springer, New York, pp 447–450

Garrity GM, Holt JG (2001f) Phylum BVIII. *Nitrospirae phy. nov*. In: Boone DR, Castenholz RW (eds) Bergey's manual of systematic bacteriology: the Archaea and the deeply branching and phototrophic Bacteria, vol 1. Springer-Verlag, New York, pp 451–464

Garrity GM, Holt JG (2001g) Phylum BXI. *Chlorobi phy. nov*. In: Boone DR, Castenholz RW (eds) Bergey's manual of systematic bacteriology: the Archaea and the deeply branching and phototrophic Bacteria, vol 1. Springer-Verlag, New York, pp 601–623

Garrity GM, Holt JG (2001h) The road map to the manual. In: Boone DR, Castenholz RW, Garrity GM (eds) Bergey's manual of systematic bacteriology: the Archaea and the deeply branching and phototrophic Bacteria, vol 1, 2nd edn. Springer, New York, pp 119–166

Garrity GM, Boone DR, Castenholz RW (2001) Bergey's manual of systematic bacteriology. Springer-Verlag, New York

Gauthier S, Tetu A, Himaya E, Morand M, Chandad F, Rallu F, Bujold E (2011) The origin of *Fusobacterium nucleatum* involved in intra-amniotic infection and preterm birth. J Matern Fetal Neonatal Med 24:1329–1332

Geissinger O, Herlemann DP, Morschel E, Maier UG, Brune A (2009) The ultramicrobacterium "*Elusimicrobium minutum*" gen. nov., sp. nov., the first cultivated representative of the termite group 1 phylum. Appl Environ Microbiol 75:2831–2840

Ghai R, Mizuno CM, Picazo A, Camacho A, Rodriguez-Valera F (2013) Metagenomics uncovers a new group of low GC and ultra-small marine Actinobacteria. Sci Rep 3:2471

Gharbia SE, Shah HN, Edwards KJ (2010) Genus I. *Fusobacterium* Knorr 1922, 4AL. In: Krieg NR, Staley JT, Brown DR, Hedlund BP, Paster BJ, Ward NL, Ludwig W, Whitman WB (eds) Bergey's manual of systematic bacteriology, vol 4. Springer, New York/Dordrecht/Heidelberg/London, pp 747–774

Glaeser SP, Kampfer P (2015) Multilocus sequence analysis (MLSA) in prokaryotic taxonomy. Syst Appl Microbiol 38:237–245

Goris J, Konstantinidis KT, Klappenbach JA, Coenye T, Vandamme P, Tiedje JM (2007) DNA-DNA hybridization values and their relationship to whole-genome sequence similarities. Int J Syst Evol Microbiol 57:81–91

Gtari M, Ghodhbane-Gtari F, Nouioui I, Ktari A, Hezbri K, Mimouni W, Sbissi I, Ayari A, Yamanaka T, Normand P et al (2015) Cultivating the uncultured: growing the recalcitrant cluster-2 *Frankia* strains. Nature Sci Rep 5:13112

Gupta RS (2004) The phylogeny and signature sequences characteristics of *Fibrobacteres, Chlorobi,* and *Bacteroidetes*. Crit Rev Microbiol 30:123–143

Gupta RS, Bao B (2012) Phylogenetic framework and molecular signatures for the main clades of the phylum Actinobacteria. Microbiol Mol Biol Rev 76:66–112

Gupta RS, Lali R (2013) Molecular signatures for the phylum *Aquificae* and its different clades: proposal for division of the phylum *Aquificae* into the emended order *Aquificales*, containing the families *Aquificaceae* and *Hydrogenothermaceae*, and a new order *Desulfurobacteriales* ord. nov., containing the family *Desulfurobacteriaceae*. Antonie Van Leeuwenhoek 104:349–368

Gupta RS, Chander P, George S (2013) Phylogenetic framework and molecular signatures for the class *Chloroflexi* and its different clades; proposal for division of the class *Chloroflexia* class. nov. [corrected] into the suborder *Chloroflexineae* subord. nov., consisting of the emended family *Oscillochloridaceae* and the family *Chloroflexaceae* fam. nov., and the suborder *Roseiflexineae* subord. nov., containing the family *Roseiflexaceae* fam. nov. Antonie Van Leeuwenhoek 103:99–119

Gutierrez T, Berry D, Teske A, Aitken MD (2016) Enrichment of Fusobacteria in sea surface oil slicks from the Deepwater Horizon oil spill. Microorganisms 4:24

Guy L, Ettema TJ (2011) The archaeal 'TACK' superphylum and the origin of eukaryotes. Trends Microbiol 19:580–587

Hamdi O, Ben Hania W, Postec A, Bouallagui H, Hamdi M, Bonin P, Ollivier B, Fardeau ML (2015) Aminobacterium thunnarium sp. nov., a mesophilic, amino acid-degrading bacterium isolated from an anaerobic sludge digester, pertaining to the phylum *Synergistetes*. Int J Syst Evol Microbiol 65:609–614

Hedlund BP, Gosink JJ, Staley JT (1997) *Verrucomicrobia* div. nov., a new division of the bacteria containing three new species of *Prosthecobacter*. Antonie Van Leeuwenhoek 72:29–38

Horn M (2008) *Chlamydiae* as symbionts in eukaryotes. Annu Rev Microbiol 62:113–131

Hovanec TA, Taylor LT, Blakis A, Delong EF (1998) *Nitrospira*-like bacteria associated with nitrite oxidation in freshwater aquaria. Appl Environ Microbiol 64:258–264

Huber H, Hohn MJ, Rachel R, Fuchs T, Wimmer VC, Stetter KO (2002) A new phylum of Archaea represented by a nanosized hyperthermophilic symbiont. Nature 417:63–67

Hugenholtz P, Stackebrandt E (2004) Reclassification of *Sphaerobacter thermophilus* from the subclass *Sphaerobacteridae* in the phylum *Actinobacteria* to the class *Thermomicrobia* (emended description) in the phylum *Chloroflexi* (emended description). Int J Syst Evol Microbiol 54:2049–2051

Ji Q, Luo ZX, Yuan CX, Wible JR, Zhang JP, Georgi JA (2002) The earliest known eutherian mammal. Nature 416:816–822

Johnson JL, Ordal EJ (1968) Deoxyribo- nucleic acid homology in bacterial taxonomy: effect of incubation temperature on reaction specificity. J Bacteriol 95:893–900

Jumas-Bilak E, Roudiere L, Marchandin H (2009) Description of '*Synergistetes*' phyl. nov. and emended description of the phylum '*Deferribacteres*' and of the family *Syntrophomonadaceae*, phylum '*Firmicutes*'. Int J Syst Evol Microbiol 59:1028–1035

Kadnikov VV, Mardanov AV, Beletsky AV, Shubenkova OV, Pogodaeva TV, Zemskaya TI, Ravin NV, Skryabin KG (2012) Microbial community structure in methane hydrate-bearing sediments of freshwater Lake Baikal. FEMS Microbiol Ecol 79:348–358

Kashefi K, Holmes DE, Reysenbach AL, Lovley DR (2002) Use of Fe(III) as an electron acceptor to recover previously uncultured hyperthermophiles: isolation and characterization of *Geothermobacterium ferrireducens* gen. nov., sp. nov. Appl Environ Microbiol 68:1735–1742

Kielak AM, Barreto CC, Kowalchuk GA, van Veen JA, Kuramae EE (2016) The ecology of Acidobacteria: moving beyond genes and genomes. Front Microbiol 7:744

Kishimoto N, Tano T (1987) Acidophilic heterotrophic bacteria isolated from acidic mine drainage, sewage, and soils. J Gen Appl Microbiol 33:11–25

Krieg NR, Ludwig W, Whitman WB, Hedlund BP, Paster BJ, Staley JT, Ward N, Brown D (2010) Bergey's manual of systematic bacteriology. Springer, New York

Kunnimalaiyaan M, Stevenson DM, Zhou Y, Vary PS (2001) Analysis of the replicon region and identification of an rRNA operon on pBM400 of *Bacillus megaterium* QM B1551. Mol Microbiol 39:1010–1021

Lagkouvardos I, Jehl MA, Rattei T, Horn M (2014) Signature protein of the PVC superphylum. Appl Environ Microbiol 80:440–445

Lapage S, Sneath P, Skerman E, Lessel H, Seeliger R, Clark W (1992) International code of nomenclature of bacteria: bacteriological code, 1990 revision. ASM Press, Washington, DC

Lassalle F, Campillo T, Vial L, Baude J, Costechareyre D, Chapulliot D, Shams M, Abrouk D, Lavire C, Oger-Desfeux C et al (2011) Genomic species are ecological species as revealed by comparative genomics in *Agrobacterium tumefaciens*. Genome Biol Evol 3:762–781

Ley RE (2010) Obesity and the human microbiome. Curr Opin Gastroenterol 26:5–11

Lindström K, Young JPW (2011) International committee on systematics of prokaryotes subcommittee on the taxonomy of *Agrobacterium* and *Rhizobium*. Int J Syst Evol Microbiol 61:3089–3093

Linnaeus C (1753) Species Plantarum. Stockholm

Lopez-Garcia P, Moreira D (2008) Tracking microbial biodiversity through molecular and genomic ecology. Res Microbiol 159:67–73

Macy JM, Nunan K, Hagen KD, Dixon DR, Harbour PJ, Cahill M, Sly LI (1996) *Chrysiogenes arsenatis* gen. nov., sp. nov., a new arsenate-respiring bacterium isolated from gold mine wastewater. Int J Syst Bacteriol 46:1153–1157

Maiden MC, Bygraves JA, Feil E, Morelli G, Russell JE, Urwin R, Zhang Q, Zhou J, Zurth K, Caugant DA et al (1998) Multilocus sequence typing: a portable approach to the identification of clones within populations of pathogenic microorganisms. Proc Natl Acad Sci U S A 95:3140–3145

Margulis L (1981) Symbiosis in cell evolution. W.H. Freeman & Co, San Francisco

Mayr E (1963) Animal species and evolution. Belknap Press of Harvard University Press, Cambridge, MA

Meier-Kolthoff JP, Auch AF, Klenk HP, Goker M (2013) Genome sequence-based species delimitation with confidence intervals and improved distance functions. BMC Bioinformatics 14:60

Miller JM, Rhoden DL (1991) Preliminary evaluation of Biolog, a carbon source utilization method for bacterial identification. J Clin Microbiol 29:1143–1147

Miroshnichenko ML, Lebedinsky AV, Chernyh NA, Tourova TP, Kolganova TV, Spring S, Bonch-Osmolovskaya EA (2009) *Caldimicrobium rimae* gen. nov., sp. nov., an extremely thermophilic, facultatively lithoautotrophic, anaerobic bacterium from the Uzon Caldera, Kamchatka. Int J Syst Evol Microbiol 59:1040–1044

Mohamad KY, Kaltenboeck B, Rahman Kh S, Magnino S, Sachse K, Rodolakis A (2014) Host adaptation of *Chlamydia pecorum* towards low virulence evident in co-evolution of the *ompA*, *incA*, and ORF663 Loci. PLoS One 9:e103615

Mori K, Yamaguchi K, Sakiyama Y, Urabe T, Suzuki K (2009) *Caldisericum exile* gen. nov., sp. nov., an anaerobic, thermophilic, filamentous bacterium of a novel bacterial phylum, *Caldiserica* phyl. nov., originally called the candidate phylum OP5, and description of *Caldisericaceae* fam. nov., *Caldisericales* ord. nov. and *Caldisericia classis* nov. Int J Syst Evol Microbiol 59:2894–2898

Mougel C, Thioulouse J, Perriere G, Nesme X (2002) A mathematical method for determining genome divergence and species delineation using AFLP. Int J Syst Evol Microbiol 52:573–586

Moussard H, L'Haridon S, Tindall BJ, Banta A, Schumann P, Stackebrandt E, Reysenbach AL, Jeanthon C (2004) *Thermodesulfatator indicus* gen. nov., sp. nov., a novel thermophilic chemolithoautotrophic sulfate-reducing bacterium isolated from the Central Indian Ridge. Int J Syst Evol Microbiol 54:227–233

Murray RGE (1984) Validation of the publication of nex names and new combinations previously effectively published outside the IJSB. Int J Syst Bacteriol 34:355–357

Murray RG, Stackebrandt E (1995) Taxonomic note: implementation of the provisional status Candidatus for incompletely described procaryotes. Int J Syst Bacteriol 45:186–187

Oba T, Andachi Y, Muto A, Osawa S (1991) CGG: an unassigned or nonsense codon in *Mycoplasma capricolum*. Proc Natl Acad Sci U S A 88:921–925

Ochman H, Wilson AC (1987) Evolution in bacteria: evidence for a universal substitution rate in cellular genomes. J Molec Evol 26:74–86

Oren A, da Costa MS, Garrity GM, Rainey FA, Rossello-Mora R, Schink B, Sutcliffe I, Trujillo ME, Whitman WB (2015) Proposal to include the rank of phylum in the International Code of Nomenclature of Prokaryotes. Int J Syst Evol Microbiol 65:4284–4287

Oyaizu H, Debrunner-Vossbrinck B, Mandelco L, Studier JA, Woese CR (1987) The green nonsulfur bacteria: a deep branching in the eubacterial line of descent. Syst Appl Microbiol 9:47–53

Patel BKC (2011) Phylum XX. *Dictyoglomi* phyl. nov. In Krieg NR, Staley JT, Brown DR, Hedlund BP, Paster BJ, Ward NL, Ludwig W, Whitman WB (eds) Bergey's manual of systematic bacteriology, 2nd edn, vol 4 (The Bacteroidetes, Spirochaetes, Tenericutes (Mollicutes), Acidobacteria, Fibrobacteres, Fusobacteria, Dictyglomi, Gemmatimonadetes, Lentisphaerae, Verrumicrobia, Chlamydiae, and Planctomycetes. Springer, New York, p 775

Persson T, Battenberg K, Demina IV, Vigil-Stenman T, Vanden Heuvel B, Pujic P, Facciotti MT, Wilbanks EG, O'Brien A, Fournier P et al (2015) *Candidatus* Frankia datiscae Dg1, the actinobacterial microsymbiont of *Datisca glomerata*, expresses the canonical *nod* genes *nodABC* in symbiosis with its host plant. PLoS One 10:e0127630

Pettengill EA, Pettengill JB, Binet R (2015) Phylogenetic analyses of *Shigella* and enteroinvasive *Escherichia coli* for the identification of molecular epidemiological markers: whole-genome comparative analysis does not support distinct genera designation. Front Microbiol 6:1573

Polonia AR, Cleary DF, Freitas R, Coelho FJ, de Voogd NJ, Gomes NC (2016) Comparison of archaeal and bacterial communities in two sponge species and seawater from an Indonesian coral reef environment. Mar Genomics 29:69–80

Rainey FA, Ray K, Ferreira M, Gatz BZ, Nobre MF, Bagaley D, Rash BA, Park MJ, Earl AM, Shank NC et al (2005) Extensive diversity of ionizing-radiation-resistant bacteria recovered from Sonoran Desert soil and description of nine new species of the genus *Deinococcus* obtained from a single soil sample. Appl Environ Microbiol 71:5225–5235

Ransom-Jones E, Jones DL, McCarthy AJ, McDonald JE (2012) The *Fibrobacteres*: an important phylum of cellulose-degrading bacteria. Microb Ecol 63:267–281

Ray J (1686) History of plants (Historia plantarum species hactenus editas aliasque insuper multas noviter inventas & descriptas complectens: in qua agitur primò De plantis in genere). Clark, London

Reysenbach A-L (2001a) Phylum BI, *Aquificae phy. nov.* In: Boone DR, Castenholz RW (eds) Bergey's manual of systematic bacteriology : the Archaea and the deeply branching and phototrophic bacteria, vol 1. Springer, New York, pp 359–367

Reysenbach A-L (2001b) Phylum BII, *Thermotogae phy. nov*. In: Boone DR, Castenholz RW (eds) Bergey's manual of systematic bacteriology : the archaea and the deeply branching and phototrophic bacteria, vol 1. Springer, New York, pp 369–387

Rinke C, Schwientek P, Sczyrba A, Ivanova NN, Anderson IJ, Cheng JF, Darling A, Malfatti S, Swan BK, Gies EA et al (2013) Insights into the phylogeny and coding potential of microbial dark matter. Nature 499:431–437

Rogers MJ, Simmons J, Walker RT, Weisburg WG, Woese CR, Tanner RS, Robinson IM, Stahl DA, Olsen G, Leach RH et al (1985) Construction of the mycoplasma evolutionary tree from 5S rRNA sequence data. Proc Natl Acad Sci U S A 82:1160–1164

Rooney AP, Price NP, Ehrhardt C, Swezey JL, Bannan JD (2009) Phylogeny and molecular taxonomy of the *Bacillus subtilis* species complex and description of *Bacillus subtilis* subsp. *inaquosorum* subsp. nov. Int J Syst Evol Microbiol 59:2429–2436

Saiki T, Kobayashi Y, Kawagoe K, Beppu T (1985) *Dictyoglomus thermophilum* gen. nov., sp. nov., a chemoorganotrophic, anaerobic, thermophilic bacterium. IJSEM 35:253–259

Saitou N, Nei M (1987) The neighbor-joining method: a new method for reconstructing phylogenetic trees. Mol Biol Evol 4:406–425

Schleifer K-H (2001) Phylum XIII, *Firmicutes* Gibbons and Murray 1978, 5 (Firmacutes [sic] Gibbons and Murray 1978, 5). In: De Vos P, Garrity GM, Jones D, Krieg NR, Ludwig W, Rainey FA, Schleifer K-H, Whitman WB (eds) In Bergey's manual of systematic bacteriology: the Firmicutes, vol 3. Springer, New York, pp 19–1422

Schopf JW (2014) Geological evidence of oxygenic photosynthesis and the biotic response to the 2400-2200 ma "great oxidation event". Biochemistry (Mosc) 79:165–177

Sen A, Daubin V, Abrouk D, Gifford I, Berry AM, Normand P (2014) Phylogeny of the class Actinobacteria revisited in the light of complete genomes. The orders '*Frankiales*' and *Micrococcales* should be split into coherent entities: proposal of *Frankiales* ord. nov., *Geodermatophilales* ord. nov., *Acidothermales* ord. nov. and *Nakamurellales* ord. nov. Int J Syst Evol Microbiol 64:3821–3832

Siqueira FM, Cibulski SP, Teixeira TF, Mayer FQ, Roehe PM (2017) Draft genome sequence of *Acholeplasma laidlawii*, a common contaminant of cell cultures. Genome Announc 5:e01578–e01516

Skennerton CT, Haroon MF, Briegel A, Shi J, Jensen GJ, Tyson GW, Orphan VJ (2016) Phylogenomic analysis of *Candidatus* 'Izimaplasma' species: free-living representatives from a *Tenericutes* clade found in methane seeps. ISME J 10:2679–2692

Slobodkin AI, Reysenbach AL, Slobodkina GB, Baslerov RV, Kostrikina NA, Wagner ID, Bonch-Osmolovskaya EA (2012) *Thermosulfurimonas dismutans* gen. nov., sp. nov., an extremely thermophilic sulfur-disproportionating bacterium from a deep-sea hydrothermal vent. Int J Syst Evol Microbiol 62:2565–2571

Sokal RR, Sneath PHA (1963) Principles of numerical taxonomy. W.H. Freeman, San Francisco

Soo RM, Woodcroft BJ, Parks DH, Tyson GW, Hugenholtz P (2015) Back from the dead; the curious tale of the predatory cyanobacterium *Vampirovibrio chlorellavorus*. PeerJ 3:e968

Sorokin DY, Muyzer G (2010) *Desulfurispira natronophila* gen. nov. sp. nov.: an obligately anaerobic dissimilatory sulfur-reducing bacterium from soda lakes. Extremophiles 14:349–355

Sorokin DY, Foti M, Tindall BJ, Muyzer G (2007) *Desulfurispirillum alkaliphilum* gen. nov. sp. nov., a novel obligately anaerobic sulfur- and dissimilatory nitrate-reducing bacterium from a full-scale sulfide-removing bioreactor. Extremophiles 11:363–370

Spang A, Saw JH, Jorgensen SL, Zaremba-Niedzwiedzka K, Martijn J, Lind AE, van Eijk R, Schleper C, Guy L, Ettema TJ (2015) Complex archaea that bridge the gap between prokaryotes and eukaryotes. Nature 521:173–179

Spring S, Bunk B, Sproer C, Schumann P, Rohde M, Tindall BJ, Klenk HP (2016) Characterization of the first cultured representative of *Verrucomicrobia* subdivision 5 indicates the proposal of a novel phylum. ISME J 10:2801–2816

Sridhar S, Sharma A, Kongshaug H, Nilsen F, Jonassen I (2012) Whole genome sequencing of the fish pathogen *Francisella noatunensis* subsp. orientalis Toba04 gives novel insights into *Francisella* evolution and pathogenecity. BMC Genomics 13:598

Stackebrandt E, Goebel B (1994) Taxonomic note: a place for DNA-DNA reassociation and 16S rRNA sequence analysis in the present species definition in bacteriology. International Journal Systematic Bacteriology 44:846–849

Stackebrandt E, Rainey FA, Ward-Rainey NL (1997) Proposal for a new hierarchic classification system, actinobacteria classis nov. Int J Syst Bacteriol 47:479–491

Strous M, Fuerst JA, Kramer EH, Logemann S, Muyzer G, van de Pas-Schoonen KT, Webb R, Kuenen JG, Jetten MS (1999) Missing lithotroph identified as new planctomycete. Nature 400:446–449

Suehiro Y, Sakai K, Nishioka M, Hashimoto S, Takami T, Higaki S, Shindo Y, Hazama S, Oka M, Nagano H et al (2017) Highly sensitive stool DNA testing of *Fusobacterium nucleatum* as a marker for detection of colorectal tumours in a Japanese population. Ann Clin Biochem 54:86–91

Szeszak F, Skripeczky K, Princzinger A, Szabo G (1973) Functional units of *Streptomyces* hyphae, equivalent with prokaryotic cells. Zentralbl Bakteriol Parasitenkd Infektionskr Hyg 128:243–251

Tamaki H, Tanaka Y, Matsuzawa H, Muramatsu M, Meng XY, Hanada S, Mori K, Kamagata Y (2011) *Armatimonas rosea* gen. nov., sp. nov., of a novel bacterial phylum, *Armatimonadetes* phyl. nov., formally called the candidate phylum OP10. Int J Syst Evol Microbiol 61:1442–1447

Thrash JC, Coates JD (2011) Phylum XVII. Acidobacteria phyl. nov.. In Krieg NR, Staley JT, Brown DR, Hedlund BP, Paster BJ, Ward NL, Ludwig W, Whitman WB (eds) Bergey's manual of systematic bacteriology, 2nd edn, vol 4 (The Bacteroidetes, Spirochaetes, Tenericutes (Mollicutes), Acidobacteria, Fibrobacteres, Fusobacteria, Dictyglomi, Gemmatimonadetes, Lentisphaerae, Verrumicrobia, Chlamydiae, and Planctomycetes). Springer, New York, pp. 725–727

Tyson GW, Lo I, Baker BJ, Allen EE, Hugenholtz P, Banfield JF (2005) Genome-directed isolation of the key nitrogen fixer *Leptospirillum ferrodiazotrophum* sp. nov. from an acidophilic microbial community. Appl Environ Microbiol 71:6319–6324

Vos P, Garrity G, Jones D, Krieg NR, Ludwig W, Rainey FA, Schleifer K-H, Whitman WB (2009) Bergey's manual of systematic bacteriology. Springer, New York

Vulic M, Lenski RE, Radman M (1999) Mutation, recombination, and incipient speciation of bacteria in the laboratory. Proc Natl Acad Sci U S A 96:7348–7351

Wang NF, Zhang T, Yang X, Wang S, Yu Y, Dong LL, Guo YD, Ma YX, Zang JY (2016) Diversity and composition of bacterial community in soils and lake sediments from an arctic lake area. Front Microbiol 7:1170

Ward NL (2011) Phylum XXV. Planctomycetes Garrity and Holt 2001, 137 emend. Ward (this volume). In Krieg NR, Staley JT, Brown DR, Hedlund BP, Paster BJ, Ward NL, Ludwig W, Whitman WB (eds) Bergey's manual of systematic bacteriology, 2nd edn, vol 4 (The Bacteroidetes, Spirochaetes, Tenericutes (Mollicutes), Acidobacteria, Fibrobacteres, Fusobacteria, Dictyglomi, Gemmatimonadetes, Lentisphaerae, Verrumicrobia, Chlamydiae, and Planctomycetes). Springer, New York, p 879

Wayne LG, Brenner DJ, Colwell RR, Grimont PAD, Kandler O, Krichevsky MI, Moore LH, Moore WEC, Murray RGE, Stackebrandt E et al (1987) International committee on systematic bacteriology. Report of the ad hoc committee on reconciliation of approaches to bacterial systematics. Int J Syst Bacteriol 37:463–464

Weisburg WG, Giovannoni SJ, Woese CR (1989) The *Deinococcus-Thermus* phylum and the effect of rRNA composition on phylogenetic tree construction. Syst Appl Microbiol 11:128–134

Whitman WB, Goodfellow M, Kämpfer P, Busse H-J, Trujillo ME, Ludwig W, Suzuki K-i (2012) Bergey's manual of systematic bacteriology. Springe, New York

Woese CR (1987) Bacterial evolution. Microbiol Rev 51:221–271

Woese CR, Fox GE (1977) Phylogenetic structure of the prokaryotic domain: the primary kingdoms. Proc Natl Acad Sci U S A 74:5088–5090

Woese CR, Stackebrandt E, Ludwig W (1984) What are mycoplasmas: the relationship of tempo and mode in bacterial evolution. J Mol Evol 21:305–316

Woese CR, Stackebrandt E, Macke TJ, Fox GE (1985) A phylogenetic definition of the major eubacterial taxa. Syst Appl Microbiol 6:143–151

Woese CR, Kandler O, Wheelis ML (1990) Towards a natural system of organisms: proposal for the domains archaea, Bacteria, and Eucarya. Proc Natl Acad Sci U S A 87:4576–4579

Wolk CP (2000) Heterocyst formation in *Anabaena*. In: Brun YVS, Shimkets LJ (eds) Prokaryotic development. ASM, Washington, DC, pp 83–104

Wong MT, Wang W, Lacourt M, Couturier M, Edwards EA, Master ER (2016) Substrate-driven convergence of the microbial community in lignocellulose-amended enrichments of gut microflora from the Canadian beaver (*Castor canadensis*) and north american moose (*Alces americanus*). Front Microbiol 7:961

Wu D, Raymond J, Wu M, Chatterji S, Ren Q, Graham JE, Bryant DA, Robb F, Colman A, Tallon LJ et al (2009) Complete genome sequence of the aerobic CO-oxidizing thermophile *Thermomicrobium roseum*. PLoS One 4:e4207

Yamao F, Muto A, Kawauchi Y, Iwami M, Iwagami S, Azumi Y, Osawa S (1985) UGA is read as tryptophan in *Mycoplasma capricolum*. Proc Natl Acad Sci U S A 82:2306–2309

Yap WH, Zhang Z, Wang Y (1999) Distinct types of rRNA operons exist in the genome of the actinomycete *Thermomonospora chromogena* and evidence for horizontal transfer of an entire rRNA operon. J Bacteriol 181:5201–5209

Young JM, Kuykendall LD, Martínez-Romero E, Kerr A, Sawada H (2001) A revision of *Rhizobium* Frank 1889, with an emended description of the genus, and the inclusion of all species of *Agrobacterium* Conn 1942 and *Allorhizobium* undicola de Lajudie et al. 1998 as new combinations: *Rhizobium radiobacter, R. rhizogenes, R. rubi, R. undicola* and *R. vitis*. Int J Syst Evol Microbiol 51:89–103

Yu XL, Chan Y, Zhuang LF, Lai HC, Lang NP, Lacap-Bugler DC, Leung WK, Watt RM (2016) Distributions of *Synergistetes* in clinically-healthy and diseased periodontal and peri-implant niches. Microb Pathog 94:90–103

Zahradka K, Slade D, Bailone A, Sommer S, Averbeck D, Petranovic M, Lindner AB, Radman M (2006) Reassembly of shattered chromosomes in *Deinococcus radiodurans*. Nature 443:569–573

Zaremba-Niedzwiedzka K, Caceres EF, Saw JH, Backstrom D, Juzokaite L, Vancaester E, Seitz KW, Anantharaman K, Starnawski P, Kjeldsen KU et al (2017) Asgard archaea illuminate the origin of eukaryotic cellular complexity. Nature 541:353–358

Zhang H, Sekiguchi Y, Hanada S, Hugenholtz P, Kim H, Kamagata Y, Nakamura K (2003) *Gemmatimonas aurantiaca* gen. nov., sp. nov., a gram-negative, aerobic, polyphosphate-accumulating micro-organism, the first cultured representative of the new bacterial phylum *Gemmatimonadetes* phyl. nov. Int J Syst Evol Microbiol 53:1155–1163

Zuo G, Xu Z, Hao B (2013) *Shigella* strains are not clones of *Escherichia coli* but sister species in the genus *Escherichia*. Genomics Proteomics Bioinformatics 11:61–65

Chapter 3
Importance of Prokaryotes in the Functioning and Evolution of the Present and Past Geosphere and Biosphere

Bernard Ollivier, Nina Zeyen, Gregoire Gales, Keyron Hickman-Lewis, Frédéric Gaboyer, Karim Benzerara, and Frances Westall

Abstract On a volcanic and anaerobic planet characterized by abundant hydrothermal activity, physicochemical gradients and disequilibria at the local scale would have been fundamental for the emergence of life on Earth. Unfortunately, the early rock record pertaining to this existential process no longer exists, and, while chemists attempt to recreate life in a test tube, two other approaches can provide some information about early life and its metabolic processes. In the first place, phylogenetic, geological, thermodynamic, and microbiological settings suggest that disproportionation of reduced sulfurous compounds might have been essential for microbial evolution by delivering both sulfide and sulfate on Earth's surface. These processes would have been fueled by serpentinization reactions in hydrothermal systems. Another approach is to study ancient and modern microbialites in order to better understand primitive microbial metabolic traits that occurred more than 3 billion years ago. The combination of all of these approaches to understanding early terrestrial life is of relevance to the emergence of life on other planets and satellites in the solar system, especially, for example, Mars.

B. Ollivier (✉) · G. Gales
Aix Marseille Université, Université de Toulon, CNRS, IRD, MIO UM 110, Marseille, France
e-mail: bernard.ollivier@mio.osupytheas.fr

N. Zeyen · K. Benzerara
Centre National de la Recherche Scientifique, Institut de Minéralogie, de Physique des Matériaux et de Cosmochimie, Sorbonne Universités, Paris, France

UM9R 750, Muséum National d'Histoire Naturelle, Université Pierre et Marie Curie, Paris, France

K. Hickman-Lewis · F. Gaboyer · F. Westall
CNRS, Centre de Biophysique Moléculaire, UPR 4301, Rue Charles Sadron, Orléans CEDEX, France

J.-C. Bertrand et al. (eds.), *Prokaryotes and Evolution*,
https://doi.org/10.1007/978-3-319-99784-1_3

Keywords Primitive metabolisms · Past geosphere · Past biosphere · Hydrothermal systems · Serpentinization · Sulfur compound disproportionation · Stromatolites · Microbialites

3.1 Evolution of the Geosphere and Associated Prokaryotic Metabolisms

The aim of this chapter is to highlight the available materials and types of systems on the early Earth which may have facilitated the emergence of life through in situ, thermodynamically favorable physicochemical reactions, taking into account the role played by possible biotic partnerships in the use of abiotically produced molecules (e.g., syntrophic associations, use of minerals as terminal electron acceptors, inorganic fermentation, etc.). Here, we will consider the geochemical functioning of specific ecosystems together with associated ancient anaerobic metabolisms conditioned by the availability of the main mineral and organic compounds that prevailed on the early Earth, respecting the laws of thermodynamics associated with their use, and how these processes might have been of significance in the origin of life.

3.1.1 General Considerations

Nevertheless, we need first to place early life in a coherent environmental setting, and for this we will first look at the development of the early Earth as a habitable planet. It should be stated at the outset that elucidating the history of life on our planet is beset with problems, principally because we are lacking the rock record (the history books of the Earth) from the critical period covering the early evolution of the planet after its consolidation about 4.56 Ga (Patterson et al. 1955; Manhes et al. 1980, see Zhang 2002 for another view) and the emergence of life sometime before 4 Ga. Nevertheless, we can construct our hypotheses from certain basic principles, and, extrapolating from relevant geological and geochemical information from the oldest preserved rocks (4–3.5 Ga), as well as modeling, we can arrive at a coherent geological and geochemical scenario for the context of the origin and early evolution of life.

Terranes older than 3.85 Ga are rare, the oldest dated rocks (4.031 Ga) belonging to the Acasta Gneiss in the Northwest Territories of Canada (Bowring and Williams 1999). These are highly metamorphosed rocks. There are other ancient rock exposures dating from ~3.8 to 3.6 Ga from Greenland (Isua and Akilia, Furnes et al. 2007) and again from Canada, all highly metamorphosed. The rarity of ancient outcrops can be attributed to two important phenomena, e.g., destruction due to a peak in meteorite impact occurrence (the Late Heavy Bombardment, LHB), a period of huge and frequent destructive meteorite impacts that affected all the inner planets of the solar system, and dynamics of inner and outer shells of the Earth. LHB occurred ~700 Ma after planet accretion and differentiation. This event has been linked to destabilization of the planetesimal disk outside the orbits of the planets, which caused a sudden massive delivery of planetesimals to the inner solar system (Gomes et al. 2005). This 20–200 Myr cataclysm likely resurfaced most of the Earth and, in the severest modeled scenario, may even have partially vaporized the oceans (no such event occurring in the gentlest scenario, Sleep et al. 1989).

In addition, the Earth has always been a tectonically dynamic body, and its fluid envelopes such as atmosphere and hydrosphere erase continually Earth's surface with processes known as alteration and erosion. Tectonic recycling of the crust was also a contributing factor to the destruction of the early crust (Kamber 2015). For these reasons also, crustal remnants older than 3.5 Ga are highly metamorphosed: the Acasta Gneisses represent completely altered, reconstituted, and metamorphosed ancient hydrated crust. Other crustal remnants dating from 3.7 to 3.6 Ga do contain some recognizable sediments (Isua, e.g., Rosing 1999), but they have also been metamorphosed to amphibolite/granulite facies (i.e., the rocks have been subjected to temperatures up to 800 °C or more and pressures up to 10 kbar or more). Thus, it is difficult to decipher information from these rocks about the environmental conditions of the early Earth that were the context for the emergence of life. Under these extreme conditions (by today's standards), the extraordinary odyssey of life may have started more than 4 billion years ago as the result of myriad thermodynamically favorable, abiotic physicochemical reactions essential for cell biosynthesis and division (Dass et al. 2016).

3.1.2 *Life-Essential Components*

Life as we know is based on carbon molecules with liquid water as the accompanying solvent. Certain transition elements are also important to the structure of molecules and metabolic functioning. These are the basic ingredients of life, together with the availability of energy sources. Thus, to propose consistent models for the emergence of life, understanding the physicochemical conditions prevailing on the young Earth's fluid envelopes is necessary.

Liquid Water Among its numerous characteristics, Earth is unique in the solar system due to the presence, over long time scales, of liquid water at its surface, which is thought to be required for life. Because of its mass, it has a permanent atmosphere and hydrosphere, contrary to Mars or the Moon. However, according to the standard model of stellar evolution (Gough 1981), the Sun was 20–30% weaker during the Hadean and the Archaean. With the present atmospheric composition, such a weak solar luminosity would lead to a global icehouse. Nevertheless, the presence of early liquid water (and protocontinental crust) is attested by the occurrence of zircon crystals formed between 4.4 and 4.3 Ga through the reworking and fractionation of hydrated crust and found reworked in younger sediments in Jack Hills, Western Australia (Wilde et al. 2001; Harrison et al. 2005). The presence of liquid water from between 4.4 and 4.3 Ga and throughout the Archaean Eon (Feulner 2012) means that the atmosphere had to have been above freezing, probably due to the presence of greenhouse gases.

Atmosphere Major greenhouse gases at this time were carbon dioxide and methane with some water (Kasting and Catling 2003) but not ammonia, as this photolyzes rapidly in the absence of UV screening by atmospheric ozone (Kasting 1982). N_2 and CO_2 were the main constituents of the atmosphere of the early Earth. Other components included H_2, CO, CH_4, NH_3, H_2O, and sulfurous gases comprising H_2S and SO_2 originated from volcanic exhalations and sulfides. By contrast, free oxygen was absent (Kasting 1993; Amend et al. 2013), although small amounts of the gas would have been produced by abiotic reactions including photolysis of water vapour in the upper atmosphere, radiolysis of water at the surface of the ocean since there was no protective ozone layer and therefore radiation levels reaching the surface of the Earth were much higher (Cockell and Raven 2004) as well as dissociation of boiling water exiting shallow hydrothermal vents. Apart from sulfurous compounds, there is evidence that the presence of elemental sulfur (S°) as a product of photolysis of volcanic gases might have been significant on the early Earth (Philippot et al. 2007).

Organic Molecules It appears that the great majority of organic molecules contributing to the prebiotic processes leading to the origin of life could be of extraterrestrial origin. We know that the flux of extraterrestrial materials to the early Earth was orders of magnitude higher than at present Maurette and Brack (2006). Carbonaceous chondrites and comets (Altwegg et al. 2016) contain sugar alcohols, sugars, amino acids, and nucleobases, which might have been important for the emergence of life (Cooper et al. 2001; Burton et al. 2011; Amend et al. 2013; McCaffrey et al. 2014). Much of the organic matter delivered to the early Earth arrived in the form of micrometeorites, tiny fragments of cometary dust in the size range of 50–200 μm, that survived entry into the Earth's atmosphere (Maurette and Brack 2006). A wide range of organic molecules has been identified in comets and cometary materials. The Rosetta mission to comet Churyumov-Gerasimenko detected the amino acid glycine (Altwegg et al. 2016), as well as complex macromolecular organics in the grains emitted by the cometary nucleus, similar to the insoluble organic matter found in carbonaceous chondrites (Fray et al. 2016), and a suite of organic compounds, including methyl isocyanate, acetone, propionaldehyde, and acetamide, kicked up from the dust at the comet's surface (Goesmann et al. 2015).

A number of relevant prebiotic molecules have been also detected in the interstellar medium (Nuevo et al. 2010). Among them, glycolaldehyde has been postulated as a precursor of ribose (backbone of RNA) and is known to lead to the generation of other sugars like threose and erythrose together with alcohols (e.g., ethylene glycol) at pressures ranging from 5 to more than 25 GPa (McCaffrey et al. 2014).

Despite the importance of these phenomena in the interstellar medium, a small number of organic compounds were already present on Earth, thanks to hydrothermal activity in the submarine realm, and a range of chemical reactions in the atmosphere and at the interface between rocks and water. In a now famous experiment, Stanley Miller in 1953 made a successful experiment to determine whether organic molecules could have been formed in a reducing atmosphere of methane, ammonia, hydrogen, water, and silent electric discharge. However, since this early groundbreaking experiment, it has now been determined that the atmosphere was less reducing than Miller initially believed and that NH_3 was not a major constituent of the atmosphere (Kasting 1982). More recent work has shown that localized prebiotic synthesis could have been effective in volcanic plume steam submitted to spark discharges (Johnson et al. 2008).

The early Earth was characterized by a highly active, global hydrothermal system in which seawater circulation in the upper crust was driven by higher heat flow from the mantle (Grigné et al. 2005). Thus little amounts of small organic molecules (e.g., methane, ketones etc., Shock et al. 1998) were produced by hydrothermal processing of the hydrated crust (a process also giving rise to serpentinization and producing large amounts of metabolically useful H_2).

3.1.3 Last Universal Common Ancestor (LUCA)

Before talking of the metabolic pathways possibly used by the first living organisms, we insert here a short paragraph regarding the hypothesized nature of LUCA because it follows on logically from the description of the hydrothermally active early Earth. Phylogenetic studies have yielded different hypotheses regarding the identity of the last universal common ancestor (LUCA) in terms of physiological traits (mesophilic/moderate thermophilic, hyperthermophilic) (Woese 1987; Brochier and Philippe 2002; Di Giulio 2003), but in order to have a better understanding, it is necessary to take into account the thermodynamic laws and the early geochemical conditions existing on our planet. In addition, we should keep in mind that LUCA was not the first biological entity. There were probably many unsuccessful attempts before an effective genetic process was put into place allowing for the emergence of a living, independent cell preceding LUCA. LUCA itself was probably not a well-defined species but rather an assemblage of microorganisms characterized by a low or nonexistent species barrier that rapidly and efficiently exchanged genetic information (Bayman et al. 2002). Given the anoxic, hot, reducing, and toxic (due to the presence of reduced sulfurous gases, Kasting 1993) early environment described above, it is likely that LUCA had a strict anaerobic metabolism based on existing prebiotic molecules and was probably a (hyper)thermophile thriving in the hydrothermally dominated niches of the early Earth (Weiss et al. 2016). This idea was reinforced by the phylogenetic tree of life proposed by Woese, which placed hyperthermophilic anaerobic microorganisms within the domains *Bacteria* and *Archaea*, specifically at the deepest phylogenetic branches of this tree (Woese 1987). It has been proposed that the thermophilic nature of LUCA could have been an artifact of the fact that the surface of the Earth was severely affected by the LHB to the extent that microbial life, outside protective hydrothermal vents, may have been annihilated (Sleep et al. 1989, 2001; Weiss et al. 2016). However, recent work has highlighted the fact that the habitable environment of the early Earth was, in any case, hot and strongly influenced by hydrothermal activity (Hofmann and Harris 2008; Kamber 2015; Westall et al. 2018), thus reinforcing the evidence from Woese's phylogenetic tree.

Note that there is no fossil record of the emergence and early evolution of life because of the destruction and or severe metamorphism of the rock record prior to 3.5 Ga. As we will see below, the earliest fossil record documents an already thriving and global biosphere that had evolved far beyond the first protocells and, probably, well beyond LUCA.

3.1.4 Electron Donors and Acceptors on Early Earth

The rocks and geochemical environment of the early Earth clearly conditioned the delivery of transition metals and various electron donors and electron acceptors used by the first microorganisms (Charlou et al. 2002; McCollom and Seewald 2007). Hot, hydrothermal fluids flushing transition element-rich mafic and ultramafic rocks brought metals important to metabolism and the structure of many biomolecules, such as Fe, Ni, Co, Zn, As, Cu, and more, through the vent edifices into the seawater. Other elements and compounds that could be used as electron donors and acceptors are detailed below.

Although ammonia was in low concentrations in the Hadean-Archaean atmosphere, its protonated form, NH_4^+, was a priori present in the primitive oceans. NH_4^+ in this environment may have originated from nitrogen reduction catalyzed by hydrothermal circulation (Brandes et al. 1998). Since the early oceans were probably acidic as a consequence of the CO_2-rich atmosphere (pH 5.5, Grotzinger and Kasting 1993; Pinti 2005) and had temperatures ranging between 20 °C and 80 °C (Knauth and Lowe 2003; Robert and Chaussidon 2006; Tartèse et al. 2017), and since the ammonium acidic dissociation constant lies between 10 and 8, as calculated at temperatures ranging from 0 to 50 °C and pKa and decreasing as a function of temperature (Bates and Pinching 1949) [NB thermodynamical calculation extending to 200 °C, performed by Amend and Shock 2001, leads to the same assessment (pKa = 6 at 200 °C)], it can be hypothesized that the early oceans contained substantial ammonium concentrations.

N_2 and CO_2 were the main constituents of the atmosphere of the early Earth. Other components included H_2, CO, CH_4, H_2O, and sulfurous compounds comprising H_2S and SO_2 originated from volcanic exhalations, and S° (see S° formation above). While H_2, CH_4, NH_4^+ (see above regarding the presence of NH_4^+ in the primitive oceans), CO, and H_2S may serve as electron donors, CO_2 may be used only as a terminal electron acceptor during oxidative processes and should be considered as a powerful major oxidant that could be used by microorganisms. SO_2 and S° may be used both as electron donors and electron acceptors thus delivering in each case a more oxidized (sulfate) and a more reduced chemical form (sulfide). These reactions are known under the term of "disproportionation," and we will see below that some of them (e.g., elemental sulfur disproportion) could have been part of the metabolisms that have been carried out by first living cells (Finster 2008). Among all these available compounds in the atmosphere, SO_2 has not yet been reported to be used either as an electron acceptor or an electron donor by prokaryotes. By contrast, there is evidence that, at high temperatures between 200 and 330 °C, SO_2 may have been disproportionated to sulfate and elemental sulfur (Kusakabe et al. 2000) (Eq. 3.1), thus reinforcing the idea that this latter compound was usable by primitive thermophilic life. Whether primordial or not, we will discuss it below.

$$3SO_2 + 3H_2O \rightarrow 2HSO_4^- + S^{\circ} + 2H^+ \quad (3.1)$$

Apart from what is known of the main components in the gas atmosphere and what has been provided during the LHB on the early Earth, a number of mineral and organic compounds have been reported to be produced by geochemical reactions in fluids emitted by various hydrothermal systems, including terrestrial hot springs and deep and shallow marine vents (Amend et al. 2013). Mafic-hosted vents related to basaltic effusion were discovered in the late 1970s in the Atlantic Ocean (e.g., Lucky Strike, Menez Gwen, Broken Spur, etc.). Hot fluids emitted from black smoker chimneys were found to contain H_2, CH_4, H_2S, metals, and metalloids under their reduced forms due to precipitation with sulfide together with alkanes and carboxylic acids (Amend et al. 2013). In addition to H_2, CH_4, alkanes, carboxylic acids, and aromatic compounds were also detected from geological alkaline (pH up to 12) fluids originating from ultramafic-hosted vents (e.g., Rainbow, Lost City) (Konn et al. 2009). In these geological settings, H_2 is known to be generated by the oxidation of Fe(II)-bearing minerals such as olivine contained in mantle rocks. In this process, the rock is hydrated in contact with water and produces serpentinite and ferrobrucite as an intermediate (serpentinization reaction; Eq. 3.2). Thereafter, H_2 production results from the oxidation of ferrobrucite (Eq. 3.3) (Konn et al. 2009).

$$2\underset{\text{olivine}}{\left[Mg_{1.8}Fe_{0.2}SiO_4\right]} + 3H_2O \rightarrow \underset{\text{serpentine}}{\left[Mg_{2.85}Fe_{0.15}Si_2O_5(OH)_4\right]} + \underset{\text{ferrobrucite}}{\left[Mg_{0.75}Fe_{0.25}(OH)_2\right]} \quad (3.2)$$

$$\underset{\text{Ferrobrucite}}{\left[Mg_{0.75}Fe_{0.25}(OH)_2\right]} \rightarrow \underset{\text{magnetite}}{Fe_3O_4} + 9\underset{\text{brucite}}{Mg(OH)_2} + H_2 + 2H_2O \quad (3.3)$$

Methane and other hydrocarbons may result from the abiotic reduction of CO_2 by the H_2 produced during olivine oxidation by what are called the Fischer-Tropsch-type reactions. This is exemplified by Eq. 3.4, corresponding to CH_4 production (Konn et al. 2009). It is noteworthy that this reaction may occur biotically, therefore delivering sufficient energy for the growth of hydrogenotrophic methanogens of the domain *Archaea* (see below).

$$4H_2 + CO_2 \rightarrow CH_4 + 2H_2O \qquad \Delta G^{\circ\prime} = -135\ \text{kJ/mol} \quad (3.4)$$

Pasini et al. (2013) suggested these hydrocarbons and/or carboxylic acids may originate from high-temperature cracking of buried organic matter in a similar manner to that reported in oil formation from organic matter in petroleum systems. The production of H_2 and low molecular weight acids by the oxidation of ferrobrucite in peridotites has also been confirmed by laboratory experiments very recently (Miller et al. 2017). Besides a substantial content of H_2 and organic compounds generated by hydrothermal alkaline serpentinized systems, significant quantities of iron in its reduced form [Fe(II)], a common constituent of mantle minerals, olivine and pyroxene, which can be leached by supercritical hydrothermal fluids, may act as a potential electron donor for microorganisms living under anaerobic conditions. In addition, while these extreme environments are poorly distributed in modern terrestrial and marine environments, with a hotter mantle and a higher heat flux during the Hadean and Archaean (Turcotte 1980; Grigné et al. 2005; Kamber 2015), their distribution was more important on the early Earth. In this respect, hydrothermal systems and particularly those depending on serpentinization reactions may have delivered valuable carbon, including recycled extraterrestrial carbon, and provided energy sources which could have been used by early microorganisms, all within a disequilibrium scenario between alkaline hydrothermal fluids exiting into acidic seawater (cf. Russel et al. 2013; Westall et al. 2018) (Fig. 3.1).

Fig. 3.1 The "Aiguille de Prony" chimney within the submarine serpentinizing hydrothermal field of Prony Bay in New Caledonia. (Photography: courtesy of Roy Price)

Interestingly, serpentines have been discovered within Mars' oldest crust and have, as a consequence, opened a debate with regard to the origin of life in a similar context on the Red Planet (Ehlmann et al. 2010). Numerous compounds such as amino acids, sugars, and nucleobases have been formed experimentally under hydrothermal conditions reflecting early Earth scenarios (Amend et al. 2013). However, the abiotic generation of these compounds in modern hydrothermal ecosystems is still a matter of debate (see above discussion of Fischer-Tropsch-type processes).

Since early life has often been hypothesized to have started by lithoautotrophic growth fostered by the delivery of organic compounds (lipids, sugars, proteins, etc.) through anabolic processes, the main molecules which could have been used as early electron donors, as noted above, were H_2, CO, H_2S, S°, and NH_4^+. On the other hand, the possible electron acceptors that could have been used by primitive microorganisms comprised S°, and CO_2, all potential early oxidants for microbial oxidation of hydrogen. The hydrogenotrophic-based subsurface lithoautotrophic microbial ecosystem (SLIME) has been proposed as one plausible scenario for the earliest life forms (Chapelle et al. 2002). The thermodynamically favorable reactions related to anaerobic hydrogen oxidation include methanogenesis, acetogenesis, and sulfur reduction.

3.1.4.1 Hydrogen Oxidation Coupled to CO_2 Reduction

Methanogens belong to the phylum *Euryarchaeota* and represent a deep phylogenetic lineage within the domain *Archaea*. They are known to use a wide range of substrates including acetate and formate and methylated compounds (methanol, methylamines, etc.), with most members within the orders *Methanobacteriales*, *Methanococcales*, *Methanopyrales*, and *Methanomicrobiales* having both the ability to oxidize H_2 and to reduce CO_2 to methane (Garcia et al. 2000; Liu et al. 2012) (see Eq. 3.4). Another way for some of these methanogens to oxidize H_2 requires the presence of methanol and/or methylated compounds which can be used as terminal electron acceptors to produce methane (Garcia et al. 2000) (Eq. 3.5). They pertain to the genus *Methanosphaera* and to the recently described genus *Methanomassiliicoccus* (Dridi et al. 2012). Since the oxidation state of early hydrothermal fluids could have differed from that in present-day systems (Amend et al. 2013), the production of alcohols, such as methanol, cannot be ruled out. Interestingly, one *Methanobacterium* sp. has been isolated using metallic iron as the only electron donor (Dinh et al. 2004). Its growth was faster than that obtained via hydrogen consumption, and this was attributed to the use of electrons originating directly from iron oxidation. More attention should therefore be paid to this process when performed by a methanogen since methane is produced without the involvement of any hydrogenase (Dinh et al. 2004). The direct use of electrons by methanogens (interspecies electron transfer) was also shown when *Methanothermobacter thermautotrophicus* was cocultivated with *Pelotomaculum thermopropionicum* in the form of aggregates using particularly propionate as a substrate (Choi and Sang 2016).

$$H_2 + CH_3OH \rightarrow CH_4 + H_2O \qquad \Delta G^{\circ\prime} = -112.5\,\text{kJ/mol} \qquad (3.5)$$

Most acetogens so far isolated and studied belong to the domain *Bacteria* and consist of numerous species within the genera *Clostridium*, *Moorella*, and *Sporomusa*, phylum *Firmicutes*, the genus *Acetomicrobium*, phylum *Bacteroidetes*, the genus *Desulfotignum*, and phylum *Proteobacteria*. More than 100 acetogens have been isolated to date (Ragsdale and Pierce 2008). They have the ability to oxidize hydrogen and reduce CO_2 to acetate following Eq. 3.6:

$$4H_2 + 2CO_2 \rightarrow CH_3COO^- + H^+ + 2H_2O \qquad \Delta G^{\circ\prime} = -107.1\,\text{kJ/mol} \qquad (3.6)$$

It has been recently established, as reported above for methanogens, that a close phylogenetic relative of the acetogenic *Sporomusa sphaeroides* used iron [Fe(0)] as the sole electron donor to reduce CO_2 to acetate (Choi and Sang 2016). Numerous other acetogens including members of the genera *Sporomusa*, *Clostridium*, and *Moorella* were also shown to directly use electrons from a cathode for producing acetate. This suggests that the direct use of electrons by both methanogens and acetogens should be considered as a common microbial evolutionary trajectory for further study (e.g., the direct use of electrons instead of hydrogen).

Interestingly, in a manner similar to methanogens, the Wood-Ljungdahl pathway suggested to be prevalent in LUCA (Weiss et al. 2016) is used in a reductive direction for energy conservation and autotrophic carbon assimilation in acetogens. This complies with the two fundamental requirements to sustain life: energy conservation and biomass production (Ragsdale and Pierce 2008). At low pressures of hydrogen, methanogens are known to outcompete acetogens in the use of the gas, but when the latter is readily accessible in the environment (e.g., in hydrothermal serpentinizing systems), its use by both prokaryotic lineages is probably of similar significance. Acetate delivered by acetogens may become an important carbon source for other microorganisms, second only to CO_2. Although H_2 oxidation by methanogenic archaeons has been identified as a possible primitive metabolism with methanogenesis being traced back to the Late Archaean era (2.8–3.2 Ga) (Blank 2009), it is still questioned whether they could have operated this way on the early Earth due to the small amount of redox energy available to them at that period (Russel et al. 2013).

3.1.4.2 Hydrogen Oxidation Coupled to Sulfur Reduction

Elemental sulfur (S°) might have been a dominant sulfur species in marine anoxic hydrothermal environment of the Archaean (Philippot et al. 2007; Finster 2008) and therefore represented a potential electron acceptor to oxidize hydrogen. Hedderich et al. (1999) showed that numerous microorganisms have the ability to store energy (ATP) by lithotrophic sulfur reduction following Eq. 3.7:

$$H_2 + S \rightarrow H_2S \qquad \Delta G^{\circ\prime} = -27.9\,\text{kJ/mol} \qquad (3.7)$$

This is particularly the case for sulfur-dependent hyperthermophiles of the phylum *Crenarchaeota*, such as the *Acidianus*, *Thermoproteus*, *Thermococcus*, *Pyrococcus*, *Pyrodictium*, *Desulfurococcus*, *Pyrobaculum*, or *Stygioglobus* genera, as well as for thermophilic (e.g., *Ammonifex*, *Desulfurobacterium* spp., etc.) or mesophilic bacteria (e.g., *Desulfuromonas*, *Desulfovibrio*, *Desulfomicrobium*, *Desulfosporosinus*, *Shewanella*, *Sulfurospirillum*, *Wolinella* spp., etc.) (Stetter et al. 1986; Hedderich et al. 1999; Amend and Shock 2001; Fauque and Barton 2012). This reductive process is therefore widely distributed among prokaryotes. It is noticeable that many prokaryotes within the *Bacteria* and *Archaea* domains may use elemental sulfur as an electron acceptor, thus avoiding the accumulation of reducing equivalents, as is the case for many members of the order *Thermotogales* (Cappelleti et al. 2014). This situation has also been observed in numerous mesophilic and thermophilic/hyperthermophilic methanogens (e.g., *Methanococcus*, *Methanothermococcus*, *Methanobacterium*, *Methanopyrus*) which, in the presence of hydrogen as electron donor and CO_2 and S° as electron acceptors, were shown to produce both methane and sulfide with energy recovered only from CO_2 reduction (i.e., not from sulfur reduction) (Stetter and Gaag 1983; Liu et al. 2012). This reductive process of elemental sulfur performed by methanogens was regarded as a "window into ancient sulfur metabolism" (Liu et al. 2012). Because elemental sulfur has low solubility in water, the direct use of this electron acceptor via a sulfur reductase by sulfur reducers has been questioned (Fauque and Barton 2012). In contrast to sulfate, S° is known to be relatively chemically reactive, and its solubilization by the cleaving of the S_8-ring via a nucleophilic attack of HS^- in an aqueous solution may result in the formation of polysulfide ($S_{(n+1)}^{2-}$) (Eq. 3.8). Polysulfide has been suggested as being preferably used as a terminal electron acceptor when compared to elemental sulfur (Fauque and Barton 2012). This leads us to rewrite this important reaction in the context of primitive life since both HS^- and S° and consequently $S_{(n+1)}^{2-}$ might have prevailed on the early Earth with elemental sulfur and polysulfide possibly acting as potential oxidants to hydrogen and other available substrates. In this respect, most known sulfur-reducing microorganisms (with a peculiar emphasis for (hyper)thermophilic *Archaea*) might in fact be polysulfide reducers following Eq. 3.9 (Hedderich et al. 1999).

$$nS^{\circ} + HS^{-} \rightarrow S_{(n+1)}^{2-} + H^{+} \quad (3.8)$$

$$H_2 + S_{(n+1)}^{2-} \rightarrow HS^{-} + S_n^{2-} + H^{+} \quad (3.9)$$

3.1.4.3 CO Utilization as Electron Donor

CO was present in the early atmosphere as a result of volcanic emissions at concentrations of up to 100 ppm (versus 0.05–0.35 ppm nowadays in nonurban environments) (Jeoung et al. 2014). Despite its low solubility in water and its known toxicity, numerous anaerobic microorganisms use it as a source of carbon and energy. Accordingly, CO may have participated in early metabolisms (Miyakawa et al. 2002). Anaerobic conversion of CO may be performed by a wide range of microorganisms including sulfate reducers, hydrogenogens, phototrophs, homoacetogens, and methanogens (Mörsdorf et al. 1992; Davidova et al. 1994; Parshina et al. 2005). The enzyme which converts CO to CO_2 and electrons as products is called carbon monoxide dehydrogenase (CODH) (Jeoung et al. 2014). Thereafter, the presence of these reducing equivalents allows sulfurous compounds and protons to be reduced to sulfide and hydrogen, respectively (see Eq. 3.10 for production of hydrogen from CO). While some microorganisms can take energetic advantage in producing hydrogen from CO (e.g., *Carboxydothermus hydrogenoformans*, *Rhodospirillum rubrum*, *Thermosinus carboxydivorans*) (Jeoung et al. 2014), others, like acetogens and methanogens, use hydrogen as an electron source for reducing CO_2 to acetate and CH_4, respectively, in a similar manner to that described above. The use of CO by acetogens and methanogens using the reductive acetyl-COA pathway (also called the Wood-Ljungdahl pathway) to deliver energy and generate biomass from CO (Jeoung et al. 2014) may be compared to the production of methane from hydrogen by reducing CO_2, which is only possible through the activity of the CODH, and should therefore be considered as a supplementary step with regard to metabolic evolution in early life. Despite its toxicity, CO is used by some sulfate-reducing bacteria (SRB) as their sole carbon and energy source at concentrations of up to 100% in the gas phase (e.g., *Desulfotomaculum carboxydivorans*) (Parshina et al. 2005).

$$CO + H_2O \rightarrow CO_2 + H_2 \qquad \Delta G^{\circ\prime} = -28.5\,kJ/mol \tag{3.10}$$

3.1.4.4 Disproportionation of Inorganic Sulfur Compounds

Definition Disproportionation corresponds to a chemical/biological reaction where the same mineral or organic compound serves as electron donor and electron acceptor. In the case of inorganic sulfur compounds such as thiosulfate ($S_2O_3^{2-}$), sulfite (SO_3^{2-}), and elemental sulfur (S°), each one may be oxidized to $SO_4^{2-}$and reduced to HS^-or in other terms disproportionated to $SO_4^{2-}$and to HS^-(Fig. 3.2).

Working in an anoxic artificial seawater medium containing elemental sulfur (S°), Belkin et al. (1985) demonstrated that at temperatures above 80 °C, sulfide was produced abiologically at linear rates and was most probably the result of the disproportionation of S°. This oxidoreductive process of S° leading to the production of sulfate (SO_4^{2-}) and sulfide (HS^-) was confirmed by Smith (2000) at temperatures ranging from 50 to 200 °C, with the presence of sulfide-scavenging cations such as copper catalyzing this reaction. Other inorganic sulfur compounds, including thiosulfate and sulfite, can also be disproportionated by a wide range of anaerobic bacteria, especially within the *Deltaproteobacteria* (Finster 2008). This metabolic feature is assimilated to a fermentative process and therefore termed "mineral fermentation" or "inorganic fermentation" (Fig. 3.2).

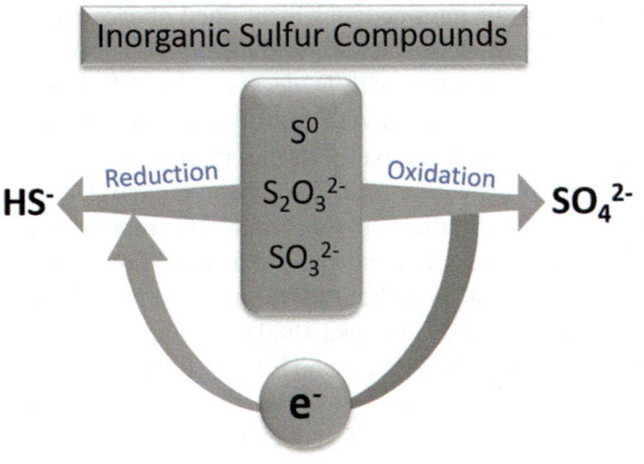

Fig. 3.2 Disproportionation of thiosulfate ($S_2O_3^{2-}$), sulfite (SO_3^{2-}), and elemental sulfur (S°) to sulfide (HS^-) and sulfate (SO_4^{2-})

While thiosulfate (Eq. 3.11) and sulfite (Eq. 3.12) disproportionation are thermodynamically favorable under standard conditions, disproportionation of elemental sulfur (Eq. 3.13) requires the scavenging of sulfide to become exergonic. This is made possible by the presence of ferric iron as a sulfide chemical-scavenging agent as shown in Eq. 3.14. The overall reaction stoichiometry is found in Eq. 3.15. Moreover, FeS, pyrite, a mineral linked quite closely to the origin of life in multiple geological scenarios, formed when elemental sulfur was disproportionated by microorganisms. The presence of the microorganisms catalyzed the rate at which pyrite was formed as compared to what is conventionally known from reported kinetics of natural abiotic pyrite formation processes (Finster 2008). In this sense, pyrite thus formed can be considered as a biomineral.

$$S_2O_3^{2-} + H_2O \rightarrow SO_4^{2-} + HS^- + H^+ \qquad \Delta G^{\circ\prime} = -21.9\,\text{kJ/mol}\left(\text{per}\,S_2O_3^{2-}\right) \tag{3.11}$$

$$4SO_3^{2-} + H^+ \rightarrow 3SO_4^{2-} + HS^- \qquad \Delta G^{\circ\prime} = -58.9\,\text{kJ/mol}\left(\text{per}\,SO_3^{2-}\right) \tag{3.12}$$

$$4S^{\circ} + 4H_2O \rightarrow SO_4^{2-} + 3HS^- + 5H^+ \qquad \Delta G^{\circ\prime} = +10.2\,\text{kJ/mol}\left(\text{per}\,S^{\circ}\right) \tag{3.13}$$

$$3HS^- + 2FeOOH + 3H^+ \rightarrow S^{\circ} + 2FeS + 4H_2O \qquad \Delta G^{\circ\prime} = -143.9\,\text{kJ/mol}\left(\text{per}\,S^{\circ}\right) \tag{3.14}$$

The overall reaction stoichiometry is thus (Eqs. 13 plus 14):

$$3S^{\circ} + 2FeOOH \rightarrow SO_4^{2-} + 2FeS + 2H^+ \qquad \Delta G^{\circ\prime} = -30\,\text{kJ/mol}\left(\text{per}\,S^{\circ}\right) \tag{3.15}$$

Among the oxidized sulfur intermediates mentioned above, thiosulfate is disproportionated by a *Desulfovibrio* species, *D. sulfodismutans* (Finster 2008), as well as sulfite. *D. sulfodismutans* together with *D. desulfuricans* CSN can produce ATP by substrate-level phosphorylation via ATP sulfurylase when grown on thiosulfate (Krämer and Cypionka 1989), probably in the absence of electron transport and proton gradients as reported earlier for *D. sulfodismutans* (Bak and Cypionka 1987; Bak and Pfennig 1987). Studies on thiosulfate disproportionation in cell-free extracts of *D. desulfuricans* CSN indicated that the initial step of this reaction started with the cleavage of thiosulfate by the thiosulfate reductase enzyme into sulfite and sulfide (Krämer and Cypionka 1989) in a similar manner as reported for *Desulfocapsa sulfoexigens* (Frederiksen and Finster 2004; Finster 2008). Thereafter sulfite can be oxidized to sulfate along two parallel pathways. One depends on the reverse reduction of sulfate with APS reductase, ATP sulfurylase, and with the involvement of pyrophosphatase, whereas the second depends only on the activity of the sulfite oxidoreductase (Frederiksen and Finster 2003; Finster 2008). The first demonstration of elemental sulfur disproportionation was made by Thamdrup et al. (1993) when studying anaerobic marine enrichment cultures containing S° and ferric hydroxide (FeOOH) or manganese oxide (MnO_2) as sulfide-scavenging minerals. Interestingly, growth of these disproportionating microorganisms was dependent on a chemolithotrophic process with oxidized sulfurous compounds and CO_2 acting as energy and carbon sources, respectively. The results obtained with *Desulfocapsa sulfoexigens* indicated that the enzymes involved in elemental sulfur disproportionation were the same as those expressed in the presence of thiosulfate with the exception of the thiosulfate reductase (Finster 2008). The final step of S° reduction is still a matter of debate, with the possible involvement of a sulfite reductase or a sulfite oxidoreductase, but further experiments are needed (Finster 2008). After these important discoveries regarding the metabolism of sulfurous compounds by anaerobes, numerous other SRB pertaining to the *Deltaproteobacteria* were shown to be able to also use them as electron donors and electron acceptors. They include members of the genera *Desulfobacter*, *Desulfococcus*, *Desulfomonile*, *Desulfobulbus*, and *Desulfofustis* (Finster 2008). However, within the *Deltaproteobacteria*, this metabolic ability using sulfurous compounds was not restricted to SRB. The hydrogenotrophic thiosulfate-, elemental sulfur-, polysulfide-reducing alkaliphilic bacterium *Desulfurivibrio alkaliphus*, isolated from soda lakes, was shown to use elemental sulfur as its sole energy source (Poser et al. 2013). Other isolates including *Dissulfurimicrobium hydrothermale*, and *Dissulfuribacter thermophilus* originating from hydrothermal ponds, and deep-sea hydrothermal vents have been also recognized as growing by disproportionation of elemental sulfur. These represent deep phylogenetic lineages within the *Deltaproteobacteria* (Slobodkin et al. 2013, 2016). However, disproportionation of sulfurous compounds is not necessarily coupled to growth as is the case for *Desulfovibrio* species, *Desulfomonile tiedje,* and other species mentioned above. In addition to thiosulfate and sulfite, energetically coupling growth on elemental sulfur was also observed particularly in members of the family *Desulfobulbaceae* (e.g., *Desulfobulbus*, *Desulfofustis*, and *Desulfocapsa* spp.) (Finster 2008). This oxidoreductive process was found to be constitutive for *Desulfovibrio sulfodismutans* or *D. desulfuricans* CSN (Krämer and Cypionka 1989).

In addition to *Deltaproteobacteria*, representatives of the genus *Desulfotomaculum* (e.g., *Desulfotomaculum thermobenzoicum, Desulfotomaculum nigrificans*) in class *Clostridia*, phylum *Firmicutes*, were found to disproportionate thiosulfate (Finster 2008; Aüllo et al. 2013). One representative of this class isolated from alkaline lakes (*Dethiobacter alkaliphilus*) grew by disproportionation of elemental sulfur (Poser et al. 2013). Experiments that have been performed by molecular approaches (clone libraries of 16S rDNA genes) on elemental sulfur-disproportionating enrichment cultures revealed the predominance of members of *Desulfobulbaceae*, a family comprising only strict anaerobes (Finster 2008). It was therefore quite surprising when Obraztsova et al. (2002) reported that a facultative anaerobe pertaining to the *Gammaproteobacteria*, *Pantoea agglomerans*, disproportionated also elemental sulfur. This observation confirmed that this sulfur metabolism was not restricted to (i) strict anaerobes or (ii) only *Deltaproteobacteria*, as demonstrated by some members of the *Clostridia* and *Gammaproteobacteria* classes mentioned above. This raises the question: Are these poorly studied metabolisms much more widespread within the prokaryotes than previously believed? What is their importance with regard to the origin of life, and to what extent might this metabolism have been conserved in prokaryotic microorganisms throughout microbial evolution? Exciting answers come from recent microbial investigations of deep-sea hydrothermal vents as well as studies of the anaerobic oxidation of methane (AOM). Slobodkin et al. (2012) isolated from a deep-sea hydrothermal vent an extremely thermophilic non-SRB named *Thermosulfurimonas dismutans* belonging to the class *Thermodesulfobacteria*, one of the deepest branches within the phylogenetic tree comprising numerous sulfate reducers (e.g., *Thermodesulfobacterium, Thermodesulfatator* spp.) (Garrity and Holt 2001; Hatchikian et al. 2001) that grows by disproportionation of elemental sulfur (Slobodkin et al. 2012). Under optimal growth conditions, the doubling time of *Thermosulfurimonas dismutans* was 45 min. Despite its inability to reduce sulfate, its genome contained the set of genes involved in sulfate reduction (Mardanov et al. 2016), thus suggesting that, similar to *Desulfocapsa sulfexigens* and *Desulfovibrio sulfodismutans*, these genes might be used by this extremophilic thermophilic bacterium for elemental sulfur disproportionation following the reverse reduction of sulfate. Interestingly, the genome content of *Thermosulfurimonas dismutans* indicates that it may perform autotrophic carbon fixation by the Wood-Ljungdahl pathway and it possesses all genes required for nitrogen fixation (Mardanov et al. 2016). Moreover, experiments with sulfate-reducing *Thermodesulfobacterium* spp. isolated from oil field waters led us to the conclusion that these microorganisms had also the ability to disproportionate thiosulfate (B. Ollivier, unpublished results). Further experiments are now clearly needed to be performed with members of the class *Thermodesulfobacteria* including *Thermodesulfatator*, *Caldimicrobium*, and *Thermosulfidibacter* spp. to confirm whether disproportionation of sulfurous compounds is indeed deeply anchored in the phylogenetic tree, thus signifying that such metabolisms might have been one of the earliest metabolisms in the Hadean and Early Archaean. In addition to various studies which have been undertaken with regard to the use of sulfurous compounds by chemolithoautotrophic microorganisms pertaining to different phylogenetic lineages, there is recent evidence that such reactions were of primary importance in the global carbon cycle. This has been highlighted by Milucka et al. (2012) with work focusing on anaerobic oxidation of methane (AOM). Indeed, AOM, when coupled to sulfate reduction, is known to be of ecological significance in the carbon and sulfur cycling of marine environments (Eq. 3.16).

$$CH_4 + SO_4^{2-} \rightarrow HS^- + HCO_3^- + H_2O \qquad \Delta G^{o\prime} = -34\,kJ/mol \qquad (3.16)$$

CH_4 is oxidized by methanotrophic *Archaea* (ANME-1, ANME-2, and ANME-3 groups) performing the reverse reaction of methanogenesis in the presence of SRB pertaining to the class *Deltaproteobacteria* (Knittel and Boetius 2009; Timmers et al. 2017). None of these partners have been cultivated in pure culture to date, suggesting that this oxidative process resulted from syntrophic associations, with different scenarios being hypothesized. In contrast, during continuous cultivation over 8 years of an AOM enrichment from sediments of the Mediterranean mud volcano Isis, Milucka et al. (2012) provided evidence that both the large and irregularly shaped ANME-2 and SRB belonging to the *Desulfosarcina-Desulfococcus* cluster performed unexpected metabolisms at least for the ANME-2 group of microorganisms. Indeed, this latter group was shown to form zero-valent sulfur compounds during AOM through a new pathway for dissimilatory sulfate reduction, yet to be fully understood, while SRB were shown to disproportionate disulfide, an important component used by methanogenic archaea for respiration (Hedderich et al. 1999), into sulfate and sulfide. Despite the claim that AOM might not be an obligate syntrophic process, there is as yet no evidence that these microorganisms grow isolated in culture. In this respect, it is tempting to think that a syntrophic association is still necessary for energy conservation during AOM. Nevertheless these results show that, within the AOM consortium, SRB may have played a crucial role by disproportionating sulfurous compounds during the AOM process. Thus, if we consider inorganic sulfur compound disproportionation to be truly ancestral, as hypothesized above, could anaerobic methanotrophy by the archaea involved in the AOM have been anterior, at geological scales, to methanogenesis from acetate? This hypothesis is in agreement with the deep phylogenetic position of ANME groups within the order *Methanosarcinales* comprising acetoclastic representatives. Even if other hypotheses may be postulated with regard to the evolution of AOM process during time, this latter viewpoint should merit further attention.

Generally, biological sulfurous compound disproportionation has been reported for neutrophilic conditions. Pikuta et al. (2003), and more recently Sorokin et al. (2008, 2011), successfully enriched and isolated SRB with the ability to perform thiosulfate and sulfite disproportionation from soda lakes. Thereafter, Poser et al. (2013) reported elemental sulfur disproportionation by alkaliphilic anaerobic bacteria including *Desulfovibrio alkaliphilus*, class *Deltaproteobacteria*, and *Dethiobacter alkaliphilus*, class *Clostridia*. In both cases, culture experiments were performed at pH 10. Under these alkaline conditions, it is known that elemental sulfur may be present as soluble polysulfides after reacting with sulfide and may have different chain lengths (see Eq. 3.9). In contrast to neutrophilic disproportionating bacteria, Poser et al. (2013) also obtained growth of the two alkaliphilic bacteria tested in the absence of ferric iron, thus suggesting that the use of elemental sulfur as the sole energy source was favored under such conditions as compared to neutral ones. The authors pointed out the formation of polysulfides having chain lengths ranging from S_2^{2-} to S_8^{2-} during this process with more polysulfide being produced in the absence of iron. In addition, while in the presence of iron, dominant polysulfide chain lengths ranged from S_3^{2-} to S_4^{2-}, those obtained in the absence of iron ranged from S_4^{2-} to S_6^{2-}. All these results are in agreement with polysulfides being significant intermediates during elemental sulfur disproportionation especially in alkaline conditions. Apart from the fact that the result of elemental sulfur reduction may be that of polysulfide reduction (but possibly not in neutral conditions), the involvement of polysulfide disproportionation in that of elemental sulfur is difficult to discount, especially at high pH values, as noted by Poser et al. (2013). This observation made under alkaline conditions has to be clearly taken into account if considering this inorganic sulfur fermentation as a primitive, if not the most primitive, metabolism on Earth.

Based on sulfur isotopic studies ($^{34}S/^{32}S$ ratios), Shen et al. (2001) reported that formation of microscopic sulfides in about 3.47-Gyr-old barites from the North Pole in the Pilbara region of Western Australia could have resulted from sulfate reduction. However, multiple sulfur isotope analyses (^{34}S, ^{32}S, but also ^{33}S) from drill-core samples from the chert-barite deposit at the North Pole (Philippot et al. 2007) instead concluded that sulfide production resulted from elemental sulfur disproportionation rather than by sulfate reduction. The conclusion drawn by Philippot et al. (2007) is more consistent with what we know of the oxidation state of the atmosphere, the lithosphere, and the hydrosphere as being free of oxygen in the Early Archaean. However, with regard with what has been discussed above for alkaliphilic disproportionating microorganisms, and the fact that sulfide was present during this early geological period, we cannot exclude that polysulfides possibly reacting with sulfides and elemental sulfur might have been the preferred electron donor sources for producing sulfide and sulfate. Whatever the sulfur sources which were used for disproportionation (either elemental sulfur or polysulfides), it is clear that low concentrations of sulfate were delivered to the primitive oceans as a result of inorganic sulfur disproportionation and thus constituted a new potential oxidant to oxidize substrates by microorganisms in a strictly anaerobic atmosphere (Ollivier and Guyot 2009). This may have given an opportunity for the first sulfate reducers to express their presence, with more resounding activity occurring alongside the Great Oxidation Event some 2.5–2.2 billion years ago and rendering possible the use of a wide range of electron acceptors by anaerobes (sulfate, nitrate, nitrite, oxidized forms of metals or metalloids) (Sessions et al. 2009).

There is additional evidence in the geological record for SRB activity. Wacey et al. (2011) proposed that micrometer-sized pyrite crystals co-occurring with spheroidal and ellipsoidal cells in 3.43 Ga rocks of Western Australia were formed by the activity of sulfate-reducing and sulfur-disproportionating microorganisms, thus supporting that sulfur-based metabolisms were present in early life.

Here again, elemental sulfur/polysulfide disproportionation is at the center of the debate around primitive life and might have been essential in producing sulfate to be subsequently used by SRB. Indeed, since early conditions on Earth were anoxic and highly reductive, the possible availability of sulfate to be used as electron acceptors by microorganisms was made more probable by disproportionation processes involving more reduced sulfurous compounds. Briefly, it is thus quite plausible to conclude that sulfur disproportionation and sulfate reduction might have occurred at the type of biotope reported in the above geological studies in the Early Archaean sediments of Western Australia and South Africa, even if the former metabolism (sulfur disproportionation) seems to us more likely as a pioneering one.

Interestingly, biochemical studies reporting sulfurous compound disproportionating SRB demonstrated that enzymes involved in sulfate reduction (APS reductase, ATP sulfurylase, pyrophosphatase) were used in the reverse reduction of sulfate during these inorganic oxidoreductive processes (Finster 2008). This was confirmed with cultures of *Desulfocapsa sulfoexigens* which demonstrably disproportionated elemental sulfur. After these experiments conducted with SRB, two questions were posed by Finster (2008): (i) *Why do not all sulfate-reducers possess the capacity to disproportionate, assuming that they all carry the required enzyme machinery?; and (ii) why do some of them disproportionate, but do not grow?* The recent isolation and cultivation of *Dissulfurimicrobium hydrothermale*, *Dissulfuribacter thermophilus* (Slobodkin et al. 2013, 2016), and *Dissulfurirhabdus thermomarina* (Slobodkina et al. 2016) recognized that elemental sulfur-disproportionating, non-sulfate-reducing bacteria are located deep in the phylogenetic branches of the class *Deltaproteobacteria*. This complicates the situation as the activity of these microorganisms might have preceded that of SRB. This hypothesis is supported by the genome analysis of *Thermosulfurimonas dismutans* which also performed elemental sulfur disproportionation, but not sulfate reduction, despite possessing all genes for performing this latter dissimilatory process. The phylogenetic position of *Thermosulfurimonas dismutans*, class *Thermodesulfobacteria*, within the deepest branches of the domain *Bacteria* reinforces this hypothesis. It would be of great interest to analyze the genome of *Dissulfurimicrobium hydrothermale*, *Dissulfuribacter thermophilus*, and *Dissulfurirhabdus thermomarina* together with that of *Dethiobacter alkaliphilus*, class *Clostridia*, and that of *Pantoea agglomerans*, class *Gammaproteobacteria*, which have the same metabolism as *Thermosulfurimonas dismutans* regarding sulfurous compound disproportionation, although without having the ability to use sulfate as a terminal electron acceptor. Such information will be helpful to better understand the metabolic functioning and evolutionary history of elemental sulfur disproportionation by these microorganisms and concomitantly attempt to link it to early life on Earth. Future studies will be necessary to determine whether this metabolism has been shared by members of the domain *Archaea*.

3.1.4.5 H_2S and NH_4^+ Oxidation

Compounds that were present in the Early Archaean, either in the atmosphere, lithosphere, or hydrosphere, included H_2S, NH_4^+ (see above comments on NH_4^+ production in the primitive oceans), and Fe^{2+}. The biological oxidative process of H_2S or NH_4^+ might not have been relevant because of poor or non-availability of potential oxidants (e.g., nitrate and nitrite). They may, however, have been of significance as sulfur and nitrogen sources for biosynthesis of primitive microorganisms. The case of ferrous iron is questionable as its oxidation may be possible via the direct use of electrons by CO_2-reducing microorganisms, as noted above (e.g., methanogens and acetogens). This metabolism could thus also be a primitive one.

3.1.4.6 Hydrocarbon and Carboxylic Acid Oxidation

Under anaerobic conditions, microbial hydrocarbon or carboxylic acid oxidation requires the presence of terminal electron acceptors. Sulfate resulting from elemental sulfur disproportionation might have been an excellent acceptor for oxidizing these organic compounds. However, in the absence of sulfate, hydrogenotrophic methanogens may also serve as alternative H_2/e- scavengers (in syntrophic associations), thus rendering the organic oxidative process thermodynamically favorable (Stams and Plugge 2009). In terms of microbial evolution, these obligate syntrophic associations, and particularly those involving direct electron transfers, would have been highly beneficial since they propagated genetic exchange between microorganisms. Going further in this direction, it has been proposed that obligate syntrophic communities involving bacteria together with archaeons might have been part of the pathway to the eukaryotic cell (Martin and Müller 1998).

Finally, existing physicochemical gradients and disequilibria were probably characteristics of the geological setting for the origin of life on Earth (Russel et al. 2013; Westall et al. 2018), allowing for a wide range of thermodynamically favorable chemical reactions and for primitive microorganisms to take advantage of these opportunities to develop energy-yielding metabolisms. Despite several scenarios not being be ruled out (e.g., hydrogen oxidation via CO_2 or S°/polysulfide reduction, etc.), based on phylogenetic, geological, and microbiological arguments and obeying thermodynamic laws, data suggest that primitive and possibly earliest living elements might have been acting anaerobically using CO_2 as a carbon source and elemental sulfur and/or polysulfides as the sole energy source. The hypothesis of disproportionation of these sulfurous compounds to sulfate and sulfide in a hot environment by first microorganisms is increasingly convincing. This is in accordance with recent microbial investigations, which have demonstrated that these oxidoreductive processes were performed by a wider range of microorganisms than previously believed, including deep phylogenetic lineages within the domain *Bacteria* (e.g., class *Thermodesulfobacteria*) (Garrity and Holt 2001; Hatchikian et al. 2001). Whether this metabolism is shared by members within the domain *Archaea* has yet to be determined. Surprisingly, inorganic sulfur compound disproportionation has been poorly studied to date, and there are still unknown biochemical mechanisms involved in the process. In this respect, further attention should be paid by biochemists, microbiologists, and phylogeneticians to these biological reactions to better understand the pivotal role that the disproportionation of sulfurous compounds might have played in early life.

It is noteworthy that the conversion of S° and polysulfides into sulfide and sulfate allowed the creation of significant redox energy and, thus, the possibility for sulfate-reducing bacteria to progressively adapt to using sulfate as a terminal electron acceptor on geological time scales. In addition, S° and polysulfide disproportionation is thermodynamically facilitated by alkaline conditions and does not require ferric iron as a sulfide-scavenging mineral, contrary to the situation in neutral conditions. Interestingly, during this oxidoreductive process at high pH, polysulfides are produced in greater quantity and are more chemically stable.

Given the highly hydrothermally influenced marine environment of the Early Archaean, we may expect basaltic- and/or ultramafic-hosted vents to have actively participated both in the emergence of life and in its metabolic evolution. The alkaline conditions that prevailed in fluids emitted from serpentinized hydrothermal systems, in particular, might have been of primary importance to reinforce the capacity of the first living microorganisms to chemolithotrophically use S° and polysulfides by disproportionation. Moreover, apart from appropriate pH conditions, these fluids might have fueled early ecosystems with important quantities of H_2 and CH_4 which were then further used as electron donors by microorganisms (e.g., hydrogenotrophic sulfate-reducing or methane-producing microorganisms, together with anaerobic methane oxidizers). The delivery of alkaline fluids into a slightly acidic, early ocean could have also encouraged the creation of a proton gradient (Russel et al. 2013; Westall et al. 2018) which might have been helpful to establish the "proton-motive force" as an essential mechanism for energy conservation and, consequently, microbial evolution (Mitchell 1961).

The involvement of sulfur-disproportionating microorganisms pertaining to the *Deltaproteobacteria* in AOM is interesting. Taking into account the metabolic antiquity of sulfurous compound disproportionation and the importance of this anaerobic oxidative process in carbon and sulfur cycling in modern marine environments, we may expect this process to have had a similar importance in the anoxic Archaean oceans. Such anaerobic oxidation would have been of significance in removing CH_4 from the atmosphere, thus participating in the progressive cooling of the planet. The increasing availability of sulfate as a terminal electron acceptor would be pivotal to this. An acceleration of all of these oxidative processes occurred with the progressive appearance of oxygen on Earth, especially during the "Great Oxidation Event" at around 2.45–2.32 Ga (Bekker et al. 2004). Apart from increasing levels of oxygen, this event paved the way to enhance the number of mineral and organic components on the Earth which could be used as terminal electron acceptors, thus widening the potential lineages of evolution (Sessions et al. 2009; Hazen and Sverjensky 2010). Finally, it has been reported that sulfurous compounds (e.g., sulfate) (Aubrey et al.

2006; Rettberg et al. 2016) and "serpentinites" (Ehlmann et al. 2010) have been found on Mars. If sulfate possibly resulted from inorganic sulfurous oxidoreductive abiotical processes, was the combination of both of these chemical factors relevant for emergence of life on this planet and possibly others? That remains a matter of debate for the coming decades.

3.2 Early Traces of Life: The Fossil Record

The fossil record of early life starting at ~3.5 Ga documents a flourishing biosphere in which anaerobic life forms appeared to have colonized all of the preserved, available habitats on the early Earth. There are, however, no traces of the emergence of life from abiotic components nor of what must have been the very first cells, owing to the abovementioned destruction of the early crust due to impacts and tectonic activity. Thus, the oldest, well-preserved crustal rocks still in existence that host traces of life date back only as far as the Early to Mid-Archaean era (~3.8–3.3 Ga).

3.2.1 The Environment of the Early Earth from a Microbial Point of View

Since the nature and distribution of life is strictly related to its habitat, it is pertinent here to provide a brief overview of the habitats on the early Earth, especially because the environmental conditions were very different to those at the Earth's surface today (Westall 2012). The Earth was an essentially volcanic planet with a more or less global ocean; continents as we know them did not exist, and exposed land masses were represented only by the exposed portions of volcanoes or oceanic plateau formed atop mantle plumes, not dissimilar to the Hawaiian Islands or Iceland today. The Earth's mantle was much warmer, thus driving intensified volcanic activity, as well as volcanism-associated hydrothermal activity. The crust-seawater interface was therefore warm to hot, with hot hydrothermal fluids emerging into somewhat cooler seawater (Van den Boorn et al. 2007; Tartèse et al. 2017). This is the critical interface from which we find most of the early traces of life. In addition to being warm to hot in temperature, the environment was anaerobic with only very small traces of abiotically produced oxygen, the main atmospheric gas being carbon dioxide. Acid rain and CO_2 dissolved in the seawater provoked acidic conditions, as did the influx of hot, acidic hydrothermal fluids. The early Earth can be viewed therefore as a hot and acidic water world (Figs. 3.3 and 3.4).

There have been some suggestions, however, that our young planet could have been cold and maybe even partially frozen because of the reduced illumination and heat emanating from the early Sun (the "faint young Sun paradox", Sagan and

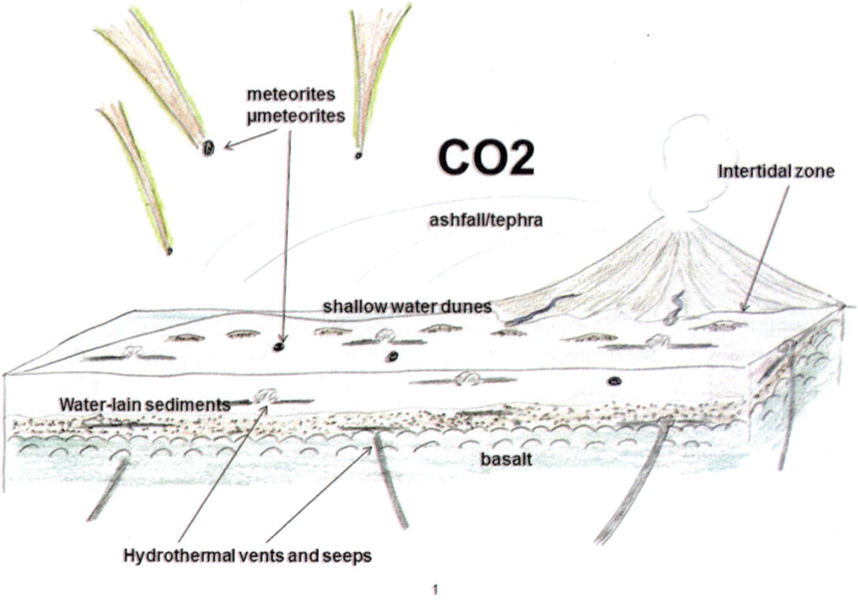

Fig. 3.3 Simplified sketch of the early Earth environment, highlighting some of the defining processes of the Hadean-Archaean era. The planet was more extreme than at present: it suffered near-constant bombardment by extraterrestrial material, volcanism was widespread, and hydrothermal activity was present over a much greater area than on the modern planet. Here we have highlighted a shallow-water "volcanic shelf" environment at the interface of these processes. It is such environments that preserved the oldest remaining traces of life. (From Westall et al. 2015b)

Mullen 1972). The preserved crustal fragments, however, do not support a cold Earth (e.g., Hofmann and Harris 2008).

We have painted a global to regional view of the environment of the early Earth and will now address the microbial-scale habitats. In this volcanic and hydrothermal world, the primary habitats would have been intimately linked to these kinds of substrates with potential habitats occurring on and within the volcanic sediments, in crevasses and fractures in the volcanic rocks and the crust, and in association with hydrothermal vent environments themselves (Figs. 3.3 and 3.4). On the modern Earth, such environments are colonized by a wide variety of chemotrophic microorganisms, i.e., those which do not rely on sunlight for energy production but are instead able to harness energy from sources in their direct environment or substrate. While primarily subaqueous, the habitable environments existed theoretically also on exposed landmasses, although sediments formed on the latter are very rarely preserved. The subsurface environment would also have been potentially habitable; indeed, microbes occur in the modern crust down to several kilometers (Onstott 2016), although on the early Earth, fractures and faults in the subsea crust would have been rapidly sealed by hydrothermal silica. Consequently, subsurface habitats

Fig. 3.4 Modern high-temperature hydrothermal vent (black smoker) from the "Snake Pit" locality, Mid-Atlantic Ridge. Hot (100–400 °C), mineral-rich effluent pours out of the vent into high-pressure ocean floor environments. A multitude of elements are leached from the vent edifice by the fluid, including Fe, Co, Zn, As, and Cu. Image from Woods Hole Oceanographic Institution. (www.whoi.edu)

would have been restricted to the upper meters of sediments and inactive fractures and hydrothermal vents.

Other types of habitats within reach of sunlight were available, where phototrophic microorganisms could thrive. These included shallow seas, the uppermost meters of the water column, as well as the beach environment. One intriguing aspect of the existence of phototrophs on the early Earth is that UV radiation levels must have been very high compared to those today by a factor of 50–1000 (Cockell and Raven 2004) owing to the lack of atmospheric oxygen and the consequent lack of a protective ozone layer. This means that the early phototrophs needed to have access to radiation mitigation strategies, such as rapid gene repair, pigmentation, thick EPS coatings, etc.

The salient message here is that the early Earth can be considered as an extreme environment compared to the modern Earth: hot, anoxic, generally acidic, high radiation, etc. Thus, any organisms living in this kind of environment can be considered to be polyextremophilic. Note, however, that these kinds of environmental conditions would be common on most habitable rocky planets and satellites in the universe. What are not normal are the conditions reigning at the surface of the Earth today!

This brief environmental overview shows that there was a range of environments that could have hosted and, as we will describe below, apparently did host chemotrophic and phototrophic life forms. In the above section, we extensively addressed the potential electron donors and acceptors that would have been available to the primitive life forms, as listed on Table 3.1. Note that, while there was almost no oxygen in the atmosphere, small amounts produced by abiotic reactions (via photolysis/radiolysis of water and dissociation of the H_2O molecule in boiling hydrothermal fluids) would have been sporadically present and potentially available. Thus many oxidized species, such as nitrate and sulfate, would have been available in only infinitesimally small amounts and probably very locally, for instance, nitrate formed by abiotic processes, such as lightning and impacts (Stern et al. 2015). The variety of electron donors and acceptors potentially available could have supported a relatively wide variety of chemolitho- and chemoorganotrophic metabolisms, as well as fermenters and phototrophs (Table 3.1).

Table 3.1 List of potential electron donors and acceptors on the early Earth

Electron donors	Electron acceptors	Metabolisms	Comments
Chemolithotrophs			
S°	S°	Disproportionation of S°	S°, $S_2O_3^{2-}$, SO_3^{2-} derived from sulfur species of the atmosphere and of precipitated polysulfide
$S_2O_3^{2-}$	$S_2O_3^{2-}$	Disproportionation of $S_2O_3^{2-}$	
SO_3^{2-}	SO_3^{2-}	Disproportionation of SO_3^{2-}	
S_n^{2-} (polysulfide)	S_n^{2-}	Disproportionation of S_n^{2-}	
H_2S, S°, $S_2O_3^{2-}$	O_2[bs], NO_3^-, Fe^{3+}	Oxidation of reduced sulfur species	
H_2	CO_2, O_2[b], NO_3^-, MnO_2,$Fe(OH)_3$,SO_4^{2-}	H_2 oxidation coupled to O_2, Fe^{3+}, MnO_2, NO_3^-, and SO_4^{2-}/S° reduction. Methanogenesis. Acetogenesis	H_2 produced by serpentinization
CH_4	O_2[b], NO_3^-, MnO_2, SO_4^{2-}, S°[a]	Methanotrophy, coupled to O_2, MnO_2, NO_3^-, and SO_4^{2-} reductions	CH_4 produced by (i) Fisher-Tropsch reaction from CO_2 and H_2 (serpentinization) and (ii) methanogens. Anaerobic oxidation of methane (AOM): AOM is performed by cocultures associating methanogens to sulfate reducers
CO	CO_2, O_2[b], SO_4^{2-}	Carbon monoxide oxidation	
Fe^{2+} dissolved in water and contained in rocks	CO_2, O_2[b], NO_3^-, MnO_2, SO_4^{2-}, S°	Iron oxidation	May result from volcanic exhalation and from alteration of siderite ($FeCO_3$), magnetite (Fe_3O_4), pyrite (FeS_2), basalt…
NH_4^+, NO_2^-	NO_2^-, O_2[b]	Oxidation of reduced nitrogen species	Anaerobic ammonium oxidation (ANAMMOX). NH_4^+ from Archean oceans. NO_2^- can be produced by nitrate reduction

(continued)

Table 3.1 (continued)

Electron donors	Electron acceptors	Metabolisms	Comments
Chemoorganotrophs			
Oxidation/Respiration			
Amino acids, organic acids, alcohols, fatty acids, sugars…	Fe^{3+}, SO_4^{2-}, NO_3^-, NO_2^-, O_2[b]…	Fe^{3+}/nitrate/sulfate reduction	Organics can be of abiotic (extraterrestrial and/or hydrothermal) or of biotic origin
Interspecies H_2 transfer	Hydrogenotrophic microorganisms	H_2 is produced by a chemoorganotroph to be used by a hydrogenotrophic partner	In this case the hydrogenotrophic partner (e.g., sulfate reducer or methanogen) can be considered as a terminal biological electron acceptor
Fermentation			
Amino acids, sugars, organic acids...	Small organics, typically pyruvate (glycolysis)	Production of hydrogen, volatile fatty acids, alcohols (end products of organic acid fermentation)...	Organics can be of abiotic origin (extraterrestrial and/or hydrothermal) or of biotic origin. Methanogens ferment acetate and methylated compounds to produce methane
Phototrophs			
H_2S, $S_2O_3^{2-}$, S°, Fe^{2+}, NO_2^-	Quinones, cytochromes of the electron transport system	Anoxygenic photosynthesis	Purple (i.e., *Rhodobacter* sp.), green sulfur (i.e., *Chlorobium* sp.) or green non-sulfur (i.e., *Chloroflexus* sp.) bacteria and *Heliobacterium* sp. (*Firmicutes*)

[a]S° is not properly an electron acceptor but is an intermediate in anaerobic oxidation of methane (AOM)

[b]Although the atmospheric oxygenation only occurred ~2.5 Ga ago, oxygen and oxidizing conditions could occur in the Archean on local scales (notably with the radiolysis of water)

Of interest with respect to early life is that there appears to have been a strong control on the distribution of chemotrophic microbial colonies by hydrothermal activity (Westall et al. 2015a, b). Close to hydrothermal vents, chemotrophic biomass is high, and colonization is relatively dense; further away in oligotrophic waters, only delicate monolayers of lithotrophs colonize the surfaces of volcanic particle. Of course, the phototrophs were more widespread and not spatially limited to the vicinities of hydrothermalism. They could colonize oligotrophic environments also, such as sediment surfaces away from hydrothermal nutrient sources.

3.2.2 Microbial Biosignatures and Their Preservation

Our knowledge of early life is based on the signatures preserved in ancient rocks. We noted above the rapid silicification characteristic of the Early Archaean (4.0–3.3 Ga) sediments. Indeed, silicification was so rapid that its precipitation around the microbial colonies inhabiting the volcanic substrates preserved individual cells, cell colonies, biofilms, or even degraded, disseminated organic molecules. The rapidity of the preservation is demonstrated by the manner and fidelity of preservation: colonies and biofilms have been preserved in their life positions. For instance, colonies of probable lithotrophic, coccoidal microbes are known to be preserved in multiple stages of the cell cycle, including cell division and lysis (Westall et al. 2006a, 2011b). In another case, delicate filaments forming a phototrophic biofilm have been bent over by flowing water and preserved in that position (Westall et al. 2006b). Where no cells are preserved, amorphous carbonaceous matter is disseminated in the sediment, especially where colonies were degraded by high-temperature hydrothermal fluids. The fossil record is patchy, however, and not all microorganisms are preserved physically. An experiment to silicify two, similar, hyperthermophilic archaeal species, *Pyrococcus abyssi* and *Methanocaldococcus jannaschii*, demonstrated that the latter lysed rapidly during silicification and consequently only its abundant extracellular materials (EPS, extracellular polymeric substance) were fossilized, while *P. abyssi* fossilized very well as a morphological signature (Fig. 3.5). Very recently, it has also been shown that amide groups derived from protein compounds can still be detected in silicified microbial cells as old as 1.88 Ga, supporting further this idea of a high preservation potential of silicification (Alleon et al. 2016).

Thus, organisms preserved in the rock record represent only a fraction of the microbial diversity and biomass that must have been present in the Early Archaean era. Similarly, since there is only a small amount of Archaean and older rock preserved, we must assume that our estimates of biological diversity are further underestimated: life, even on the early Earth, would have been more diverse, at least at a metabolic level, than we can ever realize.

A brief overview of the preservation mechanisms and resultant biosignatures is opportune here (cf. Orange et al. 2009; Westall and Cavalazzi 2011). Microorganisms and their colonial communities and associations are characterized by their morphology, the molecular, elemental, and isotopic composition of their organic components, as well as by their direct and indirect effects on the immediate environment. Table 3.2 presents a resume of these biosignatures. The latter includes corrosion and leaching features in mineral substrates (so-called trace fossils); biofilm stabilization of sediment bedding planes through a network of EPS or contribution to the formation of vertical, laminated structures, such as stromatolites, through the trapping of detrital particles on and between the sticky EPS laminae of biofilms and microbial mats; as well as microbial influence on the precipitation of minerals. Microbial mineral precipitation may be controlled (i.e., involving specific genes), as in the case of magnetosomes in magnetotactic bacteria, or induced (i.e., following modifications of the chemical conditions but not involving specific genes), such as the precipitation of calcium carbonate, for example, by sulfate-reducing bacteria.

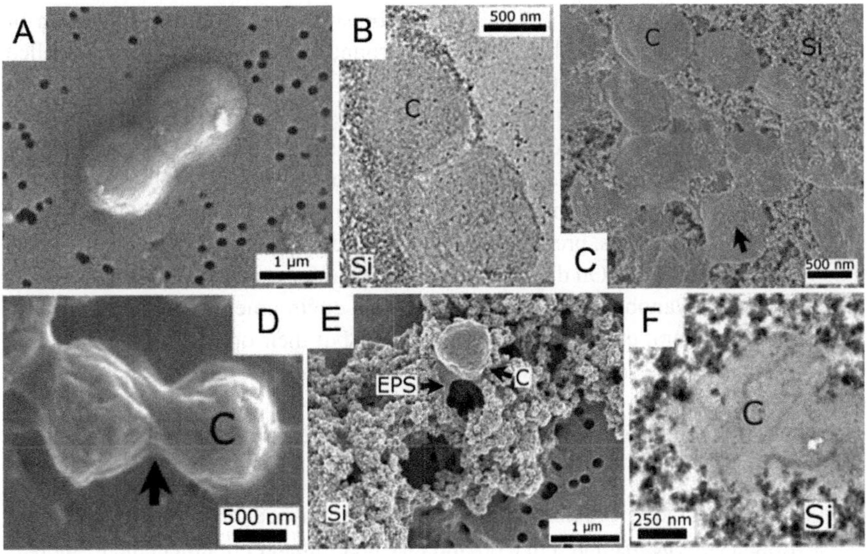

Fig. 3.5 Experimental fossilization of cells exposed to a 500 ppm silica solution. (**a**) SEM image of fresh, dividing, unsilicified *Pyrococcus abyssi* cells. (**b**) Fossilization of *P. abyssi* after 1 month; note the deposition of silica (thin dark coating) around the cell (C) envelope. (**c**) Fossilization of *P. abyssi* after 9 months; note the visible silica (Si) granules precipitating in the cell (C) wall. The cell is now trapped within silica. (**d**) SEM image of fresh, dividing, unsilicified *Methanocaldococcus jannaschii* cells. (**e**) Precipitation of silica on the EPS (*left arrow*) of the cells after 24 h; only 1 remnant cell (C) remains (*right arrow*). (**f**) TEM of lysed cell (C) in silica precipitates after 24 h

Table 3.2 List of potential biosignatures of past (fossil) life

Major microbial components	Biosignature	Specific component or structure
Cell colonies, biofilms, mats, EPS	***Fossil cells, colonies, mats***	Clotted fabrics
		Microbial mounds
		Biolaminites, e.g., stromatolites, MISS[a]
Organic components	***Carbon molecules (kerogen)***	Composition
		Structure, e.g., odd/even C number
		µ-structure
Cell metabolic activity	***Biominerals***	Direct (e.g., magnetite)
		Indirect (e.g., microcrystalline calcite, aragonite, dolomite)
	Elements	Concentration, e.g., Ni, Fe, Co, As, Mg
		Isotopic ratios, e.g., C, S, N, Fe
	Microbial influence	Mineral composition, dissolution, habit, size

[a]MISS: microbially influenced sedimentary structures (cf. Noffke 2009)

Induced mineral precipitation of calcium carbonate can lead to the formation of decimeter-sized, irregularly shaped features around nutrient-rich cold seeps called bioherms or also to the formation of stromatolites.

Perhaps the most important aspect in recognizing potential biosignatures involves their propensity for preservation over long geological periods in the rock record. For this, the morphological, organic, elemental, isotopic, and mineral biosignatures need to be rapidly cemented in a nonporous medium, i.e., permineralized or covered by fine-grained sediments, preferably in an anoxic environment in order to conserve the organic components. On the modern, oxic Earth, for example, the morphologies of thermophilic cyanobacteria living in hot spring environments may be beautifully preserved by mineral encasement (fossilization), but their organic components are destroyed by thermal and oxidative degradation (Cady and Farmer 1996; Jones et al. 2001).

Preservation mechanisms of biosignatures can be divided into three types (Westall et al. 2015b):

(1) Generic, degraded organic molecules can be trapped in fine-grained anoxic sediments or encapsulated in a chemically precipitated mineral cement, such as hydrothermal silica.
(2) The remains of microorganisms can be sedimented together with other detrital matters onto the bottom of a water body (sea, lake, etc.) and then encased either in a mineral cement or simply by the surrounding, anoxic sediments.
(3) Cells, colonies, biofilms, or mats may be directly encased in a mineral crust and preserved as two- or three-dimensional morphological fossils.

Numerous experimental studies have investigated the mechanisms of fossilization. Of interest here are the mechanisms by which microbial features have been silicified, since it is this process that led to the preservation of the various types of biosignatures from the Early Archaean. Briefly, silicification of these microbial biosignatures occurred through passive precipitation of silica from supersaturated, hydrothermally influenced seawater (or from hydrothermal fluids themselves) onto organic surfaces. The silica ion (Si^{4+}) attaches itself to negatively charged functional groups, such as hydroxyls (OH), carboxyls (COOH), or phosphoryls (PO), exposed on the surface of the organic substrate (Monty et al. 1991).

Metal cation bridging also aids silicification (Beveridge and Fyfe 1985; Orange et al. 2011). Iron is particularly effective; by chelating to functional groups (carboxyls or phosphoryls in lipid membranes), Fe effectively inhibits early cell degradation, thus promoting silicification by increasing the amount of time that the cells are exposed to the fossilizing solution. This was experimentally demonstrated with cultures of *M. jannaschii* which, without Fe, lysed before the cells could be silicified but, after exposure to Fe in solution, was preserved by a silica crust (Fig. 3.5). After initial fixation to the organic template, the natural polymerization process of silica leads to the formation of an encapsulating crust and, in some cases, complete pervasive impregnation of the cell. Silica as a fossilizing medium has the advantage of having a very small crystal lattice and can therefore infiltrate the cells in a very delicate manner, reproducing the original cell morphology perfectly (Westall et al. 2011b). Thus, even the delicate S-layer can be silicified (Westall et al. 1995). While silica has the advantage of being an unparalleled agent of preservation, its most pervasive infiltration, including the development of proliferating spherulites, can occasionally obfuscate fabrics (Brasier et al. 2006; Orberger et al. 2006).

Another factor influencing the fossilization process is cell wall composition (Westall 1997; Orange et al. 2009). Gram-positive bacteria, with their thick, peptidoglycan-rich outer envelopes, produce a thick, robust crust, while Gram-negative bacteria, which have thin layers of peptidoglycan, and *Archaea*, which have none, produce delicate crusts.

Mineral encapsulation leads to the death of the cell, if it is not already dead, and gradual degradation of its organic components. The degraded components are entrapped within the fossilized cell and also within the cementing mineral matrix either around the fossil cells or, if cells are not preserved as intact fossilized forms, within the encasing anoxic sediments and cementing minerals. This has been demonstrated experimentally: Orange et al. (2012) documented changes in the composition and quantity of amino acids, monosaccharides, and fatty acids over the period of 1 year during silicification of *M. jannaschii* (Fig. 3.6). While the cells themselves did not fossilize, the organic components initially present in cells and accompanying EPS were preserved in the precipitated silica, although with change ratios, amino acids, and fatty acids being the best-preserved compounds in this experiment.

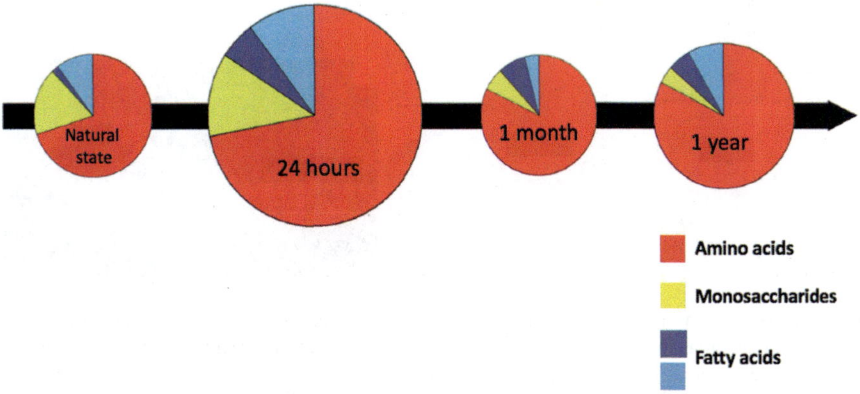

Fig. 3.6 Evolution of relative proportions of different types of organic molecules during experimental fossilization of *M. jannaschii*. Areas are proportional to the masses analyzed. Note that the fatty acids were analyzed using two different methods, thermochemolysis (*dark blue*) and simple hydrolysis (*light blue*)

3.2.3 Determining Biogenicity and Syngenicity

The earliest traces of life were formed under environmental conditions that no longer exist on the present-day Earth, and, having been submitted to several billion years of processes that could change them beyond recognition, their identification and analysis can be fraught with difficulties. The first task is to determine whether the purported biogenic signature is really of biogenic origin and not an abiotic look-alike (biomorph) or artifact (Fig. 3.7). Microbial structures have very simple shapes that can be imitated by abiotic processes: spheroidal microfossils may be easily confused with spheroidal minerals, such as silica, while a sheetlike concentration of extraterrestrial organics could imitate a biofilm. Disseminated organic matter in sediments, especially if significantly degraded, needs to be distinguished from abiotic organic matter of hydrothermal or extraterrestrial origin. A noteworthy case study of mistaken biogenicity is the pseudofossils of the 3.46 Ga "Apex Chert," Western Australia. Although initially interpreted as organisms with a cyanobacterial affinity (Schopf and Packer 1987; Schopf 1993), later studies of the same material gradually unravelled the case for their biogenicity. Though of superficially microfossil-like morphology (filamentous, apparently septate), high-resolution FIB-SEM work demonstrated that they were in fact aluminous clay minerals onto

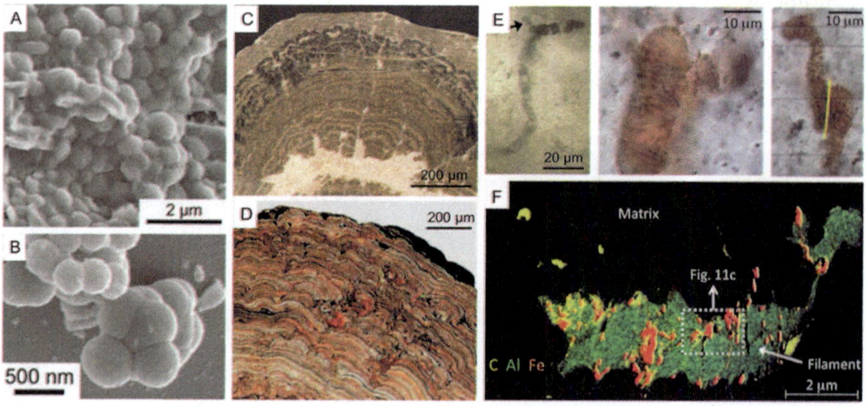

Fig. 3.7 Biomorphs and false-positive detections of biosignatures from the rock record. (**a**) Colony of coccoidal cells forming a biofilm on a volcanic grain, from Kitty's Gap Chert, Pilbara. (From Westall et al. 2011b). (**b**) Agglomerated silica spheres, a crystalline biomorph for coccoidal cells, which even bear hallmarks such as pseudo-division. (From Westall et al. 2011b). (**c**) Biogenic stromatolite, preserved from an ancient carbonate platform. (**d**) The infamous "Taylor stromatolite," texturally similar to the biological conical-domical stromatolites but formed abiotically during a paint-spraying process. (**e**) Three examples of filamentous, pseudoseptate Apex chert microfossil-like objects. Arrow indicates a side branch (a feature incompatible with prokaryotic morphology); yellow line indicates plane of the FIB-SEM image in **f**). (From Schopf 1993 and Wacey et al. 2016b). (**f**) FIB-SEM slice showing the aluminosilicate composition of the pseudofossil. Al, green (proxy for phyllosilicate); C, yellow (adhered only to the outermost extremities); Fe, red. (From Wacey et al. 2016b)

which carbon had become fortuitously adhered (Brasier et al. 2015a, b; Wacey et al. 2016a, b). As described in the next section, such cases of mistaken biogenicity plague biosignatures of all sizes, up to and including stromatolites. A famous extreme example of such a feature is the "Taylor stromatolite," a complex laminar-domical structure closely resembling a modern stromatolite but having been created, seemingly coincidentally, during paint spraying in the mid-twentieth century. Similarly convincing but supposed abiological examples are known from the geological record, and especially ancient stromatolitic occurrences, such as the 3.49 Ga Dresser Formation stromatolite, are regularly subject to strong criticism; for these, the less commonly used term "stromatoloid" (of stromatolitic form) is perhaps more appropriate.

Having established the biogenicity of the feature, the second task is to establish whether it formed at the same time as its host rock, i.e., is it syngenetic? Microbes today happily infiltrate cracks and fissures in rocks of various ages (as chasmoliths or endoliths) and can even become fossilized in their endo-/chasmolithic habitats. For instance, Westall and Folk (2003) described silicified endolithic cyanobacteria (< 8000 years old) within rocks ~3.8 billion years from the Isua complex in Greenland. Without conducting tests for syngenicity, younger microorganisms could be mistaken for ancient fossils and, indeed, were (Pflug and Jaeschke-Boyer 1979). We will not dwell on these topics, and the interested reader is referred to other texts documenting the procedures to be undertaken in the determination of biogenicity (e.g., Westall and Cavalazzi 2011; Westall et al. 2015b; Wacey 2009; Noffke 2010).

A final detail of prime importance for the study of purported traces of fossil life is the environment of formation, i.e., does the purported biosignature occur in a geological context consistent with a microbial habitat? For example, while cyanobacteria can be thermophilic and are found around hot springs (Campbell et al. 2015), they do not occur within hydrothermal veins, but this type of niche is colonized by chemotrophic organisms (at temperatures <119 °C). As another example, organotrophs need direct access to a carbon source and would not inhabit an oligotrophic, carbon-poor environment, though lithotrophs and phototrophs could.

3.2.4 The Early Record of Life

We will not present an exhaustive description of all the purported traces of ancient life but instead provide descriptions of key discoveries to illustrate the diversity and distribution of microbial remains in the rock record. We will address the ancient traces of life according to their habitat and presumed nutrient sources. It is important to understand how we attempt to identify these parameters. We noted above that study of the environment of deposition is as important as determination of the biogenicity and syngenicity of the purported fossils. Certain types of biosignature are more readily identifiable because of either features that are difficult to imitate by abiological processes or because of their relatively large size. For instance,

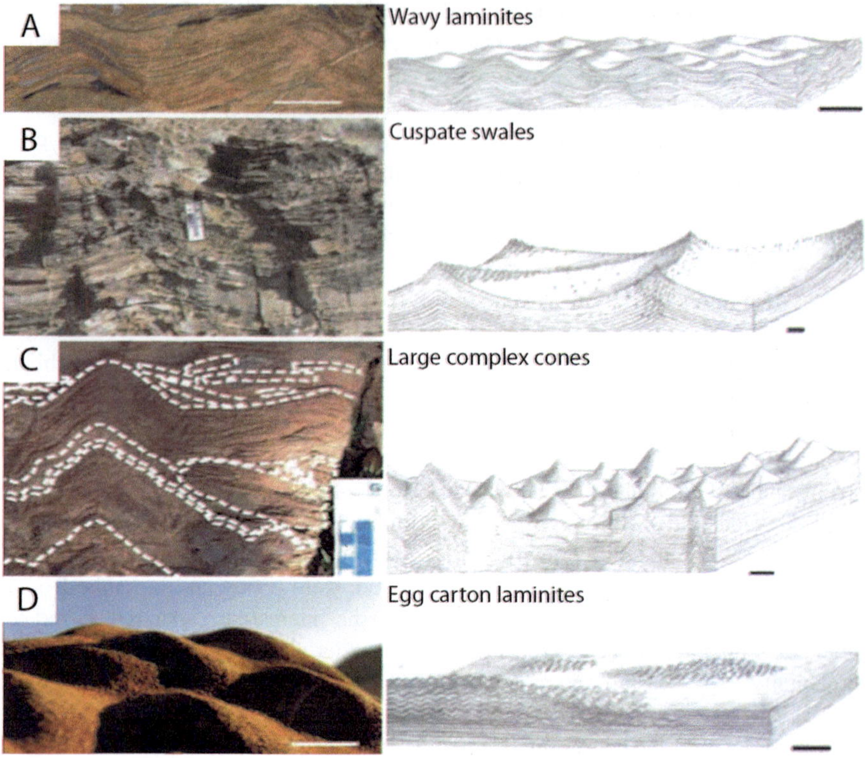

Fig. 3.8 Stromatolite morphologies from the Strelley Pool Formation, Pilbara. Each example shows a field photograph from outcrops and an interpretive sketch. (**a**) Wavy laminites, a traditional carbonate platform stromatolite morphology. This is interpreted as the morphology developing in subtidal, current-agitated environments. (**b**) Cuspate swales, large-scale peaked laminations; microbial textures are concentrated at the peaks. These are part of a typical platform carbonate transgression sequence. (**c**) Large complex cones with individual apices. These occur in the carbonate platform sequence, together with cuspate swales and egg-carton laminae. (**d**) Egg-carton laminae, a more curious zonal morphology. These occur in the carbonate platform sequence, alongside complex cones and cuspate swales. Scale bar in photograph of (**d**), 1 cm; all other scale bars, 5 cm. (From Allwood et al. 2006)

phototrophic microorganisms generally produce biofilm or mat-like structures that, due to their filamentous form and sticky, EPS-rich compositions, directly interact with their sedimentary substrates, to form what are known as microbially induced sedimentary structures (MISS, cf. Noffke 2009). These create characteristic anastomosing, weblike features in sediments that can be observed either by the naked eye or in thin sections of rocks observed by optical microscopy (Figs. 3.8 and 3.9). MISS occur at a range of scales, from the microscopic to over several kilometers of outcrop. Phototrophic mats today, through a combination of the entrapment of detrital particles and precipitation of calcium carbonate, can form vertical, domical, or columnar structures called stromatolites. Phototrophic microorganisms are not only

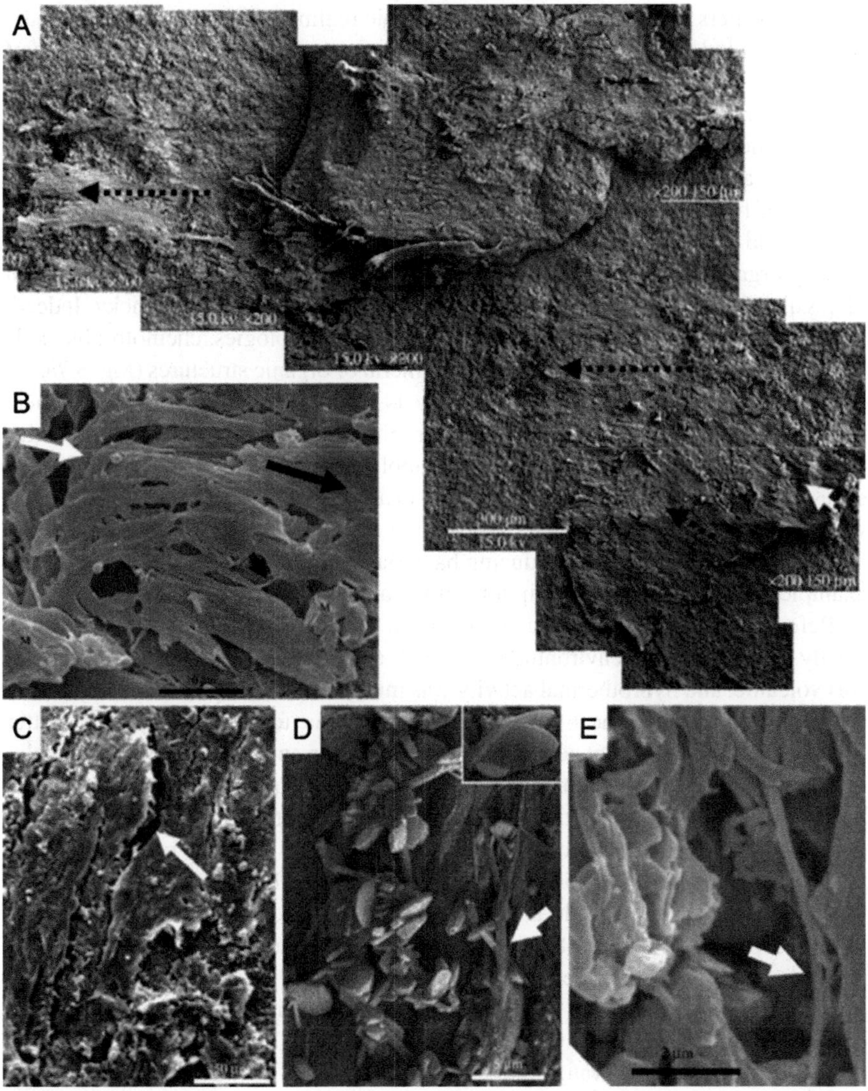

Fig. 3.9 The exceptional preservation of a phototrophic microbial mat in the 3.33 Ga Josefsdal Chert. (**a**) Plan view montage of SEM micrographs for half of the mat exposed on the surface of a bedding plane. Larger black arrows represent current direction. (**b**) Parallel and overturned filaments of the microbial mat (white arrow) relative to current flow (black arrow). (**c**) Desiccation cracks (white arrow), one indicator of subaerial exposure. (**d**) Evaporite minerals, such as gypsum and aragonite, are a further indicator of subaerial exposure. White arrow indicates a strand of mat covering the grains. (**e**) Rare, isolated turgid filament from the mat. These are usually deeply embedded in polymer film. (From Westall et al. 2006b)

bottom dwellers but also occur in the planktonic realm, sedimenting out with detrital particles and depositing at the bottom of the water body before becoming a part of the rock record.

Chemotrophs, on the other hand, tend to be smaller in size and do not necessarily produce such characteristic; often macroscopic features, although relatively thick biofilms, can form where and when there are abundant nutrients, such as in the vicinity of hydrothermal vents. Direct association of a biofilm with a mineral substrate could indicate lithotrophic behavior, but lithotrophs could also be colonized later by organotrophs, compounding the difficulty of their detection: the distinction of fossil colonies as having a specific chemotrophic metabolism is tricky. Indeed, because of their generally small size and simple morphologies, chemotrophic cells could be confused with abiotic, spherical mineral or organic structures (Fig. 3.7a, b).

Sometimes the evidence of biogenicity is of an indirect nature, for instance, related to corrosion of a substrate by lithotrophs or precipitation of secondary minerals due to the influence of microbial metabolisms on the physical-chemical properties of the immediate environment. An example of this could be calcification within a phototrophic mat due to degradation of the primary carbon dead mats by organotrophs, such as sulfate-reducing bacteria. We will address these aspects with examples in the following descriptions of the ancient traces of life.

Before continuing to specific studies from the Early Archaean record, we will briefly recapitulate the environment in which early life thrived. The global habitat was volcanic, and hydrothermal activity was much more widespread than at present. Coupled with increased heat flux through the crust due to the ancient Earth's hotter mantle temperatures, latent heat of planetary accretion and crystallization, and volatile radionuclide decay, Earth surface temperatures would have been quite warm. Seawater has been estimated, from oxygen isotope data in Archaean cherts and complementary computed global climate models (GCMs), to have been between 50 and 80 °C, while fluid inclusion studies yield an acidic pH of around 6–6.5 (though this would be locally variable around the aforementioned hydrothermal systems). However, pH levels were probably even lower, down to about pH 4, due to the high partial pressure of CO_2 in the atmosphere. UV radiation levels reaching the Earth's surface were high due to the lack of oxygen and ozone in the atmosphere (Westall et al. 2006b; Cockell and Raven 2004; Cnossen et al. 2007). Most preserved environments were shallow water, ranging from depths up to wave base (and rarely deeper) to the beach environment. That most organisms appear to have survived in relatively high radiation conditions at the Earth's surface implies that some evolutionary mechanism was in place to protect them; certainly, radiation seems not to have been prohibitive to their development. In summary, the environment of the early Earth was extreme by modern standards.

We note also that study of the early traces of life is constrained by the small amount of preserved ancient crust older. Although rocks as old as 4 Ga or more exist, such as the Acasta Gneisses (Bowring et al. 1989), the majority of the rare terranes more than 3.5-Ga-old consist of gneisses and granites, i.e., fractionated portions of crust that do not record habitable environments. Sediments as old as ~3.7 Ga have been recorded from Greenland (Rosing 1999) and also Northern

Canada (Dodd et al. 2017), but they are highly metamorphosed. While it is likely that life was present at this period in time, purported biosignatures from such highly altered rocks are highly controversial, being drawn from extremely modified, undependable geochemistry. The oldest well-preserved sediments, i.e., formations recording decodable potentially habitable environments, and thus the oldest possible instances of definitive biosignatures come from only two locations on Earth. Both formed a billion years after the consolidation of the planet: the $\leq$ 3.5 Ga Pilbara Greenstone Belt in Australia and the Barberton Greenstone Belt in South Africa. They record predominantly volcanic rocks (extrusive and intrusive), thin layers of sediments deposited on what appear to be protocontinental, oceanic plateaus (Lowe and Byerly 1999; Westall 2011), and some fractionated plutonic rocks (primitive granites). The majority of the sedimentary deposits represent shallow-water environments, i.e., foreshore to beach environments.

3.2.4.1 Phototrophic Remains

Phototrophic fossils constitute the majority of the convincing evidence for the early evolution of life on Earth. This is because, as noted above, the most readily identifiable microbial remains are those of phototrophic biofilms and mats. Phototrophic biofilms display a wide variety of morphologies, and multiple studies have noted a correlation between mat morphology and sedimentary environment of deposition (Allwood et al. 2006; Westall et al. 2006b; Tice 2009; Homann et al. 2015). One recurring theme in these interrelationships is that environmental stresses, e.g., strong, pluridirectional currents, elevated levels of toxic elements, or vigorous inputs of clastic and volcaniclastic material, seem to force the evolution of mat morphology to more complex forms. Throughout both the Barberton and Pilbara regions, mat morphology complexifies in accordance with hydraulic energy levels. In addition to mats and stromatolites, microfossils of suspected phototrophic affinity are proposed, but the proof of their biogenicity is more challenging. Examples of phototrophic biosignatures of known biogenicity from the Pilbara and Barberton regions are described in the following.

3.43 Ga Strelley Pool Formation

The Strelley Pool Formation is a sequence of sedimentary rocks (generally silicified peritidal carbonates and sandstones) sandwiched between two volcanic successions. Stromatolite-like structures are found within laminated carbonates and a bedded chert layer (Fig. 3.8). These shallow-water, carbonate platform sediments are important in that they enable decoding of the history of phototrophic organisms based on different morphotypes (Allwood et al. 2006). These are thus Earth's oldest widely accepted stromatolites.

First described almost 40 years ago (Lowe 1980, 1983) as possible microbial sediments, further analyses raised questions regarding this interpretation (Lowe 1994). However, the multi-scale observations of Allwood et al. (2006, 2009) strongly favored the biological origin. This demonstrates the changing interpretations which make appraisal of the ancient fossil record challenging. The stromatolites are environmentally controlled, and begin to grow immediately in the carbonate facies, after the cessation of siliciclastic input. Their seven described morphologies are as follows:

1. Encrusting, domical laminites, which form coatings over bedded conglomerates.
2. Small, crested or conical laminites, a high-frequency, low-amplitude form which occurs in very shallow waters, of little more than 1 m depth.
3. Cuspate swales, whose laminae form highly complex, meter-scale networks of undulating crested ridges with intersections adorned by smaller cones, directly comparable to known stromatolite facies such as *Thesaurus* (Fig. 3.8b).
4. Decimeter-scale large complex cones, most plausibly aligned with known conical stromatolites such as *Conophyton* (Fig. 3.8c).
5. "Egg-carton"-type conical stromatolites, an unusual wavy morphotype interpreted as a subtype of a *Conophyton*-style stromatolite (Fig. 3.8d).
6. Extensive wavy laminite stromatolites, a conical version of a typical carbonate platform facies, the high growth angles of which are interpreted to necessitate biological accretionary mediation (Fig. 3.8a).
7. Iron-rich laminites, continuous wavy-undulose laminations which occur at irregular intervals in the stratigraphy. These could plausibly be a poorly developed equivalent to one of the lower amplitude stromatolitic fabrics.

That there are seven distinct, syngenetic stromatolite fabrics, for which a purely mechanical depositional process is untenable in explaining their inherent heterogeneity, denotes their probable biological origin. At the microscopic scale, lamination fabrics are seen to covary with environmental stresses (represented by sedimentary facies changes), which is further evidence for biological processing at the microbial scale (Allwood et al. 2009). These morphologies are also very closely correlated to water depth. In addition to the geometric arguments against their being mechanically deposited, the paleoenvironmental outcrop evidence, combined with geochemical (REE+Y) analyses (Allwood et al. 2010), argues against their being abiotic hydrothermal precipitates. REE analyses of both the carbonate and chert laminae demonstrate a 250-fold enrichment in REE abundance in the former relative to the latter, which corresponds to similar measurements in known microbial carbonate against abiotic diagenetic counterparts (Webb and Kamber 2000; Kamber and Webb 2001).

3.45–3.41 Ga Buck Reef Chert

The Buck Reef Chert is a diverse stratigraphic suite, consisting of a basal evaporitic facies, a middle platform facies, and an upper deep basin facies (i.e., below wave base), that represents a progressively deepening marine environment lasting around 400 Ma (Lowe and Byerly 1999; de Vries et al. 2010). This is particularly significant from the geological point of view, since it is one of startlingly few (relatively) deeper-water environments preserved from the Archaean record (note also that water depths in the Early Archaean, by virtue of the globality of the ocean, would not have exceeded ~2 km).

The platform facies contains anastomosing, carbonaceous laminations, locally composed of filamentous structures with dimensions of 1–1.5 μm by 100 μm, entraining and draping over detrital grains, which are interpreted as phototrophic microbial mats (Walsh and Lowe 1999; Tice and Lowe 2004). Additional support for the biogenicity of these structures comes from the finding of an exceptionally preserved, mm-scale roll-up structure, which is clear evidence or the current-driven erosion and reworking of the microbial mat. Since, in life, the mat filaments would have been bound by glutinous, glue-like EPS, continuous water currents could have deformed this packet of laminae plastically following their erosion from the microbial mat (Tice and Lowe 2004). Carbon isotopic fractionation measurements of −35 to −20‰ within the microbial fabric are compatible with biological processing (Walsh and Lowe 1999).

3.33 Ga Josefsdal Chert

The effects of silicification in the ancient rock record are well-documented; the overwhelming majority of our clearest windows into the delicate biosphere of the early Earth are cherts. Petrographic, geochemical, and tomographic studies clearly demonstrate the abilities of silica to encapsulate and safeguard organic matter. In the vicinities of rapidly silicifying hydrothermal systems, chemically precipitated cherts can host examples of remarkably high-fidelity preservation. The Josefsdal Chert in the Barberton Greenstone Belt is one location where this phenomenon is well demonstrated. The geological setting of the Josefsdal Chert is, in many ways, similar to that of the Strelley Pool Chert: silicified volcanic sediments sandwiched between thick successions of submarine basalt lava flows. These sediments were deposited in shallow-water depths ranging from upper offshore to shoreface (tidal), i.e., well within the photic zone.

The Josefsdal Chert conserves an example of the exceptional preservation, in three dimensions, of a filamentous phototrophic microbial mat (Fig. 3.9) (Westall et al. 2001, 2006b, 2011a). The ~10-μm-thick, silicified biofilm still contains degraded organic matter (i.e., kerogeneous) and is composed of multiple laminae of filaments and their associated EPS forming a thin biofilm which covers an exposed bedding plane. The biofilm is both impregnated with silica and thickly coated with it, thus, preserving fine microbial filaments (<0.5 μm in thickness and up to 10s μm in length) probably representing the primary phototrophic producers that created the biofilm in the first place, as well as a thick coating of EPS. The filamentous texture of the surface of the biofilm is oriented in the same direction, as are portions of torn and rolled-up sections, suggesting the influence of unidirectional water flow over the biofilm. In addition, the biofilm appears to have been exposed subaerially, as evidenced by its desiccated surface. Further support for subaerial exposure comes from a suite of authigenic evaporitic minerals, including aragonite, gypsum, and halide (all coated with silica), which are intimately associated with the laminations. Subaerial exposure implies that these organisms would have needed to be radiation-tolerant, possessing mechanisms to counteract radiation damage, such as rapid genetic repair, pigmentation, and a thick, external coat of EPS.

While the surface of the film and its constituent microbial filaments is extremely well preserved, the internal structure of the film is amorphous, consisting of network textured degraded carbon molecules that have been impregnated not only with silica but also contain tiny crystallites (5–10 nm) of calcium carbonate (aragonite), a by-product of microbial metabolism, such as sulfur- or sulfate-reducing microorganisms that degraded the underlying dead part of the biofilm (only the upper layer being the phototrophically active layer) (Westall et al. 2011a).

The Josefsdal Chert microbial biofilms are an example of a community of microorganisms exhibiting characteristics and living in an environmental environment that can be used to decode their ecology. Anaerobic phototrophs were the primary producers that supported secondary heterotrophic activity in an environment characterized by a suite of extreme conditions. The anaerobic microbial communities exhibited high UV radiation tolerance, tolerance to acidic seawater, halotolerancy, and thermophily. This community thus represents one of the earliest examples of polyextremotolerant organisms. Such organisms would have been widespread and eminently more successful on the young planet.

3.22 Ga Moodies Group

The 3.22 Ga Moodies Group is a lithologically diverse sequence: quartzitic sandstones, volcanic tuffaceous deposits, siltstones, and localized conglomerates constitute its stratigraphy. Recent descriptions have highlighted that these represent Earth's oldest well-preserved siliciclastic, tidal-deltaic environment, i.e., a dynamic shallow-water coastal environment in the Archaean. Laminated sedimentary structures fitting the description of MISS have been traced over more than 15 km of outcrop (i.e., over the shoreline of an Archaean paleocontinent): morphologically, they are anastomosing, tufted, and bulbous wrinkle structures that colonized and stabilized sedimentary bedding surfaces (Noffke et al. 2006; Heubeck 2009; Homann et al. 2015). Their in vivo plastic, viscid quality is denoted by oriented, entrained grains which stick to individual laminae and, by roll-up structures, similar in origin to those described in the Buck Reef Chert. The ability of microbial mats to stabilize sediment is evident here: above the microbial mat horizons are subaerially exposed, desiccated, and fragmented layers of sediment. The appearance, survival, and preservation of microbial mats seem to be related to depositional environment. The highest abundances of microbial mats are found in high-energy upper intertidal to supratidal facies sandstones, although tidal coast facies sandstones also contain mats. Morphological diversity in the upper intertidal and supratidal facies is high: mat morphologies are not limited to planar laminations, but also feature an intriguing suite of features hinting at the life processes of the mats. Among these are gas and fluid escape structures, which deform individual laminae; the bubbles which would have formed these are, in some cases, silicificed in situ (Homann et al. 2015).

Finally, large, organic-walled, cohesive, spherical structures within the microbial mat horizons of the Moodies Group have been interpreted as isolated unicells (Javaux et al. 2010). Having sizes ranging from 50 to 300 μm, with most such structures being between 50 and 200 μm, their morphology, featuring cell-lumen-like features contained within taphonomically wrinkled spheroids, aligns them with robust organisms from the younger fossil records of the Proterozoic and Phanerozoic. Since organisms of this size are uncommon within even advanced archaea, protists and bacteria are used for comparison; Javaux et al. (2010) suggest that they may be stem clades of cyanobacteria, but their true biological affinity (and indeed biogenicity, given that taphonomic processes have significantly altered their surface textures) remains unclear.

3.2.4.2 Chemotrophic Remains

The appraisal of phototrophic biosignatures is helped by the macroscopic sizes of their biosignatures; stromatolites, for instance, can be readily observed at the outcrop scale, and even individual complex phototrophic organisms are generally of large size. Chemotrophs, by contrast, leave more enigmatic traces in the rock record: their simple prokaryotic morphologies are not always immediately indicative of biology, and their preservation, though often also exceptional in the vicinity of hydrothermal vents, is typically poorer due to the extreme environmental conditions in which they frequently lived. Where proven and putative chemotrophic biosignatures are preserved, as described below for the Josefsdal Chert, Kitty's Gap Chert, and possibly the Dixon Island Formation, they demonstrate some of the most primitive extremophiles known and are thus not only the most compelling explanations for the earliest evolution of life on Earth but also the most applicable for use in the search for life on other solar system bodies, for example, Mars, as these bodies tend to have been even more extreme in terms of environmental conditions than the early Earth.

3.446 Ga Kitty's Gap Chert

Kitty's Gap Chert in the Pilbara, Australia, is part of a volcano-sedimentary regime of volcanic sediments formed in a shallow water to tidal environment intercalated with thick lava flows (Westall et al. 2006a; Westall et al. 2011b; de Vries et al. 2006, 2010). As with the previously described Early Archaean (4.0–3.3 Ga) sediments, Kitty's Gap sediments have been silicified. These volcanic sediments host chemotrophic biosignatures described by Westall et al. (2006a, 2011b, 2015b).

Kitty's Gap volcanic particles are often coated with a thin layer of carbon, which, under scanning electron microscopy imaging, reveals monolayer biofilms of submicron-sized coccoidal organisms coated with EPS, which binds them to one another and to the substrate. Their morphologies can be distinguished from silica biomorphs (Fig. 3.7) in that they still contain organic carbon, show evidence of multiple life stages (cell division, life, and lysis), and are coated with an EPS-like film (Fig. 3.10). The preservation of dividing cells is testament to the rapidity of fossilization (silicification). Lysed cells having deflated and wrinkled morphology are also preserved. The direct association of the monolayer colonies with the surfaces of the volcanic grains indicates a likely lithotrophic metabolism. Indeed, this interpretation is enhanced by the finding that some of volcanic particles have tunnels leading from their exterior surfaces inward that are filled with an amorphous organic substance similar to EPS (Foucher et al. 2010).

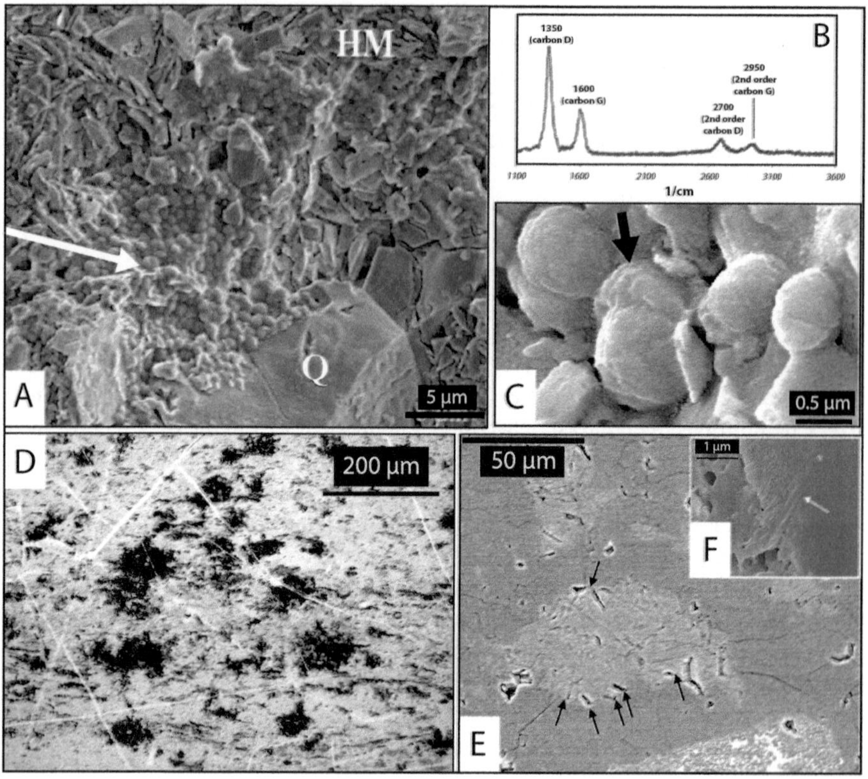

Fig. 3.10 Chemotrophic biosignatures from Early Archaean cherts. (**a**) Monolayer biofilm of micron-scale, coccoidal, chemolithotrophic cells (arrow) which coat a volcanic particle in the Kitty's Gap Chert. (**b**) Raman spectra documenting the presence of carbonaceous matter in the microfossils. (**c**) Cells from Kitty's Gap Chert biofilm in multiple stages of their life cycle: cell division and lysis (arrow). (**d**) Irregularly shaped black carbonaceous clots of up to 2 mm size in almost pure hydrothermal silica (pale background) from the Josefsdal Chert; these reflect chemotrophic colonies flourishing in the vent fluids. (**e**) Backscatter electron micrograph of small tunnel-like structures (arrows) at the margins of a pseudomorphed volcanic clast from Kitty's Gap Chert, evidence of microbial corrosion. (**f**) High-resolution SEM image of one such tunnel-like structure, filled with thick EPS. This forms a filamentous structure which eventually disappears into the matrix. (**a–c** from Westall et al. 2006a; **d** from Westall et al. 2015a; **e–f** from Foucher et al. 2010)

3.33 Ga Josefsdal Chert

The Josefsdal Chert exhibits one of the most diverse microbial consortia of the Early Archaean. In the last section, we described the variety of phototrophic microbial mats found in this unit. Here we describe chemotrophic microfossils whose remains are varied in morphology, taphonomy, and environmental context and exhibit exceptional preservation by virtue of their proximity to a rapidly silicifying hydrothermal system.

These chemotrophic remains (termed "clots") can be broadly characterized into two populations:

1. Rounded, carbon-coated sediment particles intimately associated with phototrophic mat fabrics
2. Spiky, irregular, isolated clots within the hydrothermal effluent itself, which may be up to 500 μm in size (Westall et al. 2015a, b).

The carbon-coated sediment particles up to several hundred microns in size occur in layers up to several centimeters in thickness. They co-occur with intercalated phototrophic biofilms that also draped the layers of chemotrophic clots. The latter may also be "trapped" within phototrophic laminae. The carbon coatings can be quite thick, up to several tens of microns, and are generally irregular in three dimensions, in this respect differing from the monolayer coatings seen in Kitty's Gap Chert. These carbonaceous-volcanic clots occur in close vicinity to hydrothermal effluents. The carbon coatings are interpreted as chemotrophic colonization of the volcanic grains. Possibly the initial colonizers were lithotrophs, as in the case of Kitty's Gap Chert, but the relative thickness of the carbonaceous coatings and their proximity to hydrothermal activity suggest that the lithotrophs may have subsequently been colonized by organotrophs.

The second type of clot, irregular structures with a spiky morphology (Fig. 3.10d), has a slightly different origin. These carbonaceous clots are preserved in rocks consisting of more than 99% silica which, from their location within the stratigraphy, are clearly genetically linked to hydrothermal vents and conduits. Volcanic detritus is more or less completely lacking in these deposits; consequently, this silica is more or less pure hydrothermal effluent in which individual spiky clots, up to 500 μm in diameter, are preserved exceptionally in three dimensions. The very irregular morphologies of these clots suggest that they formed in situ. This observation, together with their carbonaceous character and association with hydrothermal fluids, could imply that they are the remains of chemotrophic colonies. However, it is not possible to distinguish between a lithotrophic and organotrophic mode of life or even both as in the previous example. The encapsulation and concomitant high fidelity of conservation signifie particularly rapid silicification. By extension, it appears that these organisms were colonial chemotrophic colonies, which flourished in the nutrient-rich hydrothermal vent effluent itself. Indeed, there appears to be a strong correlation between interpreted chemotrophs and hydrothermal activity the Josefsdal Chert environment and, consequently, a hydrothermal control on such metabolisms (Westall et al. 2015a, b).

As with the above-described phototrophs, these chemotrophic microorganisms must have been polyextremophiles inhabiting environments characterized by stresses due to high temperature, high salinity, and likely regular influxes of semitoxic elements of hydrothermal origin detrimental to microbial growth, e.g., Cu.

3.22 Ga Moodies Group
In addition to the phototrophic remains described above, the Moodies Group portrays a more diverse biosphere. The recent discovery of coelobiontic (cavity-dwelling) ecosystems intercalated with microbial mats has unveiled another habitable niche in these rocks; due to the aforementioned high radiation levels on the early Earth, this concealed habitat is advantageous in that it circumvents the apparent threat. Homann et al. (2016) describe multiple cavities, directly overlain by microbial mats, in which are found downward-growing protrusions and pendant columns of laminated kerogen. They also highlight threadlike, filamentous microstructures, which are interpreted as bona fide microfossils by virtue of their regularity of subdivision (septate structure) and intimate, syngenetic association with the kerogenous cavity fabrics. Without advanced DNA repair mechanisms, the surface of the Archaean Earth may have dosed organisms with such levels of radiation that they would have been unable to survive for more than days (Cnossen et al. 2007); coeolobionts and other endolithic organisms could have avoided this danger.

Finally, the challenge of understanding the Earth's most ancient biosphere is the subject of much ongoing research. Due to the small amount of crust preserved from that period, possibilities for study are limited. Nonetheless, we are able to decode many aspects of the Archaean Earth's surface environment: we know that the Earth was an anaerobic, broadly thermophilic, dominantly oceanic planet, with widespread volcanism and hydrothermalism. Its surface, land and water alike, was bathed in UV radiation. Consequently all forms of life were what are today considered as polyextremophiles, but for the early Earth, the extremophilic ecology was the norm!

Although formed a billion years after the consolidation of the Earth, the Early Archaean terranes of the Pilbara in Australia and Barberton in South Africa preserve diverse traces of life including both chemotrophs (lithotrophs and organotrophs) and phototrophs. They were fueled by a more reduced variety of electron donors and acceptors owing to the overall anoxic conditions. Microorganisms inhabited a wide range of ecological niches from partially exposed beach surfaces, rock cavities, subsurface sediments, hydrothermal environments, the seafloor in shallow-water environments within reach of sunlight, and possibly also the planktonic realm.

While it is clear that the fossil record is not complete and represents only a fraction of the variety of microorganisms that must have lived on the early Earth, one of the remarkable aspects of these earliest traces of life is their degree of preservation. Seawater saturated with silica led to rapid fossilization, often with a high degree of morphological fidelity.

These earliest preserved biosignatures document a world in which all habitable niches had been colonized and where life had reached a relatively high degree of evolution, having reached the stage of oxygenic photosynthesis. However, this means that the earliest stages of the development of life and, indeed, its emergence from inanimate components are missing from the terrestrial fossil record. Life likely appeared as soon as habitable conditions were present, i.e., the presence of organic constituents, liquid water at temperatures <100 °C, and availability of nutrients, possibly around 4.3 Ga (Westall 2011). If we accept the hypothesis that life is a cosmic phenomenon with biological processes being simply an extension of chemistry (De Duve 1995), life could have emerged on other rocky planets and satellites in the solar system and elsewhere. The planet Mars is an example. Although inhospitable on its surface today, the Red Planet was habitable until between about 4.0 and 3.8 Ga, i.e., around the same time as the simple chemotrophic life described herein flourished on the early Earth. Unlike the Earth, Mars did not have crust-destroying tectonic activity, and thus, if life did appear there, the very early traces of its evolution could still be present in the ancient terranes (Westall et al. 2015b). For this reason a series of future astrobiological missions, the European-Russian ExoMars 2020 and NASA Mars 2020 missions, will be searching for fossil biosignatures.

3.3 Modern and Fossil Stromatolites

Stromatolites are emblematic rocks for geobiologists for several reasons: (1) they record information on old traces of life and are found throughout the geological record and (2) they might have had a major impact on the development of the early atmosphere, through, for example, photosynthetic consumption of the greenhouse gas CO_2 and production of free oxygen. In this section, we review the current knowledge on past and modern stromatolites and detail how they are recognized and which microbial groups contribute to their formation.

3.3.1 Definitions and Diversity of Modern and Ancient Stromatolites

As detailed by Hofmann (1973), stromatolites have received attention from the scientific community for a long time. The first reports of stromatolites date back to 1825 (Steel 1825). Hall (1883) first gave a species name to some of them (*Cryptozoon proliferum*), showing that a biological origin was already considered. The word stromatolite (stromatolith) was first coined by Kalkovsky in 1908.

The definition of the term "stromatolite" is debated (e.g., Krumbein 1983; Grotzinger and Knoll 1999). On the one hand, Semikhatov et al. (1979) proposed a descriptive/textural definition of stromatolites as "attached, laminated, lithified sedimentary growth structures, accretionary away from a point or limited surface of initiation". Within this definition, the term stromatolite also includes some abiogenic rocks. In contrast, the term "microbialites" received a genetic definition, i.e., based on how they formed. Burne and Moore (1987) defined them as "organosedimentary deposits that accrete as a result of a benthic microbial community trapping and binding detrital sediment and/or forming the locus of mineral precipitation". Based on their internal fabric, i.e., the arrangement of mineral grains in these rocks, microbialites are then classified into three main categories: (1) stromatolites, which are microbialites with a laminated macrofabric. This definition excludes abiotic deposits: (2) thrombolites, displaying a clotted macrofabric, and (3) leiolites, structures without a well-defined macrofabric (Dupraz et al. 2009). In all cases, lamination is one of the main attributes of stromatolites, but the detailed mechanisms governing their formation remain obscure (Grotzinger and Knoll 1999; Bosak et al. 2013). The morphologies of microbialites/stromatolites are varied and include flat laminated, oncolitic, pseudo-columnar, branching and non-branching columnar, conical or domal, as illustrated by Walter et al. (1992) (Fig. 3.11). The recognition of these morphologies and macrofabrics has often been the basis of attributing a biogenic origin to these sedimentary structures.

Many ancient marine and lacustrine rocks interpreted as stromatolites have been found in the geological record (Table 3.3). Some examples of ancient lacustrine microbialites are the 16 Ma tufa mounds from the Miocene Barstow Formation, California (Cole et al. 2004), the stromatolites and thrombolites from the 49 to 53 Ma Eocene Green River Formation in Wyoming (Seard et al. 2013), or the stromatolites of the 2.72 Ga Tumbiana Formation in Australia (Lepot et al. 2009; Stüeken et al. 2015).

Ancient marine stromatolites are very abundant, especially in the early Proterozoic when they formed massive carbonate reefs, similar in size and distribution to present coral reefs (Grotzinger and Knoll 1999; Awramik 1984). The oldest rocks interpreted as marine stromatolites were described from the 3.700 Ga Isua Greenstone Belt in Greenland (Nutman et al. 2016), the 3.481 Ga stromatolites from the Dresser Formation in the Pilbara Craton in Australia (Fig. 3.12) (e.g., Hofmann et al. 1999; Van Kranendonk et al. 2008). It should be noted that all stromatolites older than 2.3 Ga formed on a planet that was globally anoxic. This suggests that environmental conditions and biological communities associated with microbialites were likely very different to those we observe at present (e.g., Bosak et al. 2013).

Time variation in the diversity and "abundance" of stromatolites has been addressed (e.g., Awramik 1971; Walter and Heys 1985). However, "abundance" measured the spatial distribution of different types of stromatolites. The volume or mass of stromatolitic rocks on carbonate platforms, i.e., what would be a "true" abundance measurement, has not been clearly assessed for different geological periods (Grotzinger 1989). Over geological eras, stromatolites increased in "abundance" from the late Archaean to around 2.30 Ga; they remained at this level until 1.45 Ga and then increased to a peak at 1.35–1.10 Ga (Awramik and Sprinkle 1999; Riding 2006a). Proterozoic stromatolite diversity declined from 1.35–1.10 Ga to 550 Myr ago (Fig. 3.13). This decline has been classically attributed to the advent of grazing and burrowing metazoan organisms (Grotzinger 1990). This theory may explain why recent microbial carbonate platforms are restricted to locations where predators are rare, such as in saline alkaline lakes or some oligotrophic marine environments. However, it has been noted that there is a gap between the decline of stromatolite "abundance" and the first evidence of metazoans at 600 Ma (Knoll and Carroll 1999; Riding 2006a). Consequently, Grotzinger (1990) suggested an alter-

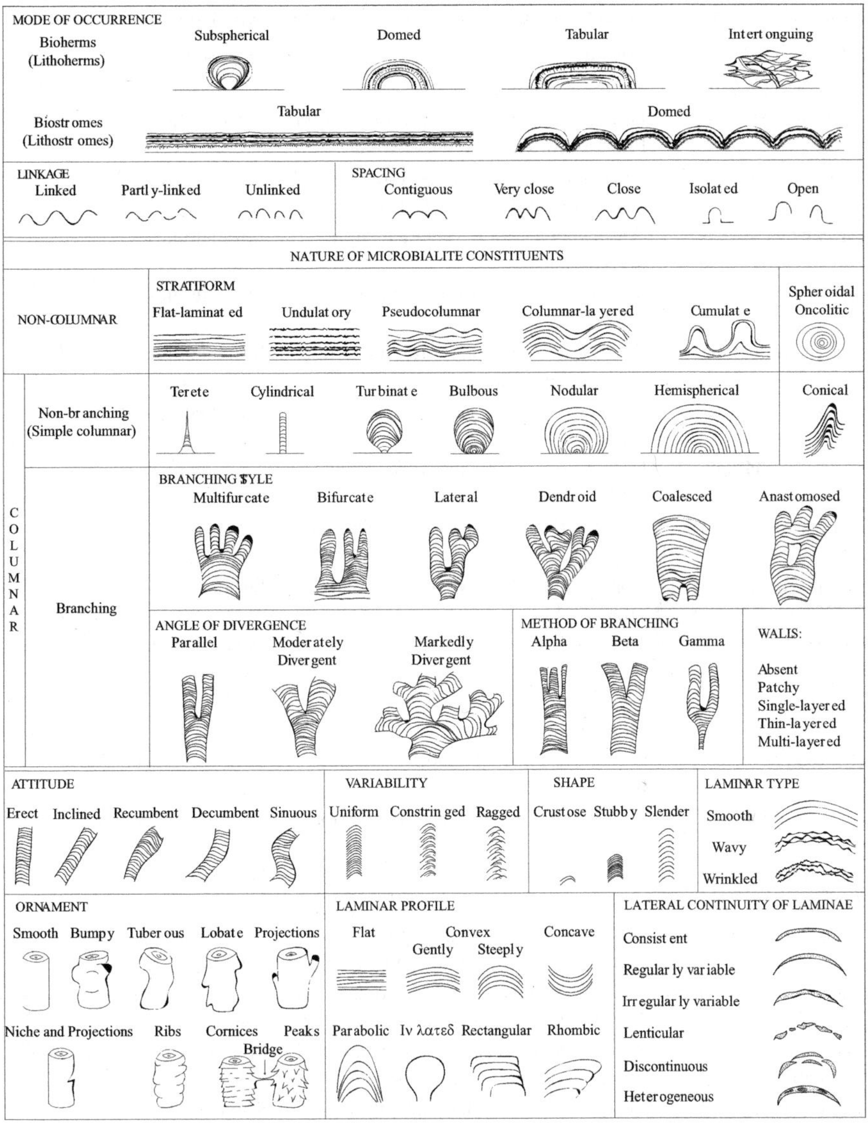

Fig. 3.11 Morphological features commonly used to characterize microbialites. (Modified after Walter et al. 1992)

native explanation for the decline of stromatolites, due to a decrease in the chemical saturation of seawater with Ca-carbonates during the Proterozoic. Overall, several questions remain unsolved, such as how to recognize the biogenicity of ancient stromatolites or what are the biological and environmental conditions conducive to their formation? Answers call for an increased knowledge of how modern stromatolites form.

Table 3.3 Summary of ancient marine and lacustrine microbialites reported in the literature

System	Locality	Deposit type
Isua supracrustal belt	Southwest Greenland	Stromatolites
North Pole Chert	Dresser Formation, Warrawoona Group, Pilbara Craton, Western Australia	Stromatolites
Barberton chert	Onverwacht Group, Barberton Greenstone Belt, South Africa	Stromatolites
Strelley Pool Chert	Warrawoona Group, Pilbara Craton, Western Australia	Stromatolites
Mushandike	Masvingo Greenstone Belt, Southern Zimbabwe	Stromatolites
Lake Meentheena	Tumbiana Formation, Fortescue Group, Western Australia	Stromatolites
Rocknest Formation	Wopmay Orogen, Northwest Territories, Canada	Thrombolites-stromatolites
Sulky Formation	Hornby Bay and Dismal Lake groups, Arctic Canada	Microbialites
Gaoyuzhuang Formation	Near the Pangjapu Iron Mine, Northern China	Stromatolites
Copper Harbor Conglomerate	Upper Peninsula of Michigan, Lake Superior, USA	Stromatolites
Diabaig	Torridon Group, Northwest Scotland	Microbially induced sedimentary structures (MISS)
Gourma Basin	Mali, Western Africa	Stromatolites-oolites
Talchir	Giridih Basin, Jharkhand, Orissa state, India	Stromatolites-oolites
Lake T'oo'dichi	Morrison Formation, Western Interior, USA	Microbialites-oolites
Mercia Mudstone Group	Southern Britain	Oolites
Toca Formation	Offshore, Congo Basin, Popular Republic of Congo and Cabinda, Angola	Coquinas-microbialites-stromatolites-oncolites
Lagoa Feia Formation	Pre-salt Campos Basin, Offshore, South Atlantic rift-sag lacustrine basins, Brazil	Coquinas-stromatolites-microbialites-oolites
Barra Velha Formation	Pre-salt Santos Basin, South Atlantic rift-sag lacustrine basin, Brazil	Microbialites-stromatolites
Rayoso Formation	Neuquén Basin (oil bearing), West-Central Argentina	Stromatolites
El-Molino Formation	Potosí Basin, El-Molino Formation, Bolivia	Stromatolites-thrombolites
Eastern Ebro Basin	Alcorisa, Vilanova de Prades, Igualada and Montserrat areas, Northeastern Spain	Stromatolites
Gosiute Lake	Laney Member, Green River Formation, Utah-Wyoming, USA	Stromatolites-thrombolites-coquinas-tufas-oolites
Bahariya Depression	Western Desert, Egypt	Microbialites-stromatolites-ooids

(continued)

Table 3.3 (continued)

System	Locality	Deposit type
Campins Basin	Northeastern Spain	Microbialites-stromatolites
Montaigu-le-Blin	Bassin des limagnes, Allier, France	Stromatolites
Manuherikia Group	Vinegar Hill, Lauder and Fiddler Flat, Central Otago, New Zealand	Stromatolites-oncoids-ooids
Barstow Formation	Mojave Desert, Southern California, USA	Tufas
Ries Lake	Impact-crater, Southern Germany	Stromatolites-microbialites-thrombolites
Duero Basin	Segovia Province, Northwestern region of the Iberian Peninsula, Spain	Stromatolites
Mediterranean Pelagian Platform	Beni Khiar and Oued el Bir formations, Tunisia, Spain, Italia, Maroc	Stromatolites-thrombolites-microbialites-oolites
Ridge Route Formation	Ridge Basin, California, USA	Stromatolites
Lake Lisan	Dead Sea, Israël	Speleothems-cave stromatolites
Sarliève	Bassin des Limagnes, Allier, France	Microbialites
Lake Rudolf	Modern Lake Turkana, Koobi Fora Formation, Kenya	Stromatolites-microbialites-thrombolites

System	Age (Myr)	Paleoenvironment	References
Isua supracrustal belt	3700	Marine	Nutman et al. (2016)
North Pole Chert	3481	Marine	Van Kranendonk et al. (2008)
Barberton chert	3400–3300	Marine	Byerly et al. (1986)
Strelley Pool Chert	3430	Marine	Allwood et al. (2006)
Mushandike	2828	Enclosed basin isolated from open seawater	Kamber et al. (2004)
Lake Meentheena	2720	Alkaline lake	Stüeken et al. (2015)
Rocknest Formation	1900	Marine	Kah and Grotzinger (1992)
Sulky Formation	1663	Marine	Bartley et al. (2014)
Gaoyuzhuang Formation	1400	Marine	Golubic and Seong-Joo (1999)
Copper Harbor Conglomerate	1090	Alluvial fan, fluviolacustrine	Fedorchuk et al. (2016)
Diabaig	994	Fluviolacustrine	Callow et al. (2011)
Gourma Basin	650–560	Marine	Bertrand-Sarfati and Moussine-Pouchkine (1983)
Talchir	~ 300	Lacustrine with influx of saline water	Das and Tripathi (2009)
Lake T'oo'dichi'	230–203	Wetland-lacustrine	Dunagan and Turner (2004)
Mercia Mudstone Group	Upper Triassic	Playa-lacustrine	Milroy and Wright (2002)

(continued)

Table 3.3 (continued)

System	Age (Myr)	Paleoenvironment	References
Toca Formation	130–125	Lacustrine	Thompson et al. (2015)
Lagoa Feia Formation	125.8–120	Lacustrine	Thompson et al. (2015)
Barra Velha Formation	125–112	Lacustrine	Wright and Barnett (2015)
Rayoso Formation	129–113	Lacustrine	Zavala et al. (2006)
El-Molino Formation	83–53	Playa-lake	Camoin et al. (1997)
Eastern Ebro Basin	66–34	Fluviolacustrine	Zamarreño et al. (1997)
Gosiute Lake	53–49	Lacustrine-playa	Seard et al. (2013)
Bahariya Depression	Middle Eocene	Marine	Salama et al. (2013)
Campins Basin	34–23	Lacustrine	Anadón and Utrilla (1993)
Montaigu-le-Blin	23–20	Lacustrine	Wattinne et al. (2003)
Manuherikia Group	Miocene	Lacustrine	Lindqvist (1994)
Barstow Formation	16	Alkaline lake and ground or spring waters	Cole et al. (2004)
Ries Lake	15	Soda lake	Arp (1995)
Duero Basin	15–009	Lacustrine	Sanz-Montero et al. (2008)
Mediterranean Pelagian Platform	Messinian ~ 6	Marine	Moissette et al. (2010)
Ridge Route Formation	Pliocene	Lacustrine	Link et al. (1978)
Lake Lisan	0.075–0.017	Lacustrine	Sorin et al. (2010)
Sarliève	0.008460–0.001910	Lacustrine	Bréhéret et al. (2008)
Lake Rudolf	0.0045	Lacustrine	Abell et al. (1982)

Modern stromatolites are rarely found in coastal marine environments, while many are found in freshwater, mostly lacustrine systems (Fig. 3.14). Modern marine stromatolites have been extensively studied in Hamelin Pool at Shark Bay in Australia (Logan 1961) and on the island of Highbourne Cay in the Bahamas (Reid et al. 2003; Khodadad and Foster 2012). It has usually been assumed that they are better analogues for ancient stromatolites, which are often considered as marine. However, this assumption should be moderated since (1) there are many ascertained ancient lacustrine stromatolites and (2) the growth of modern marine stromatolites seems to involve a lot of trapping and binding, with little in situ precipitation in contrast to lacustrine stromatolites and many ancient stromatolites (Grotzinger 1990). Modern stromatolites exhibiting a broad range of morphologies and microfabrics have been reported from more than 50 lakes around the world. The occurrence of these different morphologies has been shown to be related to various external environmental conditions such as the strength of waves, water depth, and/or the distance to the shore (Jahnert and Collins 2012).

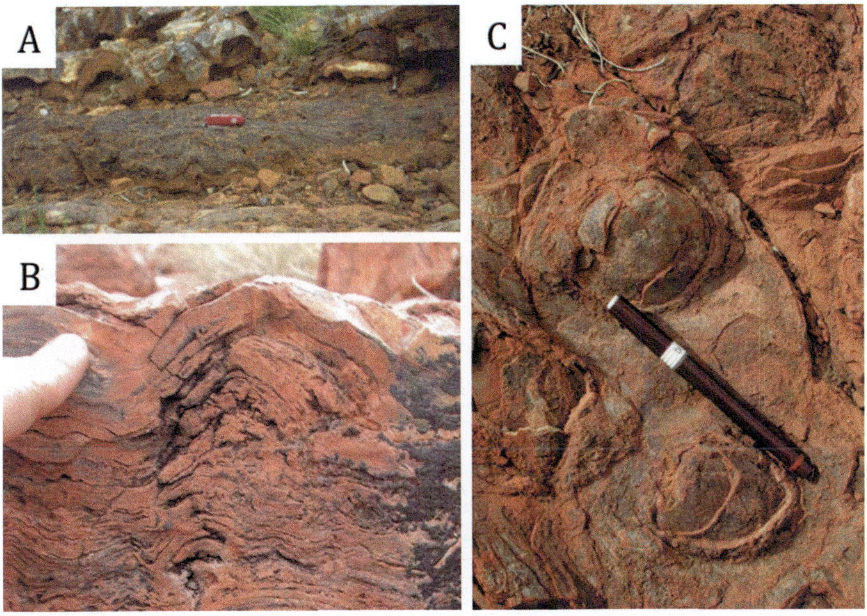

Fig. 3.12 Ancient stromatolites from the 3481 ± 2 Myr old Dresser Formation (Pilbara Craton, Australia) (Photo credits: Martin Van Kranendonk). (**a**) Bedding plane with broad domical stromatolites. (**b**) Cross-sectional outcrop view of wrinkly laminated carbonate and coniform stromatolite. (**c**) Plan outcrop view of bedding plane with coniform stromatolites

3.3.2 Geomicrobiological Processes Involved in the Formation of Modern Microbialites

Modern microbialites are associated with very diverse microbial populations thriving in oxic and anoxic microenvironments developing within the biofilms. Classically, studies have stressed the role of cyanobacteria in the formation of microbialites, because they can be studied easily by optical microscopy and taxonomically affiliated according to their morphology. Moreover, they are known to induce carbonate precipitation (Merz 1992; Gérard et al. 2013; Shih et al. 2013; Fig. 3.15).

However, the use of culture-independent approaches has revealed the existence of a much wider microbial diversity associated with microbialites (Fig. 3.16). For example, in microbialites from Lake Alchichica (Mexico), Saghaï et al. (2015) have detected in addition to cyanobacteria members of *Chlorobi*, *Chloroflexi*, and several *Alphaproteobacteria* (e.g., *Rhodobacterales*) and *Gammaproteobacteria* (*Chromatiales*) which can perform anoxygenic photosynthesis, partly coupled with sulfur oxidation but also members of several deltaproteobacterial orders such as *Desulfobacterales*, *Desulfuromonadales*, and *Desulfovibrionales* and some *Firmicutes* involved in sulfate reduction. In addition to oxygenic photosynthesis,

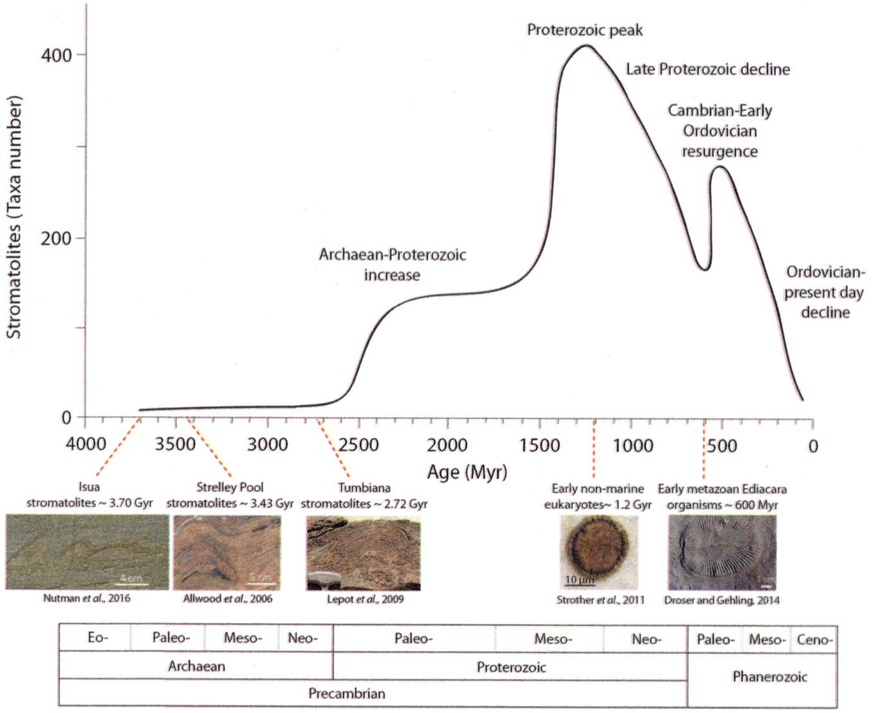

Fig. 3.13 Stromatolite diversity over geological time (modified after Riding 2006a)

several of these metabolisms such as anoxygenic photosynthesis (Bundeleva et al. 2012), sulfate reduction (Gallagher et al. 2012), or anaerobic methane oxidation coupled to sulfate reduction (Michaelis et al. 2002) are also able to induce carbonate precipitation. Saghaï et al. (2016) also showed that heterotrophs carrying out aerobic and anaerobic respiration and fermentation were more abundant than phototrophs in Alchichica microbialites. Extracellular polymeric substances (EPS) are an important structural component of the biofilms covering microbialites. They contain abundant chemical functional groups, such as carboxylic groups which can complex Ca^{2+}. This may favor precipitation of Ca-carbonates when EPS are hydrolyzed by heterotrophic EPS-degrading bacteria (Dupraz et al. 2009) such as some members of the *Bacteroidetes* phylum (Ben-Hania et al. 2016). Alchichica microbialites as many other modern microbialites contained a relatively minor, but not negligible, fraction of eukaryotes dominated by photosynthetic lineages such as diatoms and green algae. This was not the case for the most ancient microbialites, illustrating some limits of actualism. For decades ribosomal RNA genes have been analyzed to document the taxonomic diversity of microbialites (e.g., López-García et al. 2005; Centeno et al. 2012); metagenomes now provide more direct evidence of the metabolic capabilities of biofilms covering microbialites. The meta-comparison of microbialite metagenomes from several locations is progressing. It has been sug-

Fig. 3.14 Modern living microbialites from Mexican crater lakes. (**a**) Overview of Lake Atexcac. (**b**) Microbialites forming a steep wall superimposed on the Atexcac crater walls. (**c**) Microbialite from the Lake Alchichica showing white and orange colors and a green biofilm at its surface. (**d**) Microbialites from Lake Aljojuca, encrusting a basaltic rock and characterized by a green color and a high porosity

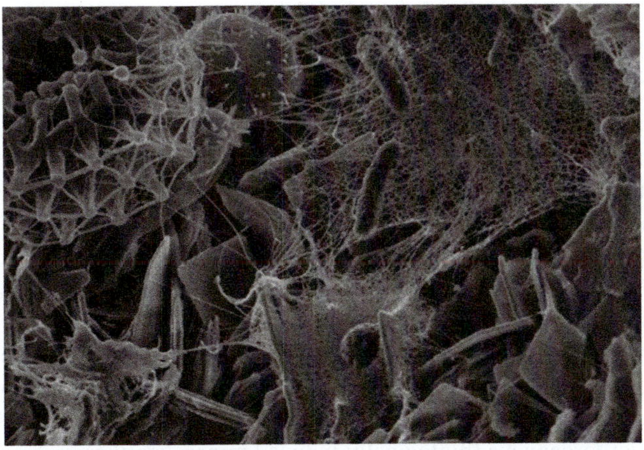

Fig. 3.15 Scanning electron microscopy image of a modern biofilm associated with microbialites from Lake Alchichica (Photo credits: Sebastien Charron). The biofilm is composed of EPS forming a kind of web in the center of the image, possible prokaryotic cells on the EPS, and larger cells with wall ornament which are possibly eukaryotic microorganisms. The biofilm is intimately associated with hydromagnesite ($Mg_5(CO_3)_4(OH)_2{\cdot}4H_2O$) crystals appearing with a platy crystal morphology

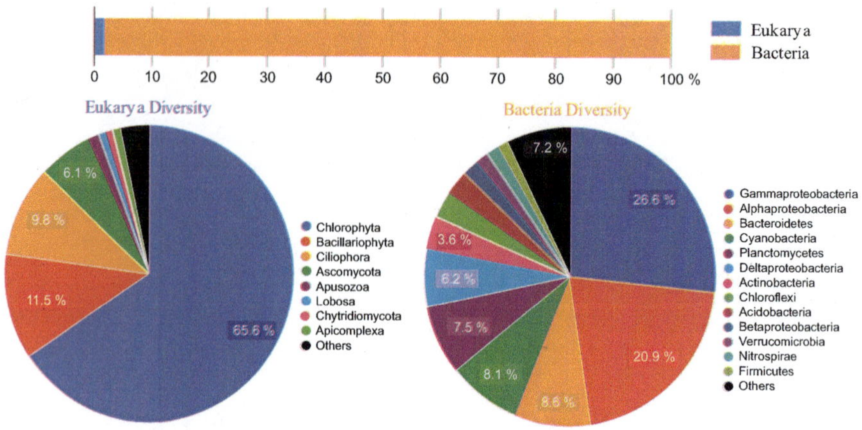

Fig. 3.16 Microbial diversity associated with a modern microbialite from Lake Alchichica, determined by metagenomics-based diversity analyses (modified after Saghaï et al. 2015). Top – Histogram showing the relative abundance of eukaryotes and bacteria. Bottom left – Pie graph representing the relative abundance of phylogenetic clades of microbial eukaryote showing the domination of *Chlorophyta* and *Bacillariophyta*. Bottom right – Pie graph illustrating the relative abundance of bacterial phyla or classes (for *Proteobacteria*) showing the dominance of *Gammaproteobacteria* and *Alphaproteobacteria*

gested that, despite differences in geographic location and salinity, significant functional similarities exist between the microbial communities of diverse microbialites that are not found in other communities from soils and oceans (Saghaï et al. 2016; Casaburi et al. 2015). Moreover, there are significant differences in the taxonomic composition of these biofilms suggesting that "who" is in the community may matter less than "what" they can do.

How do these microbial populations impact the formation of microbialites? Four processes can contribute to the formation of modern microbialites (Burne and Moore 1987) (Fig. 3.17).

(1) Trapping and binding of detrital sediments, a process sometimes termed agglutination, which seems to be a prominent mechanism in modern coastal marine environment (Riding 1991; Corkeron et al. 2012).
(2) In situ precipitation of authigenic minerals under the influence of microorganisms, a major process, in particular in modern lacustrine microbialites. This process has been termed organomineralization by Dupraz et al. (2009) to encompass passive mineralization of organic matter in supersaturated solutions (biologically influenced mineralization), as well as mineral precipitation resulting from the increase of the saturation of a solution by the metabolic activity of microorganisms (biologically induced mineralization).
(3) In situ abiotic precipitation of authigenic minerals, a process that may be difficult to distinguish from the previous one.

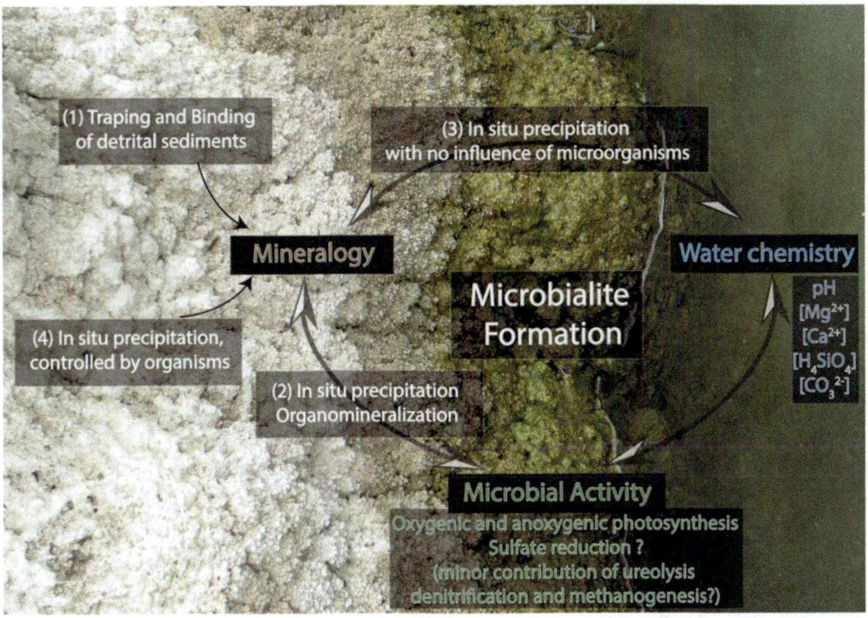

Fig. 3.17 Schematic representation of the main processes involved in the formation of microbialites. Image of a microbialite from Lake Atexcac, Mexico

(4) In situ precipitation of minerals controlled by organisms, e.g., the skeletons of eukaryotic microorganisms and intracellular formation of minerals in some prokaryotes (e.g., Couradeau et al. 2012).

Microbialites are mainly composed of Ca- and/or Mg-carbonates. The propensity of a solution to precipitate Ca-carbonates can be assessed by calculation of the saturation index (SI) with respect to a given mineral phase. The SI is defined as the decadic logarithm of the ratio of the ion activity product (IAP) over the solubility constant of the corresponding mineral (Ks).

$$\mathrm{IAP} = \left(\mathrm{Ca}^{2+}\right) \times \left(\mathrm{CO}_3^{\ 2-}\right)$$

$$\mathrm{SI} = \log\left(\mathrm{IAP} / \mathrm{Ks}\right),$$

where (Ca^{2+}) and (CO_3^{2-}) denote the activities of calcium and carbonate ions, respectively. The solution is oversaturated with a mineral phase when SI is positive. In this case, thermodynamics predicts that there should be precipitation. However, even in this case, precipitation may be very slow and may therefore not happen. The higher the oversaturation is, the faster the precipitation is (De Yoreo and Vekilov 2003). Some microorganisms can increase (CO_3^{2-}) and therefore SI and promote carbonate precipitation. Dupraz et al. (2009) listed primary metabolisms such as oxygenic and anoxygenic photosynthesizers as well as sulfate reducers, which tend to increase (CO_3^{2-}), i.e., promote precipitation, and others, such as aerobic heterotrophs, sulfide oxidizers, and fermenters, which tend to decrease (CO_3^{2-}), i.e., promote dissolution (Visscher and Stolz 2005; Dupraz et al. 2009). Other metabolisms, such as methanogenesis or ureolysis, may also promote carbonate precipitation and therefore contribute to the formation of microbialites. Overall, the capability of a phototrophic microbial mat or biofilm to promote the formation of a microbialite results from the balance between the activities of precipitation and dissolution promoters. This is why it is important to have a better assessment of the functional diversity of microbial populations living at the surface of microbialites. Metabolisms such as anoxygenic photosynthesis and/or methanogenesis may have been particularly important several billions years ago before the surface of the Earth was oxygenated.

However, this paradigm may be too simplistic for the following reasons. First, the induction of Ca-carbonate precipitation by sulfate reduction, for example, depends on the type of substrate that is used by microorganisms. Gallagher et al. (2012) showed that extracellular pH increases when bacteria use hydrogen and/or formate as electron donors, while it decreases when they use lactate, ethanol, or glycolate. In the first case, sulfate reducers favor carbonate precipitation, while, in the second case, they favor carbonate dissolution. Secondly, not only primary metabolisms but also some secondary metabolisms can play a role in the increase of pH, as shown by Barabesi et al. (2007) and Marvasi et al. (2010). Finally, in the case of cyanobacteria, it seems that not all of them have the same capability to induce carbonate precipitation. For example, some cyanobacteria such as the members of the *Pleurocapsales* order are more active than others (Couradeau et al. 2013; Gérard et al. 2013; Saghaï et al. 2015). In contrast, the cyanobacterium *Gloeomargarita lithophora* which forms intracellular Ca-carbonates (Couradeau et al. 2013) may hinder extracellular calcium carbonate precipitation (Cam et al. 2018). On the other end of the spectrum, some cyanobacteria such as *Mastigocoleus testarum* dissolve calcium carbonate instead of precipitating them (Guida and Garcia-Pichel 2016). Last but not least, microorganisms do not impact mineral precipitation only by changing the saturation index of a solution. They can also decrease (or increase) the thermodynamic barrier of the nucleation of mineral phases and thus accelerate (or decelerate) the kinetics of mineral precipitation (Giuffre et al. 2013). In this case, it is more the surface chemistry of the cells and their extracellular polymers that matter than their primary metabolism.

3.3.3 Deciphering the Biogenicity of Ancient Stromatolites and Identification of Associated Microorganisms

The oldest well-documented stromatolites are those from the Warrawoona Group (Pilbara Craton, Australia) and the Onverwacht Group (Barberton Greenstone Belt, South Africa), as described above. The 3.43-Ga-old Strelley Pool Chert located in the Warrawoona Group was deposited in a quiet coastal context and contains stromatolite-like structures that occur as pseudo-columnar structures (Hofmann et al. 1999; Van Kranendonk et al. 2003; Allwood et al. 2006). The Warrawoona Group also harbors the 3.481-Gyr-old stromatolites of the North Pole Chert forming domal structures (Walter et al. 1980; Van Kranendonk et al. 2008). In the Onverwacht Group, the pseudo-columnar stromatolites 3.4–3.3 Ga were described by Byerly et al. (1986).

The question of biogenicity of ancient stromatolites has fostered much debate. Buick et al. (1981) listed a series of criteria to decipher their biogenic origin: stromatolites must be found in sedimentary or meta-sedimentary rocks (i.e., geological context); they should be as old as the sediments in which they are deposited; and they should be formed mainly of convex-upward laminations (wavy or wrinkle) that should thicken over the crests of flexures. Moreover, microfossils or trace fossils should be present within the structures, and changes in the microfossil assemblage should be accompanied by morphological changes of the stromatolites. Finally, fossils must be organized in a way that indicates trapping, binding, or mineral precipitation induced by living microorganisms.

Grotzinger and Knoll (1999) reported that less than 1% of all stromatolites have a fossilized microbiota associated with them. This is particularly true for the most ancient stromatolites which rarely contain fossil microbes. This is why Lowe (1994) argued provocatively that stromatolites older than 3.2 Ga were probably formed through abiotic processes. He proposed that flat carbonate laminations can also form by abiotic processes such as deposition from currents, cyclic physical or chemical sedimentation, and diagenesis. For example, he proposed an abiotic formation for the Strelley Pool stromatolites by evaporative processes. In contrast, Allwood et al. (2006) described several morphotypes in the Strelley Pool Formation and attributed this diversity and complexity to paleoenvironmental responses by stromatolite-forming microbes. For the North Pole Chert, Lowe (1994) suggested that specific features such as the presence of faults may indicate that "stromatolites" formed abiotically by the deformation of a soft sediment. Finally, Byerly et al. (1986) and Lowe (1994) noted that stromatolites from Barberton are rich in microcrystalline tourmaline, a boron silicate, appearing within the laminations. Lowe (1994) interpreted these stromatolites as inorganic precipitates possibly formed around thermal springs. Alternatively, as we know, microbial communities can thrive around thermal springs. Westall et al. (2011a, 2015a, b) described in 3.33-Ga-old hydrothermally silicified sediments in the Josefsdal Chert, Barberton, carbonaceous material interpreted as remnants of silicified phototrophic biofilms with traces of in situ calcification.

This debate was further fed by the development of several types of mathematical growth models of stromatolites. The nonlinear equation of Kardar-Parisi-Zhang (KPZ equation, Kardar et al. 1986) was used to simulate stromatolite growth following several steps such as the fallout of suspended particles, the diffusion of settled particles at the surface of stromatolites, smoothing of the surface, and mineral precipitation or mat growth perpendicularly to the surface. With this model, which can be dominantly controlled by abiotic processes, Grotzinger and Knoll (1999) managed to grow some stromatolite morphologies, such as domes, suggesting that macroscopic morphologies may not be useful in determining biogenicity (Fedorchuk et al. 2016). The same models, however, could account for the involvement of microorganisms as well in the growth of various morphologies of stromatolites, including coniform stromatolites such as *Conophyton* (e.g., Batchelor et al. 2004). Another type of model known as diffusion-limited aggregation (DLA) has also been used (e.g., Verrecchia 1996). This type of model mimics a broader variety of morphologies found in the fossil record, especially in the Precambrian, with complex columnar and branching morphologies. Dupraz et al. (2006) used a combination of DLA and cellular automata to suggest that most stromatolite morphologies are produced by interactions between environmental conditions (e.g., salinity, alkalinity, hydrodynamics) and the effect of microbial mats or biofilms (e.g., stickiness, chemical control on $CaCO_3$ precipitation).

Numerous features of stromatolites have been interpreted as directly or indirectly resulting from the biological activity. As mentioned above, the search of microfossils in the geological record has been a long-term challenge causing multiple debates (Schopf and Packer 1987; Brasier et al. 2002, 2005; Schopf et al. 2002, 2010). It should be noted that detecting the past presence of microorganisms does not imply that they contributed to the formation of stromatolites. However, another part of the problem comes from the intrinsic difficulty to identify microfossils, since abiotic processes can produce objects with similar morphologies (García Ruiz et al. 2002, 2003). Since morphology is not a reliable criterion on its own, additional indications of biogenicity have been characterized (e.g., Brasier et al. 2015a, b). For example, remnants of organic molecules preserved within minerals have been detected in the 2.72-Gyr-old stromatolites of the Tumbiana Formation (Australia), similar to those found in modern microbialites (Benzerara et al. 2006; Lepot et al. 2009). Again, it should be noted that reduced carbon (referred here as organic carbon) can also be produced by abiotic processes, sometimes in large quantities (e.g., Galvez et al. 2013). The measurement of carbon isotopes in the organic matter and/or in the carbonates has also been used to infer the past presence of microorganisms based on the consideration that autotrophic organisms transform inorganic carbon into organic carbon with a more or less specific isotope fractionation favoring lighter isotope (Schidlowski 2001; Thomazo et al. 2009). However, the $^{13}C/^{12}C$ fractionation can be secondarily erased by isotopic re-equilibration during metamorphic events (Dunn and Valley 1992), and $^{13}C/^{12}C$ fractionations by abiotic processes can be similar to biogenic ones (McCollom 2011).

Biomarkers have also been looked for in stromatolites. They consist of fossil molecules, usually fossil lipids from ancient microbial cell walls that resisted aging (Summons et al. 1999; Jahnke et al. 2001; Brocks et al. 2003, 2005; Birgel et al. 2006). However, this molecular approach also has several limits. First, the phylogenetic affiliation of these molecules relies on our (limited) knowledge of the diversity of modern microorganisms that can form them. Some biomarkers have been claimed to be specific to certain taxa – before they were discovered in other taxonomical groups. For instance, Brocks et al. (2003) proposed that steranes (alteration product of sterols) reveal the presence of eukaryotes in the geological record. However, steroids can also be produced by some *Deltaproteobacteria* (Bode et al. 2003). An additional problem is the potential post-contamination of ancient rocks by (endolithic) microorganisms. For example, lipids extracted from shales of the Pilbara Craton (Australia), and first claimed as an evidence for the presence of eukaryotes in 2.7-Ga-old stromatolitic formations (Brocks et al. 1999), were later shown to be younger than the rocks enclosing them, i.e., resulting from some post-contamination (Rasmussen et al. 2008).

Beyond this need to find traces of life in ancient stromatolites, many authors have speculated on the type of microorganisms that inhabited these rocks. For example, Wacey et al. (2011) suggested the presence of sulfur-metabolizing cells in 3.430 Ga stromatolites from the Strelley Pool Formation based on the observation of putative microfossils spatially close to pyrite crystals with a particular S isotopic compositions. Similarly, Westall et al. (2011a) hypothesized the implication of SRB contributing to calcification in the 3.33 Ga Josefsdal Chert based on the observation of pyrite crystals. By measuring the isotopic composition of ancient microfossiliferous units such as the 3.481-Gyr-old Dresser Formation, Schopf (2011) proposed that RuBisCO-mediated CO_2fixation was already present. However, these results do not necessarily reflect the presence of oxygenic photoautotrophy. Indeed, several studies concluded that oxygenic photoautotrophy appeared much later than the most ancient stromatolites, possibly around 2.5–2.6 Ga (Shih et al. 2017). It is now increasingly suggested that ancient stromatolites may have been formed by anoxygenic, rather than oxygenic, photosynthesizers, in particular through the involvement of sulfur-oxidizing anoxygenic phototrophs (Bosak et al. 2007; Knoll et al. 2016). Before the great oxidation event 2.3 Ga ago, reduced chemical species such as Fe^{2+} and S^{2-} were more abundant in the ocean and may therefore have been used as electron donors by anoxygenic phototrophs. The resulting production of oxidized species may have fed a local sulfur cycle involving diverse S-oxidizing and -reducing microorganisms. In the 2.72 Ga stromatolites of the Tumbiana Formation, Thomazo et al. (2011) observed ^{15}N enrichments inversely correlated with the very low $\delta^{13}C$ values associated with organic matter. These authors attributed this correlation to the onset of nitrification coupled with a continuous removal of its derivatives (nitrite and nitrate) by denitrification.

Fossil cyanobacteria have received great interest. The identification of fossil cyanobacteria is aided by the study of modern cyanobacteria within their natural habitats (Golubic and Seong-Joo 1999) and their relative morphological complexity. Based on morphological features and modes of cell division, cyanobacteria are organized into five major phenotypes: unicellular, baeocytous (baeocytes are small, spherical reproductive cells), filamentous, heterocystous, and ramified (Rippka et al. 1979). For example, fossil filamentous cyanobacteria possibly affiliated with oscillatoriacean cyanobacteria have been described in silicified stromatolites of the 1.4 Ga Gaoyuzhuang Formation in Northern China (Golubic and Seong-Joo 1999). However, it is usually difficult to affiliate microfossils found in the fossil record with certainty to a particular taxonomic group based on their morphology only. Indeed, comparative genomics of 54 phylogenetically and phenotypically diverse cyanobacteria strains showed the lack of specific and unique genes determining unambiguously the 5 major phenotypes (Shih et al. 2013).

Modern cyanobacteria import inorganic carbon actively, mostly as HCO_3^-, through what is known as carbon concentration mechanisms or CCM (Miller and Colman 1980). Intracellular conversion of HCO_3^- to CO_2 followed by CO_2fixation by ribulose-1,5-bisphosphate carboxylase/oxygenase (RuBisCO) and the associated importation of H^+ to regulate intracellular pH results in an increase of extracellular pH, which favors $CaCO_3$ precipitation (Verrecchia et al. 1995; Badger and Price 2003; Riding 2006b). In some cases, mineral precipitation occurs on the surface of the cells, encrusting them and therefore fossilizing them (Couradeau et al. 2013). The resulting microfossils of calcified cyanobacteria have been called calcimicrobes (Riding 2006b). The earliest undisputed evidence of calcimicrobes is the filamentous morphotype *Girvanella*, dated at 750–700 Ma (Knoll et al. 1993). Although cyanobacteria were present at least since the Great Oxidation Event, dated between ~2.45 and ~2.32 Ga (Farquhar et al. 2011), calcimicrobes are scarce during the Proterozoic Eon. This paradox is called the Precambrian Enigma (Riding 1994). One of the hypotheses suggested to explain this enigma is that seawater was highly supersaturated with $CaCO_3$ phases in the Precambrian, in particular because of the high CO_2 content of the atmosphere (Knoll et al. 1993; Arp et al. 2001). This disfavored precipitation around cyanobacterial cells. Another hypothesis is that the decrease of the CO_2 content of the atmosphere in the end of the Precambrian drove the appearance of CCM by natural selection, thus leading to the appearance of calcimicrobes. The debate remains open as to whether environmental or biological changes are responsible for the hiatus between the fossil record of calcimicrobes and the appearance of cyanobacteria.

Overall, ancient stromatolites comprise one of the major and most visible records of early life on Earth, but hints about the identity of microorganisms associated with them and the environmental conditions prevailing during their formation remain scarce. Stromatolites are the result of a combination of abiotic and biotic factors. It is crucial to assess the respective roles of microorganisms and environmental conditions allowing their formation. Answers may possibly come from better understanding of how modern marine and lacustrine microbialites form. However, transformations by diagenesis and metamorphism affect the message initially carried by ancient microbialites. One key issue for future studies will be to understand how authigenic carbonate and silicate phases observed in modern microbialites transform with diagenesis and metamorphism. It will be also important to determine whether microfossils and organic polymers observed in modern microbialites can be preserved for extended periods, such as million to billions of years.

Finally, the role of eukaryotes, such as diatoms and green algae, systematically observed in association with modern microbialites in diverse modern environments, is poorly understood. While prokaryote-constructed stromatolites were abundant during the early Proterozoic, eukaryotes only became abundant later on, in the Neoproterozoic geological record. How this may have affected the formation of stromatolites remains to be further determined.

3.4 General Conclusion

Although there is still some debate, evidence from terrestrial fossil record shows that the Early Archaean Earth was inhabited by microorganisms exhibiting a chemotrophic or a phototrophic metabolism. While ancient stromatolites are the most immediately visible indication of early life, they are rare compared to the abundant but more subtle, microscopic remains of primitive chemotrophic and phototrophic life forms. Inhabiting an early Earth whose ambient conditions were very different to those of the present-day Earth, anaerobic, relatively high temperatures at the rock/seawater interface, high flux of UV radiation within a volcanic, and highly hydrothermally influenced marine environment, the first life forms to emerge would be classed today as extremophilic microorganisms. Placing recent phylogenetic, geological, and microbiological data together in the context of thermodynamic laws suggests that the disproportionation of sulfurous compounds (e.g., elemental sulfur and/or polysulfides disproportionation) may have been important as one of the first metabolisms. Such a metabolism could have taken thermodynamic advantage from the H_2- and CH_4-rich alkaline fluids emitted from abundant serpentinized hydrothermal systems. Disproportionation of sulfurous compounds into sulfide and sulfate would have allowed sulfate-reducing bacteria to develop in an anoxic atmosphere. The existence of sulfate minerals, serpentinite, and possibly past hydrothermal activity on Mars offers intriguing scenarios of ancient life on the "Red Planet".

Acknowledgments We thank Guy Fauque for revising the manuscript.

References

Abell PI, Awramik SM, Osborne RH, Tomellini S (1982) Plio-pleistocene lacustrine stromatolites from lake Turkana, Kenya: Morphology, stratigraphy and stable isotopes. Sediment Geol 32:1–26

Alleon J, Bernard S, Le Guillou C, Marin-Carbonne J, Pont S, Beyssac O, McKeegan KD, Robert F (2016) Molecular preservation of 1.88 Ga Gunflint organic microfossils as a function of temperature and mineralogy. Nat Commun 7:11977

Allwood AC, Walter MR, Kamber BS, Marshall CP, Burch IW (2006) Stromatolite reef from the Early Archaean era of Australia. Nature 441:714–718

Allwood AC, Kamber BS, Walter MR, Burch IW, Kanik I (2010) Trace elements record depositional history of early Archaean stromatolitic carbonate platform. Chem Geol 270:148–163

Allwood AC, Grotzinger JP, Knoll AH, Burch IW, Anderson MS, Coleman ML, Kanik I (2009) Controls on development and diversity of Early Archean stromatolites. Proc Natl Acad Sci USA 106:9548–9555

Altwegg K, Balsiger H, Bar-Nun A, Berthelier JJ, Bieler A, Bochsler P, Briois C, Calmonte U, Combi MR, Cottin H, De Keyser J, Dhooghe F, Fiethe B, Fuselier SA, Gasc S, Gombosi TI, Hansen KC, Haessig M, Jäckel A, Kopp E, Korth A, Le Roy L, Mall U, Marty B, Mousis O, Owen T, Rème H, Rubin M, Sémon T, Tzou CY, Hunter Waite J, Wurz P (2016) Prebiotic chemicals-amino acid and phosphorus-in the coma of comet 67P/Churyumov-Gerasimenko. Sci Adv 2. https://doi.org/10.1126/sciadv.1600285

Amend JP, LaRowe DE, McCollom TM, Shock EL (2013) The energetics of organic synthesis inside and outside the cell. Philos Trans R Soc B 368:20120255

Amend JP, Shock E (2001) Energetics of overall metabolic reactions of thermophilic and hyperthermophilic *Archaea* and *Bacteria*. FEMS Microbiol Rev 25:175–243

Anadón P, Utrilla R (1993) Sedimentology and isotope geochemistry of lacustrine carbonates of the Oligocene Campins Basin, north-east Spain. Sedimentology 40:699–720

Arp G (1995) Lacustrine bioherms, spring mounds, and marginal carbonates of the Ries-impact-crater (Miocene, Southern Germany). Facies 33:35–89. https://doi.org/10.1007/BF02537444

Arp G, Reimer A, Reitner J (2001) Photosynthesis-induced biofilm calcification and calcium concentrations in Phanerozoic oceans. Science 292:1701–1704

Aubrey A, Cleaves HJ, Chalmers JH, Skelley AM, Mathies RA, Grunthaner FJ, Ehrenfreund P, Bada JL (2006) Sulfate minerals and organic compounds on Mars. Geology 34:357–360

Aüllo T, Ranchou-Peyruse A, Ollivier B, Magot M (2013) *Desulfotomaculum* spp. and related Gram-positive sulfate-reducing bacteria in deep subsurface environments. Front Microbiol 4:362

Awramik SM (1971) Precambrian columnar stromatolite diversity: reflection of metazoan appearance. Science 1974:825–827

Awramik SM (1984) Ancient stromatolites and microbial mats. In: Cohen Y, Castenlolz RW, Halvorson HO (eds) Microbial mats: stromatolites. Alan R. Liss, New York, pp 1–22

Awramik SM, Sprinkle J (1999) Proterozoic stromatolites: the first marine evolutionary biota. Hist Biol 13:241–253

Badger MR, Price GD (2003) CO_2 concentrating mechanisms in cyanobacteria: molecular components, their diversity and evolution. J Exp Bot 54:609–622

Bak F, Cypionka H (1987) A novel type of energy metabolism involving fermentation of inorganic sulphur compounds. Nature 326:891–892

Bak F, Pfennig N (1987) Chemolithotrophic growth of *Desulfovibrio sulfodismutans* sp. nov. by disproportionation of inorganic sulfur compounds. Arch Microbiol 147:184–189

Barabesi C, Galizzi A, Mastromei G, Rossi M, Tamburini E, Perito B (2007) *Bacillus subtilis* Gene cluster involved in calcium carbonate biomineralization. J Bacteriol 189:228–235

Bartley JK, Kah LC, Frank TD, Lyons TW (2014) Deep-water microbialites of the Mesoproterozoic Dismal Lakes Group: microbial growth, lithification, and implications for coniform stromatolites. Geobiology 13:15–32

Batchelor MT, Burne RV, Henry BI, Jackson MJ (2004) A case for biotic morphogenesis of coniform stromatolites. Physica A Stat Mech Appl 337:319–326

Bates RG, Pinching G (1949) Acidic dissociation constant of ammonium ion at 0 °C to 50 °C, and the base strength of ammonia. J Res Natl Bur Stand 42:419–430

Bayman F, Lebrun E, Brugna M, Schoepp-Cotenet B, Giudici-Orticoni M-T, Nitschke W (2002) The redox protein construction kit: pre-last universal common ancestor evolution of energy-conserving enzymes. Philos Trans R Soc Lond 358:267–274

Bekker A, Holland HD, Wang PL, Rumble D III, Stein HJ, Hannah JL, Coetzee LL, Beukes NJ (2004) Dating the rise of atmospheric oxygen. Nature 427:117–120

Belkin S, Wirsen CO, Jannasch HW (1985) Biological and abiological sulfur reduction at high temperatures. Appl Environ Microbiol 49:1057–1061

Ben-Hania W, Joseph M, Bunk B, Sproer C, Klenk HP, Fardeau ML, Spring S (2016) Characterization of the first cultured representative of a *Bacteroidetes* clade specialized on the scavenging of cyanobacteria. Environ Microbiol 19:1134–1148

Benzerara K, Menguy N, López-García P, Yoon T-H, Kazmierczak J, Tyliszczak T, Guyot F, Brown GE (2006) Nanoscale detection of organic signatures in carbonate microbialites. Proc Natl Acad Sci 103:9440–9445

Bertrand-Sarfati J, Moussine-Pouchkine A (1983) Platform-to-basin facies evolution; the carbonates of late Proterozoic (Vendian) Gourma (West Africa). J Sediment Res 53:275–293

Beveridge TJ, Fyfe W (1985) Metal fixation by bacterial cell walls. Can J Earth Sci 22:1893–1898

Birgel D, Thiel V, Hinrichs K-U, Elvert M, Campbell KA, Reitner J, Farmer JD, Peckmann J (2006) Lipid biomarker patterns of methane-seep microbialites from the Mesozoic convergent margin of California. Org Geochem 37:1289–1302

Blank CE (2009) Not so old Archaea-the antiquity of biogeochemical processes in the archaeal domain of life. Geobiology 7:495–514

Bode HB, Zeggel B, Silakowski B, Wenzel SC, Reichenbach H, Müller R (2003) Steroid biosynthesis in prokaryotes: identification of myxobacterial steroids and cloning of the first bacterial 2, 3 (S)-oxidosqualene cyclase from the myxobacterium *Stigmatella aurantiaca*. Mol Microbiol 47:471–481

Bosak T, Greene SE, Newman DK (2007) A likely role for anoxygenic photosynthetic microbes in the formation of ancient stromatolites. Geobiology 5:119–126

Bosak T, Knoll AH, Petroff AP (2013) The meaning of stromatolites. Annu Rev Earth Planet Sci 41:21–44

Bowring SA, Williams IS (1999) Priscoan (4.00–4.03 Ga) orthogneisses from northwestern Canada. Contrib Mineral Petrol 134:3–16

Bowring SA, Williams IS, Compston W (1989) 3.96 Ga gneisses from the Slave province, Northwest Territories, Canada. Geology 17:971–975

Brandes JA, Boctor NZ, Cody GD, Cooper BA, Hazen RM, Yoder HS Jr (1998) Abiotic nitrogen reduction on the early Earth. Nature 395:365–367

Brasier MD, Green OR, Jephcoat AP, Kleppe AK, Van Kranendonk MJ, Lindsay JF, Steele A, Grassineau NV (2002) Questioning the evidence for Earth's oldest fossils. Nature 416:76–81

Brasier M, Green O, Lindsay J, Mcloughlin N, Steele A, Stoakes C (2005) Critical testing of Earth's oldest putative fossil assemblage from the ~3.5Ga Apex chert, Chinaman Creek, Western Australia. Precambrian Res 140:55–102

Brasier MD, McLoughlin N, Green OR, Wacey D (2006) A fresh look at the fossil evidence for early Archaean cellular life. Philos Trans R Soc B 361:887–902

Brasier MD, Antcliffe J, Saunders M, Wacey D (2015a) Earth's earliest fossils (3.5– 1.9 Ga): changing the picture with new approaches and new discoveries. Proc Natl Acad Sci USA 112:4859–4864

Brasier AT, Rogerson MR, Mercedes-Martin R, Vonhof HB, Reijmer JJG (2015b) A Test of the biogenicity criteria established for microfossils and stromatolites on Quaternary Tufa and Speleothem materials formed in the "Twilight Zone" at Caerwys, UK. Astrobiology 15:883–900

Bréhéret J-G, Fourmont A, Macaire J-J, NéGrel P (2008) Microbially mediated carbonates in the Holocene deposits from Sarliève, a small ancient lake of the French Massif Central, testify to the evolution of a restricted environment. Sedimentology 55:557–578

Brochier C, Philippe H (2002) Phylogeny: A non-hyperthermophilic ancestor for *Bacteria*. Nature 417:244

Brocks JJ, Logan GA, Buick R, Summons RE (1999) Archean molecular fossils and the early rise of eukaryotes. Science 285:1033–1036

Brocks JJ, Buick R, Summons RE, Logan GA (2003) A reconstruction of Archean biological diversity based on molecular fossils from the 2.78 to 2.45 billion-year-old Mount Bruce Supergroup, Hamersley Basin, Western Australia. Geochim Cosmochim Acta 67:4321–4335

Brocks JJ, Love GD, Summons RE, Knoll AH, Logan GA, Bowden SA (2005) Biomarker evidence for green and purple sulphur bacteria in a stratified Palaeoproterozoic sea. Nature 437:866–870

Buick R, Dunlop JSR, Groves DI (1981) Stromatolite recognition in ancient rocks: an appraisal of irregularly laminated structures in an Early Archaean chert-barite unit from North Pole, Western Australia. Alcheringa Aust J Palaeontol 5:161–181

Bundeleva IA, Shirokova LS, Benezeth P, Pokrovsky OS, Kompantseva EI, Balor S (2012) Calcium carbonate precipitation by anoxygenic phototrophic bacteria. Chem Geol 291:116–131

Burne RV, Moore LS (1987) Microbialites: organosedimentary deposits of benthic microbial communities. Palaios 2:241–254

Burton AS, Glavin DP, Callahan MP, Dworkin JP, Jenniskens P, Shadad MH (2011) Heterogeneous distributions of amino acids provide evidence of multiple sources within the Almahata Sitta parent body, asteroid 2008 TC_3. Meteorit Planet Sci 46:1703–1712

Byerly GR, Lower DR, Walsh MM (1986) Stromatolites from the 3,300–3,500-Myr Swaziland Supergroup, Barberton Mountain Land, South Africa. Nature 319:489–491

Cady SL, Farmer JD (1996) Fossilization processes in siliceous thermal springs: trends in preservation along thermal gradients. In: Bock GR, Goode JA (eds) Evolution of Hydrothermal Ecosystems on Earth (and Mars?), Proceedings of the CIBA Foundation Symposium, 202. Wiley, Chichester, pp 150–173

Callow RHT, Battison L, Brasier MD (2011) Diverse microbially induced sedimentary structures from 1Ga lakes of the Diabaig Formation, Torridon Group, northwest Scotland. Sediment Geol 239:117–128

Cam N, Benzerara K, Georgelin T, Jaber M, Lambert JF, Poinsot M, Skouri-Panet F, Raimbault E, Cordier L, Jezequel D (2018) Active biomineralization of intracellular carbonates by high Ca uptake in cyanobacteria. Geobiology 16:49–61

Camoin G, Casanova J, Rouchy JM, BlancValleron MM, Deconinck JF (1997) Environmental controls on perennial and ephemeral carbonate lakes: the central palaeo-Andean Basin of Bolivia during late Cretaceous to early Tertiary times. Sediment Geol 113:1–26

Campbell KA, Guido DM, Gautret P, Foucher F, Ramboz C, Westall F (2015) Geyserite in hot-spring siliceous sinter: window on Earth's hottest terrestrial (paleo)environment and its extreme life. Earth Sci Rev 148:44–64

Cappelleti M, Zannoni D, Postec A, Ollivier B (2014) Members of the order *Thermotogales*: from Microbiology to hydrogen production. In: Zannoni D, De Philippis R (eds) Microbial bioenergy: hydrogen production, advances in photosynthesis and respiration including bioenergy and related processes, Chapter 9, vol 38. Springer, Dordrecht, pp 197–224

Casaburi A, Piombino P, Nychas G-J, Villani F, Ercolini D (2015) Bacterial populations and the volatilome associated to meat spoilage. Food Microbiol 45:83–102

Centeno CM, Legendre P, Beltrán Y, Alcántara-Hernández RJ, Lidström UE, Ashby MN, Falcón LI (2012) Microbialite genetic diversity and composition relate to environmental variables. FEMS Microbiol Ecol 82:724–735

Chapelle FH, O'Neill K, Bradley PM, Methé BA, Ciufo SA, Knobel LL, Lovley DR (2002) A hydrogen-based subsurface microbial community dominated by methanogens. Nature 415:312–315

Charlou JL, Donval JP, Fouquet Y, Jean-Baptiste P, Holm N (2002) Geochemistry of high H_2 and CH_4 vent fluids issuing from ultramafic rocks at the Rainbow hydrothermal field (36 degrees 14 ' N, MAR). Chem Geol 191:345–359

Choi O, Sang B-I (2016) Extracellular electron transfer from cathode to microbes: application for biofuel production. Biotechnol Biofuels 9:11

Cnossen I, Sanz-Forcada J, Favata F, Witasse O, Zegers Z, Arnold NF (2007) Habitat of early life: solar X-ray and UV radiation at Earth's surface 4–3.5 billion years ago. J Geophys Res 112:E02008

Cockell CS, Raven JA (2004) Zones of photosynthetic potential on Mars and the early Earth. Icarus 169:300–310

Cole JM, Rasbury ET, Montañez IP, Pedone VA, Lanzirotti A, Hanson GN (2004) Petrographic and trace element analysis of uranium-rich tufa calcite, middle Miocene Barstow Formation, California, USA: Uranium-rich tufa deposits, California. Sedimentology 51:433–453

Cooper G, Kimmich N, Belisle W, Sarinana J, Brabham K, Garrel L (2001) Carbonaceous meteorites as a source of sugar-related organic compounds for the early Earth. Nature 414:879–883

Corkeron M, Webb GE, Moulds J, Grey K (2012) Discriminating stromatolite formation modes using rare earth element geochemistry: trapping and binding versus in situ precipitation of stromatolites from the Neoproterozoic Bitter Springs Formation, Northern Territory, Australia. Precambrian Res 212–213:194–206

Couradeau E, Benzerara K, Gerard E, Moreira D, Bernard S, Brown GE, Lopez-Garcia P (2012) An early-branching microbialite cyanobacterium forms intracellular carbonates. Science 336:459–462

Couradeau E, Benzerara K, Gérard E, Estève I, Moreira D, Tavera R, López-García P (2013) Cyanobacterial calcification in modern microbialites at the submicrometer scale. Biogeosciences 10:5255–5266

Das SS, Tripathi MK (2009) Trace fossils from Talchir carbonate concretions, Giridih basin, Jharkhand. J Earth Syst Sci 118:89–100

Dass AV, Hickman-Lewis K, Brack A, Kee TP, Westall F (2016) Stochastic prebiotic chemistry within realistic geological systems. Chem Select 1:4906–4926

Davidova MN, Tarasova NB, Mukhitova FK, Karpilova IU (1994) Carbon monoxide in metabolism of anaerobic bacteria. Can J Microbiol 40:417–425

De Duve C (1995) Vital Dust. BasicBooks, New York 362 pp

de Vries ST, Nijman W, Wijbrans JR, Nelson DR (2006) Stratigraphic continuity and early deformation of the central part of the Coppin Gap Greenstone Belt, Pilbara, Western Australia. Precambrian Res 147:1–27

de Vries ST, Nijman W, de Boer PL (2010) Sedimentary geology of the Palaeoarchaean Buck Ridge (South Africa) and Kittys Gap (Western Australia) volcano-sedimentary complexes. Precambrian Res 183:749–769

De Yoreo JJ, Vekilov PG (2003) Principles of crystal nucleation and growth. Rev Mineral Geochem 54:57–93

Di Giulio M (2003) The universal ancestor and the ancestor of bacteria were hyperthermophiles. J Mol Evol 57:721–730

Dinh HT, Kuever J, Mußmann M, Hassel AW, Stratmann M, Widdel F (2004) Iron corrosion by novel anaerobic microorganisms. Nature 427:829–832

Dodd MS, Papineau D, Grenne T, Slack JF, Rittner M, Pirajno F, O'Neil J, Little CTS (2017) Evidence for early life in Earth's oldest hydrothermal vent precipitates. Nature 543:60–64

Dridi B, Fardeau ML, Ollivier B, Raoult D, Drancourt M (2012) *Methanomassiliicoccus luminyensis* gen. nov., sp. nov., a methanogenic archaeon isolated from human faeces. Int J Syst Evol Microbiol 62:1902–1907

Dunagan SP, Turner CE (2004) Regional paleohydrologic and paleoclimatic settings of wetland/lacustrine depositional systems in the Morrison Formation (Upper Jurassic), Western Interior, USA. Sediment Geol 167:269–296

Dunn SR, Valley JW (1992) Calcite–graphite isotope thermometry: a test for polymetamorphism in marble, Tudor gabbro aureole, Ontario, Canada. J Metamorph Geol 10:487–501

Dupraz C, Pattisina R, Verrecchia EP (2006) Translation of energy into morphology: simulation of stromatolite morphospace using a stochastic model. Sediment Geol 185:185–203

Dupraz C, Reid RP, Braissant O, Decho AW, Norman RS, Visscher PT (2009) Processes of carbonate precipitation in modern microbial mats. Earth Sci Rev 96:141–162

Ehlmann BL, Mustard JF, Murchie SL (2010) Geological setting of serpentine deposits on Mars. Geophys Res Lett 37:L06201

Farquhar J, Zerkle AL, Bekker A (2011) Geological constraints on the origin of oxygenic photosynthesis. Photosynth Res 107:11–36

Fauque GD, Barton LL (2012) Hemoproteins in dissimilatory sulfate- and sulfur-reducing prokaryotes. Adv Microb Physiol 60:1–90

Fedorchuk ND, Dornbos SQ, Corsetti FA, Isbell JL, Petryshyn VA, Bowles JA, Wilmeth DT (2016) Early non-marine life: evaluating the biogenicity of Mesoproterozoic fluvial-lacustrine stromatolites. Precambrian Res 275:105–118

Feulner G (2012) The faint young Sun problem. Rev Geophys 50:RG2006

Finster K (2008) Microbiological disproportionation of inorganic sulfur compounds. J Sulfur Chem 29:281–292

Foucher F, Westall F, Brandstatter F, Demets R, Parnell J, Cockell C, Edwards H, Beny JM, Brack A (2010) Testing the survival of microfossils in artificial martian sedimentary meteorites during entry into Earth's atmosphere: the STONE 6 experiment. Icarus 207:616–630

Fray N, Bardyn A, Cottin H, Altwegg K, Baklouti D, Briois C, Colangeli L, Engrand C, Fischer H, Glasmachers A, Grün E, Haerendel G, Henkel H, Höfner H, Hornung K, Jessberger EK, Koch A, Krüger H, Langevin Y, Lehto H, Lehto K, Le Roy L, Merouane S, Modica P, Orthous-Daunay FR, Paquette J, Raulin F, Rynö J, Schulz R, Silén J, Siljeström S, Steiger W, Stenzel O, Stephan T, Thirkell L, Thomas R, Torkar K, Varmuza K, Wanczek KP, Zaprudin B, Kissel J, Hilchenbach M (2016) High-molecular-weight organic matter in the particles of comet 67P/Churyumov–Gerasimenko. Nature 538:72–74

Frederiksen TM, Finster K (2003) Sulfite-oxido-reductase is involved in the oxidation of sulfite in *Desulfocapsa sulfoexigens* during disproportionation of thiosulfate and elemental sulfur. Biodegradation 14:189–198

Frederiksen TM, Finster K (2004) The transformation of inorganic sulfur compounds and the assimilation of organic and inorganic carbon by the sulfur disproportionating bacterium *Desulfocapsa sulfoexigens*. Antonie Van Leeuwenhoek 85:141–149

Furnes H, de Wit M, Staudigel H, Rosing M, Muehlenbachs K (2007) A vestige of Earth's oldest ophiolite. Science 315:1704–1707

Gallagher KL, Kading TJ, Braissant O, Dupraz C, Visscher PT (2012) Inside the alkalinity engine: the role of electron donors in the organomineralization potential of sulfate-reducing bacteria. Geobiology 10:518–530

Galvez ME, Beyssac O, Martinez I, Benzerara K, Chaduteau C, Malvoisin B, Malavieille J (2013) Graphite formation by carbonate reduction during subduction. Nat Geosci 6:473–477

García Ruiz JM, Carnerup A, Christy AG, Welham NJ, Hyde ST (2002) Morphology: an ambiguous indicator of biogenicity. Astrobiology 2:353–369

García Ruiz JM, Hyde ST, Carnerup AM, Christy AG, Kranendonk MJV, Welham NJ (2003) Self-assembled silica-carbonate structures and detection of ancient microfossils. Science 302:1194–1197

Garcia JL, Patel BKC, Ollivier B (2000) Taxonomic, phylogenetic, and ecological diversity of methanogenic *Archaea*. Anaerobe 6:205–226

Garrity GM, Holt JG (2001) Phylum BIII. *Thermodesulfobacteria* phyl. nov. In: Boone DR, Castenholz RW (eds) Bergey's manual of systematic bacteriology, vol 1. Springer, New York/Berlin/Heidelberg, p 389

Gérard E, Ménez B, Couradeau E, Moreira D, Benzerara K, Tavera R, López-García P (2013) Specific carbonate–microbe interactions in the modern microbialites of Lake Alchichica (Mexico). ISME J 7:1997–2009

Giuffre AJ, Hamm LM, Han N, Yoreo JJD, Dove PM (2013) Polysaccharide chemistry regulates kinetics of calcite nucleation through competition of interfacial energies. Proc Natl Acad Sci 110:9261–9266

Goesmann F, Rosenbauer H, Bredehöft JH, Cabane M, Ehrenfreund P, Gautier T, Giri C, Krüger H, Le Roy L, MacDermott AJ, McKenna-Lawlor S, Meierhenrich UJ, Muñoz Caro GM, Raulin F, Roll R, Steele A, Steininger H, Sternberg R, Szopa C, Thiemann W, Ulamec S (2015) Organic compounds on comet 67P/Churyumov-Gerasimenko revealed by COSAC mass spectrometry. Science 349:6247

Golubic S, Seong-Joo L (1999) Early cyanobacterial fossil record: preservation, palaeoenvironments and identification. Eur J Phycol 34:339–348

Gomes R, Levison HF, Tsiganis K, Morbidelli A (2005) Origin of the cataclysmic Late Heavy Bombardment period of the terrestrial planets. Nature 435:466–469

Gough DO (1981) Solar interior structure and luminosity variations. Sol Phys 74:21–34

Grigné C, Labrosse S, Tackley PJ (2005) Convective heat transfer as a function of wavelength: implications for the cooling of the Earth. J Geophys Res 110(B3):B03409

Grotzinger JP (1989) Facies and evolution of Precambrian carbonate deposition systems: emergence of the modern platform archetype. In: Crevello PD, Wilson JL, Sarg JF, Read JF (eds)

Controls on carbonate platform and basin development, Special Publication, vol 44. Society of economic Paleontologists and Mineralogists, Tulsa, pp 79–106
Grotzinger JP (1990) Geochemical model for Proterozoic stromatolite decline. Am J Sci 290-A:80–243
Grotzinger JP, Kasting JF (1993) New constraints on Precambrian ocean composition. J Geol 101:235–243
Grotzinger JP, Knoll AH (1999) Stromatolites in Precambrian carbonates: evolutionary mileposts or environmental dipsticks? Annu Rev Earth planet Sci 27:313–358
Guida BS, Garcia-Pichel F (2016) Extreme cellular adaptations and cell differentiation required by a cyanobacterium for carbonate excavation. Proc Natl Acad Sci USA 113:5712–5717
Hall JD (1883) Cryptozoon (proliferum) N. G. (an sp). New York State Mus Ann Rept 36
Harrison TM, Blichert-Toft J, Müller W, Albarede F, Holden P, Mojzsis SJ (2005) Heterogeneous Hadean hafnium: evidence of continental crust at 4.4 to 4.5 Ga. Science 310:1947–1950
Hatchikian CE, Ollivier B, Garcia J-L (2001) Class I. *Thermodesulfobacteria* class. nov., Order I.*Thermodesulfobacteriales* ord. nov., Family I. *Thermodesulfobacteriaceae* fam. nov. In: Boone DR, Castenholz RW (eds) Bergey's manual of systematic bacteriology, vol 1. Springer, New York/Berlin/Heidelberg, pp 389–390
Hazen RM, Sverjensky DA (2010) Mineral surfaces, geochemical complexities, and the origins of life. Cold Spring Harb Perspect Biol 2:a002162
Hedderich R, Klimmek O, Kröger A, Dirmeier R, Keller M, Stetter KO (1999) Anaerobic respiration with elemental sulfur and with disulfides. FEMS Microbiol Rev 22:353–381
Heubeck C (2009) An early ecosystem of Archean tidal microbial mats (Moodies Group, South Africa, ca. 3.2 Ga). Geology 37:931–934
Hofmann HJ (1973) Stromatolites: characteristics and utility. Earth Sci Rev 9:339–373
Hofmann A, Harris C (2008) Silica alteration zones in the Barberton Greenstone Belt: a window into subseafloor processes 3.5–3.3 Ga ago. Chem Geol 257:221–239
Hofmann H, Grey K, Hickman A, Thorpe R (1999) Origin of 3.45 Ga coniform stromatolites in the Warrawoona Group, Western Australia. Geol Soc Am Bull 111:1256–1262
Homann M, Heubeck C, Airo A, Tice MM (2015) Morphological adaptations of 3.22 Ga-old tufted microbial mats to Archean coastal habitats (Moodies Group, Barberton Greenstone Belt, South Africa). Precambrian Res 266:47–64
Homann M, Heubeck C, Bontognali TTR, Bouvier A-S, Baumgartner LP, Airo A (2016) Evidence for cavity-dwelling microbial life in 3.22 Ga tidal deposits. Geology 44:51–54
Jahnert RJ, Collins LB (2012) Characteristics, distribution and morphogenesis of subtidal microbial systems in Shark Bay, Australia. Mar Geol 303–306:115–136
Jahnke LL, Eder W, Huber R, Hope JM, Hinrichs K-U, Hayes JM, Des Marais DJ, Cady SL, Summons RE (2001) Signature lipids and stable carbon isotope analyses of Octopus Spring hyperthermophilic communities compared with those of *Aquificales* representatives. Appl Environ Microbiol 67:5179–5189
Javaux EJ, Marshall CP, Bekker A (2010) Organic-walled microfossils in 3.2- billion-year-old shallow-marine siliciclastic deposits. Nature 463:934–938
Jeoung J-H, Fesseler J, Goetzl S, Dobbek H (2014) Carbon monoxide. Toxic gas and fuel for anaerobes and aerobes: carbon monoxide dehydrogenases. Met Ions Life Sci 14:37–69
Johnson AP, Cleaves HJ, Dworkin JP, Glavin DP, Lazcano A, Bada JL (2008) The Miller volcanic spark discharge experiment. Science 322:404
Jones B, Renaut RW, Rosen MR (2001) Taphonomy of silicified filamentous microbes—implications for identification. PALAIOS 16:580–592
Kah LC, Grotzinger JP (1992) Early Proterozoic (1.9 Ga) thrombolites of the Rocknest Formation, Northwest Territories, Canada. PALAIOS 7:305
Kalkovsky E (1908) Oolith und stromatolith im norddeutschen Buntsandstein. Zeitschrift der Deutschen Geologischen Gesellschaft 60:68–125
Kamber BS (2015) The evolving nature of terrestrial crust from the Hadean, through the Archaean, into the Proterozoic. Precambrian Res 258:48–82

Kamber BS, Webb GE (2001) The geochemistry of late Archaean microbial carbonate: implications for ocean chemistry and continental erosion history. Geochim Cosmochim Acta 65:2509–2525

Kamber BS, Bolhar R, Webb GE (2004) Geochemistry of late Archaean stromatolites from Zimbabwe: evidence for microbial life in restricted epicontinental seas. Precambrian Res 132:379–399

Kardar M, Parisi G, Zhang Y-C (1986) Dynamic scaling of growing interfaces. Phys Rev Lett 56:889–892

Kasting JF (1982) Stability of ammonia in the primitive terrestrial atmosphere. J Geophys Res 87:3091–3098

Kasting JF (1993) Earth's early atmosphere. Science 259:920–926

Kasting JF, Catling D (2003) Evolution of a habitable planet. Annu Rev Astron Astrophys 41:429–463

Khodadad CLM, Foster JS (2012) Metagenomic and metabolic profiling of nonlithifying and lithifying stromatolitic mats of Highborne Cay, The Bahamas. PLoS One 7:e38229

Knauth LP, Lowe DR (2003) High Archean climatic temperature inferred from oxygen isotope geochemistry of cherts in the 3.5 Ga Swaziland Supergroup, South Africa. Bull Geol Soc Am 115:566–580

Knittel K, Boetius A (2009) Anaerobic oxidation of methane: progress with an unknown process. Annu Rev Microbiol 63:311–334

Knoll AH, Carroll SB (1999) Early animal evolution: emerging views from comparative biology and geology. Science 284:2129–2137

Knoll AH, Fairchild IJ, Swett K (1993) Calcified microbes in Neoproterozoic carbonates; implications for our understanding of the Proterozoic/Cambrian transition. PALAIOS 8:512–525

Knoll AH, Bergmann KD, Strauss JV (2016) Life: the first two billion years. Philos Trans R Soc B-Biol Sci 371:20150493

Konn C, Charlou JL, Donval JP, Holm NG, Dehairs F, Bouillon S (2009) Hydrocarbons and oxidized organic compounds in hydrothermal fluids from Rainbow and Lost City ultramafic-hosted vents. Chem Geol 258:299–314

Krämer M, Cypionka H (1989) Sulfate formation via ATP sulfurylase in thiosulfate- and sulfite-disproportionating bacteria. Arch Microbiol 151:232–237

Krumbein WE (1983) Stromatolites – the challenge of a term in space and time. Precambrian Res 20:493–531

Kusakabe M, Komoda Y, Takano B, Abiko T (2000) Sulfur isotopics effects in the disproportionation reaction of sulfur dioxide in hydrothermal fluids: implications of the $\Delta\ ^{34}S$ variations of dissolved bisulfate and elemental sulfur from active crater lakes. J Volcanol Geotherm Res 97:287–307

Lepot K, Benzerara K, Rividi N, Cotte M, Brown GE, Philippot P (2009) Organic matter heterogeneities in 2.72Ga stromatolites: alteration versus preservation by sulfur incorporation. Geochim Cosmochim Acta 73:6579–6599

Lindqvist JK (1994) Lacustrine stromatolites and oncoids: Manuherikia Group (MIOCENE), New Zealand. In: Bertrand-Sarfati J, Monty C (eds) Phanerozoic Stromatolites II. Springer, Dordrecht, pp 227–254

Link MH, Osborne RH, Awramik SM (1978) Lacustrine stromatolites and associated sediments of the Pliocene Ridge Route Formation, Ridge Basin, California. J Sediment Res 48:143–157

Liu Y, Beer LL, Whitman WB (2012) Methanogens: a window into ancient sulfur metabolism. Trends Microbiol 20:251–258

Logan B (1961) Cryptozoon and associate stromatolites from the recent, Shark Bay, Western-Australia. J Geol 69:517–533

López-García P, Kazmierczak J, Benzerara K, Kempe S, Guyot F, Moreira D (2005) Bacterial diversity and carbonate precipitation in the giant microbialites from the highly alkaline Lake Van, Turkey. Extremophiles 9:263–274

Lowe DR (1980) Stromatolites 3,400-Myr old from the Archean of Western Australia. Nature 284:441–443

Lowe DR (1983) Restricted shallow water sedimentation of 3.4 Byr-old stromatolitic and evaporitic strata of the Strelley Pool Chert, Pilbara block, Western Australia. Precambrian Res 19:239–283

Lowe DR (1994) Abiological origin of described stromatolites older than 3.2 Ga. Geology 22:387–390

Lowe DR, Byerly GR (1999) Stratigraphy of the west-central part of the Barbeton Greenstone Belt, South Africa. Special Paper of the Geol Amer Soc 329:1–36

Manhes G, Allègre CJ, Dupré B, Hamelin B (1980) Lead isotope study of basic-ultrabasic layered complexes: speculations about the age of the earth and primitive mantle characteristics. Earth Planet Sci Lett 47:370–382

Mardanov AV, Beletsky AV, Kadnikov VV, Slobodkin AI, Ravin NV (2016) Genome analysis of *Thermosulfurimonas dismutans*, the first thermophilic sulfur-disproportionating bacterium of the phylum *Thermodesulfobacteria*. Front Microbiol 7. https://doi.org/10.3389/fmicb2016.00950

Martin W, Müller M (1998) The hydrogen hypothesis for the first eukaryote. Nature 392:37–41

Marvasi M, Visscher PT, Perito B, Mastromei G, Casillas-Martínez L (2010) Physiological requirements for carbonate precipitation during biofilm development of *Bacillus subtilis etfA* mutant. FEMS Microbiol Ecol 71:341–350

Maurette M, Brack A (2006) Cometary petroleum in Hadean time? Meteorit Planet Sci 41:5247

McCaffrey VP, Zellner NEB, Waun CM, Bennett ER, Earl EK (2014) Reactivity and survivability of glycolaldehyde in simulated meteorite impact experiments. Orig Life Evol Biosph 44:29–42

McCollom TM (2011) What can carbon isotopes tell us about sources of reduced carbon in rocks from the Early Earth. In: Golding SD, Glikson M (eds) Earliest life on Earth: habitats, environments and methods of detection. Springer, Dordrecht, pp 291–311

McCollom TM, Seewald JS (2007) Abiotic synthesis of organic compounds in deep-sea hydrothermal environments. Chem Rev 107:382–401

Merz MU (1992) The biology of carbonate precipitation by cyanobacteria. Facies 26:81–101

Michaelis W, Seifert R, Nauhaus K, Treude T, Thiel V, Blumenberg M, Knittel K, Gieseke A, Peterknecht K, Pape T, Boetius A, Amann R, Jorgensen BB, Widdel F, Peckmann J, Pimenov NV, Gulin MB (2002) Microbial reefs in the Black Sea fueled by anaerobic oxidation of methane. Science 297:1013–1015

Miller SL (1953) The production of amino acids under possible primitive Earth conditions. Science 117:528–529

Miller AG, Colman B (1980) Evidence for HCO_3^- transport by the blue-green alga (Cyanobacterium) *Coccochloris peniocystis*. Plant Physiol 65:397–402

Miller HM, Mayhew LE, Ellison ET, Kelemen P, Kubo M, Templeton AS (2017) Low temperature hydrogen production during experimental hydration of partially-serpentinized dunite. Geochim Cosmochim Acta 209:161–183

Milroy PG, Wright VP (2002) Fabrics, facies control and diagenesis of lacustrine ooids and associated grains from the Upper Triassic, southwest England. Geol J 37:35–53

Milucka J, Ferdelman TG, Polerecky L, Franzke D, Wegener G, Schmid M, Lieberwirth I, Wagner M, Widdel F, Kuypers MMM (2012) Zero-valent Sulphur is a key intermediate in marine methane oxidation. Nature 49:541–546

Mitchell P (1961) Coupling of phosphorylation to electron and hydrogen transfer by a chemiosmotic type of mechanism. Nature 191:144–148

Miyakawa S, Yamanashi H, Kobayashi K, Cleaves HJ, Miller SL (2002) Prebiotic synthesis from CO atmospheres: implications for the origins of life. Proc Natl Acad Sci USA 99:14628–14631

Moissette P, Cornée J-J, Mannaï-Tayech B, Rabhi M, André J-P, Koskeridou E, Méon H (2010) The western edge of the Mediterranean Pelagian Platform: a Messinian mixed siliciclastic–carbonate ramp in northern Tunisia. Palaeogeogr Palaeoclimatol Palaeoecol 285:85–103

Monty CLV, Westall F, van der Gaast S (1991) Diagenesis of siliceous particles in subantarctic sediments, ODP Leg 114, Hole 699a: possible microbial mediation. In: Ciesielski PF,

Kristoffersen Y, Clement B, Moore TC (eds) Proceedings of the ODP scientific research, vol 114. ODP, College Station, pp 685–710

Mörsdorf G, Frunzke K, Gadkari D, Meyer O (1992) Microbial growth on carbon monoxide. Biodegradation 3:61–82

Noffke N (2009) The criteria for the biogenicity of microbially induced sedimentary structures (MISS) in Archean and younger, sandy deposits. Earth Sci Rev 96:173–180

Noffke N (2010) Geobiology: microbial mats in sandy deposits from the Archaean Era to today. Springer, Berlin/Heidelberg

Noffke N, Eriksson KA, Hazen RM, Simpson EL (2006) A new window into Archean life: Microbial mats in Earth's oldest siliciclastic tidal deposits (3.2 Ga Moodies Group, South Africa). Geology 34:253–256

Nuevo M, Bredehöft JH, Meierhenrich UJ, d'Hendecourt L, Thiemann WH (2010) Urea, glycolic acid, and glycerol in an organic residue produced by ultraviolet irradiation of interstellar/pre-cometary ice analogs. Astrobiology 10:245–256

Nutman AP, Bennett VC, Friend CRL, Van Kranendonk MJ, Chivas AR (2016) Rapid emergence of life shown by discovery of 3,700-million-year-old microbial structures. Nature 537:535–538

Obraztsova AY, Francis CA, Tebo BM (2002) Sulfur disproportionation by the facultative anaerobe *Pantoea agglomerans* SP1 as a mechanism for chromium (VI) reduction. Geomicrobiol J 19:121–132

Ollivier B, Guyot F (2009) Sulfate-reducing bacteria: a deep biosphere-early life connection ? Environ Microbiol Rep (Crystal Ball 2009) 1:14–16

Onstott TC (2016) Deep life: the hunt for the hidden biology of earth, Mars, and beyond. Princeton University Press, 512 pp

Orange F, Westall F, Disnar J-R, Prieur D, Bienvenu N, Le Romance M, Défarge C (2009) Experimental silicification of the extremophilic Archaea *Pyrococcus abyssi* and *Methanocaldococcus jannaschii*: applications in the search for evidence of life in early Earth and extraterrestrial rocks. Geobiology 7:403–418

Orange F, Disnar JR, Westall F, Prieur D, Baillif P (2011) Metal binding by the early Earth analogue microorganism, archaea *Methanocaldococcus jannaschii* and its effects on silicification. Palaeontology 54:953–964

Orange F, Disnar J-P, Gautret P, Westall F, Bienvenu N, Lottier N, Prieur D (2012) Preservation and evolution of organic matter during experimental fossilisation of the hyperthermophilic archaea *Methanocaldococcus jannaschii*. Orig Life Evol Biosph 42:587–609

Orberger B, Rouchon V, Westall F, de Vries ST, Wagner C, Pinti DL (2006) Protoliths and micro-environments of some Archean Cherts (Pilbara, Australia). In: Reimold WU, Gibson R (eds.) Processes on the early Earth. Geological Society of America, special paper Boulder 405: 132–154

Parshina SN, Sipma J, Nakashimada Y, Meint Henstra A, Smidt H, Lysenko AM, Lens PNL, Lettinga G, Stams AJM (2005) *Desulfotomaculum carboxydivorans* sp. nov., a novel sulfate-reducing bacterium capable of growth at 100% CO. Int J Syst Evol Microbiol 55:2159–2165

Pasini V, Brunelli D, Dumas P, Sandt C, Frederick J, Benzerara BS, Ménez B (2013) Low temperature hydrothermal oil and associated biological precursors in serpentinites from Mid-Ocean Ridge. Lithos 178:84–95

Patterson C, Tilton G, Inghram M (1955) Age of the Earth. Science 121:69–75

Pflug HD, Jaeschke-Boyer H (1979) Combined structural and chemical analysis of 3,800-Myr-old microfossils. Nature 280:483–448

Philippot P, Van Zuiten M, Lepot K, Thomazo C, Farquhar J, Van Kranendonk MJ (2007) Early Archaean microorganisms preferred elemental sulfur, not sulfate. Science 317:1534–1537

Pikuta EV, Hoover RB, Bej AK, Marsi D, Whitman WB, Cleland D, Krader P (2003) *Desulfonatronum thiosdismutans* sp. nov., a novel alkaliphilic, sulfate-reducing bacterium capable of lithoautotrophic growth. Int J Syst Evol Microbiol 53:1327–1332

Pinti DL (2005) In: Gargaud M, Barbier B, Martin H, Reisse J (eds) Lectures in astrobiology, Advances in Astrobiology and Biogeophysics. Springer, Berlin, pp 83–107

Poser A, Lohmayer R, Vogt C, Knoeller K, Planer-Friedrich B, Sorokin D, Richnow H-H, Finster K (2013) Disproportionation of elemental sulfur by haloalkaliphilic bacteria from soda lakes. Extremophiles 17:1003–1012

Ragsdale SW, Pierce E (2008) Acetogenesis and the Wood-Ljungdahl pathway of CO_2 fixation. Biochim Biophys Acta 1784:1873–1898

Rasmussen B, Fletcher IR, Brocks JJ, Kilburn MR (2008) Reassessing the first appearance of eukaryotes and cyanobacteria. Nature 455:1101–1104

Reid RP, James NP, Macintyre IG, Dupraz CP, Burne RV (2003) Shark Bay stromatolites: microfabrics and reinterpretation of origins. Facies 49:299–324

Rettberg P, Anesio AM, Baker VR, Baross JA, Cady SL, Detsis E, Foreman CM, Hauber E, Ori GG, Pearce DA, Renno NO, Ruvkun G, Sattler B, Saunders MP, Smith DH, Wagner D, Westall F (2016) Planetary protection and Mars special regions--a suggestion for updating the definition. Astrobiology 16:119–125

Riding R (1991) Classification of microbial carbonates. In: Riding R (ed) Calcareous algae and stromatolites. Springer, Berlin, pp 21–51

Riding R (1994) Evolution of algal and cyanobacterial calcification. In: Bengtson S (ed) Early life on Earth. Columbia University Press, New York, pp 426–438

Riding R (2006a) Microbial carbonate abundance compared with fluctuations in metazoan diversity over geological time. Sediment Geol 185:229–238

Riding R (2006b) Cyanobacterial calcification, carbon dioxide concentrating mechanisms, and Proterozoic-Cambrian changes in atmospheric composition. Geobiology 4:299–316

Rippka R, Deruelles J, Waterbury JB, Herdman M, Stanier RY (1979) Generic assignments, strain histories and properties of pure cultures of cyanobacteria. J Gen Microbiol 111:1–61

Robert F, Chaussidon M (2006) A palaeotemperature curve for the Precambrian oceans based on silicon isotopes in cherts. Nature 443:969–972

Rosing MT (1999) 13C-Depleted carbon microparticles in >3700-Ma sea-floor sedimentary rocks from west Greenland. Science 283:674–676

Russel MJ, Nitschke W, Branscomb E (2013) The inevitable journey to being. Philos Trans R Soc 368:20120254

Sagan C, Mullen G (1972) Earth and Mars: evolution of atmospheres and surface temperatures. Science 177:52–56

Saghaï A, Zivanovic Y, Zeyen N, Moreira D, Benzerara K, Deschamps P, Bertolino P, Ragon M, Tavera R, López-Archilla AI, López-García P (2015) Metagenome-based diversity analyses suggest a significant contribution of non-cyanobacterial lineages to carbonate precipitation in modern microbialites. Front Microbiol 6:797. https://doi.org/10.3389/fmicb.2015.00797

Saghaï A, Zivanovic Y, Moreira D, Benzerara K, Bertolino P, Ragon M, Tavera R, López-Archilla AI, López- García P (2016) Comparative metagenomics unveils functions and genome features of microbialite-associated communities along a depth gradient. Environ Microbiol 18:4990–5004

Salama W, El Aref MM, Gaupp R (2013) Mineral evolution and processes of ferruginous microbialite accretion – an example from the Middle Eocene stromatolitic and ooidal ironstones of the Bahariya Depression, Western Desert, Egypt. Geobiology 11:15–28

Sanz-Montero ME, RodriGuez-Aranda JP, GarciA Del Cura MA (2008) Dolomite-silica stromatolites in Miocene lacustrine deposits from the Duero Basin, Spain: the role of organotemplates in the precipitation of dolomite: Dolomite-silica stromatolites in Miocene lacustrine deposits. Sedimentology 55:729–750

Schidlowski M (2001) Carbon isotopes as biogeochemical recorders of life over 3.8 Ga of Earth history: evolution of a concept. Precambrian Res 106:117–134

Schopf JW (1993) Microfossils of the early Archean apex chert: new evidence of the antiquity of life. Science 260:640–646

Schopf JW (2011) The paleobiological record of photosynthesis. Photosynth Res 107:87–101

Schopf JW, Packer BM (1987) Early Archean (3.3-billion to 3.5-billion-year-old) microfossils from Warrawoona Group, Australia. Science 237:70–73

Schopf JW, Kudryavtsev AB, Agresti DG, Wdowiak TJ, Czaja AD (2002) Laser-Raman imagery of Earth's earliest fossils. Nature 416:73–76

Schopf JW, Kudryavtsev AB, Sugitani K, Walter MR (2010) Precambrian microbe-like pseudofossils: a promising solution to the problem. Precambrian Res 179:191–205

Seard C, Camoin G, Rouchy J-M, Virgone A (2013) Composition, structure and evolution of a lacustrine carbonate margin dominated by microbialites: Case study from the Green River formation (Eocene; Wyoming, USA). Palaeogeogr Palaeoclimatol Palaeoecol 381–382:128–144

Semikhatov MA, Gebelein CD, Cloud P, Awramik SM, Benmore WC (1979) Stromatolite morphogenesis – progress and problems. Can J Earth Sci 16:992–1015

Sessions AL, Doughty DM, Welander PV, Summons RE, Newman DK (2009) The continuing puzzle of the great oxidation event. Curr Biol 19:567–574

Shen Y, Bulck R, Canfield DE (2001) Isotopic evidence for microbial sulphate reduction in the early Archaean era. Nature 410:77–80

Shih PM, Wu D, Latifi A, Axen SD, Fewer DP, Talla E, Calteau A, Cai F, de Marsac NT, Rippka R, others (2013) Improving the coverage of the cyanobacterial phylum using diversity-driven genome sequencing. Proc Natl Acad Sci 110:1053–1058

Shih PM, Hemp J, Ward LM, Matzke NJ, Fischer WW (2017) Crown group Oxyphotobacteria postdate the rise of oxygen. Geobiology 15:19–29

Shock EL, McCollom TM, Schulte MD (1998) The emergence of metabolism from within hydrothermal systems. In: Wiegel J, Adams MWW (eds) Thermophiles: the keys to molecular evolution and the origin of life. Taylor & Francis, London, pp 59–76

Sleep NH, Zahnle KJ, Kasting JF, Morowitz HJ (1989) Annihilation of ecosystems by large asteroid impacts on the early Earth. Nature 342:139–142

Sleep NH, Zahnle K, Neuhoff PS (2001) Initiation of clement surface conditions on the early Earth. Proc Natl Acad Sci 98:3666–3672

Slobodkin AI, Reysenbach A-L, Slobodkina GB, Baslerov RV, Kostrikina NA, Wagner ID, Bonch-Osmolovskaya EA (2012) *Thermosulfurimonas dismutans* gen. nov., sp. nov., an extremely thermophilic sulfur-disproportionating bacterium from a deep-sea hydrothermal vent. Int J Syst Environ Microbiol 62:2565–2571

Slobodkin AI, Reysenbach A-L, Slobodkina GB, Kolganova TV, Kostrikina NA, Bonch-Osmolovskaya EA (2013) *Dissulfuribacter thermophilus* gen. nov., sp. nov., a thermophilic, autotrophic, sulfur-disproportionating, deeply branching deltaproteobacterium from a deep-sea hydrothermal vent. Int J Syst Environ Microbiol 63:1967–1971

Slobodkin AI, Slobodkina GB, Panteleeva AN, Chernyh NA, Novikov AA, Bonch-Osmolovskaya EA (2016) *Dissulfurimicrobium hydrothermale* gen. nov., sp. nov., a thermophilic, autotrophic, sulfur-disproportionating deltaproteobacterium isolated from a hydrothermal pond. Int J Syst Environ Microbiol 66:1022–1026

Slobodkina GB, Kolganova TV, Kopitsyn DS, Viryasov MB, Bonch-Osmolovskaya EA, Slobodkin AI (2016) *Dissulfurirhabdus thermomarina* gen. nov., sp. nov., a thermophilic, autotrophic, sulfite-reducing and disproportionating deltaproteobacterium isolated from a shallow-sea hydrothermal vent. Int J Syst Evol Microbiol 66:2515–2519

Smith JW (2000) Isotopic fractionations accompanying sulfur hydrolysis. Geochem J 34:95–99

Sorin L, Anton V, Miryam B-M, Roi P, Amos F (2010) Late Pleistocene palaeoclimatic and palaeoenvironmental reconstruction of the Dead Sea area (Israel), based on speleothems and cave stromatolites. Quat Sci Rev 29:1201–1211

Sorokin DY, Tourova TP, Henstra AM, Stams AJM, Galinski EA, Muyzer G (2008) Sulfidogenesis under extremely haloalkaline conditions by *Desulfonatrospira thiodismutans* gen. nov., sp., nov., and *Desulfonatrospira delicata* sp. nov., a novel lineage of *Deltaproteobacteria* from hypersaline soda lakes. Microbiology 154:1444–1453

Sorokin DY, Tourova TP, Dektova EN, Kolganova TV, Galinski EA, Muyzer G (2011) Culturable diversity of lithotrophic haloalkaliphilic sulfate-reducing bacteria in soda lakes and the description of *Desulfonatronum thioautotrophicum* sp. nov., *Desulfonatronum thiautrophicum* sp.

nov., *Desulfonatronovibrio thiodismutans* sp., nov., and *Desulfonatronovibrio magnus* sp. nov. Extremophiles 15:391–401

Stams AJM, Plugge CM (2009) Electron transfer in syntrophic communities of anaerobic bacteria and archaea. Nat Rev Microbiol 7:568–577

Steel JH (1825) A description of the olitic formation lately discovered in the Country of Sratoga and State of New York. Am J Sci 9:16–19

Stern JC, Sutter B, Freissinet C, Navarro-González R, McKay CP, Archer PD Jr, Buch A, Brunner AE, Coll P, Eigenbrode JL, Fairen AG, Franz HB, Glavin DP, Kashyap S, McAdam AC, Ming DW, Steele A, Szopa C, Wray JJ, Martín-Torres FJ, Zorzano MP, Conrad PG, Mahaffy PR, Science Team MSL (2015) Evidence for indigenous nitrogen in sedimentary and aeolian deposits from the *Curiosity* rover investigations at Gale crater, Mars. Proc Natl Acad Sci 112:4245–4250

Stetter KO, Gaag G (1983) Reduction of molecular Sulphur by methanogenic bacteria. Nature 305:309–311

Stetter KO, Segerer A, Zillig W, Huber G, Fiala G, Huber R, König H (1986) Extremely thermophilic sulfur-metabolizing archaebacterial. Syst Appl Microbiol 7:393–397

Stüeken EE, Buick R, Schauer AJ (2015) Nitrogen isotope evidence for alkaline lakes on late Archean continents. Earth Planet Sci Lett 411:1–10

Summons RE, Jahnke LL, Hope JM, Logan GAN (1999) Molecular fossils for cyanobacteria recording a geological history of oxygenic photosynthesis. Nature 400:554–557

Tartèse R, Chaussidon M, Gurenko A, Delarue F, Robert F (2017) Warm Archean oceans reconstructed from oxygen isotope composition of early-life remnants. Geochem Perspect Lett 3:55–65

Thamdrup B, Finster K, Hansen JW, Bak F (1993) Bacterial disproportionation of elemental sulfur coupled to chemical reduction of iron or manganese. Appl Environ Microbiol 59:101–108

Thomazo C, Pinti DL, Busigny V, Ader M, Hashizume K, Philippot P (2009) Biological activity and the Earth's surface evolution: insights from carbon, sulfur, nitrogen and iron stable isotopes in the rock record. C R Palevol 8:665–678

Thomazo C, Ader M, Philippot P (2011) Extreme 15N-enrichments in 2.72-Gyr-old sediments: evidence for a turning point in the nitrogen cycle: Extreme 15N-enrichments in 2.72-Gyr-old sediments. Geobiology 9:107–120

Thompson DL, Stilwell JD, Hall M (2015) Lacustrine carbonate reservoirs from Early Cretaceous rift lakes of Western Gondwana: pre-salt coquinas of Brazil and West Africa. Gondwana Res 28:26–51

Tice MM (2009) Environmental controls on photosynthetic microbial mat distribution and morphogenesis on a 3.42 Ga clastic-starved platform. Astrobiology 9:989–1000

Tice MM, Lowe DR (2004) Photosynthetic microbial mats in the 3416-Myr-old ocean. Nature 431:549–552

Timmers PHA, Welte CU, Koehorst JJ, Plugge CM, Jetten MSM, Stams AJM (2017) Reverse methanogenesis and respiration in methanotrophic Archaea. Archaea 1654237, 22 pp

Turcotte DL (1980) On the thermal evolution of the Earth. EPSL 48:53–58

Van den Boorn SHJM, Van Bergen MJ, Nijman W, Strauss H (2007) Dual role of seawater and hydrothermal fluids in Early Archean chert formation: evidence from silicon isotopes. Geology 35:939–942

Van Kranendonk MJ, Webb GE, Kamber BS (2003) Geological and trace element evidence for a marine sedimentary environment of deposition and biogenicity of 3.45 Ga stromatolitic carbonates in the Pilbara Craton, and support for a reducing Archaean ocean. Geobiology 1:91–108

Van Kranendonk M, Philippot P, Lepot K, Bodorkos S, Pirajno F (2008) Geological setting of Earth's oldest fossils in the ca. 3.5Ga Dresser Formation, Pilbara Craton, Western Australia. Precambrian Res 167:93–124

Verrecchia EP (1996) Morphometry of microstromatolites in calcrete laminar crusts and a fractal model of their growth. Math Geol 28:87–109

Verrecchia E, Freytet P, Verrecchia K, Dumont J (1995) Spherulites in calcrete laminar crusts – biogenic $CaCO_3$ precipitation. J Sediment Res Sect A 65:690–700

Visscher PT, Stolz JF (2005) Microbial mats as bioreactors: populations, processes, and products. Palaeogeor Palaeoclimatol Palaeoecol 219:87–100. https://doi.org/10.1016/j.palaeo.2004.10.016

Wacey D (2009) Early life on earth, a practical guide. In: Landman NH, Harries PJ (eds) Topics in geobiology, vol 31. Springer, Heidelberg

Wacey D, Kilburn MR, Saunders M, Cliff J, Brasier MD (2011) Microfossils of sulphur-metabolizing cells in 3.4-billion-year-old rocks of Western Australia. Nat Geosci 4:698–702

Wacey D, Battison L, Garwood RJ, Hickman-Lewis K, Brasier MD (2016a) High-resolution techniques for the study of the morphology and chemistry of Proterozoic microfossils. In: Brasier AT, McIlroy D, McLoughlin N (eds) Earth system evolution and early life: a celebration of the work of Martin Brasier, Geological Society of London Special Publication, vol 448. The Geological Society, London, pp 81–104

Wacey D, Saunders M, Kong C, Brasier AT, Brasier MD (2016b) 3.46 Ga Apex chert 'microfossils' reinterpreted as mineral artefacts produced during phyllosilicate exfoliation. Gondwana Res 36:296–313

Walsh MM, Lowe DR (1999) Modes of accumulation of carbonaceous matter in the early Archaean: a petrographic and geochemical study of the carbonaceous cherts of the Swaziland Supergroup. In: Lowe DR, Byerly GR (eds) Geologic evolution of the Barberton Greenstone Belt, South Africa, Special Paper, vol 329. Geological Society of America, Boulder, pp 115–132

Walter MR, Heys GR (1985) Links between the rise of the Metazoa and the decline of stromatolites. Precambrian Res 29:149–174

Walter MR, Buick R, Dunlop JSR (1980) Stromatolites 3,400–3,500 Myr old from the North Pole area, Western Australia. Nature 284:443–445

Walter MR, Grotzinger JP, Schopf JW (1992) Proterozoic stromatolites. In: Schopf JW, Klein C (eds) The Proterozoic Biosphere, A multidisciplinary study. Cambridge University Press, Cambridge, p 254

Wattinne A, Vennin E, De Wever P (2003) Evolution d'un environnement carbonaté lacustre à stromatolithes, par l'approche paléo-écologique (carrière de Montaigu-le-Blin, bassin des Limagnes, Allier, France). Bull Soc Géol Fr 174:243–260

Webb GE, Kamber BS (2000) Rare earth elements in Holocene reefal microbialites: a new shallow seawater proxy. Geochim Cosmochim Acta 64:1557–1565

Weiss MC, Sousa FL, Mrnjavac N, Neukirchen S, Roettger M, Nelson-Sathi S, Martin WF (2016) The physiology and habitat of the last universal common ancestor. Nat Microbiol 1:16116

Westall F (1997) The influence of cell wall composition on the fossilization of bacteria and the implications for the search for early life forms. In: Cosmovici C, Bowyer S, Werthimer D (eds) Astronomical and biochemical origins and the search for life in the universe. Editori Compositrici, Bologna, pp 491–504

Westall F (2011) Early life: nature, distribution and evolution. In: Gargaud M, López-Garcìa P, Martin H (eds) Origins and evolution of life – An astrobiological perspective. Astrobiology. Cambridge, pp 391–413

Westall F (2012) The early Earth. In: Impey C, Lunine J, Funes J (eds) Astrobiology. Cambridge University Press, Cambridge, pp 89–114

Westall F, Cavalazzi B (2011) Biosignatures in rocks. In: Thiel V, Reitner J (eds) Encyclopedia of geobiology. Springer, Berlin

Westall F, Folk RL (2003) Exogenous carbonaceous microstructures in Early Archean cherts and BIFs from the Isua Greenstone Belt: implications for the search for life in ancient rocks. Precambrian Res 126:313–333

Westall F, Boni L, Guerzoni E (1995) The experimental silicification of microorganisms. Paleontology 38:495–528

Westall F, de Wit MJ, Dann J, van der Gaast S, de Ronde CEJ, Gerneke D (2001) Early Archaean fossil bacteria and biofilms in hydrothermally–influenced sediments from the Barberton greenstone belt, South Africa. Precambrian Res 106:93–116

Westall F, de Vries ST, Nijman W, Rouchon V, Orberger B, Pearson V, Watson J, Verchovsky A, Wright I, Rouzaud J-N, Marchesini D, Severine A (2006a) The 3.466 Ga "Kitty's Gap Chert," an early Archean microbial ecosystem. GSA Special Papers 405:105–131

Westall F, de Ronde CEJ, Southam G, Grassineau N, Colas M, Cockell CS, Lammer H (2006b) Implications of a 3.472–3.333 Gyr-old subaerial microbial mat from the Barberton greenstone belt, South Africa for the UV environmental conditions on the early Earth. Philos Trans R Soc B Biol Sci 361:1857–1875

Westall F, Cavalazzi B, Lemelle L, Marrocchi Y, Rouzaud JN, Simionovici A, Salome M, Mostefaoui S, Andreazza C, Foucher F, Toporski J, Jauss A, Thiel V, Southam G, MacLean L, Wirick S, Hofmann A, Meibom A, Robert F, Defarge C (2011a) Implications of in situ calcification for photosynthesis in a similar to 3.3 Ga-old microbial biofilm from the Barberton greenstone belt, South Africa. Earth Planet Sci Lett 310:468–479

Westall F, Foucher F, Cavalazzi B, de Vries S, Nijman W, Pearson V, Watson J, Verchovsky A, Wright I, Rouzaud J-N, Marchesini D, Anne S (2011b) Volcaniclastic habitats for early life on Earth and Mars: a case study from ~3.5 Ga-old rocks from the Pilbara, Australia. Planet Space Sci 59:1093–1106

Westall F, Campbell KA, Breheret JG, Foucher F, Gautret P, Hubert A, Sorieul S, Grassineau N, Guido DM (2015a) Archean (3.33 Ga) microbe-sediment systems were diverse and flourished in a hydrothermal context. Geology 43:615–618

Westall F, Foucher F, Bost N, Bertrand M, Loizeau D, Vago JL, Kminek G, Gaboyer F, Campbell KA, Bréhéret JG, Gautret P, Cockell CS (2015b) Biosignatures on Mars : what, where, and how? Implications for the search for martian life. Astrobiology 15:998–1029

Westall F, Hickman-Lewis K, Hinman N, Gautret P, Campbell KA, Bréhéret JG, Foucher F, Hubert A, Sorieul S, Dass AV, Kee TP, Georgelin T, Brack A (2018) A hydrothermal-sedimentary context for the origin of life. Astrobiology 18:259–293

Wilde SA, Valley JW, Peck WH, Graham CM (2001) Evidence from detrital zircons for the existence of continental crust and oceans on the Earth 4.4 Gyr ago. Nature 409:175–178

Woese CR (1987) Bacterial evolution. Microbiol Rev 51:221–271

Wright VP, Barnett AJ (2015) An abiotic model for the development of textures in some South Atlantic early Cretaceous lacustrine carbonates. Geol Soc Lond Spec Publ 418:SP418.3

Zamarreño I, Anadón P, Utrilla R (1997) Sedimentology and isotopic composition of Upper Palaeocene to Eocene non-marine stromatolites, eastern Ebro Basin, NE Spain. Sedimentology 44:159–176

Zavala C, Ponce JJ, Arcuri M, Drittanti D, Freije H, Asensio M (2006) Ancient lacustrine hyperpycnites: a depositional model from a case study in the Rayoso Formation (Cretaceous) of West-Central Argentina. J Sediment Res 76:41–59

Zhang Y (2002) The age and accretion of the Earth. Earth Sci Rev 59:235–263

Chapter 4
Evolutionary Success of Prokaryotes

Jean-Claude Bertrand, Patricia Bonin, Bernard Ollivier, Karine Alain, Anne Godfroy, Nathalie Pradel, and Philippe Normand

Abstract How can the evolutionary success of prokaryotes be explained? How did they manage to survive conditions that have fluctuated, with drastic events over 3.5 billion years? Which significant metabolisms and mechanisms have appeared over the course of evolution that have permitted them to survive the most inhospitable conditions from the physicochemical point of view? In a "Red Queen Race," prokaryotes have always run sufficiently fast to adapt to constraints imposed by the environment and the other living species with which they have established interactions. If the criterion retained to define the level of evolution of an organism is its capacity to survive and to yield the largest number of offsprings, prokaryotes must be considered highly evolved organisms.

J.-C. Bertrand (✉)
Unité Mixte de Service, UMS 3470, OSU Pythéas, Aix Marseille Université,
Marseille Cedex, France
e-mail: jean-claude.bertrand@univ-amu.fr

P. Bonin · B. Ollivier · N. Pradel
Aix Marseille Université, Université de Toulon, CNRS, IRD, MIO UM 110, Marseille, France

K. Alain
CNRS, Université de Bretagne Occidentale, IFREMER, Laboratoire de Microbiologie des Environnements Extrêmes – UMR6197, Ifremer, Plouzané, France

A. Godfroy
IFREMER, CNRS, Université de Bretagne Occidentale, Laboratoire de Microbiologie des Environnements Extrêmes – UMR6197, Ifremer, Plouzané, France

P. Normand
Laboratoire d'Ecologie Microbienne, UMR 5557, Université Claude Bernard Lyon 1, Villeurbanne, France

J.-C. Bertrand et al. (eds.), *Prokaryotes and Evolution*,
https://doi.org/10.1007/978-3-319-99784-1_4

Keywords Artificial minimal cell · Biogeochemical cycles · Carbon cycle · Dormancy · Extremophilic microorganisms · Horizontal gene transfers · Hypermutators · Minimal bacterial gene set · Mutations · Nitrogen cycle · Spores · Starvation · Storage of organic and inorganic compounds · Theories of evolution · Ultramicrobacteria · Universal tree of life · Viable but nonculturable state · Viable prokaryotes in geological formations

4.1 Prokaryotes: An Overview of the Theories of Evolution

In this paragraph, the concept of natural selection as first proposed by Darwin and its role in the evolution of prokaryotes will be first discussed. Studies undertaken after Darwin do not fundamentally question natural selection, but it has been demonstrated to be insufficient to explain the evolution of living beings and, in particular, that of prokaryotes. Afterward, the different theories that have been proposed after Darwin will be briefly described and their application to the prokaryotic world covered.

4.1.1 Darwinian "Natural Selection"

Natural selection, as the driving force of evolution, has been proposed simultaneously and in an independent fashion by Charles Darwin (1859) and by Alfred Russel Wallace (Wallace 1870). Charles Darwin was a naturalist who had built his theory based on the observation of the phenotype of macroorganisms (plants and animals), both living ones and fossils. The mechanisms underlying natural selection were three major ones: variation, heredity, and struggle for existence.

(i) Variation: for Darwin species are not immutable (*I am fully convinced that species are not immutable*). Species evolve "through reproduction with modification," and the species of a given genus are supposed to descend from a common ancestor, even though he considers that the exact character of the common ancestor of a group of organisms is not known (*we never know the exact character of the common ancestor*). This link between all living beings and this concept of a "common ancestor" had not been proposed before Darwin.
(ii) In the framework of natural selection, each variation, which happens in a random fashion, if it is useful for the survival and fitness (maximum number of offsprings) of a living being, is preserved and inherited; conversely, any injurious variation is rejected. Besides, successive modifications that appear are slow and slight and appear in a graduated manner; a great and sudden modification will not occur in the evolution of living beings as conceived by Darwin; for him "Natura non facit saltum" or nature does not make leaps.

(iii) "Struggle for existence": in an ecosystem, the species present (and not individuals) produce more offsprings than can survive given the resources available that are for the most part limited. A struggle will thus occur to obtain these resources. Those that will survive will not necessarily be "the strongest" but those "best adapted" to the conditions of the environment and who will have a success in leaving progeny. The phenotypes emerging at random and yielding the largest number of offsprings will survive. Random variations appear, and among these variations (mutations), those that will spread in the population are those that will result in adaptation to new environmental conditions and will ensure the maximum number of offsprings.

It must be precised that natural selection is not aiming toward "progress" but toward maintenance of species best adapted to a given ecosystem at a given moment; the biotope itself being subject to changes also modifies permanently the conditions to which the species must adapt.

Moreover, species must not only adapt to the changes occurring to their biotope but also to the changes that other species within the ecosystem undergo. Darwin was conscious of the importance of interactions with other species (*if any one species does not become modified and improved in a corresponding degree with its competitors, it will soon be exterminated*).

Is natural selection proposed by Darwin, who was completely ignorant of the microbial world, applicable to prokaryotes?

The concept of natural selection, built on observation of the phenotype of macroorganisms, is applicable to prokaryotes. Indeed, in prokaryotes, as specified by the natural selection concept, each time a "variant," within a population, gains an advantage for survival – be it for the acquisition of resources or adaptation to a natural or anthropogenic modification of the environment – it is maintained and will be propagated in the population. For instance, antibiotic resistance of bacteria and bacterial xenobiotic biodegradation are consequence of evolution via natural selection. However, for Darwin the time scale to grasp evolution was geological; in prokaryotes, the time scale is much faster: the pace of prokaryotes generation (a few hours, days) is completely different from that of macroorganisms which is set in years if not millennia.

Darwin developed the concept of natural selection based on observations, without knowing its underlying mechanisms. The study of prokaryotic evolution has played a capital role in the comprehension of mechanisms involved in the evolution of living beings. What is even more remarkable is that the study of prokaryotes has brought the experimental proof of the action of natural selection and other mechanisms involved in evolution, enabling direct observation and dissection of evolutionary processes (van Ditmarsch and Xavier 2014) (*cf.* "Evolution underway in prokaryotes").

Do discoveries that have followed Darwin's works question natural selection?

Before mentioning the other theories proposed following Darwin, this theory must be first replaced in the context of the knowledge existing at the moment of its elaboration and, second, insist on the carefulness of the author concerning the role of natural selection in the evolution of living beings.

It must be recalled that Darwin was convinced natural selection had been the main, but not the only means causing modification of species (*I am convinced that natural selection has been the main but nonexclusive means of modification*).

He knew the limits to his theory. The most important was his incapacity to find the biological mechanisms at play in descent with modifications. Darwin was convinced that a large field of research would open on knowledge of the causes and laws of variation (*a great and almost untrodden field of inquiry will be opened, on the causes and laws of variation*). It was indeed the case, and the first to propose a theory of heredity compatible with the Darwinian theory was an Augustinian monk Gregor Mendel, 6 years after the publication of *On the Origin of Species* and 17 years before the disappearance of Darwin. After the death of Darwin, a copy of a paper by Mendel was found on his bookshelf, which pages had not been detached, indicating he had not read it. It can thus be said that one answer to the questions he had wrestled with all his life was to be found at home, among his books.

4.1.2 The Synthetic Theory or Neo-Darwinian Theory

The main authors of the synthetic theory, elaborated in the 1930–1948 period, were Fisher (1930), Wright (1931), Haldane (1932), Dobzhansky (1937), Mayr (1942), Simpson (1944), and Huxley (1948). In this theory coupling classical Mendelian genetics and population genetics, biodiversity, and evolution of populations can always be explained by mechanisms of "natural selection," hence the name "neo-Darwinism." However, the scale of observation has changed. Once the laws of heredity had been discovered, the mathematical means were developed to "measure" natural selection.

Genetics was integrated into the Darwinian theory. The concept of "variant" has been replaced with that of "mutant," the result of random mutations that appear in the genetic reservoir. The selective value of a mutant is its adaptation to an environmental change, its propagation in a population, and its capacity to increase its descent (fitness). New alleles, new genotypes, and thus new phenotypes appear at random. If an allele codes for an advantageous "variant," the individuals that carry this allele will reproduce more than those carrying the wild allele that will eventually disappear from the population.

For neo-Darwinians, natural selection remains the most important mechanism to explain evolution of populations. The studies undertaken with prokaryotes have provided convincing arguments in favor of this theory.

1. The first direct experimental demonstration showing any genetic mutation is the fruit of evolutionary chance was done by Luria and Delbrück (1943) who studied resistance of the bacterium *Escherichia coli* B to a bacterial virus α. They asked whether the resistance to bacteriophage in *E. coli* was induced by the presence of phage or if instead it was due to random mutations occurring prior to phage exposure. The virus was plated first on nutrient agar plates and spread over the entire surface of the agar. A few minutes later, a bacterial suspension to be tested was spread over the central part of the plate. Luria and Delbrück noticed that bacteria resistant to the viral infection appeared after a few hours. Two hypotheses concerning the origin of the bacterial resistance could be proposed: (i) the appearance of bacteria resistant to the lytic phages would result from mutations induced in a specific fashion adapted to the presence of viruses. The mutations would thus have appeared not during the growth phase of the populations, but immediately after spreading of the bacteria onto the selectively lethal medium; (ii) the genetic mutations causing the hereditary variation linked to phage resistance would be spontaneous, produced at random during the growth phase of the bacterial population before plating on the selectively lethal containing lytic phages; the viruses would have no role in their occurrence. Luria and Delbrück made fluctuation tests to understand, on the basis of statistics on the distribution of the number of resistant bacteria, if mutations had occurred before (ii) or after (i) exposure of the bacterial population to viruses, in order to determine if the results of these experiences were conform with a Lamarckian explanation or if they were in line with the synthetic theory paradigm. They concluded that "the resistance to virus is due to a heritable change of the bacterial cell which occurs independently of the action of the virus" in accordance with the synthetic theory. This was the first direct demonstration of the random nature of mutation.
2. It is the study of prokaryotes that has permitted to understand the mechanisms underlying mutations.
3. From the studies of prokaryotes, it has been demonstrated that the acquisition of an advantageous allele can also occur through HGT (*cf.* Sect. 4.2.2).

4.1.3 The Neutralist Theory of Molecular Evolution (Genetic Drift)

According to the theory proposed by Motoo Kimura (Kimura and Ohta 1974; Kimura 1986), evolutionary changes, at the molecular level, do not always result in a progressive accumulation of favorable mutations causing, at the phenotypic level, a gradual and continuous evolution of the adaptation to the environment as supposed by Darwinian thought. In a population, not only the appearance but also the diffusion and fixation of mutations – which for this author are mainly but not only strictly neutral or nearly neutral – occur in a purely random fashion, at a regular rhythm over time, whence the notion of "molecular clock." It has nevertheless been demonstrated, on the one hand, that proteins do not all have the same rate of molecular evolution and, on the other, that the mutation rate is not the same over the genome, thus questioning the concept of molecular clock, as well as that of chance (Chattopadhyay et al. 2009).

Neutral mutations, those that cause neither advantage nor disadvantage in the population, accumulate in the genotype and are not eliminated through natural selection, but are under the control of random "drift," which is a random "drift of genetic frequencies."

The neutralist theory is not opposed to the idea that evolution of forms and functions be governed by Darwinian selection, but it reveals another facet of evolutionary processes. Darwin observed phenotypes, while Kimura studied genes.

Kimura estimated that evolutionary changes, when studied at the molecular level (proteins, DNA), are not the consequence of the action of natural selection, but that of the genetic drift of neutral or nearly mutations; to get established in a population, mutant genes do not inevitably have an advantage.

The neutral theory (or more precisely, the neutral-mutation-random-drift hypothesis) claims that the great majority of evolutionary changes at the molecular level are caused not by Darwinian selection acting on advantageous mutants, but by random fixation of selectively neutral or nearly neutral mutants. The theory does not deny the role of natural selection in determining the course of adaptive evolution, but it assumes that only a minute fraction of DNA changes are adaptive in nature (Kimura 1986).

In the neutralist theory, the size of populations plays an important role (Hallatschek et al. 2007). The smaller the population, the larger the importance of the genetic drift. However this evolution also exists in large populations if the phenomenon is followed over a large number of generations. This situation is that occurring in prokaryotes which generation times are particularly short.

Neutralist Theory and Prokaryotes

(i) Application of the neutralist theory (genetic drift) to small size populations: intracellular pathogenic bacteria? It has been noted that intracellular pathogens or symbionts have a faster mutation rate than their saprotrophic neighbors and that they accumulate deleterious mutations (Moran 1996), effects due to the lack of genetic exchange, and a regular purging mechanism called Muller's ratchet (1964).

(ii) The neutralist theory considers that mutations may affect genes without incidence on the function of the proteins encoded by these genes. Examples of neutral mutations in prokaryotes: ARNr16s, nitrate reductase, etc.

4.1.4 The Theory of Punctuated Equilibria

We will recall that for Darwin, natural selection "can never take a leap, but must advance by the shortest and slowest steps," Natura non facit saltum.

The notion of discontinuity was already present in the theory of mutations proposed by Hugo de Vries, a mutationist evolutionist, who estimated that species appeared following mutations. The theory of punctuated equilibria (Eldredge and Gould 1973) reinforces the concept of discontinuities in evolution: evolution is not always gradual, moving forward step by step, as proposed by Darwin. It may also occur stepwise, its advance being discontinuous.

Examples Chosen in the Prokaryotic World to Illustrate This Theory

(i) To explain the transition from a cellular organization of the prokaryotic type to a eukaryotic type organization, basically, three "scenarios" have been proposed: *(a)* the "endosymbiotic theory," this theory postulates that mitochondria and chloroplasts were former bacteria that, during evolution, had established a symbiotic relationship with a eukaryotic ancestor, probably heterotrophic and anaerobic (Sagan 1967; Margulis 1970); *(b)* in this scenario eukaryotes result from an association between a bacterium and an archaeon (Martin and Müller 1998; Lopez-Garcia and Moreira 2006); and *(c)* the very controversial hypothesis which suggests that Archaea and Eukarya have evolved from an actinobacterium (Cavalier-Smith 2002). For the three hypotheses proposed, the prokaryotic/eukaryotic transition did not occur progressively on geological scales, but relatively quickly with a massive number of HGTs, for instance, from cyanobacteria to the nucleus of the host (in the endosymbiotic theory) or from deltaproteobacteria to archaea accompanied by the total disappearance of the bacterial genome (hypothesis based on an association between a bacterium and an archaeon).

(ii) Resistance to antibiotics (*cf.*: biosynthesis and resistance to antibiotics) (a bacterium may acquire "at once" the genetic information coding for new properties).

(iii) The association *Wolbachia*/nematodes, arthropodes, and insects (Ioannidis et al. 2013; Moriyama et al. 2015).

(iv) DNA gyrase: the discontinuity may be explained by the acquisition of a mutation that will confer an evolutionary advantage to the mutant population in which it emerged.

4.1.5 The Theory of the Selfish Gene

This theory was proposed by Richard Dawkins (1976). As he proclaims it himself, he is an enthusiastic Darwinian. Even though he considers that much of what Darwin said is, in detail, wrong, he nevertheless estimates that his theory is a different way of seeing, not a different theory.

He considers there are two manners to consider natural selection, from the point of view of the gene and from that of the individual. Well understood, they are equivalent; these are two conceptions of the same truth. One can alternate from one to the other; they are the same neo-Darwinism (*My point was that there are two ways of looking at natural selection, the gene's angle and that of the individual. If properly understood they are equivalent, two views of the same truth. You can flip from one to the other, and it will still be the same neo-Darwinism*).

For Dawkins, the fundamental unit of selection is not the species, nor the individual, but the gene or "replicator" (*the basic unit of natural selection*) which is also the unit of heredity. Natural selection favors genes that control survival machines that develop most harmoniously and yield most offsprings in the environmental conditions present.

Richard Dawkins pushes to the extreme the role of genes in evolution. He sees the evolution of the living world only through the history of genes. For him, "We are survival machines, but 'we' does not mean just people. It embraces all animals, plants, bacteria, and viruses." Genes are selfish, living only for themselves and have an autonomous life. However, if the organism is only a means to carry genes, Dawkins admits their *survival depends on the efficiency of the bodies in which they live and which they helped to build* and that *natural selection favors replicators that are good at building survival machines*.

In prokaryotes, natural selection as defined by R. Dawkins has an important pertinence. One can consider prokaryotes, from the point of view of replication, as particularly efficient machines to transmit their genes to their offsprings.

4.1.6 Natural Selection and HGTs

The tree of life should perhaps be called the coral of life, [its] base of branches dead; so that passages cannot be seen. (*Life and Letters of Charles Darwin – Volume 1, pg 368*)

The discovery of HGTs, which are particularly intense in prokaryotes, demonstrate that novelty does not emerge only in a vertical fashion "from parents to descent" but also in a horizontal fashion "from neighbor to neighbor." Horizontal transfers of genes (HGTs) are superimposed to vertical transfers (*cf.* 4.3.2 – *Horizontal gene transfers*).

The study of prokaryotes has also revealed a mode of information transfer, unthinkable for Darwin and his successors, until the work undertaken by Griffith on the transmission of hereditary characters in bacteria. Before Griffith, transmission of hereditary characters could only occur from living being to living being. But Griffith showed that information transfer could occur from "dead being to living being" and that the information thus acquired could then be transmitted to offsprings in a stable fashion. These discoveries amplified even more the intensity and complexity of exchanges in the prokaryotes world.

Darwin wanted to propose a scheme that would have highlighted the links that unite all species, hence the notion of the "tree of life." However, the whole of post-Darwinian discoveries demonstrate that the Darwinian tree of life cannot represent completely the evolution of living beings and particularly that of prokaryotes. For that reason, it is necessary to go from the "tree of life" to the "coral of life," a metaphor initially proposed by Darwin himself in his "secret notebooks." The metaphor retained *in fine* by Darwin, that "of a tree which branches spread out through geological times" – this was his time scale – can no longer be retained. If branches spread out, they often also get closer, cross, and even fuse (Fig. 4.1) (Doolittle 1999;

Fig. 4.1 Living beings carrying a very different evolutionary history exchange continually information. If branches spread out, they often also get closer, cross, and even fuse. https://www.pinterest.fr/pin/331436853799089167/

Olendzenski and Gogarten 2009; Koonin and Wolf 2012; Daubin and Szöllősi 2016). Living beings carrying a very different evolutionary history exchange continually information and may even fuse; a strictly genealogical explanation is thus no longer sufficient; it must be coupled to the notion of relatedness between species, which was demonstrated through the study of genes phylogenies, an approach that Darwin could not even imagine.

It should also be noted that the dichotomy between trees and corals with the later fusing while the former would not is false since there are many examples of fused trees in nature.

4.2 The Mode of Reproduction of Prokaryotes: High Rate of Reproduction and Maximal Utilization of Their Genetic Information

4.2.1 Maximal Utilization of Their Genetic Information

The mode of reproduction of prokaryotes through scissiparity is the most efficient to yield, in the shortest time, the highest number of offsprings (fitness). Prokaryotes represent "life at its rawest." They use their genetic information optimally: rate of replication and translation, this last amplified through coupling of transcription and translation, and their DNA containing up to 90% of coding sequences (strong correlation between genome size and number of ORFs) (Konstantinidis and Tiedje 2004) contrary to DNA of eukaryotes that contain a high percent of noncoding DNA.

Many genomes contain a lot of redundancy, in terms of gene duplications, as well as pseudogenes that seem to have lost any function. Together with repeat sequences and parasitic DNA that seem to bear no function for the organism, the only conclusion can be that bacterial genomes are not always evolving toward optimal efficiency (Land et al. 2015). The presence of such "junk" DNA is one reason for the vast variation in genome size within the bacterial world, although the genome's size is of course also dependent on the number of functional genes and pathways that are present.

4.2.2 Generation Time

Generation times of prokaryotes (the time interval required for the cells or population) to divide are significantly lower than those of eukaryotes. They are highly variable and depend on the physicochemical conditions, available nutrients, and genetic factors.

In laboratory culture, the generation time calculated during exponential growth phase varies according to the species and is evaluated most often under optimal culture conditions. For most known cultured bacteria, generation times range from about 15 to 60 min: approximate generation times were 20, 28, 30, 35, and 50 min for *Escherichia coli*, *Bacillus subtilis*, *Staphylococcus aureus*, *Pseudomonas aeruginosa*, and *Thermus aquaticus*, respectively. However, for some prokaryotes generation times are much higher and vary within a taxon.

The mean generation time of *Rhizobium japonicum* was between 2.8 and 4.1 h for fast-growing isolate and 6.7 and 13.0 h for slow-growing isolates (Keyser et al. 1982). At a pH 5.2 and 88 °C and at optimal factor concentration, the generation time of the archaeal kingdom, *Thermofilum pendens* (order of Thermoproteales), growing in a fermentor, was 10 h (Zillig et al. 1983). *Treponema pallidum*, the syphilis spirochete, doubled every 30–33 h in vivo and 30–50 h in vitro (Lafond and Lukehart 2006). In *Syntrophomonas wolfei*, an anaerobic syntrophic bacterium, the generation times obtained by co-culturing with *Desulfovibrio* and *Methanospirillum hungatei* were 54 and 84 h, respectively (McInerney et al. 1981). It also need to be mentioned that the generation time of an obligately barophilic bacteria (Yayanos et al. 1981), designated MT41 (MT for Mariana Trench), was 25 h at 2 °C and 690 bars; at a pressure of 1035 bars (close to the depth origin), the generation time was about 33 h, and the bacteria were unable to grow at a pressure below 380 bars. All of the examples cited above, which are obtained from growth undertaken in the laboratories (under optimal growth conditions), show a great variability of growth of prokaryotes.

In natural conditions, it is difficult to accurately estimate the generation time of bacteria, because nutrients and other important environmental factors for growth (e.g., temperature, salinity, pH, pressure, etc.) vary widely in time and space. In addition, most frequently (but not always), conditions for optimal growth exist only intermittently. Most often prokaryotes face to "feast-or-famine" conditions. Thus generation times in nature are most of the time well below to those recorded in the laboratory. In addition, growing in pure culture (axenic cultures), which is "absolutely impossible in nature," "protected as it is in its vial" (Winogradsky 1949), the prokaryote is not subjected to the competitive effects of other microorganisms.

In situ, generation time can be extremely high. According to Jannasch (1969) the generation time of four aquatic bacteria calculated from experiments performed in chemostat was slow, between 20 and 200 h. For their part, Wirsen and Molyneaux (1999) studying deep-sea natural microbial populations estimate that the growth rates of natural populations were extremely low (e.g., a doubling time of 629 h). In aquifer, the doubling time of an introduced bacterial strain, *Comamonas* sp. strain S DA001 (isolated from a shallow aquifer), under in situ subsurface conditions, was approximately 15 days; the calculated growth rate can be considered as identical to that of indigenous microbial populations because the strain was injected without nutrients (Mailloux and Fuller 2003). The longer generation times described are those of the endogenous prokaryotic populations living in subsurface depth. The mean generation time of the deep subseafloor microorganisms, calculated from metabolic activity of these populations (e.g., sulfato-reduction), was estimated at ~ 1000 years (Jørgensen and D'Hondt 2006).

4.2.3 Genome Size

The size of bacterial genomes varies from a few hundred nucleotides to >14 000, and it is largely dependent on lifestyle, growth conditions, phylogenetic origins, and nutritional strategies. In Tables 4.1, 4.2, and 4.3 are summarized the main characteristics of the genomes of *Bacteria* and Archaea of the most representative species of free-living organisms and organisms which are enable to have a free-living state. Unlike eukaryotes, the genome size variation in bacteria translates almost directly into biochemical, physiological, and organismal complexity because the majority of sequences are functional protein-coding regions.

Note that some prokaryotes contain more DNA than some eukaryotes. For instance, if the average genome size of fungi is 43.30 Mbp, some strains have a genome size below this average such as *Candida caseinolytica* (9.18 Mbp), *Hansenula polymorpha* (8.97 Mbp), and *Wallemia sebi* (9.82) (Mohanta and Bae 2015). In the same way, the genome size of *Ostreococcus tauri*, the smallest free-living eukaryote, marine alga, is 12.56 Mb (Derelle et al. 2006).

Finally, let's note that genome size of several prokaryotes is smaller than the larger known viral genomes at 2.5 Mbp (Philippe et al. 2013).

Table 4.1 Comparison of genomes of *Bacteria*

Organisms	Genome size (bp)	GC content (%)	Number of coding sequences	References
Mycobacterium genitalium	580,076	31.7	475	McCutcheon and Moran (2011)
Rickettsia prowazekii	1,111,523	29.0	835	McCutcheon and Moran (2011)
Pelagibacter ubique Strain HTCC1062	1,309,000	29.7	1354	Grote et al. (2012)
Methylophilaceae Strain HIMB624	1,333,209	35.37	1381	Hugget et al. (2012)
Prochlorococcus Strain MED4	1,657,990	30.8	1,716	Rocap et al. (2003)
Prochlorococcus Strain MIT9313	2,410,873	50,7	2,275	Rocap et al. (2003)
Synechococcus Strain WH8102	2,430,000	59.4	2,525	Dufresne et al. (2005)
Prochlorococcus Strain MIT9303	2,680,000	50	3,022	Scanlan et al. (2009)
Escherichia coli	4,639,675	50.8	4,145	McCutcheon and Moran (2011)
Anabaena variabilis ATCC29413	7,105,752	41.4	5,710	Thiel et al. (2014)
Rhodococcus sp. Strain RHA1	9,702,737	67	9,145	McLeod et al. (2006)
Sorangium cellulosum Strain S_0ce56	13,033,779	71.38	9,367	Schneiker et al. (2007)

Table 4.2 Comparison of genomes of Archaea

Organisms	Genome size (bp)	GC content (%)	Number of coding sequences	References
ARMAN-4	800,887	34.8	916	Baker et al. (2010)
ARMAN-2	999,043	47.2	1,033	Baker et al. (2010)
Methanothermus fervidus strain V24S^T	1,243,342	31.64	1,311	Anderson et al. (2010)
Staphylothermus marinus	1,570,485	35.7	1,610	Anderson et al. (2009)
Methanococcus Jannaschii	1,664,976	31.4	1,682	Bult et al. (1996)
Thermofilum pendens	1,781,889	57.6	1,883	Anderson et al. (2008)
Methanosarcina acetivorans	5,751,492	42.7	4,524	Galagan et al. (2002)

Table 4.3 Comparison of genomes of host-dependent prokaryotes

Organisms	Genome size (bp)	GC content (%)	Number of coding sequences	References
Candidatus Nasuia deltocephalinicola	112,031	16.6	137	Bennett et al. (2016)
Candidatus Tremblaya princeps	138,927	58.8	121	McCutcheon and Moran (2011) McCutcheon and von Dohlen (2011)
Candidatus Hodgkinia cicadicola	143,795	58.4	169	McCutcheon and Moran (2011)
Candidatus Carsonella ruddii	159,662	16.6	182	McCutcheon and Moran (2011)
Candidatus Zinderia insecticola	208,564	13.5	202	McCutcheon and Moran (2010)
Candidatus Sulcia muelleri strain CWSS	245,530	22.4	227	McCutcheon and Moran (2011)
Nanoarchaeum equitans	490,885	31.6	552	Waters et al. (2003)
Candidatus Moranella endobia PCVAL	538,000	43.5	411	Martínez-Cano et al. (2015)
Candidatus Blochmannia floridanus	705,557	27.38	583	Gil et al. (2003)

4.2.3.1 Free-Living Prokaryotes

According to Land et al. (2015), a "typical" bacterial genome is around 5 million bp and encodes about 5000 proteins; however, different lineages of free-living bacteria – most of them in marine environments – evolved reduced genomes (Table 4.1).

Pelagibacter ubique (strain HTCC1062) is a bacteria whose genome has long been considered as the smallest genome of an autonomously replicating free-living in nature (1,308,759 base pairs, 1354 protein code genes, GC% 29,7) (Giovannoni et al. 2005; Grote et al. 2012) (*cf.* Table 4.1). This strain is an ultramicrobacteria with an average cell volume (enclosed by outer membrane) of 0.037 ± 0.011 μm^3 (Zhao et al. 2017a). This strain cultivated in the laboratory (Rappe et al. 2002) thrives at low nutrient concentrations (oligotrophic conditions) characteristic of open oceans. *P. ubique* is the most studied members of the clade SAR11. All SAR11 bacteria (α-proteobacterium) are obligate aerobic and chemoheterotrophic, with full respiratory electron transport systems and the light-dependent proton pump proteorhodopsin. They have an ancient evolutionary origin, likely a Precambrian origin (Giovannoni 2017). It is the most abundant group of heterotrophic bacteria in the oceans: in some regions SAR11 clade represents 50% of the total surface microbial community and 25% of the subeuphotic microbial community, and this global abundance in the oceans worldwide would be 2.4×10^{28} cells (Morris et al. 2002). However, presently, there is no consensus about biomass and activity of the clade in nature, largely due to variable results obtain as function of the method used to evaluate these two parameters. Today, SAR11 is one of the most successful competitors for dissolved organic compounds in euphotic zone and dark ocean and plays a major role in ocean carbon cycle. SAR11 members are have likely undergone genome streamlining, an evolution that leads to minimization of cell structure and complexity (small cell size, reduced genome) in large populations of free-living microbes in nutrient-poor environment (Giovannoni et al. 2005; Grote et al. 2012). In SAR11 large populations, genome streamlining is thought to benefit cells by lowering the cost of replication and optimization of transport systems that are essential to competition in the open ocean; in some cases surface-volume ratios might also drive streamlining. According to the streamlining hypothesis, the following characteristics are consistent with natural selection acting to economize cellular metabolism. For detailed information concerning evolution, abundance, diversity, biogeography, activity, biochemistry, and metabolism of SAR11, see Grote et al. (2012) and Giovannoni's review (Giovannoni 2017).

More recently, several species with lower genome size than *P. ubique* have been described. For instance, a novel clade of an uncultured marine *Actinobacteria* has been described by Ghai et al. (2013) using a combination of metagenomic approaches, flow cytometry, and FISH. The term of subclass "*Candidatus* Actinomarinidae" was proposed to denominate this group of prokaryotes. A member of these *Actinobacteria* named "*Candidatus* Actinomarina minuta" has an average cell volume of ~ 0.013 μm^3 and a genome size estimated in the range of 823–1029 kb.

The clade OM43, related to Type 1 methylotrophs, is another group of marine and freshwater bacteria with a small genome. The most investigated strains were HTCC2181 et HIMB624 which formed a monophylic lineage within the family *Methylophilaceae*, with a genome size of 1,304,428 bp (Giovannoni et al. 2008) and 1,333,209 bp (Huggett et al. 2012), respectively. These heterotrophic strains cannot oxidize methane but oxidize C1 compounds methanol and formaldehyde as sources of carbon and energy.

Bacteria endowed with small genome have been detected in groundwater such as the candidate phyla SR1, WWE3, TM7, and OD1 with a genome size equal to 1,777,760 bp, 878,109 bp, 845,464 bp, and 693,528 bp, respectively (Kantor et al. 2013; Luef et al. 2015).

Marine picocyanobacteria, the most abundant photosynthetic free-living organism on Earth, with two genera, *Prochlorococcus* and *Synechococcus*, have larger genomes. The genome size of *Prochlorococcus* ranges from 1.66 to 2.68 Mbp (Dufresne et al. 2003; Rocap et al. 2003; Scanlan et al. 2009). Rocap et al. (2003) describe two ecotypes within the genus: a high-light-adapted ecotype (1,657,990 bp, one of the smallest known genomes of any oxygenic phototroph), which is the most abundant in surface waters, and a low-light-adapted ecotype with a larger genome (2,410,873 bp), dominating in deeper waters. *Synechococcus* genome size ranges from 1.64 to 2.68Mb (Scanlan et al. 2009). According to Scanlan et al., the genome size of *Prochlorococcus* and *Synechococcus* is still small compared to the average genome size of other sequenced cyanobacteria (5,33 ± 3,69 Mb). For instance, the genome sizes of *Anabaena variabilis* (Thiel et al. 2014) and *Acaryochloris marina* (Swingley et al. 2008) were 7.1 and 8.3, respectively. Finally, let's note that some free-living bacteria are endowed with a largest genome such as myxobacterium *Sorangium cellulosum* strain S_0ce56 (Schneiker et al. 2007) and strain So0157-2 (Han et al. 2013) with a genome reaching 13,033,779 pb and 14,782,125 pb, respectively.

Concerning the genome size of free-living Archaea (Table 4.2), some are endowed with a small genome such as uncultivated ARMAN lineages (archaeal Richmond Mine acidophilic nanoorganisms) with a genome size ranging from 800,887 to 999,043 bp; ARMAN cells have volumes of 0.009–0.04 μm^3, as determined by cryoelectron microscopy (Comolli et al. 2009). It was suggested that these organisms were at least partially dependent on (an)other community members for basic metabolic building blocks. Other archaea have larger genome size. This is the case of hyperthermophilic strain *Thermofilum pendens* with a genome size of 1,781,889 bp (Anderson et al. 2008) or *Methanosarcina acetivorans* with a genome containing 5,751,492 bp which is by far the largest known archaeal genome and unique among the Archaea in forming distinct multicellular structures (Galagan et al. 2002).

4.2.3.2 Host-Dependent

Endosymbiotic and intracellular parasites living in a protected and nutrient-rich environment of their host have many features in common, including massive gene losses according to the Darwinian principle (many molecules can be obtained from the host) (Table 4.3). Bacteria with the smallest genomes are *Candidatus* Tremblaya princeps, *Candidatus* Hodgkinia cicadicola, and *Candidatus* Carsonella ruddii with genome sizes of 138,927, 143,795, and 159,662 bp, respectively (McCutcheon and Moran 2011). Many highly reduced genomes have been described in bacteria symbionts of several insect lineages (Martínez-Cano et al. 2015). This is the case of two bacteria that are hosted by the insect *Planococcus citri*. These two named bacteria *Candidatus* Moranella endobia (*β-Proteobacteria*) and *Candidatus* Tremblaya princeps (*γ-Proteobacteria*) live symbiotically and are associated by forming an unprecedented organization: *Ca.* M. endobia lives inside *Ca.* T. princeps (Thao et al. 2002). Another intimate and specific association deserves to be reported, the one between two Archaea, *Nanoarchaeum equitans* and *Ignicoccus hospitalis*, where *N. equitans* grows only in co-culture with *I. hospitalis* (Waters et al. 2003; Podar et al. 2008).

4.2.3.3 How Minimal Bacterial Gene Set Is Necessary to Exist?

The analysis of smallest prokaryotic genomes leads to a question of fundamental biology: what is the minimum gene number necessary for an independent life, required for reproduction and self-maintenance under given environmental conditions, for a free cell exist? This question can be approached by comparative genomics.

From the comparison of genome sequencing of *Haemophilus influenzae* and *M. genitalium*, Mushegian and Koonin (1996), the pioneers in identification of minimal gene set, estimated that 256 gene were needed to sustain the existence of a modern-type cell. Gil et al. (2004) proposed gene set composed of 206 protein-coding genes necessary, but probably not enough, to maintain a cell alive in a realistic environment. So, Gil and Peretó (2015) revisited the minimal gene set by adding gene to improve cell viability and new genes for RNA processing and metabolism. In 2016, Ye et al. (2016) proposed a simplified bacterial gene set by comparative genomics and supplemented it by neo-construction of a bacterial approximately minimal metabolic network. From this procedure, they proposed a simplified bacterial gene that preserves both self-replication and self-maintenance systems including 327 genes and requiring 431 reactions.

4.2.3.4 Artificial Minimal Cell

The simplified molecular machinery of organisms with small genomes may be used to aid in the design of live artificial minimal cells in the laboratory. An artificial cell could be considered "alive" if three criteria are met: *self-maintenance* (*metabolism*), *self-reproduction*, and *capable of Darwinian evolution*. Two approaches, conceptually different, can be implemented to synthetize artificial cell (Gil et al. 2004; Juhas 2016; Martínez-García and de Lorenzo 2016; Ye et al. 2016; Glass et al. 2017) (see also the site www.ees.lanl.gov/protocells): (i) the top-down approach (genome downsizing, reprogramming existing cell with simple genome) starts from existing organism (viable "natural" cell) and sequentially removes genes, until no more genes can be removed without severe growth impairment or loss of viability; beyond this genome simplification, growth is no longer possible; (ii) the bottom-up approach (de novo synthesis) consists of to synthesize artificial cell from scratch using non-living organic and inorganic materials.

Major results in the construction of a minimal synthetic cell, using a top-down approach, were obtained by the J. Craig Institute using *Mycoplasma*, a group of bacteria characterized by lack of cell wall, obligate parasitic lifestyle, metabolic simplicity, and small genome. Indeed, C. Venter's team designed, synthetized, and assembled the genome of *Mycoplasma mycoides* (1,078 kbp) and then transplanted it into a recipient cell, the cell of another bacteria, *Mycobacterium capricolum*. The new cell obtained, called JCVI-syn1.0, was controlled only by the synthetic chromosome (901 genes), was capable of self-replication, but was not a truly "synthetic" cell because its synthetic genome was put into an existing cell (Gibson et al. 2010). In 2016, Hutchison et al. of the C. Venter Institute designed and synthetized a genome that contains 531,560 bp and 473 genes, appointed JVCI-syn3.0 (428 genes were deleted compared to the genome of JVCI-syn1.0); this genome is smaller than that of any free-living replicating cell. For instance, *M. genitalium*, one of the smallest known natural free-living bacteria, has a genome size 580,070 bp and a total 470 predicted coding regions (Fraser et al. 1995). JVCI-syn1.0 and JVCI-syn3.0 showed that they have similar colony morphologies, with smaller colony size for JCV-3.0; moreover, their generation time was different: JCVI-syn 3.0 has a doubling time of 173 min, while it is 63 min for JVCI-syn1.0. An unexpected result emerged from the analysis of JVCI-3.0 genes: if most were assigned into different functional group (cell membrane structure and function of genome information, expression and preservation, cytosolic metabolism), 149 genes (31.5% of the genome) were of unknown biological functions, suggesting the presence of undiscovered functions that are essential for life. If the work to produce a minimal set of gene necessary for the cell to survive has been focused on mycoplasma, "the quest for the minimal bacterial genome" was also undertaken at other prokaryotes (e.g., *E. coli, Bacillus subtilis, Corynebacterium glutamicum, Pseudomonas putida, Streptomyces avermitilis, M. pulmonis, M. pneumoniae*) (Martínez-García and de Lorenzo 2016; Glass et al. 2017).

Concerning the studies devoted to construction of artificial cells using a bottom-up approach, many have been focused on self-replicating lipid vesicles (Hanczyc et al. 2003; Rasmussen et al. 2004; Chiarabelli et al. 2009). Rasmussen et al. used liposome with a lipophilic peptide nucleic acid (PNA) anchored on his surface. Rasmussen's artificial cell was capable of self-replication using light as a source of energy.

Many difficulties must be overcome in the bottom-up approach due, largely, to the complex interactions between genes such as the lack of detailed information on gene networks: gene can become essential only when isolated from other genes, not integrated into the gene network. Another problem is the high number of possible combinations of genes from a gene pool. In addition, if the artificial cell cannot support life, it is difficult to know what (s) gene (s) is (are) absent(s) in order to obtain a self-replicating system.

In conclusion, the exact composition of the universal genome remains unresolved, and we are still unable to create life in the laboratory. Furthermore, the construction of minimal cell has a potential for medical application and provide large and valuable informations in the understanding of life. Furthermore, if gene set will be able to sustain the main vital functions of a hypothetical simplest bacterial cell, according to Gil et al. (2004), "one must keep in mind that this kind of research has little relevance for the study of the origin of life." The contribution of these studies on the knowledge of the origin of life remains open.

4.3 Rate of Evolution of Prokaryotes: Intensity and Propagation of Mutations Coupled to an Extraordinary Capacity to Exchange Genetic Information

4.3.1 Mutations

Darwin saw evolution as a two-step process. The first being a random process producing "variations," favorable or unfavorable for an organism, the second maintaining the favorable variations by the process of "natural selection." The notion of "variation" was reformulated in terms of "mutations" within the framework of modern synthetic theory (synthesis of Mendelian genetics and Darwinian evolutionism). Indeed, although DNA replication is a very reliable and very effective mechanism, the prokaryotic genome still undergoes permanent and hereditary modifications: mutations. A mutation corresponds to an alteration of the genetic material, the consequence of a lesion in the DNA, which occurs randomly from an evolutionary point of view; moreover, any mutation is caused by a malfunction of one of the DNA polymerases.

The first evidence that genetic mutations occur randomly was provided by Luria and Delbrück (1943), using *E. coli* as a biological model. These authors demonstrated that *E. coli* cells in the presence of phages resisted viral infection but that the presence of phages had no role in the emergence of resistant bacteria; the genetic mutations appeared spontaneously during the exponential phase of *E. coli* before contact with the virus.

The major consequence of the rapidity of reproduction of prokaryotes is their ability to produce a large number of mutants which will be subjected to the action of "natural selection." These mutations may be favorable, unfavorable (even lethal), or neutral in terms of survival and reproduction of an organism in a given biotic and abiotic environment. Most of the errors that occur in a genome are neutral or deleterious: neutral mutations will have no effect on the reproduction of organisms; deleterious mutations will be eliminated by "natural selection." But if a mutant benefits from a favorable mutation that allows him to live in a new environment, the advantage gained is such that the mutant will dominate the rest of the population. This is why the fight against pathogenic prokaryotes is an endless struggle; every time a new antibiotic is discovered and used, a mutant will appear that will resist this new product and will be able to grow in its presence. More generally, at each environmental change, a mutant appears which will adapt, and this capacity of adaptation explains the durability of the prokaryotes during the 3.5 billion years of evolution with a conquest of all biotopes, even the most extreme ones.

There are two types of mutations: *spontaneous mutations* and *induced mutations*. These two major classes of mutations have a completely different origin

4.3.1.1 Spontaneous Mutations

Spontaneous mutations result from a natural process. Indeed, during DNA replication, DNA polymerases PolI, II, and III occasionally make incorporation errors (replication is semiconservative) that they spontaneously correct. If this is not enough, a mismatch repair system (MMR) is used. In the latter case, the spontaneous frequency of mutations is of the order of 10^{-9} to 10^{-11}. These values attest to the very high stability of the DNA molecule which is essential for the organism to be genetically stable. But mutations, which are part of the natural evolution and occur at random on the nucleotide sequence, are also essential because they allow the organism to evolve "if diversity is essential to survival, and if mutagenesis is required to generate such diversity, perhaps mutagenesis has been positively selected for throughout evolution" (Radman 1999).

4.3.1.2 Induced Mutations

The second type of mutations, induced mutations, involves the action of an exogenous agent (mutagen). In principle, whether it is of physical, chemical, or biological origin, this mutagen acts, like the spontaneous mutations, at random on the genome. The SOS system (in reference to the naval Save Our Souls distress signal) is the typical example of an inducible mutator system. It is a coordinated response to DNA damage; it ensures the repair of DNA and the survival of the bacterium in response to strong environmental stresses that alter the genome (genotoxic attacks) (d'Ari 1985; Cox 2003; Fuchs et al. 2004; Nohmi 2006; Erill et al. 2007; Żgur-Bertok 2013; Baharoglu and Mazel 2014; Madigan et al. 2015; Gillings 2017). The SOS system is triggered by UV irradiation, chemicals or oxidative compounds, acids, organic mutagens, some antibiotics, high pressure, reactive oxygen species, etc. The SOS system, which is induced when normal repair systems cannot repair DNA damage, was first described in *Escherichia coli* (Radman 1975).

How does this system work? Two key proteins govern the SOS response: RecA (an inducer) and LexA (a repressor). As soon as the DNA is damaged (Fig. 4.2), the DNA polymerase stops recopying the genome. The triggering of the SOS system is recognition by the RecA protein (Fig. 4.2) of portions of the genome in the form of single-stranded DNA, the appearance of which is caused by aggression. RecA binds to this DNA (nucleofilament) and changes its conformation. Activated RecA has new properties, notably the ability to hydrolyze LexA, which is the brake of the SOS system. The cleavage of LexA leads to the depression of the SOS genes (LexA regulon). The key element is the presence of single-stranded DNA: as long as it is present, RecA retains its LexA-degrading properties, and the intensity of the SOS response depends on the number of single-stranded DNA portions. The SOS system encodes several proteins involved in various DNA repair mechanisms, in particular polymerase V (an enzyme encoded by the *umu*C, *umu*D genes) and polymerase IV

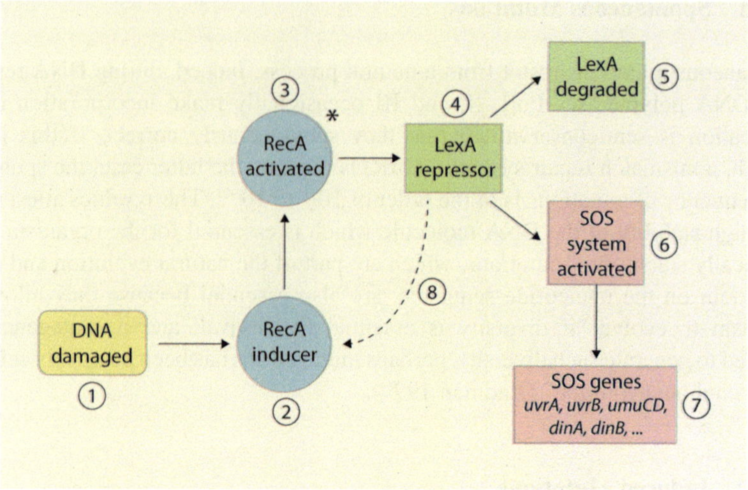

Fig. 4.2 The SOS system. DNA damage (1) activates RecA (2,3) RecA★ activated hydrolyzes LexA (4,5). Proteolysis of LexA leads to coordinate derepression of LexA regulated genes (LexA regulon) (6,7). Rec A is expressed at a low level even in the presence of the LexA repressor (8) (Modified and redrawn from Nohmi, 2006; Żgur-Bertok (2013); Baharoglu and Mazel 2014; Madigan et al. 2015)

(an enzyme encoded by *din*B). Rec A is expressed at a low level even in the presence of the LexA repressor. Polymerases V and IV have the property to bypass the lesions at the level of the DNA sequence (translesion synthesis). They continue DNA replication despite lesions and thus allow the survival of the bacterium, unlike a "normal" polymerase that no longer works when it encounters DNA damage. However, these polymerases which are not characterized by high copy fidelity make copying errors and thus produce an increase in the rate of mutations in the damaged genomic regions. These polymerases are, on the one hand, able to cope with DNA lesions and, on the other hand, to increase the frequency of mutations in the event of stress. This increase in mutations rate generates genetic diversity and adaptation (e.g., antibiotic resistance) and potentiate bacterial survival and adaptation to changing environments. The disappearance of stress causes the SOS system to stop, and the rate of mutation returns to a normal mean value.

SOS response is a widespread and ancient trait of bacteria. Indeed, almost all bacterial phyla harbor a *lexA* gene with characteristic SOS boxes (Erill et al. 2006; Žgur-Bertok 2013; Baharoglu and Mazel 2014).

In addition, potentially carcinogenic substances generally induce an SOS response in bacteria. This correlation is the basis of several genotoxicity and carcinogenicity tests, the most widely used of which is the test developed and improved by Bruce Ames and his team (Maron and Ames 1983).

4.3.1.3 Hypermutators

Another property of prokaryotes, which explains their evolutionary success, is the presence in nature of bacteria that have a rate of spontaneous mutations 100–1000 higher than a normal or wild-type cell. These bacteria are referred to as hypermutators (Taddei et al. 1997; Tenaillon et al. 1999; Oliver 2005; Denamur and Matic 2006; Woodford and Ellinton 2006; Jayarama 2009; Oliver and Mena 2010; Hammerstrom et al. 2015; Lindgren et al. 2016). This hypermutability can be perceived in real time (see Chap. 6). The appearance of mutators is rare in a stable environment but increases in stressful situations where only bacteria that rapidly acquire one or more mutations to cope with the effect of stress will be able to survive. Mutators have been found in different populations – *E. coli*, *Salmonella enterica*, *Neisseria meningitidis*, *Staphylococcus aureus*, *Streptococcus pneumoniae*, *Pseudomonas aeruginosa*, etc. – with a frequency ranging from 0.1% to 60%. The ability of hypermutability is linked to damage affecting DNA repair and replication fidelity. The hypermutability that will create biodiversity is a considerable selective advantage for the mutator when a prokaryotic population faces a change in the environment: the possibility of possessing a mutant that will resist change – for example, the arrival of an antibiotic – will be considerably higher than in a wild-type strain. Indeed, it has been demonstrated, in vivo and in vitro, that treatment with an antibiotic leads to the selection of mutators whose resistance to the antibiotic is higher than that observed in non-mutators. A large number of mutators are found in hospitals. The mutator will grow faster than wild strain that has a low rate of mutations. But the increase in mutation capacity also involves genes essential for the survival of the bacterium, i.e., the number of lethal mutations will also be higher in the mutator than in the wild-type strain. In other words, the mutator adapts faster, but it is under threat of the appearance of a lethal mutation and is condemned in the long term. What is its future? Gene transfer between the mutator and the wild-type strain (reacquisition of a wild-type allele of an anti-mutator gene from a mutator bacteria via a horizontal gene exchange) or a reversion of the mutator may occur. Another possibility exists: the mutator can transmit the favorable mutation to the wild-type strain. In conclusion, despite the innumerable deleterious mutations that occur in hypermutability, the tiny proportion of favorable mutations confers such an advantage for the mutator that it overcomes the rest of the population.

4.3.2 Horizontal Gene Transfers

Prokaryotes live and evolve in a large variety of biotopes where physical, chemical, and biological conditions change rapidly in time and space. Beyond the modulation of enzyme production and activity, mutations that are frequent in prokaryotes will permit them to cope with these permanent fluctuations. However, adaptation capabilities of prokaryotes cannot be apprehended solely as ability to mutate. Indeed, on top of mutations – which are a source of diversity at the genetic level – the evolutionary success of prokaryotes can be explained also by their extraordinary capacity to exchange genetic information through horizontal transmission (also named lateral transfer). This mode of transmission, with no relation to cell division, is a particularly efficient source of dissemination of genes or alleles carrying an evolutionary advantage. Contrary to vertical transmission of genes (from parent to offspring; from mother to daughter), there is gene transfer from a "donor cell" to a "recipient cell," cells that can have no close parenthood in horizontal transfers (HGTs). Under certain conditions, the genetic material acquired through HGTs may integrate into the genome of the "receptor" cell and become an element of its genetic heritage; it will thereafter be transmitted vertically. Such horizontal exchanges, which were certainly very common when life appeared on Earth, are still today a major mode of rapid adaptation of prokaryotes to environmental changes they face constantly in contemporary ecosystems that are very frequently perturbed by human activities (Gillings 2017). HGTs have played and will continue to play an essential role in the evolution of prokaryotes (Ochman et al. 2000; Gogarten et al. 2002; Pál et al. 2005; Boto 2010; Brochier-Armanet and Moreira 2015; Koonin 2016).

4.3.2.1 Horizontal Transfer Mechanisms

Acquisition of DNA from another organism by horizontal transfer can occur through several mechanisms. The three main ones are transformation, conjugation, and transduction (Lorenz and Wackernagel 1994; Koonin et al. 2001; Redfield 2001) (Frost et al. 2005; Daubin and Szöllősi 2016). Two other modes of transfer must also be mentioned: outer membrane vesicles (Berleman and Auer 2013; Biller et al. 2014) and nanotubes (Dubey and Ben-Yehuda 2011; Pande et al. 2015).

During *transduction* (Fig. 4.3), a bacteriophage (virus capable of infecting bacteria) causes a unidirectional transfer of DNA from one cell to another. During a normal infection cycle, the bacteriophage injects its genetic material into the bacteria and takes over the host cell replicative machinery to make new viral particles, which are liberated into the environment as a consequence of lysis of infected cells (lytic cycle). In the case of transduction, the bacteriophage will transfer DNA from one parasitized bacterium to another one. Two types of transduction exist: generalized transduction and specialized transduction.

In the first case, the phage injects its genomic DNA into the host cell where it is replicated by the parasitized cell replicative machinery, which DNA is degraded and fragmented. Such fragments of the host DNA are occasionally packaged into "phage particles," and, together with normal phages, they are liberated in the medium following lysis of the parasitized cell. Phage particles containing host DNA can then infect new bacterial cells and inject them with host DNA.

In the case of specialized transduction, "temperate" phages can integrate into the bacterial chromosome. Once integrated, the phage DNA will be replicated together with the host chromosome and be transmitted to the two daughter cells.

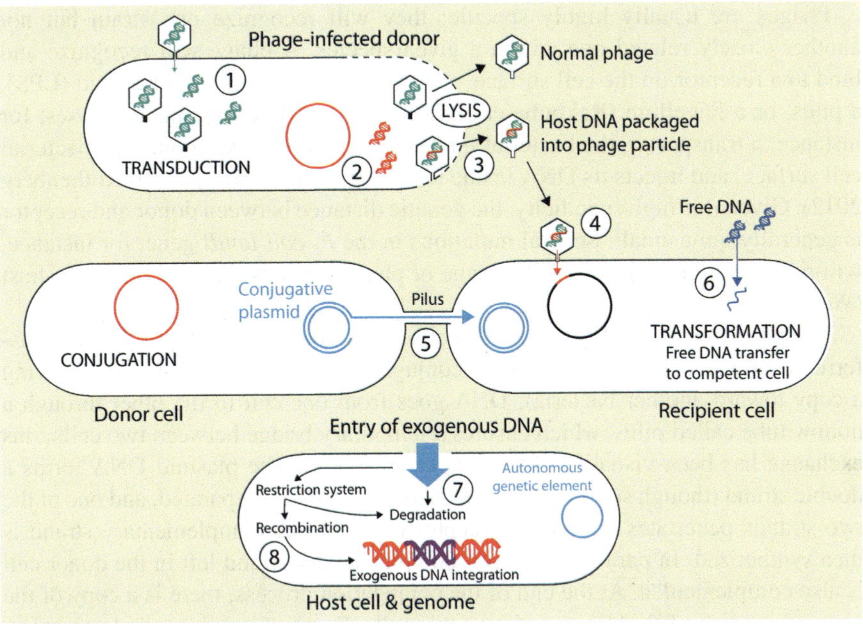

Fig. 4.3 Horizontal transfer mechanisms and the fate of transferred genetic material
DNA transfer by transduction: In bacterial cells infected by the phage, DNA fragments from the host are occasionally packaged into "particles" and following cellular lysis are transferred to new bacterial cells (1, 2, 3, 4)
DNA transfer by conjugation: DNA transfer imposes a direct cell-to-cell contact via a pilus. DNA penetrates into "recipient cells" via the pilus as a single strand (5)
DNA transfer by transformation: Competent cells can take up free DNA present in the environment (6)
Fate of the transferred DNA: The DNA that has been transferred to the host cell is either degraded or fragmented by restriction systems (7). It can integrate into the host genome or be maintained stably in the "recipient cell" as an "autonomous genetic element" (8)

In the case of transformation and conjugation, HGTs can only occur between species sharing the same biotope; on the contrary, transduction may occur between species living in spatially separated biotopes.

Restriction systems have emerged over the course of evolution because of the high risk represented by phages that can cause lysis of the bacterial host. These systems recognize a given sequence of four or more nucleotides and also whether that sequence is methylated or not. In a constant race between phages and their bacterial hosts, phages will mutate to avoid the recognized sequence, while the bacterial host will also mutate to increase the level of restriction enzymes or to change the recognized sequence. Other known defense systems are adsorption inhibition, CRISPR-Cas (clustered regularly interspaced short palindromic repeats-CRISPR-associated proteins) systems, and abortive infection (Samson et al. 2013).

Phages are usually highly specific; they will recognize one strain but not another closely related one within a given species. A phage will recognize and bind to a receptor on the cell surface, such as a protein (OmpA), or a lipid (LPS), a pilus, or a flagellum (Rakhuba et al. 2010). The well-known phage λ uses, for instance, a transporter, the bacterial maltose LamB for docking onto the bacterial cell surface, and injects its DNA inside the cytoplasm (Chatterjee and Rothenberg 2012). Given this high specificity, the genetic distance between donor and receptor is generally quite small. Several mutations in the *E. coli lamB* gene, for instance, will result in modifications of the range of phage mutants that can infect the host (Werts et al. 1994).

During *conjugation* (Fig. 4.3), two cells get in contact, and their DNA is transferred from one cell to the other via a conjugative plasmid (capable of transferring a copy toward another bacteria); DNA goes from one cell to the other through a hollow tube called pilus, which ensures a temporary bridge between two cells; this exchange has been visualized by Babic et al. (2008). The plasmid DNA forms a double strand (though some are linear), the two strands are separated, and one of the two strands penetrates within the receptor cell, and the complementary strand is then synthetized. In parallel, the DNA complementary strand left in the donor cell is also complemented. At the end of the conjugation process, there is a copy of the plasmid in each of the donor and receptor cells. Some plasmids, called episomes, after their integration to the chromosome, can cause the transfer of fragments of the chromosomal DNA. The cells containing a plasmid integrated to the chromosome are called HFR (high frequency of recombination). Some conjugative plasmids can transfer informations between phylogenetically distant organisms (e.g., from Gram-negative bacteria to Gram-positive bacteria). The size of plasmids is very variable (1 kilobase to more than 1 megabase). The main plasmids are those that confer resistance to antibiotics, virulence plasmids, and metabolic plasmids that carry genes coding for enzymes that degrade some aromatic compounds (toluene), pesticides (atrazine), etc. The conjugation process has been described by Lederberg and Tatum (1946) who wrote that "these experiments imply the occurrence of a sexual process in the bacterium *Escherichia coli*." Indeed, conjugation that implies a physical contact during the exchange of DNA has sometimes been considered as a form of sexuality; however, it is not a true one: there are no cell division and no fusion of two gametes.

Free-naked DNA is abundant in the environment, originating essentially from lysed cells from DNA excreted by living cells. This DNA can survive naked in substrates such as soil for several years, being protected by clay micelles (Paget and Simonet 1994). During natural *transformation* (Fig. 4.3), free DNA penetrates within a cell, as a single strand, inasmuch as the receptor cell is competent (presence of specialized proteins on the surface of the cell that make it competent) (Lorenz and Wackernagel 1994; Chen and Dubnau 2004; Thomas and Nielsen 2005; Johnsborg et al. 2007).

The first to demonstrate the process of bacterial transformation was Fred Griffith (Griffith 1928) who was studying virulence in "pneumococcal types." He studied two strains with different phenotypes. One strain was called "smooth" which cells were surrounded with a capsule made up of sugars and which, injected to mice, caused the death of these animals. Another strain called "rough" was not virulent and was devoid of capsule, the determinant of virulence. Griffith injected into these mice a mixture of "smooth" cells killed by a heat treatment of living "rough" cells. Following this injection, he noticed the death of his mice, and the only bacteria recovered in the mice were living "smooth" bacteria. What had occurred? The dead "smooth" bacteria had transmitted to "rough" bacteria a "transforming principle" that had resisted the heat treatment that was necessary to synthesize the capsule. The acquired "smooth" character was transmitted from generation to generation by bacteria that had inherited the "transforming principle."

From Griffith's experiences, two concepts can be drawn that were two important steps in the history of genetics. The first one is that a hereditary character can be exchanged between bacteria (e.g., the capacity to synthetize a capsule) through a transmission route different from the one known at that time (from mother cell to daughter cells): vertical transmission. He had discovered a new mode of transmission of hereditary characters: horizontal transmission. The second concept, even more surprising, was that the ability to synthesize a capsule could be transmitted from dead bacteria to living ones, a transmission from the dead! The discovery of the chemical nature of this "transforming principle" would have to wait for 16 years until Avery, MacLeod, and McCarty (1944) would demonstrate that the "transforming principle" was DNA. Dead bacteria had transmitted DNA – in other words the gene responsible for the synthesis of the capsule – to living bacteria.

Afterward, transformation was demonstrated in cells belonging to the same species but also in prokaryotes belonging to very different taxonomic (*Bacteria* and Archaea) and trophic (photolithotrophs, chemolithotrophs, heterotrophs, methylotrophs) groups. Generally the state of competence is only reached for a transitory period, and the percent of bacterial cells capable of transformation is low: *Vibrio parahaemolyticus*, 1.9×10^{-9}; *Bacillus subtilis*, 3.5×10^{-2}) (Lorenz and Wackernagel 1994). It must be underlined that it is possible to transform noncompetent cells into competent cells through artificial methods (e.g., electroporation).

HGTs can also occur through *outer membrane vesicles* (OMVs), structures that are between ~50 and 250 nm in diameter. OMVs intervene in different processes: virulence, quorum sensing, biofilm formation, redox reaction, cellular defense, and HGTs. This type of exchange, produced by a wide range of taxa, is particularly important in aquatic ecosystems. For example, a culture of *Prochlorococcus*, which is a numerically dominant marine cyanobacterium (population of ~1027 cells), releases continuously lipid vesicles that contain proteins, RNA, and DNA. Such vesicles carry DNA, from diverse bacteria, which have been identified in coastal and open-ocean samples. They could be the main vectors for horizontal gene transfer in marine systems (Biller et al. 2014).

DNA transfer can also occur between neighboring cells via *nanotubes* that permit direct cell-to-cell contact and exchange of cytoplasmic elements such as protein, metabolites, and DNA (*cf.* Chap. 1)

4.3.2.2 The Fate of Transferred Genetic Material

The DNA transferred may not be integrated into the "recipient genome," thus not resulting in a new genetic combination. It can be degraded or fragmented by endonucleases present in the cytoplasm of the receptor cell (restriction system of the host) (Fig. 4.3(7)). If the DNA escapes the host-defense systems, the exogenous DNA integrates into the host genome through recombination (homologous, illegitimate, homology-facilitated illegitimate combination) (Fig. 4.3(8)). There is then formation of a new heritable genome within the "recipient cell." It must be underlined that only a low proportion of HTG-transferred DNA is fixed in a durable manner in populations (Thomas and Nielsen 2005). The transferred genetic material will be maintained in its host only if it represents a benefit for the host.

4.3.2.3 The Prokaryotes Concerned by These Exchanges

Different techniques are implemented to detect and quantify HGTs. Such quantification is hard because the estimates obtained can be very variable from one technique to another; in particular, the different approaches do not identify the same types of HGTs (Brochier-Armanet and Moreira 2015; Daubin and Szöllősi 2016; Chan et al. 2017). Keeping in mind this problem, it is nevertheless possible to conclude that HGTs percentages can be very variable as a function of the species studied and of their function (Nakamura et al. 2004). HGTs are few in some species (0.5% in *Buchnera* sp. APS; 1.9% in *Mycoplasma genitalium* G-37); on the contrary, they are numerous in other species (10.5 in *Halobacterium* sp. NRC-1, 13% in *Synechocystis* sp. PCC6803, 24.1% in *Chlorobium tepidum* TLS, 25.2% in *Methanosarcina acetivorans* C2A) (Nakamura et al. 2004).

Besides, if horizontal transfers occur most frequently between closely related species, they are also possible between evolutionary distant organisms (Gram-negative bacteria/Gram-positive bacteria), comprising organisms belonging to different domains. In this manner, Archaea genomes code for proteins characteristic of *Bacteria* and vice versa. For example, the genome of the hyperthermophilic bacterium *Thermotoga maritima* contains around 8–11% of archaeal hyperthermophilic genes (Zhaxybayeva et al. 2009). The presence of Archaea genes in *T. maritima* would be the result of HTGs between organisms that share a given biotope, hot thermal springs. Thus, through HGTs, in a single step, barriers corresponding to million years of evolution are crossed.

4.3.2.4 Consequences on the Definition of Prokaryotes Genome: Core Genome and Pan Genome

Studies of comparative genomics have shown that the number of genes present in bacterial strains belonging to the same species may be very variable. This situation has led to distinguish in the genome of bacteria, a pan genome that comprises the

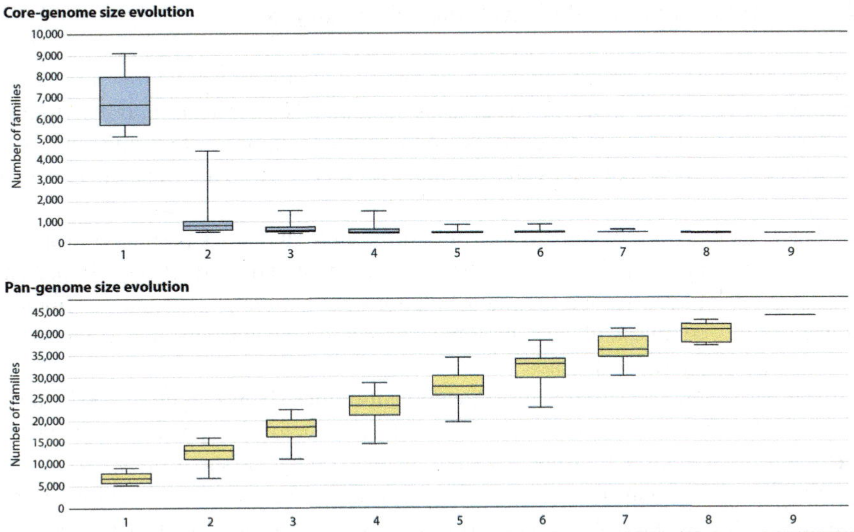

Fig. 4.4 *Frankia* core genome and pan-genome evolution according to the number of sequenced genomes
On the X-axis is the number of *Frankia* genomes analyzed in the data set. In the top "core genome," the first box represents the average number of genes (and the standard error) in the group of nine genomes studied. The second box represents the average number of genes shared (80% AA threshold) between pairs of the nine genomes studied. The third box represents the average number of genes shared (80% AA threshold) between triplets of the nine *Frankia* genomes studied, etc.
In the bottom "pan-genome" graph, the first box represents the average number of genes (and the standard error) present in the group of nine *Frankia* genomes studied. The second box represents the average number of genes (and the standard error) shared between pairs of genomes in the group of nine genomes studied plus those not shared by the pairs. (Normand P., personal communication)

whole of genes present in all individuals of a species as well as a core genome that comprises only those genes present in all strains identified in a given species.

For example, studies realized by Rasko et al. (2008), who have analyzed the genome of 17 strains of *E. coli*, have revealed a core genome of ~2200 genes and a pan genome containing more than 13,000 genes. Such elasticity can correspond to thousands of genes as in *Frankia* where genomes with less than 2% distance in the 16S rRNA genes were shown to have genomes ranging from 5.4 to 9Mb (Normand et al. 2007).

It is generally admitted that genes that constitute the core genome, which have been conserved over the course of evolution, are essential. Those genes outside the core genome should be considered dispensable. They would be necessary only to survive under certain environmental conditions particular to a group of strains.

The number of sequenced genomes is "dramatically" increasing in the last 10 years. In the case where multiple genomes of the same species are available, it is possible to calculate the pan and core genomes; in addition, if the size of the core genome remains constant, the pan genome continues to rise with the addition of more genomes (Fig. 4.4).

4.3.2.5 HTGs and Adaptive Evolution of Prokaryotes

Genes that are transferred horizontally generally code for metabolic functions other than the fundamental molecular processes such as DNA replication, transcription, and translation. There are nevertheless numerous exceptions to that rule, and, precisely, certain ribosomal genes show evident traces of horizontal transfer (Brochier et al. 2000).

The dissemination of functions acquired through HGTs has permitted prokaryotes to adapt rapidly to changes of their environment and represents an important selective advantage. Among the main functions are antibiotic resistance, virulence transmission, conquest of new biotopes (or ecological niches), and capacity to catabolize synthetic compounds.

The best known function is resistance to antibiotics (Dobrindt et al. 2004; Juhas et al. 2009). The mode of transfer varies as a function of the species (conjugation, natural transformation, transduction), but its main characteristic is the speed of dissemination of resistant mutants (*cf.* Chap. 6). Efficiency of dissemination can be explained in part only by the clustering of resistance genes. As a matter of fact, those genes acquired through HGTs are not distributed homogeneously over the length of the genome, but are clustered on plasmids and particular chromosomal regions called genomic islands (GIs) (Langille et al. 2010). Thus, through GIs, acquisition of all resistance genes can occur as a single step, and the "receiving" strain will become multiresistant. Besides, as they move from one host to another, GIs can acquire new resistance genes, and thus, bacteria that are resistant to antibiotics will become more and more efficient. Emergence of resistance to antibiotics by pathogenic bacteria has become a major public health challenge. A real "arms race" has begun between scientists attempting to elaborate new antibiotics and bacteria that develop means of resistance to these new molecules.

Another genetic transfer concerns the transfer of virulence functions by pathogenic bacteria (adhesins, toxins, invasins, protein secretion systems, iron uptake systems, modulins, etc.). The genes transferred that are responsible for virulence are localized on large regions of chromosomal and plasmidic DNA (pathogenicity islands) (Hacker and Kaper 2000). For example, the uropathogenic *E. coli* (UPEC) that is the main cause of "urinary tract infection" (UTI) represents around 40% of nosocomial UTI; UPEC virulence factors are frequently encoded on pathogenicity islands (Brzuszkiewicz et al. 2006). Usually, in closely related nonpathogenic neighbors belonging to the same genus or the same species, these genes are absent. In *Agrobacterium*, the agent of crown gall in a number of plants, the virulence genes are also plasmid-borne, but the plasmid appears to be lost frequently, in particular as a result of heat as occurs in tropical soils (Krimi et al. 2002); this plasmid that modifies the fitness of the bacterial host can later be reacquired through conjugation.

HGTs also play a role in the conquest of new biotopes through the acquisition of new metabolic capacities. For example, it has been demonstrated that a strain belonging to genus *Thermus* that grew only under strict aerobiosis had acquired the capacity to grow under anaerobiosis using nitrate as electron acceptor (acquisition of a respiratory nitrate reductase gene cluster). The capacity to respire nitrate can be transmitted through conjugation, thus perming aerobic strains to colonize anoxic biotopes containing nitrate (Ramírez-Arcos et al. 1998).

Some prokaryotes have acquired the capacity to resist, or even to use as nutrient source, man-made toxic compounds. Many examples can be cited. Bacteria have acquired gene cassettes capable to confer precipitation of heavy metals and of radionuclides that through HGTs can be disseminated within contaminated biotopes (Moënne-Loccoz et al. 2015). HGTs are also involved in degradation of lindane (Berger et al. 2015), atrazine (Soulas and Martin-Laurent 2015), etc. that represent a significant source of carbon in soil.

4.3.2.6 HGTs and the "Universal Tree of Life"

After millions of year of evolution, the genomes of prokaryotes have become true "mosaics." The extent of HGTs and their importance in the evolution of prokaryotes are still a hotly debated question: their precise identification and quantification remain always difficult, since besides HGTs, other modifications of genomes through vertical evolution occur; some authors thus tend to minimize the role of HGTs in genome phylogeny in modern cells versus primitive genomes (Kurland et al. 2003).

However, important in certain species and negligible in others – HGTs complicate the definition of "species" in prokaryotes (*cf.* Chap. 2) and also the representation of relations between species in phylogenetic trees: in such trees, if branches split, others through HGTs fuse and sometimes form new branches. Darwin himself, before adopting the vision of a "universal tree of life," had proposed a representation of evolution of species in the form of a "coral of life": "The tree of life should perhaps be called the coral of life, [its] base of branches dead; so that passages cannot be seen" *(Life and Letters of Charles Darwin – Volume 1, pg 368).* Presently, the concept of a "universal tree of life" is yielding to that of a reticulated tree, a network forming a web- or net-like pattern (Doolittle 1999; Kurland et al. 2003; Olendzenski and Gogarten 2009; Koonin and Wolf 2012). The intensification of the genomes sequencing effort should permit to build a reliable and detailed history of genomes. The concept of phylogenetic trees should not be abandoned, but one must remain aware that the evolutive history of genes can be very different from the lineal descent of the cells that carry them, "prokaryotic evolution and the tree of life are two different things" (Bapteste et al. 2009).

4.4 Advantages to Have a Small Size

4.4.1 Dissemination

Due to their small size, the spread of prokaryotes is planetary, in the four interconnected spheres of the Earth (atmosphere, hydrosphere, lithosphere, (deep)biosphere). Indeed, prokaryotes are continuously transported over long distances by winds, currents (due to the constant circulation and mixing of the oceans), the living beings (due to the migrations over thousands of kilometers for some species, including man), and the clouds. Due to this dispersion, prokaryotes are present on the whole planet Earth, in particular in some biotopes inaccessible to plants and animals: deep biosphere and clouds.

(i) *The deep biosphere* includes a variety of subsurface habitants in continental realm (mines, deep aquifer systems) and marine realm (sediments and igneous rocks) (Schrenk et al. 2010; Biddle et al. 2012; Colwell and D'Hondt 2013); deep biosphere extends down to around 2.5 km below the ocean floor (Inagaki et al. 2015). Migration of prokaryotes in deep biosphere can take place in different ways: capture in sediments during their formation, migration from the surface by natural geological process or drilling operations, and percolation from the surface through fissures in the rock. It should be noted that a transfer of microbial communities from deep subseafloor to overlying water should be considered (Hoshino et al. 2017). The biomass of deep biosphere biomass is estimated according to the authors from 3.55×10^{30} cells (Whitman et al. 1998) to 2.9×10^{29} cells (Kallmeyer et al. 2012); whatever the estimate proposed, the deep biosphere biomass is one of the largest biospheres on Earth.
(ii) *The atmosphere*. The prokaryotes (fungi and yeasts too) play an important role in the formation of clouds. They are present in the atmospheric water phase (fog and clouds), where they are capable to survive and develop in spite of harsh conditions encountered in the atmosphere (desiccation, low temperature, solar radiations, presence of oxidizing agents, oligotrophic conditions, human pathogens, etc.) (Morris et al. 2014; Pöschl and Shiraiwa 2015; Fröhlich-Nowoisky et al. 2016).

Aerosolized from virtually all surfaces (soil and water surface, vegetation), prokaryotes (and others microorganisms) are transported in altitude, where residence time in atmosphere can range from several days to weeks, long enough for cells to travel between continents (Burrows et al. 2009). The prokaryotes will then fall back to the surface of the Earth, via the rain or the snow, where they will be able to regrow. The total bacterial concentration in atmospheric waters varies from 10^3 to 10^5 cell ml^{-1} (Delort et al. 2010). The diversity of bacterial population found in clouds is great (Amato et al. 2007b; Bottos et al. 2014). For instance, from samples collected at the Puy de Dôme summit, 28 genera have been described and were found to belong to *Actinobacteria*, *Firmicutes*, *Proteobacteria* (*Alpha*, *Beta*, and *Gamma* subclasses), and *Bacteroidetes* phyla and mainly to the genera *Pseudomonas*, *Sphingomonas*, *Staphylococcus*, *Streptomyces*, and *Arthrobacter* (Amato et al. 2007b); total bacterial count were about 1×10^5 cells ml^{-1} of cloud water, and less than 1% were cultivable (Amato et al. 2005). However, the measures of the ATP concentration in cloud samples support the conclusion that a large majority of cells are likely viable but noncultivable (VBNC) and remain alive in clouds (Amato et al. 2007c). Isolated bacteria were able to degrade various organic substrates such as formate, acetate, lactate, methanol, formaldehyde, as well as H_2O_2, a precursor to oxidant species in clouds (Amato et al. 2005, 2007a; Vaïtilingom et al. 2013).

4.4.2 Surface-Volume Ratio

A small size has another advantage. The exchanges with the outside environment – the rate at which nutrient passes into and out of the cell by passive diffusion, facilitated diffusion, or active transport – are easier in a small cell than in a large cell. So, for a given amount of resource to support growth, smaller size increases the cellular metabolic rate and growth rate. Consequently the population of small cells is higher than of large cells, which in turn can affect the evolution. Indeed, a large population will have a high rate of mutations ("the raw material of evolution") which will allow a more rapid adaptation to changes in environmental conditions and better exploitation of resources. How to explain the advantage of cells with a small size? This advantage is mainly due to the fact that these cells have a surface area greater than that of large cells; they have a higher surface-to-volume ratio (Young 2006; Moya et al. 2009). For example, a coccus with a radius equal to 1 μm, its S/V ratio will be 3 (surface area ($4\pi r^2$) of 12.6 m^2/volume ($4/3\pi r^3$) of 4.2 μm^3). For a coccus with a radius equal to 2 μm, S/V ratio will be equal to 1.5 (surface area = 50.3μm^2/volume = 33.5 μm^3) (Madigan et al. 2015). For some cells with different sizes such as *Pelagibacter ubique* (S/V: 0.31/0.014), *Escherichia coli* (S/V: 6.28/1.3), and *Epulopiscium fishelsoni* (S/V: 151.000/3.10^6), S/V ratio ($\mu m^2/\mu m^3$) is equal to 22, 4.8, and 0.05, respectively (Young 2006). *P. ubique*, one of the most successful and numerous life forms on the planet, which has a very high S/V ratio (S/V: 22), fits the model in which natural selection optimizes the surface-to-volume ratio to provide appropriate transport rates in oligotrophic environment (low-nutrient conditions) (Grote et al. 2012). At the opposite end of the spectrum is the giant bacteria, *Epulopiscium fishelsoni* (S/V: 0.05), a symbiont that spends the entire of its life in the intestinal tract of a tropical marine fish, *Acanthurus nigrofuscus* (Angert et al. 1993). *E. fishelsoni* overcomes the disadvantage of a very low S/V ratio in three ways: *(i)* life in nutrient-rich environment; *(ii)* the inner membrane contains many invaginations which increases transport across the membrane; and *(iii)* the bacteria use extreme polyploidy and contain multiple copies of its genome; each cell has between 50,000 and 120,000 copies of the chromosome (Mendell et al. 2008; El-Hajj and Newman 2015).

In environment (seawater, freshwater), the consumption of prokaryotes by bacterivores varies according to the predator and the prey; however, longer and shorter cells escape more frequently when grazed by protists. So, Pernthaler et al. (1996) proposed a model that divides freshwater bacterioplankton into four size ranges of different vulnerability to size-selective protistan grazing: small cells (<0.4 μm) weakly affected by protist grazing, "grazing-vulnerable" bacteria (0.4 and 1.6 μm), "grazing suppressed" (1.6 and 2.4 μm), and "grazing-resistant" (>2.4 μm) fractions of the bacterioplankton (Young 2006; Pernthaler et al. 1996).

4.5 Through Evolution, Prokaryotes Have Developed an Array of Mechanisms to Ensure Their Survival Under Adverse Conditions

4.5.1 Life of Prokaryotes Under Dormancy and Starvation Conditions

Sergei Winogradsky (1856–1953) drew attention to the fact that in soils *la grande majorité des germes, à un moment donné, est à l'état de vie latente, une minorité seulement à l'état actif* (Winogradsky 1949). Numerous works have confirmed these observations. Indeed, when environmental conditions become unfavorable for growth, that are in *état de vie latente* are unable to grow, but are not dead can regrow with the appearance of favorable conditions. This process is known as "dormancy," a strategy to cope with environmental stress, which is defined operationally as "reversible state of metabolic shutdown" (Stevenson 1978; Roszak and Colwell 1987; Kaprelyants et al. 1993; Jones and Lennon 2010; Lennon and Jones 2011; Wang et al. 2014a; Aanderud et al. 2015). Dormancy, "a reversible state at low metabolic activity, in which cells can persist for extended periods without division", refers to two kinds of cells: cells forming specialized structures (spores, cysts) that formed the most durable dormancy present in a limited number of bacterial species (*cf.* part 4.5.4) and non-spore-forming cells which are vegetative cells which enter in a dormant state (Kaprelyants et al. 1996). Dormancy includes three steps: *(i)* initiation, in response to environmental unfavorable changes, *(ii)* microorganism at rest where dormant cell exhibits a wide range of phenotypes (spore-forming and non-spore-forming bacteria), and *(iii)* "resuscitation" from dormancy (revival of dormant cells and spores) (Lennon and Jones 2011). A portion of prokaryotic population in nature is not active, and in some soils, the active biomass may be less than 10%. For example, in soil from Argentinean Pampa, Alvarez et al. (1998) compared the active biomass (kg ha^{-1}) versus the total biomass (kg ha^{-1}) as a function of soil management: plow tillage (total biomass, 295; active biomass, 28.4), no-tillage (total biomass, 414; active biomass, 35.3), and pasture (total biomass, 1114; active biomass, 4.19). In the same way, in soils collected near Uppsala (Sweden), growing microorganisms fraction is generally small (5–20%) in soils with no recent addition of substrates (Stenström et al. 2001). In deep biosphere – proposed as the largest reservoir of biomass on planet Earth (Whitman et al. 1998) – the hypothesis proposed by Jørgensen (2011) postulates that only 1% of the cells are active and that the other cells can persist for a long time without transforming into vegetative cells.

The main reason that leads the cells to dormancy is the energy deficiency where cells enter in a physiological state known as starvation-survival as defined by Morita (1990) "physiological state resulting from the amount of nutrients available for growth and reproduction being insufficient." Microbiologists, the most of time, investigate bacterial growth under optimal conditions (carbon and energy sources, optimal pH, temperature, Eh, salinity, aeration conditions, etc.), using axenic culture that are absolutely impossible in nature (Winogradsky 1949). The cells, thus subtract from their natural biotic and abiotic environment, are placed under artificial conditions, often radically different from the natural habitat of the organisms. But, in natural environments, prokaryotes are frequently subject to alternation of feast (in rare instance) and famine (starvation stage) like in open oceans that are continuously oligotrophs with only brief periods of nutrient inputs. Indeed, the amount of dissolved organic matter in the sea is usually less than 1 mg of C/liter for surface water and 0.5 mg of C/liter for deep water (Novitsky and Morita 1977); on the other hand, standard laboratory media contain nutrients at concentrations that are very high, e.g., > 2000 mg of C/liter (Roszak and Colwell 1987). A feast and famine diet was also proposed by Jørgensen and Boetius (2007) in deep-sea bed, where bacteria live under nutrient-limiting conditions and extreme energy limitation (Hoelher and Jørgensen 2013). The period of starvation can last for long period of time. For example, Novitsky and Morita (1977) studied the survival of a psychrophilic marine *Vibrio* (Ant-300) under long-term nutrient scarce and observed that during starvation, 50% of the starved cells remained viable for 5–7 weeks and a portion of the population (10^3 cells/ml) remained viable for more than 1 year (viability was determined by the spread-plate technique). Another example of long-term starvation is this of marine bacterium *Alteromonas denitrificans* which has survived for up to 7 years in unsupplemented synthetic seawater (Nissen 1987). The main starvation responses of bacteria include formation of periplasmic spaces, decrease of size, change in shape (from a rod to a coccus), and a sharp decrease of the endogenous respiration; the endogenous respiration of the starving cells Ant-300, a psychrophilic marine vibrio, during the first days of starvation, decreased over 80%, and, after 7 days, respiration has been reduced to 0.0071% (Novitsky and Morita 1977). During starvation a part of the population survives, and the other part dies. The recycling of dead cells might play a key role in the maintenance of long-term survival (Takano et al. 2017). The survival under starvation conditions has been well documented in many different bacteria: Ant 300 *Alteromonas denitrificans*, *Escherichia coli*, *Vibrio cholerae*, *Salmonella typhimurium*, *Listeria monocytogenes*, *Staphylococcus aureus*, *Mycobacterium tuberculosis*, *Erwinia amylovora*, and *Azospirillum brasilense*. For more informations concerning the cellular (morphology, size, motility), physiological, biochemical, and genetic changes accompanying starvation, refer to previous works (Koch 1971; Kjelleberg et al. 1993; Mukamolova et al. 2003; Peterson et al. 2005; Gonzàlez-Escalona et al. 2006; Lerner et al. 2010; Santander et al. 2014).

4.5.2 Ultramicrobacteria: Prokaryotes Adapted to the Oligotrophic Way of Life

At (very) low substrate concentrations, many prokaryotes are, as described above, able to survive in such conditions by becoming inactive, and during this "rest period," the cells are nonculturable. In contrast, other prokaryotes have a lifestyle adapted to nutrient-depleted environments and are capable to grow at (very)low substrate concentrations. This is the case of different marine bacteria, called "ultramicrobacteria," which are able to grow in natural seawater which is characterized by extremely low nutrient concentrations (oligotrophic conditions), compared with media commonly used for studies of bacterial growth (Jannasch 1967). For these oligotrophic bacteria, starvation is the normal state unlike copiotroph bacteria which grow only at high concentration of nutriments and are inactive in natural seawater where they are able to survive. In 1981, Torella and Morita described two types of free-living heterotrophic bacteria in freshly collected marine water samples, which can passed through a 0.3 µm-pore-size filter used for the entrapment of bacteria. The first one, of small size and in a state of low metabolic activity, displays, upon inoculation onto a nutrient-rich agar, an increase in their size and in their growth rate. The second type of cells, designed by Torella and Morita (1981) as "ultramicrobacteria" (less than 0.1 μm^3 in volume), formed very small microcolonies and displays a slow rate; these cells do not increase in size even in the presence of high concentration of organic matter (nutrient medium). Ultramicrobacteria maintained similar size and volume regardless of their growth conditions; these bacteria were adapted to the low-nutrient conditions of the seawater. So there is a fundamental difference between ultramicrobacteria (UMB) and starvation forms: UMB retain their small size even when exposed to media with high nutrient concentration contrary to starvation forms that regain their full vegetative size and metabolic activity when flooded with nutrients. Ultramicrobacteria that can pass through a 0.2 µm filter (referred to as 0.2 µm filterable bacteria) (Hood and Macdonell 1987) belong to *femto*bacterioplankton (size: 0.02–0.2) according to the size classification of planktonic microorganisms proposed by Sieburth et al. (1978). Even if it is difficult to isolate ultramicrobacteria, different strains have been isolated and cultured: *Sphingomonas* sp. Strain RB2256 (Fegatella and Cavicchioli 2000), *Candidatus* Pelagibacter ubique (in SAR11 clade) (Rappe et al. 2002), *Polynucleobacter necessarius* (Hahn 2003), *Herminiimonas glaciei* sp. nov. (Loveland-Curtze et al. 2009), etc. Possessing a small size provides many benefits: protection against predation by bacterivorous nanoflagellates and a surface/volume reduction that optimizes exchanges with nutrient-poor environment.

4.5.3 *Viable But Nonculturable State*

4.5.3.1 Definition and Characterization

To withstand to adverse environmental conditions, in addition to spore formation, many bacteria possess the ability to enter in a state qualified of viable but noncultivable (VBNC) (Roszak et al. 1984; Roszak and Colwell 1987; Colwell 2000; Oliver 2010; Oliver and Mena 2010; Fernández-Delgado et al. 2015; Pinto et al. 2015; Ding et al. 2017; Zhao et al. 2017b). VBNC state has been described for the first time by Xu et al. (1982), from Rita Colwell's laboratory, in *E. coli* and *Vibrio cholerae*; these two bacteria incubated in sterile water at low temperature without nutriment supplements were able to persist for several days in such conditions but were unable to form colonies on agar plate, thus demonstrating that "a significant proportion of the nonculturable cells were, indeed, viable." The term VBNC was coined by Roszak and Colwell (1987) "for those bacterial cells with detectable metabolic function, but not culturable by available methods." Indeed, VBNC bacteria escape detection by standard methods for testing samples for viable bacteria (viability is equated with cultivability a single cell yielding a visible colony on the surface of nutrient agar plate). However, nonculturable on routine culture media on which they would normally grow does not mean dead, and VBNC cells demonstrate many general characteristics as a kind of viable cells. VBNC cells sustain certain functions like uptake and incorporation of amino acids into proteins, active metabolism and respiration, ATP levels and membrane potential remain high (membrane integrity), gene transcription with specific mRNA production, antibiotic resistance, plasmids are retained, etc. (Lleò et al. 2000; Yamamoto 2000; Oliver 2010; Li et al. 2014; Pinto et al. 2015).

Some authors proposed an alternative to VBNC terminology: "active but nonculturable (ABNC) state", e.g., active cells but nonculturable (ABNC) cells "exhibit measurable activity but fail to grow to a detectable level" (Kell et al. 1998), and non-growing but metabolically active (NGMA) state (Manina and McKinney 2013).

VBNC cells present some changes from culturable cells besides the inability to growth: change in shape, cell wall, and membrane composition, including proteins, fatty acids and peptidoglycan, gene expression, etc. (Pinto et al. 2013). For example, a change in the fatty acid composition of the membrane is observed in *Vibrio vulnificus* into VBNC state, suggesting a change in membrane fluidity (Day and Oliver 2004) and presence of a specific protein profile of VBNC in *Enterococcus faecalis* (Heim et al. 2002). With respect to morphology, most of the time, VBNC cells are reduced in size and changed their morphology. *Vibrio cholerae* under VBNC state change from rod to coccoid cells (Chaiyanan et al. 2007). VBNC cells can stay alive over long periods of time, even under continued stress, and many species have the ability to "resuscitate" (Cf. Sect. 4.5.3.5).

Finally, note that since the definition of the VBNC state is proposed by Xu, the interpretation of VBNC state has been controversial within the scientific community. For example, Bogosian et al. (1998) think that the return to the cultivable state is due to the presence of the few culturable cells that remain in sample and "that the nonculturable cells were dead and that the apparent resuscitation was merely due to the growth of the remaining culturable cells." Other authors propose that in adverse conditions, some cells excrete organic molecules into the surrounding medium which are used by the other members of the populations, thus ensuring cell survival until better time comes (Arana et al. 2004) (Cuny et al. 2005). Subsequently, Manina and McKinney (2013) proposed that the "death of a fraction of cells releases molecules that can be scavenged by the survivors" and the death of "altruistic" members of the population release molecules that can be used by "survivors."

Little is known about understanding the molecular control of VBNC state, and the mechanism by which bacteria adjust in the VBNC state remains unclear. Nevertheless, many proteins seem implicated into the VBNC state induction.

In *E. coli*, EnvZ/OmpR system regulates the synthesis of the major outer membrane proteins OmpF and OmpC (Li et al. 2014). EnvZ has no effect on survival (determinated by plate count), but is involved in the entry into VBNC state: indeed, *envZ* mutants were found not to enter in VBNC state (Darcan et al. 2009). Glutathione S-transferase (GST) (Abe et al. 2007) and catalase KatG (Oliver 2010) are involved in the VBNC state in *Vibrio vunificus*.

Regulators factors RpoS (Boaretti et al. 2003; Liu et al. 2010a; Kusumoto et al. 2012) and OxyR (Cuny et al. 2005; Li et al. 2014) seem to be important for the induction of VBNC state.

4.5.3.2 Detection

Various methods to bypass agar culture have been developed (Kell et al. 1998; Khan et al. 2010; Oliver 2010; Manina and McKinney 2013; Davis 2014; Ramamurthy et al. 2014; Fernández-Delgado et al. 2015; Ayrapetyan and Oliver 2016; Léonard et al. 2016; Ding et al. 2017; Zhao et al. 2017b). The most commonly used methods are as follows:

- Direct viable count (DVC) was first described by Kogure et al. (1979). The Kogure procedure is based on acridine orange staining after elongation of cells in the presence of DNA-gyrase inhibitors such as nalidixic acid. Many bacterial pathogens became resistant to nalidixic acid necessitating modification of Kogure's method. For example, Joux and Lebaron (1997) used an antibiotic cocktail instead of nalidixic acid alone, including 4 quinolones (nalidixic, piromidic, and pipemidic acids and ciprofloxacin) and one ß-lactam (cephalexin); 100 marine strains isolated from 2 coastal areas and natural marine communities were screened for their sensitivities to these antibiotic cocktail. The combination of antibiotics resulted in higher viable counts for all water samples revealing the existence of a large and unexpected number of viable cells in coastal marine areas.
- Double staining with 5-cyano-2,3-ditolyl tetrazolium chloride (CTC) and 4′,6-diamino-2-phenylindole (DAPI), this method makes it possible to count and differentiate between dead cells and living cells. The DAPI allows the detection of all the cells, the CTC, and the enumeration only of the living cells.
- LIVE/DEAD BacLight Bacterial Viability Kit (Molecular Probes) is based on double two nucleic acid stains: membrane-permeable green-fluorescent SYTO® 9 (stains both live and dead bacteria) and membrane-impermeable red-fluorescent propidium iodide that penetrates only the bacteria with damaged membranes and reduces the SYTO® 9 fluorescence when both the stains are used. Other kits are available (e.g., RedoxSensor™, Green Vitality Kit).
- BacLight Kit (molecular probes), which combines two nucleic acid stains, membrane-permeable SYTO9 (green) and membrane-impermeable propidium iodide (red), to identify "live" cells (which stain green) versus "dead" cells (which stain red).

- Multiparameter (or polychromatic) flow cytometry (MP-FCM) is a powerful tool for rapidly analyzing cell populations on a cell-by-cell basis and provides the opportunity to obtain information in real time. Light scatter and fluorescence properties are recorded simultaneously as cells cross one by one one or several laser beams. Indeed, a very large number of cells can be measured rapidly, typically up to several thousands of cells per second. Analysis of cells by MP-FCM after staining with artificial fluorescent dyes (such as DiOC6(3) as illustrated in Fig. 4.5) provides information on the cell characteristics (size, shape, granularity) and damage addressed at the single-cell level, especially for quantitative analysis of metabolism and physiological state of VBNC cells by measuring multiple cellular parameters on each cell simultaneously: membrane integrity, membrane potential, metabolic activity (respiratory activity and intracellular enzymatic activity such as dehydrogenase, esterase), and detection of damaged DNA. It assumed that cells having intact membrane are alive and those with damaged membrane potential are dead or theoretically dead. Figure 4.5 shows an example of the investigation of the membrane potential of a culture of bacteria, thanks to the carbocyanine DiOC6(3).
- Molecular Detection-Based Methods

Molecular biology techniques can also be used to determine VBN bacteria, and molecular methods for more effective detection and quantification of these cells have increased. The most used methods are the following:

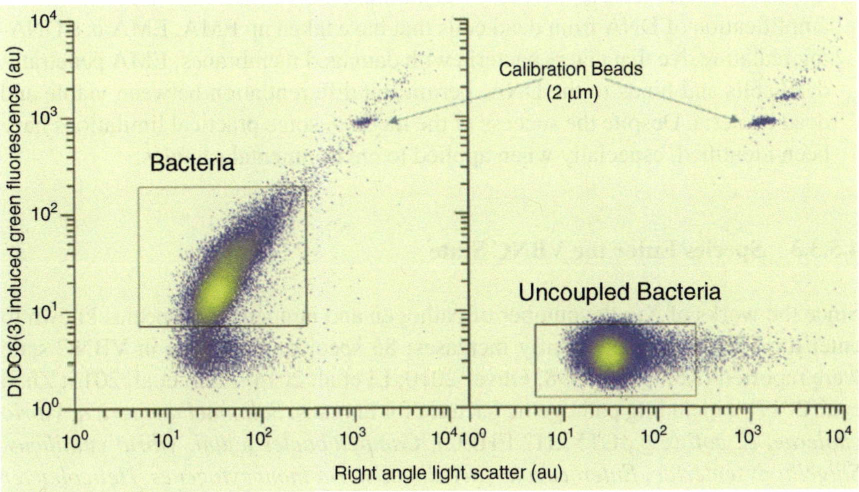

Fig. 4.5 Detection by flow cytometry
On the left, a culture of *Pseudomonas nautica* stained with carbocyanine DiOC6(3). On the right, the same culture with the membrane potential of the bacteria abolished by the uncoupler CCCP and stained with DiOC6(3). (Gregori G., personal communication)

Real-time polymerase reaction (RT-PCR) and quantitative PCR (RT-qPCR) based on detection of mRNA which have been proposed as markers for cell viability because they are very unstable molecules with very short half time inside the cells, there are only present in viable cells. Yaron and Matthews (2002) carry out gene amplification and have demonstrated that RT-PCR can serve as a detection method for *E. coli* O157:H7 cells. Using PCR, *rfbE*, *fliC*, *stx1*, *stx2*, *mobA*, *eaeA*, *hlyC*, and *A* 16S rRNA were amplified, whereas RT-PCR amplified only the 16S rRNA, *rfbE*, *stx1*, and *mobA*. Their result suggests that 16S rRNA, *rfbE*, *stx1*, and *mobA* are good targets for the detection of the presence of viable *E. coli* O157:H7 in sample containing nonculturable cells, and *rfbE* gene is the most appropriate target for detection of *E. coli* O157:H7. RT-PCR method is very specific: the targeted genes can only be used for the detection of *E. coli* O157:H7. Furthermore, 10^6 cfu were required for the detection of *rfbE*, and 10^7 cfu were needed for the detection of *fliC* and *stx1* transcripts. Using the RT-PCR microelectronic array technique, Liu et al. (2008) increased the sensitivity of the method: *rfbE* and *fliC* genes were detected with 50 VBNC cells in liter of river.

The detection of live cells can also be entreprised using viability dyes in combination with DNA amplification; the technology is based on sample treatment with the photoactivatable, and cell membrane impairment, nucleic acid-intercalating dyes ethidium monoazide (EMA) or propidium monoazide (PMA) followed by light exposure prior to extraction of DNA and amplification (Fittipaldi et al. 2012). For instance, Nogva et al. (2003) and Rudi et al. (2005) have developed a concept for quantification of viable and dead cells. The viable/dead stain ethidium monoazide (EMA) is used in combination with real-time PCR to inhibit amplification of DNA from dead cells that have taken up EMA; EMA is a DNA-intercalating dye that enters bacteria with damaged membranes; EMA penetrates dead cells and binds to the DNA, permitting differentiation between viable and dead bacteria. Despite the success of the method, some practical limitations have been identified, especially when applied to environmental samples.

4.5.3.3 Species Enter the VBNC State

Since the works of Xu, the number of pathogen and nonpathogen species known to enter the VBNC state constantly increases; 85 species of bacteria in VBNC state were reported (Kell et al. 1998; Oliver 2010; Li et al. 2014; Pinto et al. 2015; Zhao et al. 2017b), including pathogenic bacteria for humans: *Salmonella enterica*, *Vibrio cholerae*, *E. coli* (e.g., O157:H7 EHEC), *Campylobacter jejuni*, *Vibrio vulnificus*, *Shigella dysenteriae*, *Enterococcus faecalis*, *Listeria monocytogenes*, *Helicobacter pylori*, *Pseudomonas aeruginosa*, etc.

4.5.3.4 The Stressful Conditions Triggering the VBNC State

Many stressful conditions, prevalent in natural environment constantly fluctuating, can trigger transition to VBNC state (Li et al. 2014; Fernández-Delgado et al. 2015; Wu et al. 2016; Ding et al. 2017; Zhao et al. 2017b). The most stressful conditions are as follows: low nutriment concentrations, temperatures outside those that are permissive to cell growth, elevated or lowered osmotic concentrations, elevated salinity, extreme pH, oxygen stress, solar radiation, sulfur dioxide, low redox potential, heavy metal, presence of food preservatives, etc.

4.5.3.5 Return to Cultivable State: "Resuscitation"

Cells in VBNC state can recover their "culturability," in a process termed "resuscitation," by a restoration of more favorable conditions for bacterial growth (e.g., temperature upshift, rich medium). The term was first presented by Roszak et al. (1984). The definition proposed by Kell et al. (1998) will be retained, a term used "to denote transition of cells from 'nonculturable' to culturable states with respect to a given medium." The ability of a bacterium to "resuscitate" can be considered as a truly survival strategy. For instance, *Vibrio vulnificus* has an optimal growth temperature at 37 °C. In an artificial seawater microcosm incubated at 5 °C, *V. vulnificus* responds to this temperature downshift by entering to a viable but nonculturable state (Fig 4.6). The bacteria recovered culturability after a temperature increase (22 °C);

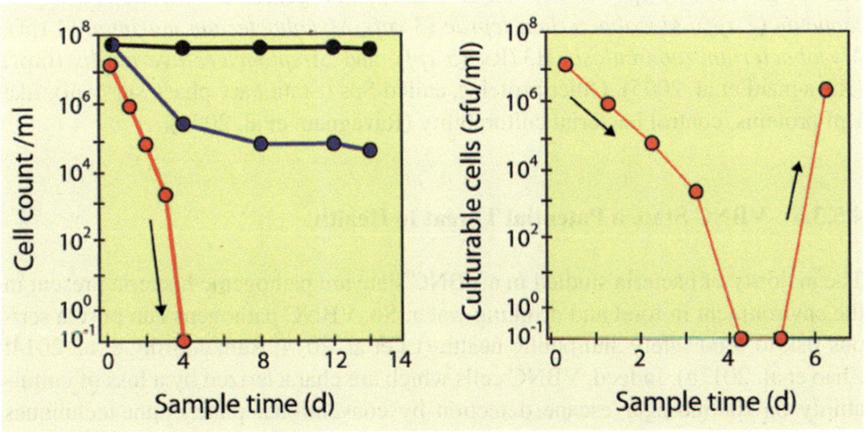

Fig. 4.6 Loss and recovery of culturability of *Vibrio vunificus*
The bacteria entered into the viable but nonculturable state in an artificial seawater microcosm incubated at 5 °C (downward arrow) and recovered culturability after a temperature increase (upward arrow). Total cell counts (acridine orange staining method), direct viable counts (method of Kogure et al. 1979), and plate counts on HI (heart infusion broth) agar in CFU per millimeter are represented by black, blue, and red circles, respectively. (Modified and redrawn from Whitesides and Oliver 1997)

the temperature upshift would release the bacteria from low-temperature stress, promoting "resuscitation" to the original cell state (Nilsson et al. 1991; Whitesides and Oliver 1997). Amel et al. (2008) claimed that VBNC cells of *Vibrio fluvialis* were able to "resuscitate" to the culturable state up to 6 years of incubation in marine sediment. Many bacteria require specific conditions to transit from a VBNC state to cultivable state (Zhao et al. 2017b). *Micrococcus luteus* cells are able to "resuscitate" from a state of nonculturability dormancy in the presence of supernatants taken from the stationary phase of batch cultures of the organism (Kaprelyants et al. 1993). *Vibrio cholerae* O1 cells in the VBNC state can be resuscitated by introduction in the intestines of human volunteers; the demonstration of a "resuscitation" in the gut supports the proposition that viable but nonculturable bacterial enteropathogens may pose a potential threat to health (Colwell et al. 1996). *Legionella pneumophila* Philadelphia JR32 in VBNC state "resuscitated" and regained pathogenic potential during intracellular residence within *Acanthamoeba castellanii* (Steinert et al. 1997), and the data of García et al. (2007) showed that *Legionella pneumophila* that became nonculturable after chlorine treatment "resuscitated" in co-culture with *Acanthamoeba polyphaga*. All the same, conversion of VBNC *V. cholerae* O139 and *V. cholerae* O1 cells to the culturable state by co-culture with eukaryotic cells (CHO, Caco-2, T84, HeLa, and intestine 407 cell lines) was reported by Senoh et al. (2010). The work undertaken on the *Micrococcus luteus* supernatant made it possible to isolate and characterize a protein (*a muralytic enzyme*), named Rpf (resuscitation-promoting factor) which is capable of very low concentration (activity at picomolar concentrations), to "resuscitate" nonculturable forms cells of *M. luteus* (Mukamolova et al. 2003; Li et al. 2014). If *M. luteus* seem to contain only one *rpf* gene, other species contain several *rpf*-like genes: *Corynebacterium glutamicum* (2 *rpf*), *Mycobacterium leprae* (3 *rpf*), *Mycobacterium marinum* (4 rpf), *Mycobacterium tuberculosis* H37Rv (5 *rpf*), and *Streptomyces avermitilis* (6rpf) (Ravagnani et al. 2005). Other proteins, called Sps (Stationary phase survival), like Rpf proteins, control bacterial culturability (Ravagnani et al. 2005).

4.5.3.6 VBNC State a Potential Threat to Health

The majority of bacteria studied in a VBNC state are pathogenic bacteria present in the environment in food and drinking water. So, VBNC pathogens can pose a serious risk to food safety and public health (Li et al. 2014; Ramamurthy et al. 2014; Zhao et al. 2017b). Indeed, VBNC cells which are characterized by a loss of culturability on routine agar escape detection by conventional plate count techniques leading to an underestimation of total viable cells in environmental, food, water, or clinical samples. VBNC cells are potentially a threat to human health because some strains remain virulent in VBNC state.

Vibrio vulnificus is an estuarine bacterium responsible for 95% of all seafood-related deaths in the United States; can enter in VBNC state and remain virulent, at least for some time; and is capable of causing fatal infections following in vivo resuscitation; however, it must be specified that virulence decreases significantly as cells enter the VBNC state mice. As *V. vulnificus* became nonculturable, virulence was determined by employing CD-1 5 to 9 weeks old (Oliver and Bockian 1995). VBNC cells of *E. coli* 0157:H7 (enterotoxigenic *Escherichia coli*), one of the most common food-borne and waterborne pathogens, retain the ability to express both stx1 (Shiga-like toxin gene1) and stx2 (Shiga-like toxin gene2) genes; Shiga-like toxins (Stx) are responsible for the major symptoms of hemorrhagic colitis and hemolytic uremic syndrome (Liu et al. 2010b). *Vibrio cholerae* can be resuscitated by introduction in the intestine of human volunteers (Colwell et al. 1996). From the analysis of all these works, it seems that in the VBNC state, the virulence is maintained in numerous pathogenic bacteria and infection can be initiated in certain strains and under certain conditions. So, VBNC state constitutes an important reservoir of pathogens. Indeed, pathogens in VBNC state have been found to be responsible for many latent infections, which may show up after months or even years.

4.5.4 Spores

Several prokaryotes belonging to the genera *Bacillus*, *Clostridium*, and *Sporosarcina* produce during a process called sporulation a highly differentiated structure named endospore, the purest form of prokaryote dormancy. In response to unfavorable conditions (starvation, desiccation, etc.), vegetative cells of spore-forming bacteria undergo an asymmetric cell division, resulting in two genetically identical daughter but morphologically distinct compartments, a larger "mother cell" and a smaller "forespore," that undergo different cell fates; the smaller of the compartment develops into a mature spore, capable of protecting the genome of the mother cell. The spore is released in the environment, when where it can persist for very long periods of time. Once favorable conditions return, the spore germinates and initiates a rapid growth, and the bacteria return to the vegetative state within minute to > 24 h (Setlow 2013, 2014a, b); spore germination required a number of spore-specific proteins (Gould 2006; Setlow 2013, 2014b). Spores, which are a long-term survival strategy, are very resistant to environmental stresses as any cell found on Earth: extreme heat, UV and γ, chemicals (acids, bases, oxidizing agents, alkylating agents, aldehydes, and organic solvents), drying, and nutrient depletion (Gould 2006; Setlow 2006, 2014a, b). Due to the ability of spores to withstand to harsh environments, the capacity of spores to survive to extraterrestrial environments has been considered, and experiments have been undertaken to test spore resistance under simulated Mars surface environment (Hagen et al. 1964; Nicholson et al. 2009; Moeller et al. 2012). Morphologically the Sps constituted by a central core containing DNA, ribosomes, and most spore enzymes (Fig. 4.7). The core is partially dehydrated (water content represents 25–50% of wet weight) and contains huge amount (~10% of the total spore dry weight) of a spore-specific molecule, the pyridine-2,6-dicarboxilic acid (dipicolinic acid, DPA) (Setlow 2014a). The core is protected by several concentrically arranged protective layers. To know the

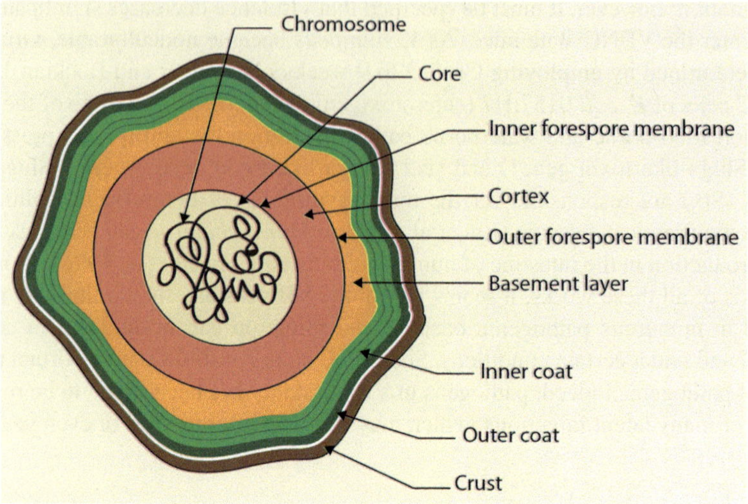

Fig. 4.7 Schematic spore structure
The multiple layers of the spore serve to protect the genome housed in the partially dehydrated central core which contains the DNA, RNA, and most enzymes. The core is surrounded by the inner forespore membrane, (ii) the protective cortex, (iii) the outer forespore membrane, and (iv) the spore coat. The spore coat consists of four layers: basement layer, inner coat, outer coat, and crust. (Modified and redrawn from McKenney et al. 2013).

physiological, physiological, and genetic mechanisms involved in the sporulation, refer to previous works (Driks 2002; Gould 2006; Galperin et al. 2012; McKenney et al. 2013; Tan and Ramamurthi 2013; Al-Hinai et al. 2015). Studies on endospores have a theoretical and applied interest. Many sporulating bacteria are pathogenic. Thus, they are of great importance in food, industrial, and medical microbiology. In addition, spore formation is an excellent model for understanding the molecular basis of developmental biology, including transcription regulation, intercellular signalling, membrane remodelling, intercellular communication, subcellular protein localization, and morphogenesis.

Other resistance structures exist in prokaryotes:

(i) Akinete, a survival structure of cyanobacteria.
(ii) *Azotobacter* and *Myxobacteria* cysts. For instance, when exposed to unfavorable conditions (e.g., starvation), cells of *Myxococcus xanthus* aggregate to form structure called fruiting bodies (multicellular structures) within which some cells differentiate into spherical, heat-resistant spores (Licking et al. 2000).
(iii) *Actinobacteria* exospores (exospore-forming filamentous *Streptomyces coelicolor*).

Spores, which are not a reproductive state but a survival state, can remain in their dormant state for years and yet return to active growth. Examples of spores able to survive for extended periods are common (Sneath 1962; Gest and Mandelstam

1987; Kennedy et al. 1994). For instance, Jacotot and Virat (1954)reported that spores of *Bacillus anthracis* from flask sealed in Louis Pasteur laboratory in 1888 were still alive 68 years later. *Clostridium aceticum* is an obligatory anaerobic spore bacteria isolated by Wieringa in 1939. The spores of this organism, which remained dormant in dry soils for more than 30 years (from 1947 to 1981), suspended in a complex medium containing fructose as energy and carbon source, were revived with growth occurring within 12 h (Braun et al. 1981). Viable *Thermoactinomyces* endospores were collected from Vindolanda site (England), where they had been deposited between 1850 and 1890 years previously (Unsworth et al. 1977).

Furthermore, increasing studies have reported the existence of prokaryotes of different origins (rocks, sediments, salt deposits, extinct animals), which could possibly have been preserved and have been survived for geological period of time. Examples are described in the following paragraph.

4.5.5 Prokaryotes Can Survive for Very Long Period Geological Time Scale (?)

The possible presence of viable prokaryotes in geological formations has been specifically studied in salt deposits (Dombrowski 1963; Kennedy et al. 1994; Grant et al. 1998; McGenity et al. 2000; Winters et al., 2015). Dombrowski (1963) was the first to announce isolation and culture from rock salt, and he describes the morphological and physiological characters of these Paleozoic bacteria. In order to avoid introduction of contaminants from surface site, he develops a rigorous method for isolation and cultivation of bacteria from Permian deposit. These results lead him to the assumption that the discovered bacteria are living representative of this geological period. Since Dombrowski's works, many studies have brought a large body of evidence that would prove the presence of long-term survival of halobacteria inside fluid inclusions of halides, and viable halophilic archaea would have been isolated from ancient salt deposit of the Permian and Triassic ages (McGenity et al. 2000; Vreeland et al. 2000; Stan-Lotter et al. 2002). For instance, Vreeland et al. (2000) announced the isolation and growth of a halotolerant spore-forming *Bacillus* species from a brine inclusion within a salt crystal of Permian Salado Formation in New Mexico. The geological age of the Salado is at least 250 Myr old. Crystal which measured 3.5 × 3.5 × 2.5 cm and contained living bacteria was collected 569 m below the surface. All precautions were taken to ensure that the samples were extracted under aseptic conditions and avoid contamination. The fluid recovered from the crystal was inoculated into casein-derived amino acid medium supplemented with 20% (W/V) NaCl. The strain isolated was designed 2-9-3; the complete 16SrRNA was sequenced and deposited in GeneBank; the sequence showed that the organism was most similar to *Bacillus marismortui*

(99% similarity) and *Virgibacillus pantothenticus* (97.5%); this ancient prokaryote is not significantly different from modern isolated prokaryote (Vreeland et al. 2000). For Vreeland et al. (2000), the strain "was present in a hypersaline during the late Permian, trapped inside a crystal at that time, and survived within the crystal until the present."

Some works are in favor of a microbial life present in the intestinal tract or in the abdomen of extent animal species. A nearly complete and remarkably well-preserved skeleton of a Pleistocene mastodon (*Mammut americanum*) was discovered in Licking County (Ohio). From a mass of plant found in the animal and interpreted as intestinal contents of the mastodon, Lepper et al. (1991) announced the isolation of enteric bacteria (*Enterobacter cloacae*). For the authors, these facultative anaerobic bacteria would be the survivors or descendants of the intestinal microflora of the mastodon.

In 1995, Cano and Borucki (Cano and Borucki 1995) using stringent, aseptic, and controlled conditions claimed that a bacterial spore isolated from abdominal content of extinct bees (*Proplebeia dominicana*) was preserved for 25–40 million years in buried Dominican amber. Spore was revived, cultured, and identified. So, a putatively ancient *Bacillus* sp. was obtained, coded BCA16. The characteristics (enzymatic, biochemical, DNA) are most closely related to extant *Bacillus sphaericus*.

Despite all the precautions taken in isolating these bacteria, all these results were viewed with scepticism, and serious arguments were made in favor of the possibility of contamination by modern bacteria during sampling and/or subsequent handling (Kennedy et al. 1994; Gutiérrez and Martin 1998; Graur and Pupko 2001). Even for samples retrieved from 2800 m depth, a very inaccessible place, the possibility that water may seep into the porous rocks during long time period, contaminating the drill sample, cannot be ruled out.

These results on the longevity of prokaryotes in geological samples – rocks, sediments, salt deposits, soils, and extinct animals – are obtained from a substantial number of samples by independent research teams; it emerges that further works are needed to validate definitively the presence of bacteria in old geological samples through the techniques described above (Parkes 2000).

The presence of microbial cells in a geological sample can also be demonstrated by the presence of a metabolic activity. Thus, Morono et al. (2011) demonstrated that carbon and nitrogen assimilation activities are maintained into microbial cells from 219-m-deep lower Pleistocene (460,000 years old); microorganisms in these old deep subseafloor sediments were alive and maintained potentials for metabolic activities and growth. For their part, Røy et al. (2012) measured an aerobic microbial respiration in 86-million-year-old deep-sea red clay, 0.001 micromole of O^2 liter^{-1}year^{-1}at 30 m depth, proving that microbial community can subsist in such sediment without fresh supply of organic matter for million years. Using a molecular technique targeting specifically rRNA, Schippers et al. (2005) demonstrated that a large fraction of the subseafloor prokaryotes are alive in very old (16 million years) and deep (> 400 m) sediments.

4.5.6 Storage of Organic and Inorganic Compounds

When resources are abundant, above what cells need to grow ("feast" periods), many prokaryotes accumulate intracellular storage polymers that will be used as endogenous energy reserves and/or sources of carbon to be used during starvation ("famine" periods) (Guerrero and Berlanga 2007; Madigan et al. 2015). Such a strategy constitutes a prokaryotic way to increase survival in the ever-changing environment and represents a Darwinian selective advantage. Endogenous organic and inorganic compounds include (i) glycogen a polymer of glucose; (ii) elementary sulfur (S_0) stored from oxidation of H_2S by sulfur chemolithotrophs to form microscopically visible granules (Fig. 4.8a) (when reduced sulfur source becomes limiting, S_0 is oxidized to sulfate (SO_4^{2-}), and the granules disappear); (iii) polyphosphate, used as sources of phosphate for nucleic acid and phospholipid biosynthesis or used directly to make ATP; (iv) polyhydroxyalkanoates (PHAs) (Fig. 4.8b–d) (poly-ß- hydroxybutyric acid (PHB), formed from ß-hydroxybutyric acid units, is probably the most common type of PHA and the best studied); (v) cyanophysin which is synthesized by cyanobacteria and some heterotrophic bacteria; and (vi) bacterioferritin, a protein that stores Fe^{2+} and thus makes it available when the supply is low and prevents the Fenton effects that causes an oxic shock.

Many prokaryotes, *Bacteria* and Archaea, produce PHAs. The occurrence of 91 different PHA constituents reflects the low substrate specificity of polyhydroxy alkanoic acid synthetases which are the key enzymes of polyhydroxyalkanoic acid biosynthesis (Steinbüchel and Valentin 1995). As a function of the number of carbon atoms of the monomer, PHAs are classified as short- (3-5 C-atoms), medium- (6 or more C-atoms), and chain-length PHAs (Mezzina and Pettinari 2016).

PHAs accumulate as intracellular granules in the form of inclusion bodies in amorphous state, as water-insoluble inclusions. Prokaryotes are able to accumulate as much as 80% of their dry weight in PHA. The size of PHA granules mostly comprises between 100 and 500 nm in diameter, and their number which varies as a function of organisms is between 5 and 10 granules per cell. Inclusion bodies contain approximately 97.5% PHA, 2% proteins, and 0.5% lipids, although some estimates of the lipid contents are considerably higher (Pötter and Steinbüchel 2005). Inclusion bodies are coated with a monolayer of phospholipids and granule-associated proteins that play a major role in biogenesis and intracellular degradation of PHAs.

Four types of granule-associated proteins are found (Steinbüchel et al. 1995; Pötter and Steinbüchel 2005; Grage et al. 2009).

(i) Polyester or PHA synthases (PhaC) which are the key enzymes of PHA biosynthesis (Rehm 2003; Peters and Rehm 2005).
(ii) PHA depolymerases (PhaZ) consist of two groups: the intracellular depolymerases (PhaZ) which degrade the amorphous PHA within granules and the extracellular depolymerases, which are secreted by most bacteria to utilize denatured PHA present in the environment from, for example, other nonliving cells.

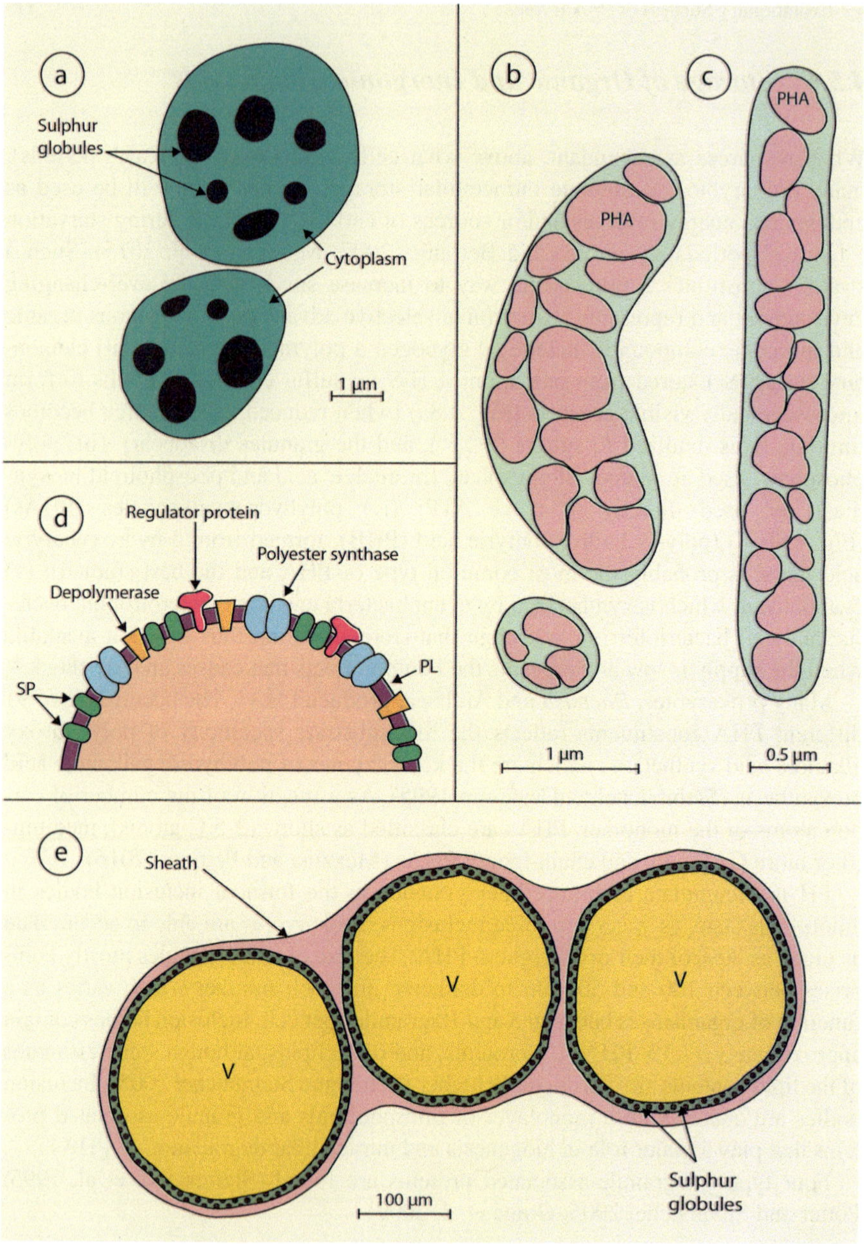

Fig. 4.8 Schematic representation of intracellular storage of organic and inorganic compounds (**a**) Bacterial intracellular sulfur globules in *Chromatium buderi* (Maki 2013); (**b**) PHA granules of the strain MAT-28 from Ebro Delta microbial mats (Guerrero and Berlanga 2007); (**c**) PHA granules of the strain *Wautersia eutropha* H16 (Tian et al. 2005): (**d**) organization of PHA granule. PHA granules are surrounded by a phospholipid membrane (PL), with embedded or attached proteins: PHA synthase, PHA depolymerase, PHA-specific regulator proteins, structural proteins (SP), additional proteins with as yet unknown functions (Rehm 2003); (**e**) *Thiomargarita namibiensis*. Most of the biovolume is occupied by a large, central vacuole (V) surrounded by internal deposit of sulfur globules (SG). Cytoplasm is restricted to a thin outer layer (green). Cells are organized in chains held together by a mucus sheath (Schulz et al. 1999). (Modified and redrawn from Schulz et al. 1999; Rehm, 2003; Tian et al. 2005; Guerrero and Berlanga 2007; Maki 2013)

(iii) Phasins (PhaPs) are noncatalytic proteins which are thought to be the major structural proteins of the membrane surrounding the inclusion, forming an interphase between the content of PHA granules (hydrophobic) and the cytoplasm content (hydrophilic) (Mezzina and Pettinari 2016).
(iv) The regulator proteins (PhaR) involved in PHA granule synthesis and phasin production (Peters and Rehm 2005; Pötter and Steinbüchel 2005).

PHAs are interest for industrial and biomedical applications (Pötter and Steinbüchel 2005; Chen 2009; Grage et al. 2009; Kourmentza et al. 2017).

A remarkable example of prokaryote able to withstand drastic changes in its environment, thanks to the accumulation of reserve substances, is that of *Thiomargarita namibiensis* (Fig. 4.8e). This giant bacterium, the largest bacterium ever discovered, visible to the naked eye, has a spherical shape and formed typical chain which contained on average 12 cells. Most cells have a diameter of between 100 and 300 μm, but some cells may have a diameter of up to 750 μm. This bacterium was discovered 1997, in Namibian shelf sediments, during a cruise aboard the R/V Petr Kottsov (Schulz et al. 1999). Each cell contains a large central vacuole, which represents about 98% of the cell volume, that it fills with high concentrations of nitrate (0.1–0.8M). The rest of the cell contains sulfide granules localized in the thin layer of cytoplasm. These sulfur globules, capable of reflecting the light, explain the name given to this bacterium "sulfur pearl of Namibia." *Thiomargarita namibiensis* that couples nitrogen cycle and sulfur cycle is able to live under oxic (O_2-rich waters) and anoxic conditions (sulfide-rich marine sediments). Its strategy is as follows: when buried in sediments, cells oxidize H_2S to S^0 anaerobically by reducing NO_3^- stored in the vacuole to ammonium (NH_4^+). They then store the S^0 as intracellular granules. When turbulent waters mix cells into the overlying oxic water column, where H_2S is lacking, they switch to the aerobic oxidation of stored S_0. The energy they gain from S^0 oxidation is used to refill their vacuole with NO_3^- from the water column so they will be able to survive under nitrate starvation for long periods (40–50 days).

4.6 The Conquest of All Biotopes by Prokaryotes and Their Role in Their Functioning

Biotopes have varied characteristics, among which varied sources of energy, various electron acceptors, and prokaryotes have evolved to colonize an amazing range of such biotopes on Earth. These biotopes undergo fluctuations, the amplitudes and periods of which vary. The microbes present have thus had to evolve to withstand the *short-term modifications* of these physicochemical parameters in these biotopes. Microbes thus have been selected to have adaptive mechanisms that allow them to face *immediate changing conditions of their habitat* (nutrient levels, thermal or osmotic shock, pH and oxygen levels, prolonged dehydration, UV irradiation, etc.).

4.6.1 *Fluctuating Parameters in the Physical World and the Need for Homeostasis*

Besides having contrasted physicochemical parameters, biotopes vary a lot in the amplitudes and rate of fluctuations for these parameters. Between a mammalian colon where temperature, acidity, salinity, and concentration of organic compounds fluctuate only marginally and a desert topsoil that sees temperature ranging between +40° and −20°, there is thus a whole array of biotopes offering contrasted conditions.

Homeostasis is the tendency of biological entities to maintain intracellular conditions as stable as possible in order to permit the hundreds of enzymes and metabolites present to interact in a sustainable fashion. All enzyme and cell constituents have optima for physicochemical parameters; these optima may not be similar for all enzymes and other determinants; therefore the cells strive to maintain these parameters close to the optima in order to permit most enzymes to function most of the time at an efficiency as high as possible.

Parameters fluctuate in general following a driver such as the sun with its diurnal and annual cycles. Other drivers are geysers and volcanoes, tides, and of course more recently anthropic inputs. Such cycles generate in general coupled fluctuations such as temperature, aridity, and UV flux, complicating the necessary biochemical adaptations. Yet bacteria have evolved to colonize and thrive in all such biotopes, even if at different concentrations.

4.6.1.1 Temperature

Temperature is the most pervasive parameter, one that microbial cells cannot escape contrary to a few large eukaryotes that can control it. Due to their small size, microbial cells will rapidly see their intracellular temperature follow that outside, and cells can only adapt the cell functioning to it. Some biotopes such as the mammal gut or the deep oceans have very stable temperature, while others such as topsoils, especially desert ones, fluctuate a lot, and those cells selected in these biotopes have evolved mechanisms to cope with it. One type of adaptation is selection of lipids, proteins, enzymes, and RNA that function better at a given temperature (Barabote et al. 2009); this process occurs over millions of years and can be accelerated by lateral gene transfer.

Temperature adaptation is also a short-term event implying HSPs (heat-shock proteins) that can renature proteins which structure has been modified by heat (Arsene et al. 2000). Cold-shock proteins are RNA chaperones that upon a cold shock will release mRNAs that will in turn modify the protein pattern of the cell (Phadtare and Inouye 2004).

Prokaryotes produce complex structures, spores, able to withstand temperatures (cf. 4.4.4). For instance, endospores present in *Firmicutes* constitute a complex structure, based on about 500 genes, many of which are conserved, that permit a bacterial culture to survive exposure to 100 °C (Galperin et al. 2012). *Actinobacteria* also produce spores which resistance to heat is modest, with mesophilic actinobacterial spores being inactivated typically by exposure to 50 °C (Fergus 1967); however, these actinobacterial spores can withstand long episodes of desiccation typical of soils.

4.6.1.2 Light

Light is a source of energy that has been harnessed by many lineages, collectively called photosynthesis. Photosynthetic lineages appeared presumably among the first microbes on Earth, and the known photosynthetic machineries are based on a conserved set of components comprising a light-harvesting pigment antenna, a reaction center, and an electron transport system (Stanier et al. 1986). Bacterial photosynthesis can be oxygenic or not, using varied chlorophyll structures, resulting in different chemistry.

The wavelength used by photosynthetic organisms is centered around 550–650 nm, but solar radiation comprises a much wider spectrum of wavelengths, many of which are hard to cope with, especially as their wavelength is close to the UV (400 nm to 100 nm, shorter than that of visible light but longer than X-rays). Such light in general accompanies the harvestable light, and so photosynthetic microbes have developed ways to protect themselves against it. UV light is absorbed by double bonds such as those present in DNA, and this causes damages like cyclobutane-pyrimidine dimers (CPDs) and 6–4 photoproducts (6–4PPs) that are cytotoxic and mutagenic (Sinha and Hader 2002). Adaptation to these challenges comprises DNA repair mechanisms that are not specific to light adaptation but are especially necessary for microbes thriving in lighted areas.

Other adaptations to light are the synthesis of pigments, such as melanins or carotenoids (Figs. 4.9 and 4.10), that absorb short wavelength light and prevent fragile cell constituents from being exposed. Melanin-pigmented microbes are often found on stone surfaces exposed to light where they form dark stains (Sghaier et al. 2015).

Light is a powerful signal that indicates the coming switch from night cold and wet conditions to hot and dry ones. Many bacteria have evolved bacteriophytochrome triggers that detect this signal and use direct protein-protein interaction to synthesize light-harvesting antennae (Giraud et al. 2002).

4.6.1.3 Nutrients (Electron Donors)

Electron donors vary a lot in the various biotopes microbes have colonized. Microbes are divided into heterotrophs that catabolize organic compounds synthesized typically by autonomous organisms and autotrophs that can synthesize compounds using energy sources such as light or chemical compounds. Those microbes able to transform light into chemical energy are called photosynthetic (above). Those able

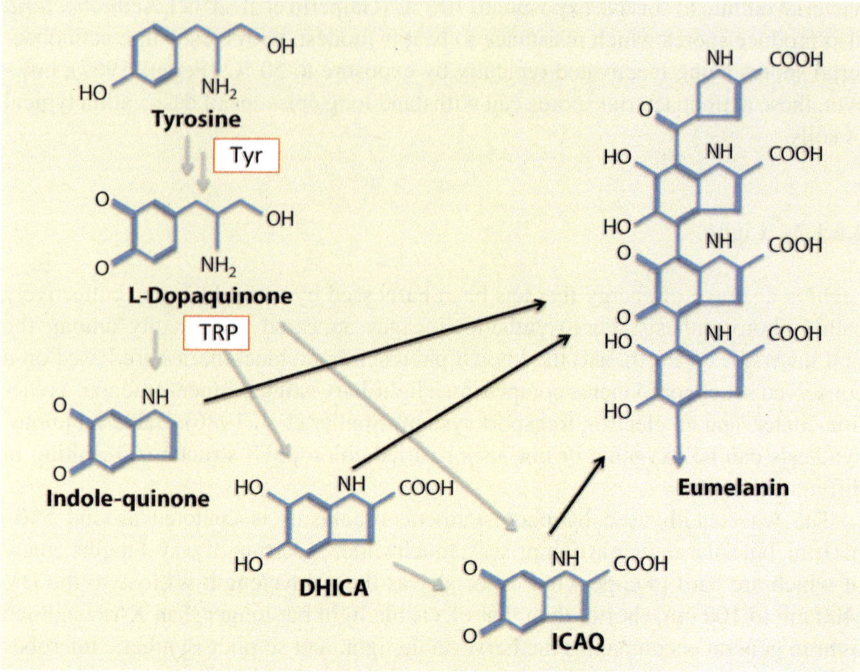

Fig. 4.9 Biosynthetic pathway for synthesis of melanin pigments
The amino acid tyrosine is transformed by tyrosinase into L-dopaquinone and DOPA, then into indole-5,6-quinone (IQ) and indole-2-carboxylic acid-5,6-quinone (ICAQ) by TRPs (tyrosine-related proteins), and finally into eumelanin by poorly characterized enzymes. Melanin is a broad term to describe brown to black pigments found in a large array of bacteria and fungi

to use energy-rich chemical compounds are autotrophs, and among the compounds used contain nitrogen (nitrate, nitrite), sulfur (sulfide), iron, hydrogen, etc. These microbes must also have ways to fix carbon to synthesize cell constituents; they do so using rbc or use organic compounds as building blocks and are then called facultative autotrophs.

The oxidation of reduced compounds has the potential to yield energy, the amount of which depends on the redox difference as indicated in Fig. 4.11.

4.6.1.4 Nutrients (Electron Acceptors)

The most prevalent electron acceptor is oxygen; it has the most positive redox and thus produces the most energy and is the preferred acceptor when present. Initially the term "oxidation" was coined by Lavoisier who recognized that compounds were chemically modified by contact with this gas. Aerobic bacteria couple redox reaction and transfer electrons toward oxygen; they thus need oxygen. Aerobic bacteria have thus developed dedicated electron transfer systems to harness the chemical

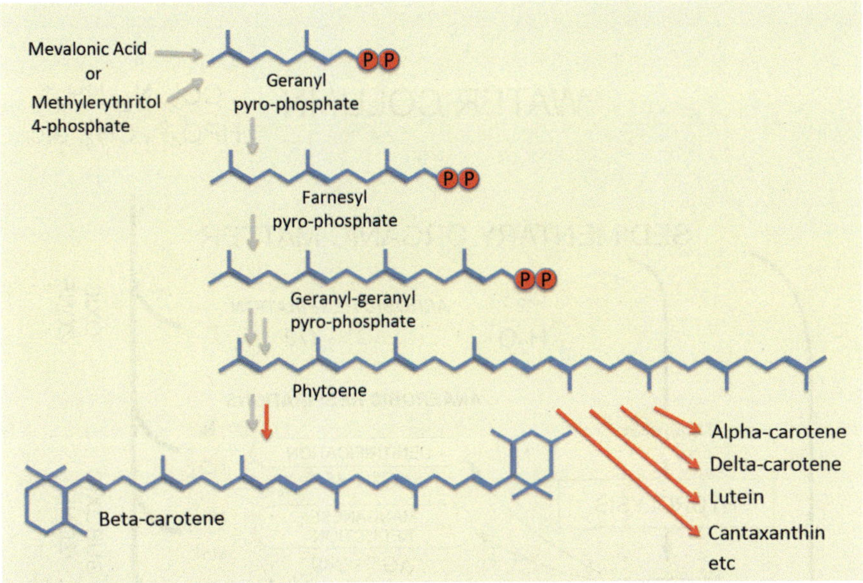

Fig. 4.10 Biosynthetic pathway for synthesis of carotenoid pigments
Two characterized pathways, the mevalonate and the methyl erythritol phosphate ones, produce dimethylallyl diphosphate, which is then transformed into geranyl-pyrophosphate, then farnesyl-pyrophosphate, then geranyl-pyrophosphate, and the 40C phytoene and beta-carotene and a range of modified derivatives, accounting for the various yellow to red hues seen in nature

energy necessary for cell functioning. Typically, cells will use organic acids or hexoses to feed the TCA cycle, which comprise dehydrogenases that oxidize organic compounds and funnel the resulting electrons to membrane-bound quinones and cytochromes. These will then bind oxygen and transform it into water.

However oxygen is a highly reactive compound, with a tendency to form daughter molecules such as peroxides, superoxide, hydroxyl radical, and singlet oxygen which together are called reactive oxygen species (ROS). These ROS interact with different cell constituents; peroxide, for instance, is a well-used antibacterial compound that modifies the membrane lipids and causes cell death. Cells exposed to peroxide, for instance, pathogens of eukaryotes, typically have developed peroxidases and catalases that transform peroxide into water. Superoxide ions also present in eukaryotic tissues can be metabolized by superoxide dismutases that yield hydrogen peroxide, which must then be dealt with by peroxidases.

The Fenton reaction involves free iron that reacts with hydrogen peroxide forming a hydroxyl radical and a hydroxide ion, which is why bacteria exposed to hydrogen peroxide have developed mechanisms to minimize the level of free iron such as bacterioferritins to store iron in an inactive form and to replace iron by other metals in proteins.

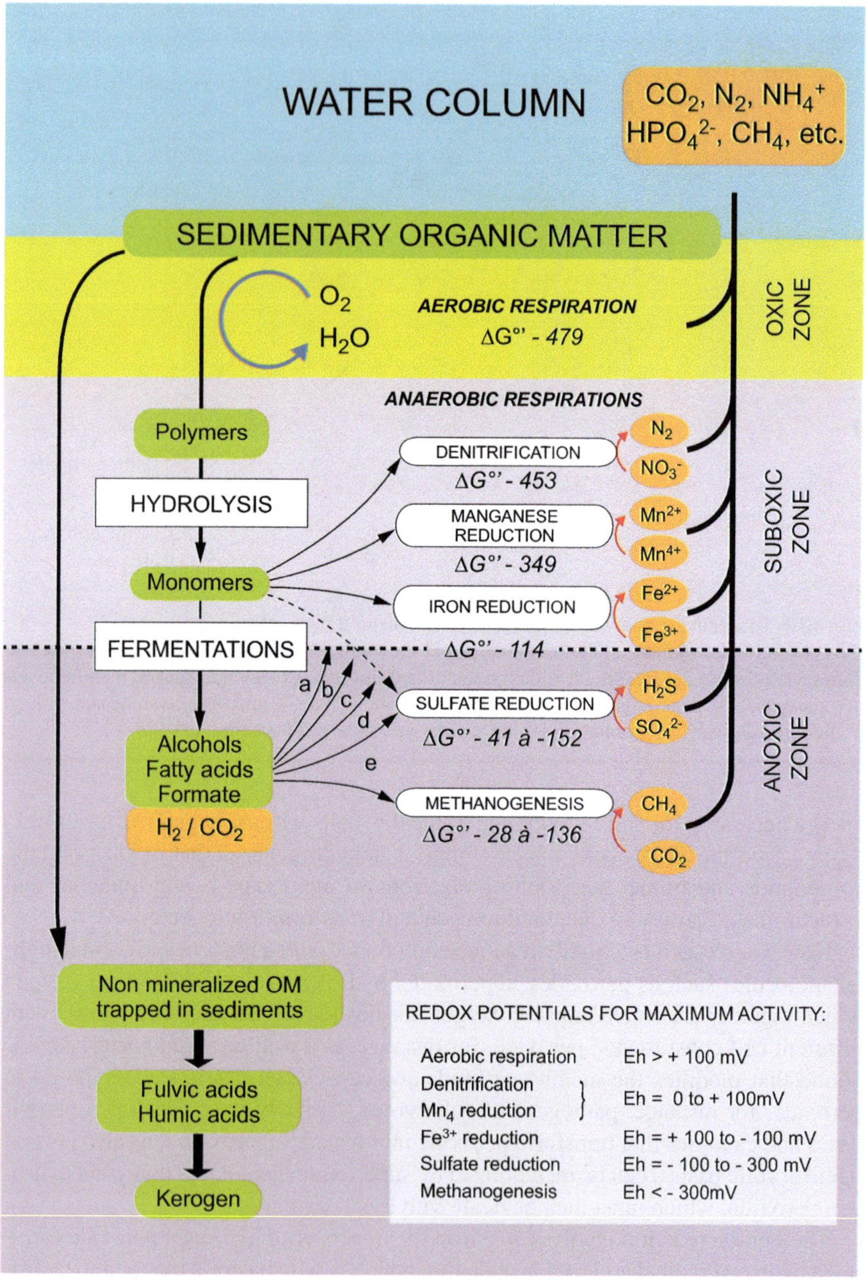

Fig. 4.11 Organic matter transformation in aquatic sediments
$\Delta G^{0'}$ values are expressed in $KJ.mole^{-1}$ of oxidized acetate (Thauer et al. 1977). For fermentations, $\Delta G^{0'}$ values range from +70 to −260 $KJ.mole^{-1}$ according to the fermented substrate. Interspecies hydrogen transfers are included. Alcohols and fatty acids are also utilized by denitrifiers (**a**), Mn (**b**), iron (**c**), and sulfate reducers (**d**) as indicated by cartouches from top to bottom. Sulfate reducers (**d**) and methanogens (**e**) use also hydrogen and formate

Anaerobic bacteria and facultative aerobic bacteria use other electron acceptors such as nitrate, nitrite, sulfate, manganese, iron oxides, carbon monoxide, or fermentation. Some of these compounds are toxic to cell constituents such as nitrite or manganese. Nitrite, which can be recovered from saltpeter on stone surfaces in caves lying next to soil or urinals, for instance, has been used for centuries to cure meats because it is a powerful antibacterial compound. The reaction starts with nitric oxide (NO) that interacts with superoxide (O2·−) to form peroxynitrite (ONOO−). The chemical basis for its toxicity is poorly known, but it is considered to stem from the formation of reactive nitrogen species (RNS) that interferes with biological processes by interacting with protein cofactors such as Fe-S clusters, heme, and lipoamides (Zhang et al. 2013).

4.6.1.5 Acidity

Several definitions of acidity exist, and most are based on the proportion of protons in a solution, a parameter that influences the proton gradient between the outside and the inside of cells. Microbial cells respire essentially by coupling oxidation of an electron donor with an electron transport chain which comprises expulsing protons outside the cells. This proton gradient is then tapped by a chemiosmotic process that allows protons back inside cells coupled with ATP synthesis. This process is in general adapted to neutral conditions, but bacteria are known to have colonized acidic and basic biotopes such as deep-sea hydrothermal vents (Bernhardt and Tate 2012).

Adaptation to an acid biotope entails either modification of the hydrogenase or local alkalinization. Local alkalinization has been described in *Helicobacter pylori* that causes stomach ulcers. This bacterium can resist highly acidic stomach content using a urease enzyme that cleaves urea, releasing ammonium that will locally neutralize hydrochloric acid (Marshall et al. 1990).

Since high pH (commonly understood as above pH 8.5) may denature DNA, destabilize membrane lipids, and denature enzymes, adaptation to an alkaline biotope is essential and has been shown to occur in two manners. Bacteria need to establish a proton gradient; therefore they may cause local acidification of the medium or bring about modification of the wall. Na+/H+ antiporters are integral membrane proteins that catalyze Na^+ uptake and proton efflux, which have been described as permitting cells to cope with high pH biotopes (Wang et al. 2014b). Use of acidic wall polymers (galacturonic, glutamic, phosphoric acids, and proteins) is another strategy that results in an acidic matrix that creates a buffer and protects the plasma membrane (Janto et al. 2011).

4.6.1.6 Water

Water is considered the best indicator for the possibility of life, and deserts are basically places where water is rare. Water is the solvent in which chemical reactions take place and osmolytes outside cells constitute a powerful force that makes water

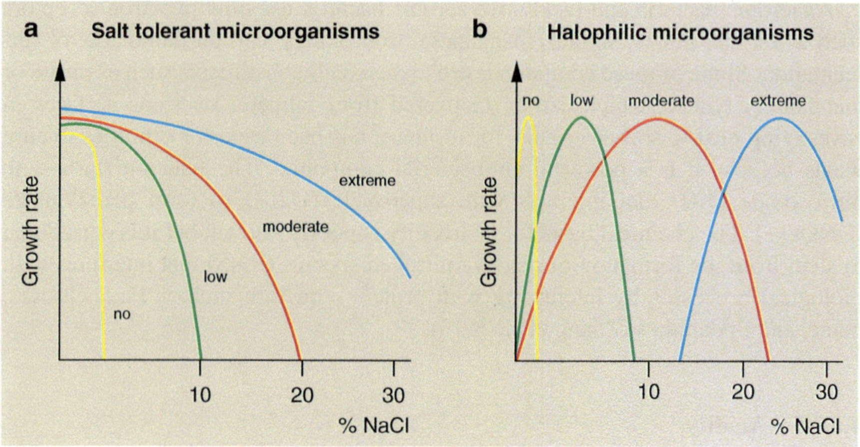

Fig. 4.12 Bacterial growth as a function of the NaCl concentration
(**a**) Salt tolerance in bacteria and archaea depends on the concentration of NaCl they can withstand: low salt tolerance, 0–8%NaCl; moderate salt tolerance, 0–20% NaCl; and extreme salt tolerance, 0–30% NaCl. (**b**) Halophilic bacteria and archaea for which are presented the lowest, highest, and optimum concentrations of NaCl in which they can develop: lowly halophilic, 1–3 – 8% NaCl; moderately halophilic, 2–3 – 10–12 – 20–25% NaCl; and extremely halophilic, 12–15 – 25–30 – 30% NaCl

difficult to get. Water abundance fluctuates typically over the day/night cycle or over the yearly cycle in terrestrial biotopes. Osmolytes that will play a major role will vary depending on the biotope, with NaCl salt being the most prevalent in seawater, but it is always accompanied by other ions, and typical seawater contains Cl^- 0.566 (mol.kg^{-1}), Na^+ 0.486, Mg^{2+} 0.055, SO_4^{2-} 0.029, Ca^{2+} 0.011, K^+ 0.011, etc. Besides seawater, there are biotopes that contain significantly higher concentration of salts such as the Dead Sea. Halophiles (salt lovers) are microbes that thrive in salted water, and they are characterized as slight, moderate, or extreme (Fig. 4.12).

4.6.1.7 Nitrogen

Nitrogen is a major constituent of cells, being part of DNA, RNA, walls, and proteins. There are animated discussions on the evolution of nitrogen in the primitive Earth atmosphere with some hypothesizing ammonia and others proposing high concentrations of cyanide (Silver and Postgate 1973; Miller and Schlesinger 1983). It is now abundant in the biosphere but in a form that most cells cannot metabolize, dinitrogen. Bacteria and archaea have developed systems for the fixation (reduction) of dinitrogen into ammonia with a very conserved protein, nitrogenase. This enzyme can in a reducing environment devoid of oxygen transform dinitrogen into ammonium ions. In order to provide this reducing environment, cells need to exclude oxygen which can be done with thick-walled cells such as heterocysts in cyanobacteria (Wolk 2000) or with multi-lamellate hopanoid walls in *Frankia* (Berry et al. 1993). Other bacteria develop clumps where cells on the outside deplete oxygen by respiring it allowing others at the center of clumps to fix nitrogen.

4.6.1.8 Other

Microbial cells need minimal amounts of oxygen, nitrogen, carbon, hydrogen, manganese, magnesium, calcium, iron, nickel, and in some cases, need maximal amounts of these. Oxygen, carbon, and nitrogen have been described above, and we will briefly describe the others below.

Iron is a constituent of some enzymes in the form of iron-sulfur clusters that are situated at the active core of nitrogenase and hydrogenase.

4.6.2 Types of Molecular Regulation and the Search for Efficiency

4.6.2.1 Transcriptional

The best understood mode of regulation is that at the transcriptional level where a transcriptional regulator, a small molecule that has affinity to a DNA sequence, has evolved to be modified by an environmental parameter, in such a way as to modify its affinity to DNA. There are negative regulators, such as Fur (ferric uptake regulator), that utilizes Fe^{2+} as a corepressor and represses siderophore (a compound to chelate iron) synthesis in pathogens as long as iron is present by binding to DNA promoters upstream of the siderophore synthesis genes and preventing their synthesis (Venturi et al. 1995). Conversely, OxyR is a positive regulator in *Escherichia coli*, negatively autoregulated, that in the presence of hydrogen peroxide will see its affinity for DNA increased and will induce resistance to subsequent lethal doses of H_2O_2 by increasing transcription of 30 proteins among which catalase, glutathione reductase, and an alkyl hydroperoxide reductase (Christman et al. 1989).

There are families of transcriptional regulators, such as the Crp, LuxR, MerR, and TetR, that vary in their structure and mode of interaction with the transcriptional apparatus. Typically bacterial genomes have hundreds of transcriptional regulators belonging to various families, the proportion of which varies as a function of the biotope (Santos et al. 2009).

Transcriptional regulation is widely represented in bacterial cells because it is an energetically cheap way to modulate cells response. However, it is a slow process, whereby the cells will see the level of a compound be modified, the transcriptional regulator will start transcription of the genes of the regulon into mRNAs, and these mRNAs will then be translated into proteins that will sometimes need to be processed to be functional and will then modify the cells physiology.

4.6.2.2 Translational

Translational regulation is the control exerted by a cell on the level of proteins from a given level of mRNA. The mechanisms involved comprise the recruitment of ribosome on initiation codons of messengers, sequestration of messengers by proteins, or modulation of elongation or termination of protein synthesis (Kozak 1999). Instances of regulation at this level are few.

4.6.2.3 Posttranslational

Posttranslational regulation refers to the control exerted on synthesized proteins by reversible or irreversible modifications. Proteins are made up of the 20 standard amino acids that are linked into chains on ribosomes using mRNA as matrix. On such raw proteins can be attached various functional groups such as phosphate, acetate, amide, methyl, glycosyl, uridyl, or lipid groups. Nonstandard amino acids can also be added such as selenocysteine, pyrrolysine, and N-formylmethionine. Maturation of enzymes, such as proteolytic cleavage, and formation of disulfide bridges will also modify the protein functions.

Nonenzymatic protein modifications can be made on proteins such as glycation (addition of a sugar molecule) and denaturation; these will modify their ability to function.

Signal transduction systems constitute the best studied part of the posttranslational control systems. These are based on proteins that are modified by environmental parameters and that can in turn modify other proteins in order to modify their function. Histidine protein kinase systems are the system most common in the bacterial world that consists of two proteins, a membrane-embedded histidine kinase protein (HK) and an intracellular response regulator (RR) that interact through the transfer of a phosphoryl group from the HK protein onto the RR protein. The phosphorylated RR protein will in turn become active until it is dephosphorylated either by the HK or by a dedicated phosphatase. Among the environment stimuli detected by HK proteins are the concentration of nutrients, the redox and the osmotic potentials, temperature, or pH. A good example is the chemotactic behavior of motile cells that can feel the concentration of nutrients. *E. coli*, for instance, has rotating flagella that switch the sense of rotation from the clockwise to the counterclockwise following a signal. Chemical compounds are sensed by transmembrane CheA proteins that autophosphorylate and then transfer the PO^{4-} to CheB and CheY that in turn interfere with the flagella Fli protein and trigger a switch in the sense of rotation which in turn modifies the direction of swimming.

4.6.3 Biodiversity and Adaptation of Extremophilic Microorganisms Inhabiting Extreme Environments

Our anthropocentric vision of life leads us to consider that extremophilic microorganisms occupy niches that are inhospitable to humans due to their harsh physicochemical conditions. These conditions are considered inhospitable nowadays, yet they may have been of primary importance for the emergence of life more than 4 billion years ago. This is true for microbes growing optimally at high (thermophiles/hyperthermophiles) or low (psychrophiles) temperatures, high (alkaliphiles) or low (acidophiles) pH, high saline conditions (hyperhalophiles), as well as high hydrostatic or lithostatic pressures (piezophiles), high levels of ionizing radiation, etc. (Cayol et al. 2015). In addition, numerous habitats may exhibit various combinations of extreme conditions, where growth of only poly-extremophiles is possible (e.g., thermohalophiles, thermoalkaliphiles, thermoacidophiles). It is quite fascinating to imagine that such microorganisms may have prevailed in the anoxic primitive atmosphere under extreme conditions with many-facetted coping strategies at the onset of life. While there are several proposed *scenarii* involving extremophiles at the origin of life, much attention has been paid to thermophilic/hyperthermophilic anaerobes as being the first microorganisms possibly occurring on Earth (Cayol et al. 2015; Westall et al. 2018). In this chapter, we will focus on microbial diversity of extremophiles and how they have adapted their physiology to their respective habitats with a peculiar emphasis on halophiles/hyperhalophiles, psychrophiles, alkaliphiles, acidophiles, piezophiles, and thermophiles/hyperthermophiles.

4.6.3.1 Halophiles/Hyperhalophiles

The ionic composition, pH, and total salt concentration reaching saturation level (34% v/v) are variable from one hypersaline ecosystem to another, whatever its origin (terrestrial, subterranean, or marine), thus allowing the establishment of a large microbial biodiversity including aerobes and anaerobes (Ollivier et al. 1994; Cayol et al. 1995; Oren 2008; Oren 2013) (Fig. 4.13). While algae (e.g., *Dunaliella* spp.), bacteria (e.g., *Halomonas* spp.), and archaea (members of the family *Halobacteriaceae*) are significant representatives of aerobes in these environments, anaerobes, which are also of importance for mineralizing organic matter in hypersaline sediments, comprise fermentative (e.g., members of the family *Haloanaerobiaceae*), homoacetogenic (e.g., *Acetohalobium arabaticum*), sulfate-reducing (e.g., *Desulfohalobium* spp.), and phototrophic bacteria (e.g., *Ectothiorhodospira* spp.), together with methanogenic archaea (e.g., *Methanohalophilus* spp.) (Ollivier et al. 1994). Since life at high salt concentrations requires a high-energy investment to counteract the effect of osmotic pressure, the free-energy changes associated with some reactions performed by anaerobes (oxidation of hydrogen or fermentation of acetate by methanogens, oxidation of acetate by sulfate-reducing bacteria) are not sufficient to allow growth, and, consequently,

Fig. 4.13 Retba Lake (Lac rose, Sénégal)
The salinity of the water is 340 g/L; it is thus a hypersaline lake in Sénégal inhabited by a wide range of aerobic and anaerobic prokaryotes

such metabolic activities do not occur *in situ* (Oren 1999). In contrast, methylotrophic methanogenesis is performed in anaerobic sediments, thus making hypersaline habitats one of the very rare environments where competition between sulfate-reducing bacteria (SRB) and methanogens does not exist since SRB are known not to use methylated compounds as electron donors (Ollivier et al. 1994). More opportunities are offered to aerobes for living in high saline conditions because of the energy provided by oxygen respiration. However, it will be seen below that energetics are not the only constraints for using one strategy instead of another in order to grow in the presence of high salt concentration.

Salts outside cells exert an osmotic pressure that attracts intracellular water from the outside to the inside and, thus, makes transport of constituents inside cells more difficult. In this respect, whatever the type of metabolism, all these microbes have to maintain cell turgor despite the high osmotic pressure exerted by the high salt concentration in the medium. This is achieved by two major strategies. One is the salting-in strategy where the uptake of potassium and chloride from the environment is followed by their accumulation inside cells, which is correlated to extracellular salt concentrations (Oren 1999; Oren 2008; Ma et al. 2010). This osmoadaptive strategy has been demonstrated many times in aerobic members of the order *Halobacteriales* (e.g., *Halobacterium*, *Haloarcula*, *Haloferax*, and *Halorubrum* spp.), as well as in anaerobic members of the order *Halanaerobiales* (e.g., *Halobacteroides* and *Halanaerobium* spp.) and in one member of the *Bacteroidetes* (e.g., *Salinibacter ruber*). The other one, called the "compatible solute strategy," consists in the accumulation of low-molecular-weight highly soluble organic solutes through uptake and/or biosynthesis in the cell cytoplasm (Oren 1999; Empadinhas and DaCosta 2008; Oren 2008; Ma et al. 2010; Shivanand and Mugeraya 2011). These organic osmolytes comprise amino acids (e.g., proline), amino acid derivatives (e.g., glycine betaine), and also sugars (e.g., trehalose) or sugar alcohols (glycerol). Glycine betaine is produced by a large range of bacteria comprising the phototrophic anaerobic *Ectothiorhodospira* spp. and cyanobacteria, domain *Bacteria*, as well as by the methylotrophic methanogenic *Methanohalophilus* spp., domain *Archaea*. Apart from glycine betaine, these methanogens may accumulate beta-glutamine, beta-glutamate, and *N*-epsilon-acetyl-beta-lysine. Other osmolytes such as sucrose and trehalose have also been shown to accumulate in osmotically stressed cyanobacteria. It has been recently demonstrated that trehalose/2-sulfotrehalose biosynthesis and glycine-betaine uptake are widespread mechanisms for osmoadaptation in the *Halobacteriales*, thus demonstrating that a given microorganism may use the two strategies mentioned above to face osmotic stress (Youssef et al. 2014). Glycerol was shown to be the major compound produced by *Dunaliella* spp. to grow under hypersaline conditions, and ectoin, a cyclic amino acid, was first identified within the cytoplasm of *Ectothiorhodospira* spp. Other organic osmolytes such as L-alpha-glutamate, L-proline, hydroxyectoine, and alpha-glucosylglycerol have been also identified in hyperhalophilic microorganisms. It is clear from these results that despite its energy cost, most hyperhalophiles use the compatible solute strategy to keep their cytoplasm isosmotic with the external medium. This may be related to the fact that cells using the salting-in strategy require extensive adaptation of intracellular enzymatic machinery and, notably, to maintain protein conformation and activity in the presence of high salt concentration (Lanyi 1974; Oren 2008), while the compatible solute strategy does not interfere with intracellular enzymatic activities and cellular processes (Grant 2004) but is most energetically costly. The resistance to or requirement for salt of *Halobacteriaceae* proteins result in an excess ratio of acidic to basic amino acids (DasSarma and Arora 2001; Ma et al. 2010) including aspartic acid and glutamic acid (Hoff 2009) that may interfere with the maintenance of physiological intracellular pH. In addition, unsaturated ether lipids have been often observed in members of this family (Ma et al. 2010), which may help in facing osmotic stress. In addition to *de novo* synthesis of compatible solutes, microorganisms have also the ability to take them up when they are present in the culture medium which is less energy costly and thus more favorable for cell growth (Oren 1999). Such solutes have been synthesized by other microbes, and their uptake by others thus constitutes a case of scavenging.

4.6.3.2 Psychrophiles

Most terrestrial and marine environments on Earth are cold, and, thus, microorganisms inhabiting them and developing under these extreme conditions of temperature, called psychrophiles, are of primary importance in governing all the existing biogeochemical cycles in nature (D'Amico et al. 2006; Cayol et al. 2015; Cavicchioli

Fig. 4.14 An example of a typical cold ecosystem in Antarctic (Photography: courtesy from Jean-Claude Marx)

2016). Psychrophiles have been recovered from various ecosystems including sea ice, snow packs and glaciers, Arctic and Antarctic Oceans, Antarctic subglacial lakes, deep-sea waters, atmosphere, clouds, and permafrost (Margesin and Miteva 2011; Cayol et al. 2015) (Fig. 4.14). More attention has been paid to them in recent decades since other planets (e.g., Mars) or satellites in our solar system (Europa), which might have been conducive to the emergence of life on them (Price 2007; Cayol et al. 2015), exhibit low-temperature conditions similar to those encountered on our planet. It is noteworthy that polyextremophily may apply to psychrophiles living at high hydrostatic pressure (piezopsychrophiles) or high salinity (halo-psychrophiles). Despite being less plausible, because most probably hot conditions have prevailed in the early Earth (Westall et al. 2018), it has even been suggested that a psychrophile was the first living microbe some 3.5 billion years ago (Price 2007).

All these microorganisms show adaptations which may be different from one another for growing at temperatures close to 0 °C with some of them having the ability to maintain activities at subzero temperatures down to – 20 °C and even less, in sea ice and possibly in permafrost (Margesin and Miteva 2011; Cayol et al. 2015; Cavicchioli 2016). The limiting factor for enzymatic activity is the availability of free water, which may occur below 0 °C depending on the presence of various salts and pressure conditions. The challenge for psychrophiles is to maintain cell integrity and functioning homeostasis particularly while water is freezing. This is possible in unfrozen water inside the permafrost soil; brine pockets resulting, for example, from marine salt increases; and channels within the ice where dissolved organic matter is available. In addition to appropriate physicochemical conditions favoring lowering of the water freezing temperature, microorganisms have also developed remarkable strategies that will be discussed below to face low temperatures when thriving in such extreme environments (for reviews, see D'Amico et al. 2006; Orfaniotou et al. 2009; De Maayer et al. 2014; Cayol et al. 2015; Cavicchioli 2016). Microbial ecology studies based on cultural and molecular approaches have reported that bacteria, archaea, yeasts, filamentous fungi, and also algae colonize these habitats. Most cold habitats are considered as highly oxygenated because of the high solubility of gases at low temperatures. In this respect, many aerobes performing either photosynthesis or oxidation of various organic electron donors are present in ice, snow, and deep-sea environments. However, anaerobic populations are also well represented, notably in permafrost which is known to transfer huge quantities of methane to the atmosphere during summer thaw, thus impacting significantly, together with anthropogenic activities, global warming. Among aerobes, the most commonly reported microorganisms include members of the *Alpha-*, *Beta-*, and *Gammaproteobacteria*, *Actinobacteria* and the *Cytophaga-Flavobacterium-Bacteroides*, and photosynthetic cyanobacteria within the domain *Bacteria*, Crenarchaeota involved in ammonium oxidation (nitrification), and at least one euryarchaeotal representative, *Halorubrum lacusprofondi* (D'Amico et al. 2006; Margesin and Miteva 2011; Cayol et al. 2015). Anaerobes comprise members of the *Deltaproteobacteria*, the *Firmicutes*, and the *Euryarchaeota*. They are found mainly in Arctic and Antarctic lakes, in cold soils, and in permafrost. As examples of oxygenic bacteria, *Alteromonas*, *Glaciecola*, *Colwellia*, and *Polaribacter* spp. were found in Arctic and Antarctic seawaters, while *Gammaproteobacteria* (e.g., *Photobacterium*, *Colwellia*, *Psychromonas*, *Moritella* spp.) occupy deep-sea waters. *Pseudomonas*, *Polaromonas*, and *Arthrobacter* spp. have been recovered from permafrost samples. Methanotrophs such as *Methylobacter*, *Methylosinus*, and *Methylomicrobium* are active in Arctic soils and permafrost where they participate in the regulation of methane emissions into the atmosphere (D'Amico et al. 2006; Margesin and Miteva 2011; Cayol et al. 2015). Surprisingly, the thermophilic aerobic hydrogen-oxidizing proteobacterium *Hydrogenophilus thermoluteolus* was detected in accretion ice in the subglacial Lake Vostok, Antarctica, where only its survival but not growth seems possible. Its presence was explained (Lavire et al. 2006) by possible seismotectonic activity episodes releasing debris along with microbial cells present within deep faults in the bedrock below the lake, where deep thermolysis of water could yield hydrogen, a source of electrons for *H. thermoluteolus*, as well as oxygen. Anaerobic activities in cold environments can be attributed to methanogens (Euryarchaeota), sulfate-reducing bacteria (*Deltaproteobacteria*), fermentative bacteria (*Firmicutes*), and possibly others which have yet to be discovered as most attention has been paid by scientists to aerobic microorganisms so far. We may expect ongoing metagenomic studies to deliver more significant information on the different anaerobic trophic groups that may participate in biogeochemical cycles in cold environments. It is known that sulfate reduction occurs in deep-sea sediments and should be mainly the result of *Desulfovibrio* spp. activity (Kaneko et al. 2007; Margesin and Miteva 2011). *Desulfotalea psychrophila*, which is able to grow by reducing sulfate to sulfide at *in situ* temperatures below 0 °C, was isolated from permanently cold Arctic sediments. In spite of the fact that *Clostridium* strains do not appear to be acclimatized to deep-sea sediments, with only spores occurring *in situ*, several of them have been isolated from Antarctic microbial mats where they may be active. *Clostridium psychrophilum* has a growth temperature optimum of 4 °C, while *C. frigoris*, *C. lacusfryxellense*, and *C. bowmani* have higher optimal growth temperatures ranging from 4 to 16 °C (Spring et al. 2003). Methanogens have been isolated from different locations in Antarctica (Ace Lake), in Alaska (Skan Bay), and in Europe (Soppen Lake, Switzerland). Their optimum growth temperature ranges from 18 to 35 °C with *Methanococcoides burtonii* being the only methanogen known so far to develop under 0 °C. They comprise hydrogenotrophic (*Methanogeniumfrigidum*, *M. marinum*), methylotrophic (*Methanococcoides burtonii*, *M. alaskense*), and acetoclastic archaea (*Methanosarcina baltica*, *M. lacustris*) (Margesin and Miteva 2011; Cavicchioli 2016). The hydrogen-oxidizing and acetate-fermenting capacities of cold-adapted methanogens suggest that complete mineralization of organic matter under anaerobic conditions producing methane may occur down to temperatures around 0 °C. The existence of such oxidative anaerobic processes at high negative temperatures needs further investigation.

All proteins have optimal temperatures at which attachment to substrate, to cofactors, and conformational changes permit them to fulfill their function. Other cell constituents, such as lipid membranes, are also capable of exchanging with proteins at temperatures that control rigidity and fluidity. Moreover the freezing of water around 0 °C makes movement of solutes impossible and generates reactive oxygen species. To grow at cold temperatures, psychrophiles have to adapt physiologically (for reviews, see D'Amico et al. 2006; De Maayer et al. 2014; Cayol et al. 2015; Cavicchioli 2016). This is achieved for most of them by an increased synthesis of membrane unsaturated fatty acids, polyunsaturated fatty acids, and branched fatty acids, as well as also sometimes by a decrease in the length of hydrocarbon chains which increase membrane fluidity. In addition, to avoid the deleterious effects of ice formation within cells, psychrophiles withstand freezing by synthesizing several types of proteins involved in various cellular processes including transcription, translation, or protein folding. Synthesis of antifreeze proteins allows lowering the freezing temperature of water. Response to cold temperature in psychrophiles may also result in the production of "cold-shock proteins" together with "cold-acclimatization proteins." Accumulation of osmolytes (sugars, polyols, amino acids, polyamines) may help microorganisms to face osmotic stress when high concentrations of salt are present in cold environments. Cell cryoprotection may be improved by the production of exopolysaccharides which have been found in Arctic and Antarctic marine bacteria and which are used for attachment of microorganisms to supports, thus facilitating in particular biofilm formation, concentration of nutrients, protection against unfavorable physicochemical conditions, and water retention. Due to the high solubility of oxygen at low temperatures, psychrophiles have also to protect themselves against reactive oxygen species that they may detoxify by a high production of antioxidant enzymes (e.g., catalase, superoxide dismutase). Finally, studies on a number of psychrophilic protein structures revealed a reduction in arginine and proline, while proteins from cold-adapted archaea have a higher content of glutamine and threonine. As demonstrated above, the strategies used by psychrophiles to thrive in a cold environment are varied and may therefore be different from one microorganism to another. Proteins play an important role within these strategies and may be the result of constitutive expression, upregulation, or downregulation of genes in response to cold exposure (De Maayer et al. 2014).

4.6.3.3 Alkaliphiles

Alkaliphilic microorganisms are divided into two groups. While obligate alkaliphiles are defined as those growing only at pH values of nine and above, facultative alkaliphiles may grow under alkaline conditions as well as at neutral pH (Yumoto et al. 2011; Cayol et al. 2015). Some of them may develop under other extreme physicochemical constraints (e.g., high temperature and/or high salinity) making them polyextremophiles (e.g., halophilic alkalithermophiles like *Natranaerobius thermophilus*) (Mesbah and Wiegel 2008; Wiegel 2011). They inhabit a wide range of environments including soda lakes (Fig. 4.15), underground alkaline waters, soil

Fig. 4.15 Alkaline crater lake (pH around 10) of Rincon de Parangueo in the state of Guanajuato, Mexico (Photography: courtesy of Manon Bartoli)

samples where ammonification is carried out, the guts of insects, alkaline industrial effluents, and the terrestrial or marine alkaline ecosystems with alkalinity resulting from serpentinization reactions from mantle rocks with water (Grant 1992; Jones et al. 1998; Grant and Sorokin 2011; Yumoto et al. 2011). Most attention has been paid by microbiologists to soda lakes, where the microbial diversity has been widely described (Grant and Sorokin 2011). However there is an increasing interest in serpentinized ecosystems (e.g., Lost City) that exhibit physicochemical conditions which might have been of primary importance for the emergence of life more than 4 billion years ago (Kelley et al. 2005; Russell et al. 2014; Westall et al. 2018). Primary production in soda lakes consists in aerobic (e.g., cyanobacteria) and anaerobic photosynthetic bacteria (e.g., *Ectothiorhodospira* spp.) which may predominate in highly saline alkaline lakes where availability of oxygen is limited due its low solubility in saturated saline waters. These primary producers deliver organic matter to the environment, which may be used by a wide range of aerobic and anaerobic microorganisms within the domains *Bacteria* and *Archaea* (Grant and Sorokin 2011; Yumoto et al. 2011; Cayol et al. 2015). Among aerobic bacteria, numerous representatives of the *Gammaproteobacteria* have been isolated from these extreme environments and include proteolytic (e.g., *Halomonas* spp.), sulfur-oxidizing (e.g., *Thioalkalivibrio* spp.), methanotrophic (e.g., *Methylomicrobium* spp.), and nitrate-reducing microorganisms (e.g., *Alkalispirillum-Alkalilimnicola* group). The *Alphaproteobacteria Nitrobacter alkalicus* was demonstrated to be involved in the nitrogen cycle by oxidizing nitrite to nitrate. A great number of aerobic archaea belonging to the order *Halobacteriales* (e.g., *Natronobacterium*, *Natronococcus*) have been identified in soda lakes where their high content in carotenoids is responsible for the red color of these lakes. Anaerobic microorganisms comprise fermentative bacteria belonging to the phylum *Firmicutes* (e.g., *Natroincola* and *Tindallia* spp.), sulfate-reducing members of the orders *Desulfovibrionales* and *Desulfobacterales*, and methylotrophic (e.g., *Methanosalsus* and *Methanohalophilus* spp.) or hydrogenotrophic methane-producing archaea (*Methanobacterium* sp.), with hydrogen being poorly or not at all used by the latter at high pH values. An anaerobic poly-extremophile, the halophilic, alkalithermophilic bacterium *Natranaerobius thermophilus*, was isolated from soda lakes of the Wadi El Natrun, Egypt. It is noteworthy that, to date, the anaerobic oxidation of acetate has never been demonstrated by any microorganism indigenous to alkaline environments, thus suggesting that complete mineralization of organic matter in the absence of oxygen still remains an enigma in soda lakes. Microbial studies that have been performed in soil samples and also in the gut of insects revealed the presence of numerous alkaliphilic *Bacillales*, while an industrial process conducted under alkaline conditions (e.g., production of the indigo fermentation liquor) identified *Alkalibacterium* spp. as the main microbial components of the liquor. Molecular analyses of alkaline groundwaters indicate also the presence of *Bacillus* spp. together with *Alkaliphilus*, *Natronoincola*, and *Anaerobranca* spp. Molecular surveys were performed in submarine alkaline hydrothermal ecosystems driven by serpentinization processes (Brazelton et al. 2010; Suzuki et al. 2013; Tiago and Verissimo 2013; Quéméneur et al. 2014; Postec et al. 2015). Such processes result in the production of hydrogen and methane by alkaline fluids originating from carbonate chimneys. These studies highlighted the presence of anaerobic members of the order *Methanosarcinales* as the main archaeal representatives possibly involved in methane production or oxidation together with sulfate-reducing bacteria. Within the *Bacteria*, evidence for the presence of anaerobic microorganisms of the *Firmicutes*, as well as aerobic *Gammaproteobacteria*-oxidizing sulfide and methane, has been provided. Few culture-dependent studies have yet been performed from serpentinite-containing environments. They have led to the isolation of heterotrophic aerobes comprising *Actinobacteria*, *Bacillus*, and *Staphylococcus* spp. (Tiago et al. 2004) and one hydrogenotroph isolated from the continental serpentinizing springs at the Cedars (CA, USA) which has been proposed as a novel genus "*Serpentinomonas*" within the *Betaproteobacteria* (Suzuki et al. 2014). It is only recently that the first anaerobic bacteria belonging to the *Firmicutes* have been isolated from such alkaline ecosystems. They include the hydrogen-producing bacteria *Alkaliphilus hydrothermalis*, *Acetoanaerobium pronyense*, and one *Clostridium* sp., all originating from the Prony hydrothermal field in New Caledonia (Mei et al. 2014; Ben Aissa et al. 2015; Bes et al. 2015).

A high external pH is destructive for cell machinery and integrity and especially for the ATP-generating proton-motive force, which is why alkaliphiles have to maintain their cytoplasmic pH at values much lower than that of the exterior environment. The mechanisms that allow microorganisms to maintain pH close to neutrality despite pH variations (acidic or alkaline pH) possibly occurring in their culture medium are known under the term "pH homeostasis" (Krulwich et al. 2011a). Such mechanisms have been discovered in aerobic *Bacillus* cells, in particular (Preiss et al. 2015). One of the major challenges for the growth of alkaliphiles is to sustain a proton-motive force in low proton concentration environments as is also the case in alkaline ecosystems (for reviews, see Krulwich et al. 2011a, b). To achieve this goal, these microorganisms increase ATP synthase activity with proton entrance in the cytoplasm being coupled to ATP generation. The energy released during the respiratory stage benefits the Na^+/H^+ or the $Na^+(K^+)/H^+$ antiporters which help also in internalizing H^+ into the cells, thus leading to acidification inside the cells. Such strategy involves cation/proton antiporters which have to allow more protons entering the cells than sodium/potassium ions, in particular, in order to maintain suitable internal pH for alkaliphiles to grow. Alkaliphiles therefore play a subtle game in the transport of cations and protons which should be favorable for proton import into the cells. Nevertheless, to ensure this bioenergetic process, the availability of cations (Na^+ or K^+) should not be limited. In this respect, the entry of Na^+, if necessary, may be activated by Na^+/soluble symporters or through the MotPS sodium ion channel that power motility. In addition, studies on aerobic alkaliphilic *Bacillus* spp. demonstrated that they possess a specific secondary cell wall polymer associated with the peptidoglycan and specific components in the cell membrane (e.g., cardiolipin, squalene) which may be helpful to trap protons at the membrane surface, thus favoring the functioning of a proton-motive force in low H^+ -containing environments, such as alkaline ones (Padan et al. 2005; Mesbah and Wiegel 2008). Experiments conducted with the anaerobic *Clostridium paradoxum* demonstrated that, in contrast to aerobic *Bacillus*, it used a Na^+-coupled ATP synthase rather than a H^+-coupled ATP synthase, thus reducing sodium toxicity and cytoplasmic proton loss that would occur in aerobic alkaliphiles using a H^+-coupled ATPase (see above).

4.6.3.4 Acidophiles

Acidophiles thrive in acidic environments with pH values ranging from 1 to 5. While extreme acidophiles grow optimally at pH below 3.0, the optimum pH for growth of moderate acidophiles is considered to be between pH 3.0 and 5.0. Acidic ecosystems occur in both subaerial and subsurface environments with low pH resulting most often from anthropogenic activities (e.g., acid mine drainage) (Johnson 1998, 2012; Cayol et al. 2015). Acidophilic microorganisms also inhabit natural ecosystems such as sulfuric springs (Fig. 4.16), volcanic ecosystems, and rivers with the Rio Tinto being one of the most studied sites from a microbiological point of view (Gonzalez-Toril et al. 2003; Sanchez-Andrea et al. 2012; Aguilera 2013; Cayol et al. 2015). In all these habitats, the acidification process mainly results from the oxidation of reduced sulfur compounds (e.g., elemental sulfur, sulfide) into sulfuric acid by a variety of aerobic or facultative anaerobic microorganisms belonging to the domains *Bacteria* and *Archaea* (Cayol et al. 2015). Interestingly, due to low pH, metal solubility is improved, and, consequently, acidophiles have been reported to resist high metal concentrations making them good candidates to be used in biotechnology (e.g., biomining, recovery of metals, bioremediation of metal-rich solid and liquid wastes) (Dopson and Holmes 2014). Among these metals, iron is recognized as the most abundant one in acidic environments, and numerous ferrous-oxidizing and ferric-reducing prokaryotes have been detected by molecular or cultural approaches in these extreme environments (Cayol

Fig. 4.16 "Los Azufres," an acidic lake (pH <3) in the state of Michoacan, Mexico (Photography: courtesy of Agnès Hirschler)

et al. 2015). It is now clearly accepted by the scientific community that, in the absence of light, reduced sulfur compounds together with ferrous iron are the primary sources of energy for providing organic matter in acidic environments. Hydrogen is also available often in these environments as an electron donor and may contribute to the overall energetic processes within such extreme habitats. Within the domain *Bacteria*, members of the genus *Acidithiobacillus* are known to oxidize both Fe(II) and sulfur compounds (e.g., *A. ferrooxidans*, *Sulfobacillus* spp.) or only sulfur compounds (e.g., *A. caldus*, *A. thiooxidans*). *Leptospirillum* spp. can grow only by oxidizing reduced iron and are considered together with *Acidithiobacillus* spp. of ecological significance in these habitats. The ability to use ferric iron as an electron acceptor has been reported for the oxidation of inorganic (e.g., elemental sulfur) or organic material (reduced carbon compounds) by *A. ferrooxidans* and *Acidiphilium* spp., respectively (Johnson 1998, 2012; Cayol et al. 2015). A moderately thermophilic weakly acidophilic anaerobic bacterium, using organic compounds as substrates and belonging to the genus *Athalassotoga,* was recently demonstrated to also reduce ferric iron (Itoh et al. 2016). Thermophilic members of the order *Sulfolobales*, domain *Archaea*, obtain energy by oxidizing elemental sulfur or sulfidic minerals (e.g., *Sulfolobus* spp.) under aerobic conditions, while others make it by reducing ferric iron (e.g., *Acidianus* spp.). Within the same domain, *Ferroplasma* spp., order *Thermoplasmatales* (e.g., *F. acidiphilum* and *F. acidarmanus*), have the ability to oxidize iron. Finally the anaerobic *Stygiolobus azoricus* and the facultative anaerobic *Acidianus* spp. were the only extreme thermoacidophiles reported to oxidize anaerobically hydrogen by reducing elemental sulfur (Johnson 1998; Cayol et al. 2015). Although these thermophilic sulfur-reducing archaea have been shown to grow under extreme acidic conditions below 3.0, cultural-dependent studies have led only to the isolation of moderate acidophilic sulfate-reducing bacteria so far. They all belong to the phylum *Firmicutes* with the exception of *Thermodesulfobium narugense* growing at pH ranging from 4.0 to 6.5 with an optimum at 5.5–6.0 (Mori et al. 2003). They include *Desulfosporosinus* spp. which incompletely oxidize their substrates. This is the case of *Desulfosporosinus acidiphilus* growing optimally at pH 5.2 using hydrogen, lactate, pyruvate, glycerol, glucose, and fructose as electron donors (Alazard et al. 2010). It is noteworthy that neither hydrogenotrophic nor acetoclastic acidophilic methanogens have been isolated to date. However recent experiments that have been performed in peatlands (pH 4.5) have demonstrated that moderate acidophilic or acidotolerant methanogens acting as acetate or hydrogen scavengers during syntrophic oxidation of butyrate, ethanol, or propionate do exist (Schmidt et al. 2016). Eukaryotes are also present in acidic environments, notably in the presence of light. However they are generally considered as acidotolerant since they do not exhibit optimum growth at low pH. They include algae, yeasts, filamentous fungi, and amoebae (Aguilera 2013; Cayol et al. 2015).

Altogether, these results demonstrate that the oxidoreductive processes making use of iron are very effective under extremely acidic conditions, while that of sulfur compounds is still a matter of debate as reduction of the latter has been mainly established only at pH values > 3.0. They also indicate that most probably complete anaerobic mineralization of organic matter in very low pH environments (< 3.0) does not occur through sulfate reduction or methanogenesis but rather through reduction of ferrous iron. However further experiments are needed to clarify the functions of the sulfur cycle in particular in acidic environments.

To face acidic conditions, acidophiles have developed different strategies with the aim of maintaining their cytoplasmic pH at near-neutral values (pH homeostasis) to avoid cell destruction (for reviews, see Baker-Austin and Dopson 2007; Krulwich et al. 2011a, b). These strategies may or may not require energy to be operational (active regulation or passive regulation, respectively). The most important mechanisms of pH homeostasis in acidophiles are to restrict the entrance of protons which could intensify cellular protonation known to be deleterious for cells. Moreover, this restriction results in a pH difference (delta) across the cytoplasmic membrane, which is essential for the functioning of the proton-motive force to produce ATP through the F_0F_1 ATPase. To sustain an appropriate proton-motive force at low pH, microorganisms possess a cell membrane highly impermeable to protons. This is the case for acidophilic archaea (e.g., *Thermoplasma*, *Ferroplasma*, *Sulfolobus* spp., etc.) which have a cell membrane composed of tetra-ether lipids. In contrast, despite the fact that tetra-ether lipids have been already detected in bacteria (e.g., diabolic acid within the membrane of *Thermotogales* (Damsté et al. 2007)), the bacterial cell membrane has been recognized to contain essentially ester lipids. Consequently, the cell membrane structure is of primary importance in allowing acidophiles to develop in their extreme habitats. Accordingly, some members of the genus *Leptospirillum* were shown to have numerous genes for cell membrane biosynthesis in their genomes. Another way to reduce the proton influx into the cells is to create a reversed membrane potential, which may be activated by the entrance of potassium ions (Baker-Austin and Dopson 2007). In this respect, many acidophilic prokaryotes were shown to have a high number of putative cation transporters that may counteract the proton influx. The excess of protons can also be removed by active pumping, as well as by using proton-driven secondary transporters. Other possibilities include the production of cytoplasmic buffer molecules to maintain the intracellular pH around 7.0. Such buffer molecules include organic compounds containing basic amino acids, and also mineral ones, such as phosphoric acid whose pH is not much affected by the addition or removal of protons. Despite organic acids are known as uncouplers to be toxic as substrates to be used under acidic conditions, heterotrophic acidophiles do exist, thus demonstrating that the latter have implemented unique strategie(s) with respect to pH homeostasis. Analysis of the genomes of some extreme acidophiles revealed the presence of several genes involved in DNA and protein repair. In addition, chaperones characterized by protein refolding were found to be highly expressed when acidophiles were exposed to a drop in pH. Finally, the pH stability of proteins at low pH may result from their content in iron. In this respect, we may expect acidophiles to use one or a combination of the strategies mentioned above to grow optimally under acidic to extreme acidic conditions, but future research in this field will provide more information on additional mechanisms that may be used by prokaryotes to circumvent the toxic effect on growth due to external low pH values.

4.6.3.5 Piezophiles

The deep biosphere includes the lithosphere, the deep oceans, and the ocean crust (Whitman et al. 1998; Cayol et al. 2015). These environments share one common feature, namely, high hydrostatic pressures (up to 1 100 atm or 110 Megapascal (MPa)). The organisms living in them and growing preferentially under high hydrostatic pressure are called "piezophiles" (defined as exhibiting optimal growth at pressures > 0.1 MPa). Hyperpiezophiles have been defined as those displaying optimal growth at pressures > 60 MPa (Allen and Bartlett 2004). The great majority of microorganisms isolated and characterized as piezophiles are psychrophilic facultative anaerobic bacteria mainly affiliated to *Gammaproteobacteria* and belonging to five genera: *Photobacterium*, *Shewanella*, *Colwellia*, *Psychromonas*, and *Moritella* (Simonato et al. 2006; Campanaro et al. 2008; Aono et al. 2010; Wang et al. 2015). These bacteria have been all isolated from deep marine sediments with the exception of the obligate piezophile *Colwellia* sp. MT-41, isolated from a shellfish collected in the Mariana Trench at 10 476 m depth, which is capable of developing only at pressures between 38 and 103 MPa (Yayanos et al. 1981). Regarding the mesophilic bacteria, they are represented by three species of sulfate-reducing bacteria belonging to the genus *Desulfovibrio* (*D. profundus*, *D. piezophilus*, and *D. hydrothermalis*) and two *Pseudomonas* strains (BT1 and MS300) (Bale et al. 1997; Kobayashi et al. 1998; Kaneko et al. 2000; Pradel et al. 2013; Amrani et al. 2014). Piezophilic thermophiles or hyperthermophiles belonging to the *Archaea* have been mainly isolated from deep-sea hydrothermal vents. They belong mainly to the *Thermococcales* order (e.g., *Thermococcus*, *Pyrococcus*, and *Palaeococcus*). A single obligate piezophile is known so far: the hyperthermophilic *Pyrococcus yayanosii* CH1, isolated from the site "Ashadze" on the Mid-Atlantic ridge at 4 100 m depth (Michoud and Jebbar 2016). Others are affiliated to the *Methanococcales* (Huber et al. 1982; Jones et al. 1983). High pressure impacts the activity of cells and cellular components and reduces the activity of numerous key processes, eventually leading to cell death of piezosensitive organisms (Simonato et al. 2006). Biochemical and genomic studies yield a fragmented view on the adaptive mechanisms in piezophiles. According to the genera investigated, piezophilic adaptation requires either the modification of a few genes, a more profound reorganization of the genome, the fine tuning of gene expression, or a stress-like physiological cell response. Global analyses, performed on few species (e.g., *Desulfovibrio hydrothermalis*, *D. piezophilus*, *Photobacterium profundum*, *Pyrococcus yayanosii*) at the genomic, transcriptomic, or proteomic levels, suggest that adaptation to high pressure is diffused in the genome and may concern only a small fraction of the genes (Le Bihan et al. 2013; Amrani et al. 2014, 2016; Michoud and Jebbar 2016). The most significant finding in studying the biochemistry of piezophilic bacteria is the discovery of *de novo* synthesis of polyunsaturated fatty acids (PFA), their proportions reaching as much as 70% of the total fatty acids in the membranes of these bacteria (Allen and Bartlett 2004; Abe 2013). The high proportion of PFA may increase the fluidity of the membrane at high hydrostatic pressures. A second major finding is the accumulation of solutes in the bacteria, which may play the expected role of "piezolyte." Similar to the mechanisms reported for microorganisms in response to osmotic or heat stresses, the piezolytes may play the role of protein-stabilizing solutes. For example, glutamate has been shown to accumulate in the piezophilic bacteria of the *Desulfovibrio* genus when pressure increases (Amrani et al. 2014, 2016). In *P. profundum*, accumulation of β-hydroxybutyrate has been reported (Martin et al. 2002).

At the genomic level, Campanaro et al. (2005) have highlighted 27 genomic regions specific to the genome of the piezophilic *P. profundum* SS9 strain, by comparison with the genome of the piezosensitive *Photobacterium* 3TCK et DSJ4 strains. These regions contain, for the greater part, genes encoding phage proteins and integrases, as well as genes for (*i*) the energy metabolism (e.g., gene encoding the NAD(P)H oxidoreductase), (*ii*) the metabolism and the transport of the tryptophan, and (*iii*) the synthesis of flagella and the assembly of pili. These specific regions may have been acquired by horizontal transfer. Comparison between 141 orthologous proteins of the piezophilic archaea *Pyrococcus abyssi* and the piezosensitive *P. furiosus* revealed a significant substitution of amino acids tyrosine and glutamine by the amino acids arginine, glycine, serine, valine, and aspartate in *P. abyssi* (Di Giulio 2005). The rate of polar amino acids was positively correlated to the pressure increase, whereas a negative relation has been established between molecular weight of amino acids and resistance to high hydrostatic pressure. Similar substitutions have been evidenced by comparing the piezophilic *D. piezophilus* to the piezosensitive *D. salexigens* (Pradel et al. 2013).

At the transcriptomic and proteomic levels, the energy metabolism has been evidenced as one of the most important cellular processes involved in high-pressure adaptation (Le Bihan et al. 2013; Pradel et al. 2013; Amrani et al. 2014, 2016). Enzymes produced by high-pressure-adapted bacteria have been shown to be more functional under high-pressure conditions than at atmospheric pressure. In this context, the expression or the structure (e.g., due to amino acids replacements) of enzymes of the metabolism may be modified in piezophilic bacteria, favoring their activity and their stability under conditions of high hydrostatic pressure. For example, a pressure-resistant terminal oxidase has been revealed in *Shewanella violacea* (Tamegai et al. 2011). In the same bacterium, pressure-regulated promoters and differentially expressed cytochromes were identified. In *P. profundum*, a TMAO reductase is more abundant in the cells at highest pressures (28 MPa) (Le Bihan et al. 2013). Finally, in the anaerobic sulfate-reducing bacterium *D. piezophilus*, transcriptomic and biochemical analyses have shown that the metabolite cycling (H_2/Formate) is an important mechanism required for energy conservation at high hydrostatic pressure (26 MPa) (Amrani et al. 2016).

4.6.3.6 Thermophiles/Hyperthermophiles

The discovery in 1965 of the first thermophilic bacterium *Thermus aquaticus* in the geothermal springs of Yellowstone marked the beginning of a new era dedicated to the study of extremophilic microorganisms, among which thermophiles and hyperthermophiles are probably the most documented ones. By definition, thermophiles grow at high temperature, with temperature optima higher than 45 °C and often above 55 °C, and hyperthermophiles thrive fastest at temperatures higher than 80 °C (Gottschal and Prins 1991; Stetter 1999). Thermophiles are found in both *Bacteria* and *Archaea*, while, to date, hyperthermophiles are mostly affiliated to the archaeal domain. The highest temperature limits for life ever recorded are held by two archaea: *Pyrolobus fumarii* exhibiting a maximal growth temperature of 113 °C (Blöchl et al. 1997) and *Methanopyrus kandleri* strain 116 proliferating at 122 °C under high pressure (20–40 MPa) where water remains liquid (Takai et al. 2008). Among *Bacteria*, the maximal growth temperatures ever documented are 95 °C and 100 °C, for *Aquifex pyrophilus* and *Geothermobacterium ferrireducens*, respectively (Huber et al. 1992; Kashefi et al. 2002). The upper temperature limit for life of prokaryotes has not yet been defined but is likely to lie in the region of 132 °C, since peptide and hydrogen bonds, two major chemical linkages governing macromolecules structures and interactions, are destabilized at this temperature. Even if life is not constrained only by temperature but rather by a range of physical and chemical parameters that act both individually and together, it is important to determine what the upper temperature for life is in order to delimit life's boundaries in hot environments and in Earth's subsurface. Such information would help in understanding in particular when and where life might have evolved on the Hadean Earth.

Thermophiles inhabit natural hot environments and may adapt to various physicochemical conditions (high hydrostatic pressure, high salinity, high pH, etc.). So far they have been recovered from water-containing natural marine and terrestrial biotopes and from artificial anthropogenic-driven systems. Hyper-/thermophiles have been recovered from marine biotopes such as shallow and deep-sea hydrothermal vents (located along the mid-ocean ridges and in back-arc basins), coastal shallow volcanic systems, marine volcanoes, and deep subseafloor sediments (the average geothermal gradient in the lithosphere is ca. 25 °C km^{-1}) (Fig. 4.17). Numerous hyper-/thermophilic taxa have also been isolated from high-temperature biotopes on land, such as hot springs, mud volcanoes, solfataric fields, petroleum

Fig. 4.17 A small hydrothermal chimney at Rainbow Hydrothermal Field (Mid-Atlantic Ridge) constituting a privileged habitat for aerobic and anaerobic hyperthermophiles (Credit photo: IFREMER – Victor 6000. EXOMAR cruise 2005)

reservoirs, and deep subterranean habitats (Cayol et al. 2015). Artificial high-temperature biotopes from which thermophiles have been isolated are various and include systems associated with human activities of extraction, transformation, and treatment of materials or wastes at elevated temperatures, such as wastewater treatment plants, smoldering coal refuse piles, sugar refineries, paper mills, hot water pipes, or domestic or industrial boilers. Surprisingly, very few thermophilic *Archaea* and *Bacteria* have been isolated from hot and arid deserts, these ecosystems being rather inhabited by mesophilic species exhibiting tolerance mechanisms to desiccation.

Thermophilic and hyperthermophilic prokaryotes available in clonal cultures are diversed in terms of phylogenetic diversity and biochemical and physiological features. However, the microorganisms cultivated so far represent only a fraction of the whole taxonomic, genomic, and phenotypic diversity of hyper-/thermophiles (Hedlund et al. 2015). Indeed, it is currently difficult to estimate the taxonomic richness and the physiological diversity of hyper-/thermophiles as most of the diversity of thermophiles is yet-uncultivated and remain to be explored. Taxa comprising hyper-/thermophilic representatives that have been detected only by molecular approaches include, for example, taxonomic groups such as "Aigarchaeota" (HWCG I, *Hot Water Crenarchaeotal Group I*), "Korarchaeota," "Calescamantes" (EM19), and "Acetothermia" (OP1) (Hedlund et al. 2015). In the past decade, previously uncultivated taxa were successfully isolated as a result of information gathered from molecular- and/or genome-based methods and/or to geochemical modelling of environmental conditions. This is notably the case for *Aciduliprofundum boonei*, an obligate thermoacidophilic sulfur- or iron-reducing heterotroph, which is the first cultured representative of the DHVE2 (*Deep-sea Hydrothermal Vent Euryarchaeotic 2*) lineage, a widespread euryarchaeotal lineage at deep-sea hydrothermal vents (Reysenbach et al. 2006). To date, less than 1000 thermophilic or hyperthermophilic species, available in axenic cultures, have been formally described (source: *List of Prokaryotic names with Standing in Nomenclature* (Parte 2014)). Efforts continue to isolate novel thermophilic taxa, even if cultivation is time-consuming. Apart from cultivation, genome sequencing efforts and progress in obtaining high-quality draft genomes from metagenomics data have impacted the taxonomic classification of these microorganisms. Some thermophile-containing lineages were reclassified or have been proposed for reclassification. This is the case for the mesophilic to moderately thermophilic *Epsilonproteobacteria* known as primary producers in deep-sea hydrothermal vent systems, where they represent most often the dominant bacterial lineage in the warm area of the deposits (Campbell et al. 2006). On the basis of a comparative genomic analysis performed on more than 600 genomes of *Bacteria*, it was recently proposed to reclassify the class *Epsilonproteobacteria* to a novel phylum named "Epsilonbacteraeota" (Waite et al. 2017). Based on current evidence with regard to archaeal and bacterial thermophiles, there are phyla containing merely thermophiles, while there are phyla containing only thermophiles and mesophiles. A great number of phylum-level lineages containing many thermophiles branch off deeply in phylogenetic trees (Stetter 2006), thus indicating that thermophilic life probably appeared early on Earth. In terms of metabolism, known hyper-/

thermophiles cover a wide range of energy-yielding reactions and can enter the food web at various levels, from the primary production to the terminal degradation of organic matter. Cultivated thermophiles involved in primary production are chemolithoautotrophs using a variety of inorganic electron donors (H_2, Fe^{2+}, S°, $S_2O_3^{2-}$, H_2S) and respiring various electron acceptors, ranging from oxygen to sulfate, and including nitrate, nitrite, ferric iron, other sulfur species, and carbon dioxide. Six different carbon fixation pathways are present among thermophiles: the reductive pentose phosphate cycle (Benson-Calvin cycle), the reductive tricarboxylic acid cycle (rTCA), the reductive acetyl-CoA pathway (Wood-Ljungdahl), the 3-hydroxypropionate bicycle (3-HP), the 3-hydroxypropionate/4-hydroxybutyrate cycle (3-HP/4-HB), and the dicarboxylate/4-hydroxybutyrate (DC/4-HB) cycle (Hügler and Sievert 2011). Several mixotrophs were also described among known thermophiles. With regard to heterotrophic hyper-/thermophiles, there are different various pathways of fermentation and types of respiration that have been reported in yet-cultivated taxa. In recent years, numerous progresses were notably done regarding the description of thermophiles with specialized metabolisms such as carboxydotrophy (Slepova et al. 2006) or sulfur compound disproportionation (Slobodkin et al. 2013), for example. The metabolic potential of yet-uncultivated thermophiles is obviously diverse as well, as revealed by metagenomics and single-genomic data analyses of hot environments, ranging from autotrophy ("Acetothermum," from a subterranean gold mine in Japan) (Takami et al. 2012) to obligate fermentation ("Korarchaeum," from Obsidian Pool, Yellowstone National Park, USA) (Elkins et al. 2008) or respiration ("Calescibacterium," from Great Boiling Spring, Nevada, USA) (Rinke et al. 2013). The discovery of an unexpected diversity of novel methyl-reducing methanogenic archaeal lineages, including those which do not belong to the Euryarchaeota (Vanwonterghem et al. 2016) and inhabit hot habitats (Sorokin et al. 2017), suggests that there are still novel lineages of thermophilic methanogens to discover.

Thermophilic and hyperthermophilic prokaryotes thrive at temperatures much higher than those tolerated by most other prokaryotes and have implemented strategies to maintain the integrity of their enzymes and genomes. A recent review on protein stability pointed out that despite much research in the field, adaptive mechanisms still remain only partially decrypted (Pucci et al. 2017). It appears that a combination of different factors that vary from one protein to another one are involved in thermostability. Nevertheless some general trends emerge clearly; they concern amino acid interactions (i.e., hydrophobic effect, salt bridges, disulfide bridges, etc.) and 3D protein structure (i.e., rigidity, loop shortening, oligomerization, etc.) (Pucci et al. 2017). These organisms can also accumulate small molecules called compatible solutes (Pais et al. 2009) that interact with the protein surface and so enhance their rigidity. Nucleic acid DNA and mostly RNA are highly sensitive to temperature. Several mechanisms that contribute to DNA stability have been identified in hyperthermophilic prokaryotes. The presence of a reverse DNA gyrase that promotes positive supercoiling of DNA (while DNA gyrase from mesophiles produce negative supercoiling) (Forterre et al. 1996; Lulchev and Klostermeier 2014; Lipscomb et al. 2017) is known to ensure a better stability of DNA to temperature.

Single-stranded DNA-binding proteins (Giaquinto et al. 2007) may contribute to the stabilization of nucleic acids and DNA-binding proteins in a similar way to that already reported for histones within the eukaryotes (histone-like proteins), thus contributing to the compaction of the DNA molecule and to its stability (Orfaniotou et al. 2009). Higher G+C% contents in tRNAs and rRNAs of many thermophiles were also reported and are likely to contribute to the thermostability of the genomes (Wang et al. 2015). Maintaining the functionality of the membranes at high temperature is also crucial. A recent review (Siliakus et al. 2017) pointed out that the presence of tetra-ether monolayers in the archaeal membrane is a highly efficient mechanism and that this feature is shared by some hyperthermophilic bacteria. Hopanoids are also found in high concentrations in membranes of thermophilic bacteria and may be involved in maintaining the integrity and fluidity of membranes at high temperatures (Caron et al. 2014).

In recent years, genetic tools have been developed for various thermophilic archaeal and bacterial models (Yu et al. 2001; Thiel et al. 2014; Lipscomb et al. 2017). In the coming years, genetic approaches (gene deletions, mutagenesis, etc.) will obviously contribute to a better understanding of life at high temperature.

Finally, mechanisms of adaptation of microorganisms to harsh and changing physicochemical conditions have in all likelihood developed early in the evolution of life as the primitive atmosphere has been recognized as being anaerobic, slightly reduced to neutral, and hot with marine ecosystems being slightly acidic (see Chap. 3). This is without considering alkaline conditions that resulted from prevailing serpentinization reactions following the contact of water with crustal rock some 3–4 billion years ago (see Chap. 3). In this respect, the first existing microorganisms already had to face extreme physicochemical conditions. We may expect that some of these mechanisms were essential to prokaryotes during the course of evolution whatever their type of metabolism whether aerobic or anaerobic. Extremophiles do not generally use a single strategy; instead they use several strategies for growth in various conditions. These strategies are different from one microorganism to another to maintain the functional integrity of their enzymes and genome. This has been made possible, in particular, by the accumulation of small molecules (organic solutes such as amino acids, cations, etc.) in the cytoplasm, quite often by modification to the lipid composition of membranes (presence of larger amounts of unsaturated fatty acids), and by the emergence of membrane carriers in order to regulate the exchange with the external environment. In some cases, there is evidence that the 3D protein structure and the amino acid content in proteins have evolved to permit cells to thrive in extreme environments. One noticeable mechanism called "pH homeostasis" is a remarkable adaptation of microorganisms to restricting the entrance of protons (acidophiles) or internalizing them in cells (alkaliphiles), thus helping in keeping the proton-motive force for ATP generation operational, despite low or high external pH conditions, respectively. Extremophilic prokaryotes may also use a set of up- and downregulation of genes to adapt to inhospitable conditions. It is noteworthy that the strategies that extremophilic microorganisms use do not allow some metabolisms due to energetic constraints (e.g., methanogenesis from $H_2 + CO_2$ or from acetate in hypersaline ecosystems, etc.). Due to the specific adaptive mechanisms that microbes have acquired, extremophiles and/or microbes that are highly resistant to unfavorable conditions will have facilities to sustain life for the next few billion years despite the important climatic changes expected to occur on our planet on geological time scales (increasing luminosity of the sun resulting in water loss and aridification of the planet).

It thus appears that microbes have colonized all regions of our planet where liquid water was present, whatever the temperature, the pH, the pressure, the presence of ionizing radiation or metal ions, or the salt concentration. It now remains to be seen whether other celestial bodies, some of which have or had liquid water, do harbor microbial life. This will likely be one of the challenges of microbial ecology for the coming century.

4.7 The Functional Redundancy in Favor of the Evolution Success of Prokaryotes

Some properties seem universally required for life. One of the essential points is the absolute necessity for all living things to transduce energy from their environment. The availability of chemical energy is controlled by the environment of an organism and by the set of potential chemical reactions to support life. On a global scale, life is sustained by a flux of energy – either visible light or from chemical sources – throughout the metabolic networks. Due to their high metabolic versatility, only the microorganisms possess such a wide range of possibilities and continues to expand as we explore more of the microbial world.

From a much-cited study from 1998 (Whitman et al. 1998), the number of prokaryotes and the total amount of their cellular carbon on Earth are estimated to be 4–63 10^{30}cells and 350–550 Pg of C, equal to between 60% and 100% of the estimated total carbon in plants, and inclusion of prokaryotic carbon in global models will almost double estimates of the amount of carbon stored in living organisms. In addition, Earth's prokaryotes contain 85–130 Pg of N and 9–14 Pg of P or about tenfold more of these nutrients than do plants and represent the largest pool of these nutrients in living organisms. There are typically 50 million bacterial cells in a gram of soil and 1 million bacterial cells in a milliliter of sea water (Whitman et al. 1998). In the global subseafloor, sedimentary microbial abundance could reach 2.9 · 10^{29}cells [corresponding to 4.1 Pg of C and~0.6% of Earth' s total living biomass]. This estimate of subseafloor sedimentary microbial abundance seems much lower than previous estimates of subsea (Kallmeyer et al. 2012). Thus, prokaryotic carbon would equal to 60–100% of the total plant carbon and potentially could double the total amount of carbon stored in living organisms.

Microorganisms represent the predominant mode of life on Earth; they obtained their energy and nutrients required for growth and maintenance from enzyme-mediated redox reactions, i.e., successive transfers of electrons and protons from a relatively limited set of chemical elements. Metabolic reactions can be classified as either assimilatory, in which chemicals from the environment are absorbed to build cellular components, or dissimilatory, in which chemicals from the environment are reacted to gain energy and the products are released back into the environment. All organisms require sources of carbon (and other structural elements) as well as sources of external energy. That is, the environment must supply a reduced compound (an electron donor) and an appropriate oxidized compound (an electron acceptor) to complete an energy-yielding reaction. Consequently, the vast majority of the fluxes between the different oxidation states of the major elements (H, C, N, O, S) are the result of microbial catalyzed redox reactions. A wide variety of redox couples can be used by various microorganisms to obtain energy, and new ones continue to be discovered periodically. The outcomes of the reactions of the five major elements connected by electron transfers are driven by thermodynamics, but selected by biologically mediated metabolisms. The biogeochemical cycles (H, C, N, O, S) are coupled via microbiologically catalyzed electron transfer reactions. Thus, biogeochemical cycles based on abiotically and biologically driven redox reactions and the feedbacks between microbial metabolisms and geochemical processes create the average redox condition of the oceans and atmosphere.

The evolution of the network of non-equilibrium redox reactions that are the sources of energy of the life on Earth remains largely unknown. However, microorganisms have probably determined the basic composition of Earth's atmosphere since the origin of life and created the breathable, O_2-rich air (Lyons et al. 2014). Indeed, 4.5 billion years ago, free oxygen was at level mostly less than 0.001% of those present in the atmosphere today. During the first half of its evolutionary history, a set of metabolic processes involving exclusively microbes has dramatically modified the surface chemistry of Earth. A most conspicuous expression of this is the accumulation of oxygen. Still, considerable controversy and debate surround when atmospheric oxygen first began to accumulate. The primitive atmosphere of Earth was mainly composed of CO_2 and N_2 with some H_2O and CH_4. Besides the oxygen production, microbes control also the atmospheric concentrations of a number of important greenhouse gases (Conrad 1996). For example, prokaryotes use CO_2 as source of carbon for their biomass synthetize or release CO_2 as a product of decomposition but are also the primary producers of CH_4 and N_2O that are strong greenhouse gases (Shoun et al. 1992). It is usually conceded that the atmospheric greenhouse effect must have been higher in the past to offset reduced solar luminosity, but the levels of atmospheric carbon dioxide and other greenhouse gases required remain speculative. A 1993 model by Kasting (1993) estimates that carbon dioxide (CO_2) levels in Earth's early atmosphere must have been 10 times to as much as 10,000 times today's level. For Kasting volcanic gases released at depth from mid-ocean ridge, hydrothermal vents could have contained appreciable concentrations of CH_4. Indeed, if the oxygen fugacity of the upper mantle was low, most of the carbon released from the mid-ocean ridge vents should have been in the form of CH_4 instead of CO_2, as it is today. The presence of this much CH_4 in the prebiotic atmosphere could have had important implications for the origin of life because it would have also permitted formation of HCN by the mechanism suggested by Zahnle (1986). This abiotic production of "organic material" fuels a long debate. Some authors argue that primeval life was based on anaerobic microorganisms able to use a wide inventory of abiotic organic materials (i.e., a heterotrophic origin), whereas others invoke an organization thrived on simple inorganic molecules such as CO_2 (i.e., an autotrophic origin).

Today, nitrogen and oxygen are by far the most common atmospheric gases; dry air is composed of about 78% nitrogen (N_2) and about 21% oxygen (O_2). Argon, carbon dioxide (CO_2, 400ppm), and many other gases are also present in much lower amounts; each makes up less than 1% of the atmosphere's mixture of gases. In the twentieth century, humans began to have an enormous impact on the global biogeochemical cycles by developing industrial processes, by implementing new agricultural practices, and by burning fossil fuels. The primary greenhouse gases in Earth's atmosphere are water vapor, carbon dioxide, methane, nitrous oxide, and ozone. The atmospheric concentration of carbon dioxide increased from 280 ppm in 1750 to 406 ppm in early 2017, and the concentration of methane has increased by about 150% during the same period. Methane is a strong greenhouse gas with a global warming potential 84 times greater than CO_2 in a 20-year time frame. Nitrous oxide has also a significant global warming potential, considered over a 100-year period; nitrous oxide has 298 times the atmospheric heat-trapping ability of carbon dioxide (CO_2).

Thus, the planet has been subjected to extraordinary environmental changes, from bolide impacts and global glaciations to massive volcanic outgassing, and despite their antiquity, microorganisms maintain their fundamental role in biogeochemical cycle functioning and in maintaining the biosphere. The origin and evolution of metabolic pathways allowed primitive cells to become more chemically independent from the prebiotic sources of essential molecules. It is reasonable to assume that during the early stages of cell evolution, the primitive metabolism was based on a limited number of rudimentary (i.e., unspecific) enzymes. Several mechanisms such as the divergence and duplication or horizontal gene transfers may account for a rapid expansion of metabolic abilities. Thus, success in the evolution of prokaryotes could be explained by a great metabolic diversity and versatility which is expressed through the biogeochemical cycles that they were the only ones initially to ensure the functioning.

4.7.1 The Carbon Cycle

4.7.1.1 Different Autotrophic CO_2 Fixation Pathways

Autotrophic CO_2 fixation represents the most important biosynthetic process in biology permitting to build all cell material solely from inorganic carbon and thus providing the organic carbon for heterotrophic organisms. Moreover, the latter oxidize organic carbon back to inorganic carbon, completing the carbon cycle.

The O_2-resistant Calvin-Benson-Bassham (*CBB*) cycle has long been considered as the only CO_2 fixation pathway, occurring primarily in the first tens of meters of the water column, with oxygenic photosynthesis as the main energy input (Bassham et al. 1950). The characteristic enzyme involved in the *CBB* cycle (or reductive pentose phosphate cycle) is ribulose 1,5-bisphosphate carboxylase/oxygenase (RubisCO) (Quayle et al. 1957), which catalyzes the primary carboxylation of ribulose 1,5-bisphosphate, yielding two molecules of 3-phosphoglycerate. The *CBB* cycle probably evolved in cyanobacteria, and it is the only carbon fixation pathway operating in eukaryotes (algae and plants) as a result of the endosymbiotic acquisition of a cyanobacterium that evolved into the chloroplasts. Overall, the phylogenetic diversity of bacterial groups using the CBB cycle is rather limited (Fig. 4.18a). Besides cyanobacteria, the CBB cycle occurs in photo- and (aerobic) chemoautotrophic *Alpha-*, *Beta-*, and *Gammaproteobacteria*. The key *CB* cycle enzyme, RubisCO, is the most abundant protein in the world (Ellis 1979), as it can comprise up to 50% of the total soluble protein in chloroplasts of a C3 plant or in bacteria using this cycle (Wildman 2002).

Five alternatives to the CBB pathway have been discovered in the last 60 years for CO2 fixation (Fig. 4.18a): the reductive tricarboxylic acid (*rTCA*) cycle, the reductive acetyl-CoA (or Wood-Ljungdahl, *WL*) pathway, the 3-hydroxypropionate (*3-HP*) bicycle, and the recently described 3-hydroxypropionate/4-hydroxybutyrate (*3-HP/4-HB*) and dicarboxylate/4-hydroxybutyrate (*DC/4-HB*) cycles (Ljungdahl 1986; Buchanan and Arnon 1990; Berg et al. 2007; Huber et al. 2008; Zarzycki et al. 2009). The importance of these CO_2 fixation pathways has only started being recognized during the last decade (Raven 2009; Berg 2011; Hügler and Sievert 2011), and our knowledge of dark CO_2 fixation outside hydrothermal vents is still in its infancy. Only an in-depth study to track these alternative pathways in the dark pelagic ocean will unequivocally change our understanding of the carbon cycle and energy flow in the ocean.

Different pathways are expected to be found in the O_2-rich seawater or in particle O_2 gradients. The *rTCA* cycle is essentially a reversal of the oxidative TCA cycle or Krebs cycle (Fig. 4.18b). The *rTCA* cycle is present in quite diverse groups of bacteria; however, due to the oxygen sensitivity of the enzymes 2-oxoglutarate and pyruvate synthase, the cycle appears to be restricted to anaerobic or microaerophilic bacteria (Fig. 4.18). Some pathways indeed harbor O_2-sensitive enzymes (*WL*, *rTCA*, *DC/4-HB*), while the others also function under fully aerobic conditions (*CBB*, *3-HP*, *3-HP/4-HB*) (Hügler and Sievert 2011). The anaerobic WL, rTCA, and DC/4-HB require 1-5 ATPs for the synthesis of pyruvate, against 7–9 ATPs with the *CBB*, *3-HP*, and *3-HP/4-HB* pathways. It is likely that dark ocean organisms adapt to use the most energetically efficient pathway, depending on their immediate physical-chemical environmental conditions. A partition of CO_2 fixation pathways is indeed observed at hydrothermal vent sites according to temperature, which is inversely proportional to the O_2 concentration (Hügler and Sievert 2011).

Even the assimilation of the simplest carbon molecule such as CO_2 leads to the introduction of a large metabolic redundancy. The further distribution of life drove the first organisms into diverse ecological niches, confronting them with different problems; this led to the appearance of new metabolic strategies. Not only were the conditions different, the organisms diversified as well, thus creating additional factors that influenced their metabolism. Life at the thermodynamical limit does not favor the usage of metabolic schemes with high-energy demands. The reversibility

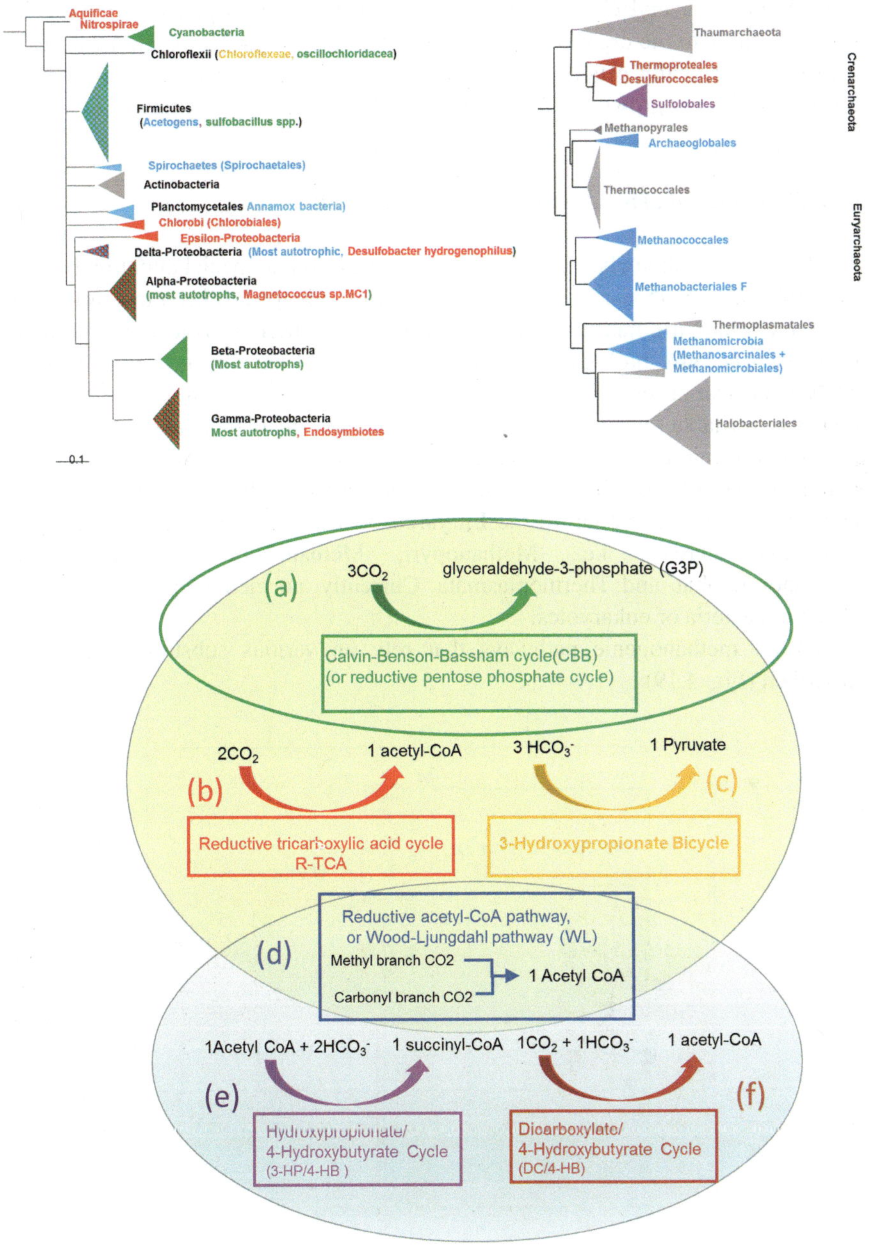

Fig. 4.18 The six pathways of autotrophic CO_2 fixation
(**a**) CBB, (**b**) rTCA, (**c**) 3-HP, (**d**) WL, (**e**) 3-HP/4-HB, and (**f**) DC/4-HB. Carbon balances are reported for each metabolic pathway. The same color is used to write the phylum and its associated metabolic patway, (Adapted from Hugler and Sievert (2011) and Berg (2011))

of the reductive acetyl-CoA pathway and the *rTCA* cycle may give additional metabolic plasticity to the organisms possessing these pathways. However, the usage of a more energetically efficient pathway is not necessarily advantageous, since it ensures the higher growth yields but may result in slower growth rates.

4.7.1.2 Methanogenesis and Methane Oxidation

Methane and its photochemical products deserve also our special attention because their roles as greenhouse gases may very well have helped to keep the early Earth habitable. Once life had arisen on Earth, the atmospheric CH_4 abundance should have risen because biological sources of CH_4 would have been available. In the absence or near absence of oxygen and sulfate, a greater amount of labile organic matter is available for microbial methane production (methanogenesis).

Methanogenic archaea are strict anaerobes that produce methane (CH_4) as the major product of their energy-conserving metabolism. All methanogenic archaea characterized so far belong to the Euryarchaeota and are distributed among five taxonomic classes, i.e., Methanopyri, Methanococci, Methanobacteria, Methanomicrobia, and Thermoplasmata. Currently, no methanogenesis has been found in bacteria or eukaryotes.

Several methanogenic pathways that rely on various substrates have been described (Fig. 4.19):

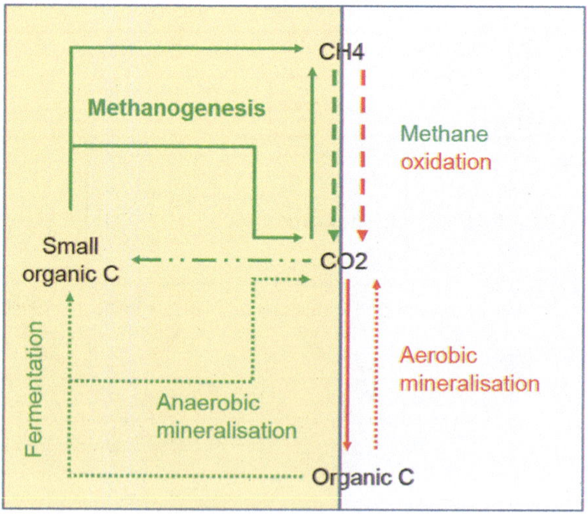

Fig. 4.19 Schematic representation of the prokaryotic carbon cycle
Red and green arrows indicate aerobic and anaerobic pathways, respectively; (1) hydrogenotrophic methanogenesis is a lithotrophic process resulting from the reduction of CO_2 with H_2 as electron donor; (2) formatotrophic methanogenesis, an organotrophic process supported by degradation of formate, acetate, and methylated compounds; (3) methane oxidation can occur via the aerobic or anaerobic pathway

(a) CO_2 reduction with hydrogen (hydrogenotrophic methanogens) or formate (formatotrophic methanogens) as electron donors or methanol reduction with hydrogen. Thermoplasmata reduce methanol with hydrogen (Paul et al. 2012).
(b) Fermentation of acetate (acetotrophic methanogens).
(c) Dismutation of methylated compounds (methylotrophic methanogens) such as methanol, methylamines, dimethyl sulfide (DMS), or methanethiol (Liu and Whitman 2008; Ferry 2010).

Whereas most cultivated methanogens reduce CO_2 with hydrogen, only members of Methanosarcinales have the ability to produce methane from the fermentation of acetate and the dismutation of methylated compounds and might also use methylamines as methanogenic substrate (Poulsen et al. 2013). Although acetate fermentation is performed by only a few cultivated methanogens, this process could account for up to two-thirds of the methane released to the atmosphere by archaeal methanogenesis; the reduction of CO_2 accounts for the rest of the archaeal contribution to atmospheric methane, with minor amounts of methane produced by the dismutation of methyl compounds (Ferry 2010).

Methanogens have been isolated from various anoxic environments (e.g., rice paddies and peat bogs; freshwater, marine, and hypersaline sediments; hydrothermal vents; deep subsurface habitats; the gastrointestinal tract of various animals) and are usually abundant where electron acceptors such as NO_3^-, Fe_3^+, and SO_4^{2-} are in short supply. Hydrogenotrophic methanogens and acetogenic bacteria have similar requirements, including anoxic conditions, a source of H2 as electron donor, and a source of CO_2 as electron acceptor. But methanogenesis occurs preferentially at low H_2 concentrations and at a pH lower than 7; it can also be performed at high temperatures (Thauer et al. 2008).

Syntrophic interactions enable methanogenesis when methanogenic substrates are limiting and methanogenic archaea established in various syntrophic partnerships. This process involves the transfer of electrons from a fermentative organism to the methanogen via a carrier molecule, such as H_2 or acetate. This transfers between two organisms enabling growth on otherwise thermodynamically unfavorable reactions. The methanogens use the carrier molecule as electron donor for energy conservation, and the fermentative organism gains energy from the redox reaction that produces the electron carrier only if the methanogens oxidize the carrier molecule, keeping the carrier at a low concentration.

Methanogenic archaea are a major source of CH_4 emissions, but some of their closest relatives in turn play a critical role in controlling these emissions by oxidizing CH_4 back to CO_2. Methane-oxidizing archaea (AMNE) are strict anaerobes, all belonging to single taxonomic class, the Methanomicrobia of the Euryarchaeota. Twenty years later, AOM was suggested to be a cooperative metabolic process mediated in marine environments by associations between anaerobic methanotrophic archaea (ANMEs) and sulfate-reducing bacteria (Knittel and Boetius 2009).

Today, at least four new ways in which microorganisms achieve anaerobic oxidation of methane (AOM) have been described (Fig. 4.20). Three of these implying two or more syntrophic partners (a, b); in the other two cases, a single microorganism performs both reactions (c,d).

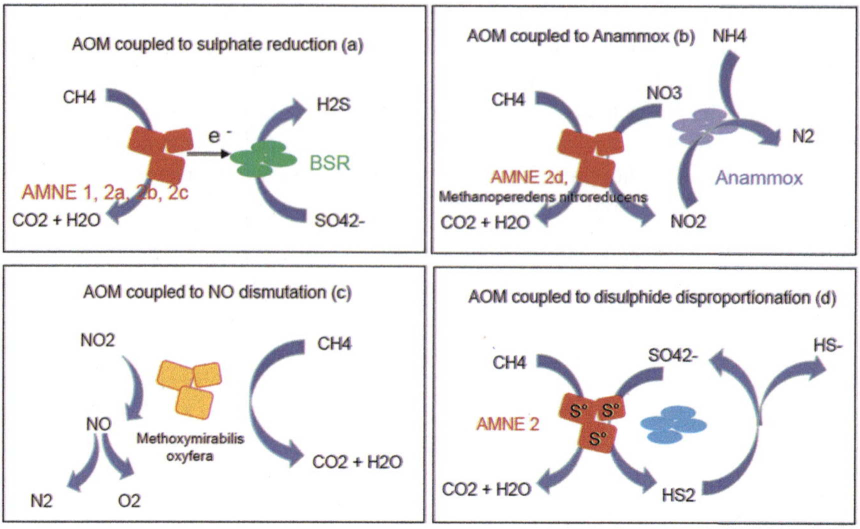

Fig. 4.20 The four known pathways of microbial anaerobic methane oxidation
(**a**) Obligate associations between two or more microbial partners, one performing oxidation of methane into carbon dioxide and water (ANME) and the other performing the reduction (here, a sulfate-reducing bacteria which convert sulfate (SO_4^{2-}) to hydrogen sulfide (H_2S))
(**b**) *Candidatus* "Methanoperedens nitroreducens" can oxidize methane anaerobically through reverse methanogenesis by using nitrate as terminal electron acceptor. Nitrite produced by ANME is reduced to dinitrogen gas through a syntrophic relationship with an anaerobic ammonium-oxidizing bacterium (anammox), with both reactions performed by a single microorganism
(**c**) The bacterium *Methylomirabilis oxyfera* converts nitrite (NO2) to nitric oxide (NO) and then dismutates NO into nitrogen and oxygen as diatomic gases. The bacterium then uses the resulting O_2 to support methane oxidation
(**d**) ANMEs oxidizes methane (as in a) but also reduces sulfate to disulfide which can be used by *Deltaproteobacteria* to yield sulfide (HS^-) and sulfate. (Modified and redrawn from Joye 2012 and Haroon et al. 2013)

(a) The anaerobic oxidation methane (CH_4) into CO2 by methanotrophic archaea (ANMEs) occurs in cooperation with sulfate-reducing bacteria, which convert sulfate (SO_4^{2-}) to hydrogen sulfide (H_2S). The mechanism of energy exchange between the ANMEs and the sulfate-reducing bacteria is unknown (Knittel and Boetius 2009).
(b) The independent AOM by *Candidatus* "Methanoperedens nitroreducens" through reverse methanogenesis by using nitrate as the terminal electron acceptor. Nitrite produced by ANME-2d is reduced to dinitrogen gas through a syntrophic relationship with an anaerobic ammonium-oxidizing bacterium (Haroon et al. 2013).
(c) The O2 produced by the dismutation of NO into diatomic gases by the bacterium *Methylomirabilis* is used by the same microorganism to support methane oxidation. Indeed, "*M. oxyfera*" bypassed the denitrification intermediate

nitrous oxide by the conversion of two nitric oxide molecules to dinitrogen and oxygen which was used to oxidize methane (Ettwig et al. 2010).

(d) Some ANMEs oxidize methane but also reduce sulfate to disulfide (HS_2^-). Zerovalent sulfur compounds (S^0) are formed during AOM through a new pathway for dissimilatory sulfate reduction performed by the methanotrophic archaea. Hence, AOM might not be an obligate syntrophic process but may be carried out by the ANME alone. Furthermore, we show that the produced S^0 – in the form of disulfide – is disproportionated by the *Deltaproteobacteria* associated with the ANME (Milucka et al. 2012).

Anaerobic methane oxidizers exert a strong control over ocean CH_4 emissions, in the oceans more than 50% of the gross annual production of CH_4 could be consumed by anaerobic methanotrophs before CH_4 is even released to ocean waters (Reeburgh 2007).

4.7.2 *The Nitrogen Cycle*

The form(s) of nitrogen, the rate of accretion, and the secondary atmosphere arising from volcanism controlled the prebiotic nitrogen cycle. Planetary accretion models generally assume that nitrogen was delivered to the protoplanet as solid (ice) NH_3, amino acids, and other simple organics. Because ultraviolet oxidation of atmospheric NH_3 (in equilibrium with NH_4^+ in the oceans) would have formed N_2 (Kasting 1982), N_2 gas remained the dominant form of nitrogen in the atmosphere (Canfield et al. 2010). Oxygen rose to its modern levels over the last 550 million years (Berner 2006). With the oxygenation of the ocean, NO_3 became the dominant nitrogen species, with minor concentrations of NH_4^+ and NO_2^- in the water column.

The nitrogen cycle is comprised of a series of redox reactions for transforming nitrogen compounds. Dinitrogen is relatively inert and may be directly used only by some microorganisms in a process called nitrogen fixation that converts inorganic nitrogen into biologically available substrates: ammonium (NH_4^+) and its conjugate acid, ammonia (NH_3). This is the main mechanism for the introduction of nitrogen into the biosphere. Ammonium and ammonia can also be converted to nitrate (NO_3^-) and nitrite (NO_2^-) during the nitrification, an aerobic process which is performed by specialized microorganisms. During denitrification, the nitrate is transformed into gaseous compounds: nitric oxide (NO), nitrous oxide (N_2O), and finally dinitrogen (N_2) which is quickly released back to the atmosphere. It is important to note that different processes can lead to the same product (Herbert 1999). Although many of the pathways in the microbial nitrogen cycle were described more than one century ago, additional fundamental pathways have been only recently discovered due to the rapid advances in molecular ecology and isotopic approaches (Fig. 4.21).

A curious feature of the modern terrestrial nitrogen cycle is that nitrogen fixation and nitrogen loss are largely balanced. This issue deals only with the nitrogen fixation and processes by which dinitrogen fixation is lost from the ecosystems.

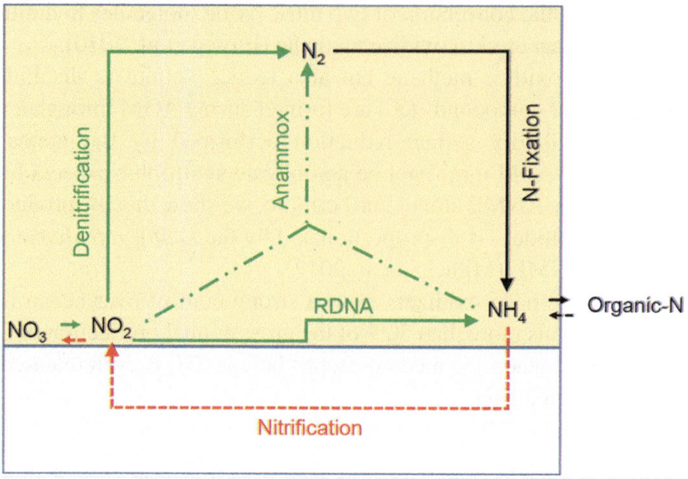

Fig. 4.21 Schematic representation of the nitrogen cycle

4.7.2.1 Dinitrogen Fixation

N_2 fixation is the process by which N_2 gas is reduced into two molecules of ammonia. The reaction consumes cellular energy, and the overall reaction is:

$$N_2 + 8H^+ + 16ATP + 8e^- \rightarrow 2NH_3 + H_2 + 16ADP + 16\,Pi$$

This reaction is catalyzed by the nitrogenase enzyme complex, which is extremely oxygen sensitive.

Nitrogen limitation would have provided strong selection pressure for the evolution of biological N_2 fixation to allow anoxygenic phototrophs to fully make use of the reducing substrates available to them in the environment. Indeed, the ability to fixN_2 evolved early in biological evolution, possibly in anaerobic photoautotrophs, all of which are in the domain *Bacteria* (Canfield et al. 2010)

The capacity for N_2 fixation occurs in bacteria and archaea of diverse physiologies (including anaerobes, facultative aerobes, aerobes, and phototrophs), and although nitrogenase is thought to be an ancient enzyme, it is not uniformly distributed and is present in perhaps a few hundred cultivated species. Cyanobacteria (mainly *Trichodesmium* species) have long been considered the main N_2 fixers in the open ocean and in many marine ecosystems. However, the importance of other diazotrophic groups such as gamma- and deltaproteobacteria has been now demonstrated. These microorganisms are diversed in terms of phylogeny but also of metabolism. Indeed, multiple sources of carbon and energy can be used by diazotrophs, which can be autotrophic or heterotrophic, phototrophic (anoxygenic photosynthetic cyanobacteria or bacteria), or chemotrophic.

Many diazotrophs have had to develop various strategies to protect the dinitrogenase complex of the oxygen. Photosynthesis and N_2 fixation can be uncoupled during the daily cycle, with fixation occurring at night. In this case, ATP required for fixation is provided by respiration and not through photosynthesis. This process has been demonstrated in *Gloeothece* sp., *Synechococcus*, *Microcoleus chthonoplastes*, and *Oscillatoria* under natural conditions; another mechanism exists in some filamentous cyanobacteria that have specialized cell structures for N_2 fixation, the heterocysts, which promote a spatial uncoupling of the two types of metabolism. The formation of heterocysts requires a significant change in vegetative cells. Indeed, the cell wall is modified, and narrow junctions appear between heterocysts and vegetative cells.

To prevent the irreversible inactivation of enzymes, other strategies consist to increase the respiratory activity, to inactivate of the enzyme when the partial pressure of oxygen becomes too high. This changed the conformation of the FeS-protein which binds to dinitrogenase in the presence of oxygen (known as conformation protection or "switch-off-switch-on" phenomenon).

The majority of marine N_2 fixation has historically been attributed to the filamentous, non-heterocystous cyanobacteria *Trichodesmium* spp. resident in the warm, stratified, and nutrient-depleted regions of the surface ocean (Carpenter 1983; Capone et al. 1997). Additionally, indirect evidence such as teledetection (Westberry and Siegel 2006) and geochemical modelling (Deutsch et al. 2007) describes geographic distributions of N_2 fixers, including *Trichodesmium* spp., that differ from our expectation of oligotrophic dominance. Finally, a number of both in situ and culture-based studies raise the possibility that sensitivities of marine diazotrophs may be different than previously thought. So, the possibility that N_2 fixation occurs in environments beyond the surface waters of the oligotrophic gyres cannot be excluded.

4.7.2.2 Dinitrogen Production

Low-oxygen environments are of particular interest for nitrogen transformations because they are the sites of fixed N loss via denitrification and anammox that are important ecological processes to maintain nitrogen equilibrium.

4.7.2.2.1 Denitrification

Denitrification is an anaerobic respiration process where nitrate is used as the terminal acceptor of electrons. The energy efficiency of denitrification is lower compared to aerobic respiration. Denitrification yields 24 molecules of ATP for one molecule of reduced. During denitrification, the transformation of nitrate to dinitrogen is carried out by four steps during which the initial nitrate (NO3) is successively reduced to nitrite (NO_2^-), nitric oxide (NO), and nitrous oxide (N_2O), up to the production of molecular nitrogen (N_2). The process involves four enzymes: dissimilatory nitrate reductase, nitrite reductase, nitric oxide reductase, and nitrous oxide reductases. For each step, there are several possibilities.

The first step of denitrification, dissimilatory reduction of nitrate to nitrite, can be performed by two types of dissimilatory nitrate reductases which have a different location in the cell: one is periplasmic (Nap) and the other is membranous (Nar). One or both enzymes are present without phylogenetic determination. The role of the periplasmic nitrate reductase (Nap) is poorly understood, and different physiological functions have been proposed. Periplasmic nitrate reductase may help to maintain the redox balance in cells when adapting to some environmental changes, including the transition between aerobic and anaerobic conditions. Because Nap is not sensitive to oxygen, some denitrifiers can realize aerobic denitrification by coupling this enzyme to nitrite reductase or to nitric oxide reductase. This is particularly favorable for microorganisms that thrive in natural environments subject to changing conditions in oxygen availability.

The nitrite transferred into the periplasmic space can get in contact with nitrite reductase, a periplasmic enzyme which catalyzes the second step of denitrification that is the reduction of nitrite to nitric oxide. There are also two types of nitrite reductases which are very different in their structure, but have a similar function: one is made of cytochrome cd1 (Nir-cd1) and the other of copper (Nir-Cu). Until now, these two types of nitrite reductases were never found simultaneously in the same organism.

Nitric oxide reductase, membranous cytochrome bc complex, and nitrous oxide reductase were responsible of the two final steps of denitrification allowing the reduction of nitric oxide (NO) to nitrous oxide (N_2O) and then to dinitrogen (N_2).

In addition to the unconventional activities by classical wellknown nitrifiers and denitrifiers and the discovery of novel N metabolic pathways in new organisms, a short circuit of the nitrification/denitrification has also been proposed recently. In marine sediments, which typically contain relatively high manganese levels, N_2 gas can be produced by the oxidation of ammonia by manganese oxide (Fernandes et al. 2015; Aigle et al. 2017).

4.7.2.2.2 Anammox

A completely novel process, in which nitrite is used as the electron acceptor for the anaerobic ammonium oxidation in nitrogen gas as the final product in the presence of ammonia, has been more recently described. Next to ammonium, organic and inorganic compounds can be used as alternative electron donors, e.g., propionate, acetate, and formate. At least for "*Candidatus* Kuenenia stuttgartiensis" and "*Candidatus* Scalindua" spp., it has been shown that, besides nitrite, iron and manganese oxides can also be use as electron acceptor.

Anammox was recently recognized as one of the major sinks for fixed inorganic nitrogen in coastal sediments and the anoxic waters of basins isolated from oxygenated deep circulation. Globally, 30–50% of the total nitrogen loss occurs in oxygen-minimum zones (OMZs) and is commonly attributed to heterotrophic denitrification.

There has been, until now, no published direct evidence for anammox in OMZs. However, the extremely low concentration of ammonium could indicate that anammox bacteria also play an important role in the nitrogen removal from OMZ waters (Kuypers et al. 2005).

All organisms responsible for this novel metabolism have been identified as relatives of *Planctomyces*. So far, the capability of anammox is limited to a very specific order of the Brocadiales, while denitrification occurs in bacteria, archaea, and even eukaryotes (van Niftrik and Jetten 2012). Anammoxifiers are anaerobic chemolithoautotrophic microorganism with an unusual morphology.

The cytoplasm in anammoxifiers was thus proposed to be divided into three cytoplasmic compartments separated by single bilayer membranes. The outermost compartment, the paryphoplasm, occurs as an outer rim, defined on its outer side by the cytoplasmic membrane and cell wall and on the inner side by the intracytoplasmic membrane. The middle compartment, the riboplasm, contains ribosomes and the nucleoid. Finally, the innermost ribosome-free compartment, the anammoxosome, occupies most of the cell volume and is bound by the anammoxosome membrane. In addition to the cell plan, anammox bacteria contain also atypical membrane lipids named ladderanes.

Anammox bacteria do not conform to the typical characteristics of bacteria but instead share features with all three domains of life, *Bacteria*, Archaea, and Eukarya, making them extremely interesting from an evolutionary perspective.

Anoxic ammonia oxidation, whether it results directly in N2 formation (as in anammox) or in nitrate production (when linked to manganese reduction), would introduce new links into the aquatic and sediment N cycle.

4.8 Conclusion

The study of the evolution of prokaryotes confirms that all discoveries done after Darwin validate his natural selection theory. The post-Darwinian theories – neo-Darwinism, neutralist evolution, punctuated equilibria, and selfish gene – are not fundamentally opposed to Darwin's proposals; they enrich it and lead to incremental revisions. For instance, the discovery of mutations and horizontal gene transfers (HGTs), and in a more general fashion of mobile elements, contradict both a progressive vision of evolution and the notion of heredity as conceived by Darwin: a purely vertical heredity (our classifications will come to be, as far as they can be so made, genealogies). As Albert Jan Kluyver famously wrote, "From elephant to butyric acid bacterium – it is all the same" to underline the profound unity of life's biochemical mechanisms, and since it is simpler to work with bacteria than mammals, evolution will continue to be most profitably studied in prokaryotes.

In this chapter were presented all the characteristics that explain the evolutionary success of prokaryotes, their role in the evolution of the geosphere, the biosphere and its functioning, and their ability to colonize all biotopes, including the most extreme ones:

- A mode of reproduction with a very high efficiency: a high rate of reproduction and an optimal utilization of their genetic information.
- A short generation time within large populations results in a high mutation rate, and haploidy makes each mutation the origin of a new lineage, thus permitting a rapid expression of mutations. The importance of mutations in prokaryotes is amplified by the presence in populations of constitutive and inducible mutators.
- An extraordinary capacity to exchange genetic information and integrate in their genome genetic information originating from other organisms through horizontal transmission. These horizontal gene transfers (HGTs) – through conjugation, transduction, and transformation – often very important may occur between evolutionarily very distant species, even belonging to different domains of life (*Bacteria*, Archaea, Eukarya). HGTs will result in the acquisition of new adaptive capacities: resistance to antibiotics, capacity to deal with artificial substances such as xenobiotics that never existed in nature before the advent of man, etc.
- The small size of their cells and their large populations have allowed prokaryotes to conquer more biotopes than other domains. Presently, they represent the most important biomass in the living world.
- Through evolution, prokaryotes have developed an array of mechanisms to ensure their survival. The survival capacity under unfavorable conditions is not limited to the emergence of mutators (constitutive or inducible). Prokaryotes possess adaptive mechanisms that allow them to face the constantly and rapidly changing conditions of their habitat (nutrients levels, thermal or osmotic shock, increase or drop of pH and oxygen concentration, prolonged dehydration, UV irradiation, etc.).
- Certain prokaryotes live under oligotrophic conditions which is their normal lifestyle (ultramicrobacteria), especially in the open ocean.
- Moreover, under adverse conditions, some prokaryotes are able to survive, sometimes for very long periods of time: (I) by utilizing precious intracellular reserves, (II) by the emergence of forms of resistance (spores, cysts, fruiting bodies), or (III) by undergoing conversion to a dormancy or viable but noncultivable (VBNC) state. If dormancy is a reversible state ("resuscitation"), the notion of viable but noncultivable cellular forms is still a matter of debate; only some strains of VBNC are known to recover a normal physiology when environmental conditions become favorable again.
- Prokaryotes have also evolved to be able to live continually under the most extreme life conditions using various strategies to adapt to in situ harsh physicochemical conditions with the aim to maintain cell integrity and functioning. They include (i) the accumulation of small molecules (organic solutes such as amino acids, cations, etc.) in the cytoplasm, (ii) a higher content of unsaturated fatty acids within the membrane lipid composition, (iii) the use of membrane carriers for regulating exchanges with the outside environment, (iv) the modification of the 3D protein structure and the protein amino acid content as well, (v) the setting up of up- and downregulation of genes to face such drastic conditions, and possibly (vi) other unknown mechanisms to be discovered. Such strategies might have been of first importance to facilitate the emergence of life on the planet where extreme physicochemical conditions prevailed some 3.7–3.8 billion years ago.

Over geological times, prokaryotes have acquired all the functions that are necessary to biogeochemical cycles functioning, enabling them to ensure all transformations of chemical elements, and, in the past and present, almost all reactions on Earth's surface are catalyzed by prokaryotes. Furthermore, their biogeochemical functions have created environmental conditions in terrestrial habitats in which eukaryotic life forms could appear, grow, and diversify, thus showing the essential and indispensable role of prokaryotes in the evolution of the living world. If all living eukaryotes should disappear following a natural or man-made catastrophe, from protists to *Homo sapiens*, biogeochemical cycles should continue to function with prokaryotes. Life would continue, and a new evolutionary cycle involving microbes would then start. On the other hand, in a bacteria- and archaea-free world, most biogeochemical cycling would cease, and "Life would not longer remain possible in the absence of microbes" (Louis Pasteur).

References

Aanderud Z, Jones S, Fierer N, Lennon J (2015) Resuscitation of the rare biosphere contributes to pulses of ecosystem activity. Front Microbiol 6:24

Abe A, Ohashi E, Ren H, Hayashi T, Endo H (2007) Isolation and characterization of a cold-induced nonculturable suppression mutant of *Vibrio vulnificus*. Microbiol Res 162:130–138

Abe F (2013) Dynamic structural changes in microbial membranes in response to high hydrostatic pressure analyzed using time-resolved fluorescence anisotropy measurement. Biophy Chemistry 183:3–8

Aguilera A (2013) Eukaryotic organisms in extreme acidic environments, the Rio Tinto case. Life (Basel) 3:363–374

Aigle A, Bonin P, Iobbi-Nivol C, Méjean V, Michotey V (2017) Physiological and transcriptional approaches reveal connection between nitrogen and manganese cycles in *Shewanella algae* C6G3. Sci Rep 7:44725

Al-Hinai M, Jones S, Papoutsakis E (2015) The *Clostridium* sporulation programs: diversity and preservation of endospore differentiation. Microbiol Mol Biol Rev 79:19–37

Alazard D, Joseph M, Battaglia-Brunet F, Cayol J, Ollivier B (2010) *Desulfosporosinus acidiphilus* sp. nov.: a moderately acidophilic sulfate-reducing bacterium isolated from acid mining drainage sediments : New taxa: Firmicutes (Class Clostridia, Order Clostridiales, Family Peptococcaceae). Extremophiles 14:305–312

Allen E, Bartlett D (2004) Piezophiles: microbial adaptation to the deep-sea environment. In: Gerday C, Glansdorff N (eds) Extremophiles, vol 3. Eolss Publishers Co Ltd., Oxford, pp 231–255

Alvarez C, Alvarez R, Grigera M, Lavado R (1998) Associations between organic matter fractions and the active soil microbial biomass. Soil Biol Biochem 30:767–773

Amato P, Ménager M, Sancelme M, Laj P, Mailhot G, Delort A-M (2005) Microbial population in cloud water at the Puy de Dôme: implications for the chemistry of clouds. Atmos Environ 39:4143–4153

Amato P, Demeer F, Melaouhi A, Fontanella S, Martin-Biesse A-S, Sancelme M, Laj P, Delort A-M (2007a) A fate for organic acids, formaldehyde and methanol in cloud water: their biotransformation by microorganisms. Atmos Chem Phys 7:4159–4169

Amato P, Parazols M, Sancelme M, Laj P, Mailhot G, Delort AM (2007b) Microorganisms isolated from the water phase of tropospheric clouds at the Puy de Dome: major groups and growth abilities at low temperatures. FEMS Microbiol Ecol 59:242–254

Amato P, Parazols M, Sancelme M, Mailhot G, Laj P, Delort A-M (2007c) An important oceanic source of micro-organisms for cloud water at the puy de Dôme (France). Atmos Environ 41:8253–8263

Amel B-N, Amine B, Amina B (2008) Survival of *Vibrio fluvialis* in seawater under starvation conditions. Microbiol Res 163:323–328

Amrani A, Bergon A, Holota H, Tamburini C, Garel M, Ollivier B, Imbert J, Dolla A, Pradel N (2014) Transcriptomics reveal several gene expression patterns in the piezophile *Desulfovibrio hydrothermalis* in response to hydrostatic pressure. PLoS ONE 9:e106831

Amrani A, van Helden J, Bergon A, Aouane A, Ben Hania W, Tamburini C, Loriod B, Ollivier B, Imbert J, Ollivier B et al (2016) Deciphering the adaptation strategies of *Desulfovibrio piezophilus* to hydrostatic pressure through metabolic and transcriptional analyses. Environ Microbiol Rep 8:520–526. https://doi.org/10.1111/1758-2229.12427

Anderson I, Rodriguez J, Susanti D, Porat I, Reich C, Ulrich L, Elkins J, Mavromatis K, Lykidis A, Kim E et al (2008) Genome sequence of *Thermofilum pendens* reveals an exceptional loss of biosynthetic pathways without genome reduction. J Bacteriol 190:2957–2965

Anderson I et al (2009) The complete genome sequence of *Staphylothermus marinus* reveals differences in sulfur metabolism among heterotrophic Crenarchaeota. *BMC Genomics* 10:145

Anderson I et al (2010) Complete genome sequence of *Methanothermus fervidus* type strain (V24S^T). Stand Genomic Sci 3:315–324

Angert ER, Clements KD, Pace NR (1993) The largest bacterium. Nature 362:239–241

Aono E, Baba T, Ara T, Nishi T, Nakamichi T, Inamoto E, Toyonaga H, Hasegawa M, Takai Y, Okumura Y et al (2010) Complete genome sequence and comparative analysis of *Shewanella violacea*, a psychrophilic and piezophilic bacterium from deep-sea floor sediments. Mol Biosyst 6:1216–1226

Arana I, Seco C, Epelde K, Muela A, Fernández-Astorga A, Barcina I (2004) Relationships between *Escherichia coli* cells and the surrounding medium during survival processes. Antonie Van Leeuwenhoek 86:189–199

Arsene F, Tomoyasu T, Bukau B (2000) The heat shock response of *Escherichia coli*. Int J Food Microbiol 55:3–9

Avery OT, MacLeod CM, McCarty M (1944) Studies on the chemical nature of the substance inducing transformation of pneumococcal types. Induction of transformation by a desoxyribosenucleic acid fraction isolated from pneumococcus type III. J Exp Med 79:137–158

Ayrapetyan M, Oliver JD (2016) The viable non-culturable state and its relevance in food safety. COFS. https://doi.org/10.1016/j.cofs.2016.1004.1010

Babic A, Lindner A, Vulic M, Stewart E, Radman M (2008) Direct visualization of horizontal gene transfer. Science 319:1533–1536

Baharoglu Z, Mazel D (2014) SOS, the formidable strategy of bacteria against aggressions. FEMS Microbiol Rev 38:1126–1145

Baker-Austin C, Dopson M (2007) Life in acid: pH homeostasis in acidophiles. Trends Microbiol 15:165–171

Baker BJ et al (2010) Enigmatic, ultrasmall, uncultivated Archaea. Proc Natl Acad Sci U S A 107:8806–8811

Bale S, Goodman K, Rochelle P, Marchesi J, Fry J, Weightman A, Parkes R (1997) *Desulfovibrio profundus* sp. nov., a novel barophilic sulphate-reducing bacterium from deep sediment layers in the Japan Sea. Int J Syst Bacteriol 47:515–521

Bapteste E, O'Malley M, Beiko R, Ereshefsky M, Gogarten J, Franklin-Hall L, Lapointe F, Dupré J, Dagan T, Boucher Y et al (2009) Prokaryotic evolution and the tree of life are two different things. Biol Direct 4:34

Barabote RD, Xie G, Leu DH, Normand P, Necsulea A, Daubin V, Medigue C, Adney WS, Xu XC, Lapidus A et al (2009) Complete genome of the cellulolytic thermophile *Acidothermus cellulolyticus* 11B provides insights into its ecophysiological and evolutionary adaptations. Genome Res 19:1033–1043

Bassham JA, Benson AA, Calvin M (1950) The path of carbon in photosynthesis. J Biol Chem 185:781–787

Beal E, House C, Orphan V (2009) Manganese- and iron-dependent marine methane oxidation. Science 325:184–187

Ben Aissa F, Postec A, Erauso G, Payri C, Pelletier B, Hamdi M, Fardeau M, Ollivier B (2015) Characterization of *Alkaliphilus hydrothermalis* sp. nov., a novel alkaliphilic anaerobic bacterium, isolated from a carbonaceous chimney of the Prony hydrothermal field, New Caledonia. Extremophiles 19:183–181

Bennett GM, Abbà S, Kube M, Marzachì C (2016) Complete genome sequences of the obligate symbionts "*Candidatus* Sulcia muelleri" and "*Ca.* Nasuia deltocephalinicola" from the Pestiferous Leafhopper *Macrosteles quadripunctulatus* (Hemiptera: Cicadellidae). Genome Announc 4(1):e1604–e1615

Berg I (2011) Ecological aspects of the distribution of different autotrophic CO_2 fixation pathways. Appl Environ Microbiol 77:1925–1936

Berg I, Kockelkorn D, Buckel W, Fuchs GA (2007) 3-hydroxypropionate/4-hydroxybutyrate autotrophic carbon dioxide assimilation pathway in Archaea. Science 31:1782–1786

Berger S, David M, Vogel TM, Simonet P (2015) Involvement of horizontal gene transfer in chlorinated compounds biological degradation: example of the biological degradation of lindane. In: Bertrand J, Caumette P, Lebaron P, Matheron R, Normand P, Sime-Ngando T (eds) Environmental microbiology: fundamentals and applications. Springer, Dordrecht/Heidelberg/New York/London, pp 705–707

Berleman J, Auer M (2013) The role of bacterial outer membrane vesicles for intra- and interspecies delivery. Environ Microbiol 15:347–354

Berner RA (2006) GEOCARBSULF: a combined model for Phanerozoic atmospheric O_2 and CO_2. Geochim Cosmochim Acta 70:5653–5664

Bernhardt HS, Tate WP (2012) Primordial soup or vinaigrette: did the RNA world evolve at acidic pH? Biol Direct 7:4

Berry AM, Harriott OT, Moreau RA, Osman SF, Benson DR, Jones AD (1993) Hopanoid lipids compose the *Frankia* vesicle envelope, presumptive barrier of oxygen diffusion to nitrogenase. Proc Natl Acad Sci U S A 90:6091–6094

Bes M, Merrouch M, Joseph M, Quéméneur M, Payri C, Pelletier B, Ollivier B, Fardeau M, Erauso G, Postec A (2015) *Acetoanaerobium pronyense* sp. nov., an anaerobic alkaliphilic bacterium isolated from a carbonate chimney of the Prony Hydrothermal Field (New Caledonia). Int J Syst Evol Microbiol 65:2574–2580

Biddle J, Sylvan J, Brazelton W, Tully B, Edwards K, Moyer C, Heidelberg J, Nelson W (2012) Prospects for the study of evolution in the deep biosphere. Front Microbiol 2:285

Biller SJ, Schubotz F, Roggensack SE, Thompson AW, Summons RE, Chisholm SW (2014) Bacterial vesicles in marine ecosystems. Science 343:183–186

Blöchl E, Rachel R, Burggraf S, Hafenbradl D, Jannasch H, Stetter K (1997) *Pyrolobus fumarii*, gen. and sp. nov., represents a novel group of archaea, extending the upper temperature limit for life to 113°C. Extremophiles 1:14–21

Boaretti M, Lleò MM, Bonato B, Signoretto C, Canepari P (2003) Involvement of *rpoS* in the survival of *Escherichia coli* in the viable but non-culturable state. Environ Microbiol 5:986–996

Bogosian G, Morris P, O'Neil J (1998) A mixed culture recovery method indicates that enteric bacteria do not enter the viable but nonculturable state. Appl Environ Microbiol 64:1736–1742

Boto L (2010) Horizontal gene transfer in evolution: facts and challenges. Proc R Soc Lond B Biol Sci 277:819–827

Bottos E, Woo A, Zawar-Reza P, Pointing S, Cary S (2014) Airborne bacterial populations above desert soils of the McMurdo Dry Valleys, Antarctica. Microb Ecol 67:120–128

Braun M, Mayer F, Gottschalk G (1981) *Clostridium aceticum* (Wieringa), a microorganism producing acetic acid from molecular hydrogen and carbon dioxide. Arch Microbiol 128:288–293

Brazelton W, Ludwig K, Sogin M, Andreishcheva E, Kelley D, Shen C, Edwards R, Baross J (2010) Archaea and bacteria with surprising microdiversity show shifts in dominance over 1,000-year time scales in hydrothermal chimneys. Proc Natl Acad Sci U S A 107:1612–1617

Brochier C, Philippe H, Moreira D (2000) The evolutionary history of ribosomal protein RpS14: horizontal gene transfer at the heart of the ribosome. Trends Genet 16:529–533

Brochier-Armanet C, Moreira D (2015) Horizontal gene transfer in microbial ecosystems. In: Bertrand J, Caumette P, Lebaron P, Matheron R, Normand P, Sime-Ngando T (eds) Environmental microbiology: fundamentals and applications. Springer, Dordrecht/Heidelberg/New York/London, pp 471–512

Brzuszkiewicz E, Brüggemann H, Liesegang H, Emmerth M, Olschläger T, Nagy G, Albermann K, Wagner C, Buchrieser C, Emody L et al (2006) How to become a uropathogen: comparative genomic analysis of extraintestinal pathogenic *Escherichia coli* strains. Proc Natl Acad Sci U S A 103:12879–12884

Bult CJ et al (1996) Complete genome sequence of the methanogenic archaeon, *Methanococcus jannaschii*. Science 273:1058–1073

Buchanan B, Arnon D (1990) Reverse KREBS cycle in photosynthesis: consensus at last. Photosynth Res 24:47–53

Burrows S, Butler T, Jöckel P, Tost H, Kerkweg A, Pöschl U, Lawrence M (2009) Bacteria in the global atmosphere – Part 2: Modeling of emissions and transport between different ecosystems. Atmos Chem Phys 9:9281–9297

Campanaro S, Treu L, Valle G (2008) Protein evolution in deep sea bacteria: an analysis of amino acids substitution rates. BMC Evol Biol 8:313

Campanaro S, Vezzi A, Vitulo N, Lauro F, D'Angelo M, Simonato F, Cestaro A, Malacrida G, Bertoloni G, Valle G et al (2005) Laterally transferred elements and high pressure adaptation in *Photobacterium profundum* strains. BMC Genomics 6:122

Campbell B, Engel A, Porter M, Takai K (2006) The versatile ε-proteobacteria: key players in sulphidic habitats. Nat Rev Microbiol 4:458–468

Canfield D, Glazer A, Falkowsli P (2010) The evolution and future of Earth's nitrogen cycle. Sciences 330:192–196

Cano R, Borucki M (1995) Revival and identification of bacterial spores in 25- to 40 million-year-old Dominican amber. Science 268:1060–1064

Capone DG, Zehr JP, Paerl HW, Bergman B, Carpenter EJ (1997) *Trichodesmium*, a globally significant marine cyanobacterium. Science 276:1221–1229

Caron B, Mark A, Poger D (2014) Some like it hot: the effect of sterols and hopanoids on lipid ordering at high temperature. J Phys Chem Lett 5:3953–3957

Carpenter EJ (1983) Nitrogen fixation by marine *Oscillatoria (Trichodesmium)* in the world's oceans. In: Carpenter EJ, Capone DG (eds) Nitrogen in the marine environment. Academic, NY, pp 65–104

Cavalier-Smith T (2002) The neomuran origin of archaebacteria, the negibacterial root of the universal tree and bacterial megaclassification. Int J Syst Evol Microbiol 52:7–76

Cavicchioli R (2016) On the concept of a psychrophile. ISME J 10:793–795

Cayol J, Ollivier B, Patel B, Ageron E, Grimont P, Prensier G, Garcia J (1995) *Haloanaerobium lacusroseus* sp. nov., an extremely halophilic fermentative bacterium from the sediments of a hypersaline lake. Int J Syst Bacteriol 45:790–797

Cayol J-L, Ollivier B, Alazard D, Amils R, Godfroy A, Piette F, Prieur D (2015) The extreme conditions of life on the planet and exobiology. In: Bertrand J-C, Caumette P, Lebaron P, Matheron R, Normand P, Sime-Ngando T (eds) Environmental microbiology: fundamentals and applications: microbial ecology. Springer, Dordrecht, pp 353–394

Chaiyanan S, Chaiyanan S, Grim C, Maugel T, Huq A, Colwell R (2007) Ultrastructure of coccoid viable but non-culturable *Vibrio cholerae*. Environ Microbiol 9:393–402

Chan C, Beiko R, Ragan M (2017) Scaling up the phylogenetic detection of lateral gene transfer events. In: Keith J (ed) Method in molecular biology bioinformatic, data, sequence analysis, and evolution, vol 1. Springer, New York

Chatterjee S, Rothenberg E (2012) Interaction of bacteriophage 1 with its *E. coli* receptor, LamB. Viruses 4:3162–3178

Chattopadhyay S, Weissman SJ, Minin VN, Russo TA, Dykhuizen DE, Sokurenko EV (2009) High frequency of hotspot mutations in core genes of Escherichia coli due to short-term positive selection. Proc Natl Acad Sci U S A 106:12412–12417

Chen G-Q (2009) A microbial polyhydroxyalkanoates (PHA) based bio- and materials industry. Chem Soc Rev 38:2434–2446
Chen I, Dubnau D (2004) DNA uptake during bacterial transformation. Nat Rev Microbiol 2:241–249
Chiarabelli C, Stano P, Luisi P (2009) Chemical approaches to synthetic biology. Curr Opin Biotechnol 20:492–497
Christman MF, Storz G, Ames BN (1989) OxyR, a positive regulator of hydrogen peroxide-inducible genes in *Escherichia coli* and *Salmonella typhimurium*, is homologous to a family of bacterial regulatory proteins. Proc Natl Acad Sci U S A 86:3484–3488
Colwell FS, D'Hondt S (2013) Nature and extent of the deep biosphere. In: Hazen RM, Jones AP, Baross JA (eds) Reviews in mineralogy and geochemistry. Mineralogical Society of America, Chantilly, pp 547–574
Colwell R, Brayton P, Herrington D, Tall B, Huq A, Levine M (1996) Viable but non-culturable *Vibrio cholerae* 01 revert to a cultivable state in the human intestine. World J Microbiol Biotechnol 12:28–31
Colwell RR (2000) Viable but nonculturable bacteria: a survival strategy. J Infect Chemother 6:121–125
Comolli L, Baker B, Downing K, Siegerist C, Banfield J (2009) Three-dimensional analysis of the structure and ecology of a novel, ultra-small archaeon. Isme J 3:159–167
Conrad R (1996) Soil microorganisms as controllers of atmospheric trace gases (H_2, CO, CH_4, OCS, N_2O, and NO). Microb Rev 60:609–640
Cox M (2003) The bacterial RecA protein as a motor protein. Annu Rev Microbiol 57:551–577
Cuny C, Dukan L, Fraysse L, Ballesteros M, Dukan S (2005) Investigation of the first events leading to loss of culturability during *Escherichia coli* starvation: future nonculturable bacteria form a subpopulation. J Bacteriol 187:2244–2248
d'Ari R (1985) The SOS system. Biochimie 67:343–347
D'Amico S, Collins T, Marx J, Feller G, Gerday C (2006) Psychrophilic microorganisms: challenges for life. EMBO Rep 7:385–389
Damsté J, Rijpstra W, Hopmans E, Schouten S, Balk M, Stams A (2007) Structural characterization of diabolic acid-based tetraester, tetraether and mixed ether/ester, membrane-spanning lipids of bacteria from the order *Thermotogales*. Arch Microbiol 188:629–641
Darcan C, Özkanca R, Idil O, Ken P, Flint K (2009) Viable but non-culturable state (VBNC) of *Escherichia coli* related to EnvZ under the effect of pH, starvation and osmotic stress in sea water. Pol J Microbiol 58:307–317
Darwin C (1859) The origin of species by means of natural selection, or the preservation of favoured races in the struggle for life. John Murray, London
DasSarma S, Arora P (2001) Halophiles. In: Encyclopedia of life sciences. Nature Publishing Group, London, pp 1–9
Daubin V, Szöllősi G (2016) Horizontal gene transfer and the history of life. Cold Spring Harb Perspect Biol 8:a018036
Davis C (2014) Enumeration of probiotic strains: review of culture-dependent and alternative techniques to quantify viable bacteria. J Microbiol Methods 103:9–17
Dawkins R (1976) The selfish gene. Oxford University Press, London
Day A, Oliver J (2004) Changes in membrane fatty acid composition during entry of *Vibrio vulnificus* into the viable but nonculturable state. J Microbiol 42:69–73
De Maayer P, Anderson D, Cary C, Cowan D (2014) Some like it cold: understanding the survival strategies of psychrophiles. EMBO Rep 15:508–517
Delort A-M, Vaïtilingom M, Amato P, Sancelme M, Parazols M, Mailhot G, Laj P, Deguillaume L (2010) A short overview of the microbial population in clouds: potential roles in atmospheric chemistry and nucleation processes. Atmos Res 98:249–260
Denamur E, Matic I (2006) Evolution of mutation rates in bacteria. Mol Microbiol 60:820–827
Derelle E, Ferraz C, Rombauts S, Rouzé P, Worden AZ, Robbens S, Partensky F, Degroeve S, Echeynié S, Cooke R et al (2006) Genome analysis of the smallest free-living eukaryote *Ostreococcus tauri* unveils many unique features. Proc Natl Acad Sci U S A 103:11647–11652

Deutsch C, Sarmiento J, Sigman DM, Gruber N, Dunne JP (2007) Spatial coupling of nitrogen inputs and losses in the ocean. Nature 445:163

Di Giulio M (2005) A comparison of proteins from *Pyrococcus furiosus* and *Pyrococcus abyssi*: barophily in the physicochemical properties of amino acids and in the genetic code. Gene 346:1–6

Ding T, Suo Y, Xiang Q, Zhao X, Chen S, Ye X, Liu D (2017) Significance of viable but nonculturable *Escherichia coli*: induction, detection, and control. J Microbiol Biotechnol 27:417–428

Dobrindt U, Hochhut B, Hentschel U, Hacker J (2004) Genomic islands in pathogenic and environmental microorganisms. Nat Rev Microbiol 2:414–424

Dobzhansky T (1937) Genetics and the origin of species. Columbia University Press, New York

Dombrowski H (1963) Bacteria from paleozoic salt deposits. Ann N Y Acad Sci 108:453–460

Doolittle W (1999) Phylogenetic classification and the universal tree. Science 284:2124–2129

Dopson M, Holmes D (2014) Metal resistance in acidophilic microorganisms and its significance for biotechnologies. Appl Microbiol Biotechnol 98:8133–8144

Driks A (2002) Overview: development in bacteria: spore formation in *Bacillus subtilis*. Cell Mol Life Sci 59:389–391

Dubey G, Ben-Yehuda S (2011) Intercellular nanotubes mediate bacterial communication. Cell 144:590–600

Dufresne A, Salanoubat M, Partensky F, Artiguenave F, Axmann I, Barbe V, Duprat S, Galperin M, Koonin E, Le Gall F et al (2003) Genome sequence of the cyanobacterium *Prochlorococcus marinus* SS120, a nearly minimal oxyphototrophic genome. Proc Natl Acad Sci U S A 100:10020–10025

Dufresne A, Garczarek L, Partensky F (2005) Accelerated evolution associated with genome reduction in a free-living prokaryote. *Genome Biology* 6:R14

El-Hajj ZW, Newman EB (2015) How much territory can a single E. coli cell control? Front Microbiol 6:309

Eldredge N, Gould S (1973) Punctuated equilibria: An alternative to phyletic gradualism. In: Schopf TJM (ed) Models in paleobiology. Freeman, Cooper & Co, San Francisco

Elkins J, Kunin V, Anderson I, Barry K, Goltsman E, Lapidus A, Hedlund B, Hugenholtz P, Kyrpides N, Graham D et al (2008) A korarchaeal genome reveals insights into the evolution of archaea. Proc Natl Acad Sci U S A 105:8102–8107

Ellis RJ (1979) The most abundant protein on. Earth Trends Biochem Sci 4:241–244

Empadinhas N, DaCosta M (2008) Osmoadaptation mechanisms in prokaryotes: distribution of compatible solutes. Int J Microbiol 11:151–161

Erill I, Campoy S, Barbé J (2007) Aeons of distress: an evolutionary perspective on the bacterial SOS response. FEMS Microbiol Rev 31:637–656

Erill I, Campoy S, Mazon G, Barbe J (2006) Dispersal and regulation of an adaptive mutagenesis cassette in the bacteria domain. Nucleic Acids Res 34:66–77

Ettwig K, Butler M, Le PD, Pelletier E, Mangenot S, Kuypers M, Schreiber F, Dutilh B, Zedelius J, de Beer D et al (2010) Nitrite-driven anaerobic methane oxidation by oxygenic bacteria. Nature 464:543–548

Fegatella F, Cavicchioli R (2000) Physiological responses to starvation in the marine oligotrophic ultramicrobacterium *Sphingomonas* sp. strain RB2256. Appl Environ Microbiol 66:2037–2044

Fergus CL (1967) Resistance of spores of some thermlophilic actinomycetes to high temperature. Mycopathol Mycol Appl 32:205–208

Fernandes SO, Javanaud C, Aigle A, Michotey V, Guasco S, Deborde J, Deflandre B, Anschutz P, Bonin PC (2015) Anaerobic nitrification–denitrification mediated by Mn-oxides in meso-tidal sediments: implications for N_2 and N_2O production. Journal of Marine Systems 144:1–8

Fernández-Delgado M, García-Amado M, Contreras M, Incani R, Chirinos H, Rojas H, Suárez P (2015) Survival, induction and resuscitation of *Vibrio cholerae* from the viable but noncultivable state in the Southern Caribbean sea. Rev Inst Med Trop Sao Paulo 57:21–26

Ferry J (2010) How to make a living by exhaling methane. Annu Rev Microbiol 64:453–473

Fisher RA (1930) The genetical theory of natural selection. Clarendon Press, Oxford

Fittipaldi M, Nocker A, Codony F (2012) Progress in understanding preferential detection of live cells using viability dyes in combination with DNA amplification. J Microbiol Methods 91:276–289

Forterre P, Bergerat A, Lopez-Garcia P (1996) The unique DNA topology and DNA topoisomerases of hyperthermophilic archaea. FEMS Microbiol Rev 18:237–248

Fraser C, Gocayne J, White O, Adams M, Clayton R, Fleischmann R, Bult C, Kerlavage A, Sutton G, Kelley J et al (1995) The minimal gene complement of *Mycoplasma genitalium*. Science 270:397–403

Fröhlich-Nowoisky J, Kampf CJ, Weber B, Huffman JA, Pöhlker C, Andreae MO, Lang-Yona N, Burrows SM, Gunthee SS, Elbert W et al (2016) Bioaerosols in the Earth system: Climate, health, and ecosystem interactions. Atmospheric Res 182:346–376

Frost L, Leplae R, Summers A, Toussaint A (2005) Motile genetic elements : the agents of open source of evolution. Nat Rev Microbiol 3:722–732

Fuchs R, Fujii S, Wagner J (2004) Properties and functions of *Escherichia coli*: Pol IV and PolV. Adv Protein Chem 69:229–264

Galagan J, Nusbaum C, Roy A, Endrizzi M, Macdonald P, FitzHugh W, Calvo S, Engels R, Smirnov S, Atnoor D et al (2002) The genome of *M. acetivorans* reveals extensive metabolic and physiological diversity. Genome Res 12:532–542

Galperin MY, Mekhedov SL, Puigbo P, Smirnov S, Wolf YI, Rigden DJ (2012) Genomic determinants of sporulation in Bacilli and Clostridia: towards the minimal set of sporulation-specific genes. Environ Microbiol 14:2870–2890

García M, Jones S, Pelaz C, Millar R, Abu Kwaik Y (2007) *Acanthamoeba polyphaga* resuscitates viable non-culturable *Legionella pneumophila* after disinfection. Environ Microbiol 9:1267–1277

Gest H, Mandelstam J (1987) Longevity of microorganisms in natural environments. Microbiol Sci 4:69–71

Ghai R, Mizuno CM, Picazo A, Camacho A, Rodriguez-Valera F (2013) Metagenomics uncovers a new group of low GC and ultra-small marine Actinobacteria. Sci Rep 3:2471

Giaquinto L, Curmi P, Siddiqui K, Poljak A, DeLong E, DasSarma S, Cavicchioli R (2007) Structure and function of cold shock proteins in archaea. J Bacteriol 189:5738–5748

Gibson D, Glass J, Lartigue C, Noskov V, Chuang R, Algire M, Benders G, Montague M, Ma L, Moodie M et al (2010) Creation of a bacterial cell controlled by a chemically synthesized genome. Science 329:52–56

Gil R, Peretó J (2015) Small genomes and the difficulty to define minimal translation and metabolic machineries. Front Ecol Evol 3:123

Gil R, Silva F, Peretó J, Moya A (2004) Determination of the core of a minimal bacterial gene set. Microbiol Mol Biol Rev 68:518–537

Gil R et al (2003) The genome sequence of *Blochmannia floridanus*: comparative analysis of reduced genomes. Proc Natl Acad Sci U S A 100:9388–9393

Gillings M (2017) Lateral gene transfer, bacterial genome evolution, and the Anthropocene. Ann N Y Acad Sci 1389:20–36

Giovannoni S (2017) SAR11 bacteria: the most abundant plankton in the oceans. Ann Rev Mar Sci 9:231–255

Giovannoni SJ, Hayakawa D, Tripp H, Stingl U, Givan S, Cho J-C et al (2008) The small genome of an abundant coastal ocean methylotroph. Environ Microbiol 10:1771–1782

Giovannoni SJ, Tripp HJ, Givan S, Podar M, Vergin KL, Baptista D, Bibbs L, Eads J, Richardson TH, Noordewier M et al (2005) Genome streamlining in a cosmopolitan oceanic bacterium. Science 309:1242–1245

Giraud E, Fardoux J, Fourrier N, Hannibal L, Genty B, Bouyer P, Dreyfus B, Vermeglio A (2002) Bacteriophytochrome controls photosystem synthesis in anoxygenic bacteria. Nature 417:202–205

Glass J, Merryman C, Wise K, Hutchison C, Smith H (2017) Minimal cells-real and imagined. Cold Spring Harb Perspect Biol 9:pii: a023861

Gogarten JP, Doolittle WF, Lawrence JG (2002) Prokaryotic evolution in light of gene transfer. Mol Biol Evol 19:2226–2238

Gonzàlez-Escalona N, Fey A, Höfle M, Espejo R, Guzmán C (2006) Quantitative reverse transcription polymerase chain reaction analysis of *Vibrio cholerae* cells entering the viable but non-culturable state and starvation in response to cold shock. Environ Microbiol 8:658–666

Gonzalez-Toril E, Liobet-Brossa E, Casamayor E, Amann R, Amils R (2003) Microbial ecology of an extreme acidic environment, the Tinto river. Appl Environ Microbiol 69:4853–4865

Gottschal J, Prins R (1991) Thermophiles – a life at elevated temperatures. Trends Ecol Evol 6:157–162

Gould G (2006) History of science – spores. In Lewis B Perry Memorial Lecture 2005

Grage K, Jahns A, Parlane N, Palanisamy R, Rasiah I, Atwood J, Rehm B (2009) Bacterial polyhydroxyalkanoate granules: biogenesis, structure, and potential use as nano-/micro-beads in biotechnological and biomedical applications. Biomacromolecules 10:660–669

Grant W (1992) Alkaline environments. In: Lederberg J (ed) Encyclopedia of microbiology, vol 2. Academic Press, New York, pp 73–84

Grant W (2004) Life at low water activity. Philos Trans Roy Soc B 359:1249–1267

Grant W, Gemmell R, McGenity T (1998) Halobacteria: the evidence for longevity. Extremophiles 2:279–287

Grant W, Sorokin D (2011) Distribution and diversity of soda lake alkaliphiles. In: Horikoshi K (ed) Extremophiles handbook. Springer, New York, pp 28–54

Graur D, Pupko T (2001) The Permian bacterium that isn't. Mol Biol Evol 18:1143–1146

Griffith F (1928) The significance of pneumococcal types. Journal of Hygiene 27:113–159

Grote J, Thrash JC, Huggett MJ, Landry ZC, Carini P, Giovannoni SJ, Rappe SJ (2012) Streamlining and core genome conservation among highly divergent members of the SAR11 Clade. MBio 3:e00252–e00212

Guerrero R, Berlanga M (2007) The hidden side of the prokaryotic cell: rediscovering the microbial world. Int Microbiol 10:157–168

Gutiérrez G, Martin A (1998) The most ancient DNA recovered from amber-preserved specimen may not be as ancient as It seems. Mol Biol Evol 15:926–929

Hacker J, Kaper J (2000) Pathogenicity islands and the evolution of microbes. Annu Rev Microbiol 54:641–679

Hagen C, Hawrylewicz E, Ehrlich R (1964) Survival of microorganisms in a simulated martian environment. I. *Bacillus subtilis* var. *globigii*. Appl Microbiol 12:215–218

Hahn M (2003) Isolation of strains belonging to the cosmopolitan *Polynucleobacter necessarius* cluster from freshwater habitats located in three climatic zones. Appl Environ Microbiol 69:5248–5254

Haldane JBS (1932) The causes of evolution. Princeton University Press, Princeton

Hallatschek O, Hersen P, Ramanathan S, Nelson DR (2007) Genetic drift at expanding frontiers promotes gene segregation. Proc Natl Acad Sci U S A 104:19926–19930

Hammerstrom T, Beabout K, Clements T, Saxer G, Shamoo Y (2015) *Acinetobacter baumannii* repeatedly evolves a hypermutator phenotype in response to tigecycline that effectively surveys evolutionary trajectories to resistance. PLoS One 10:e0140489

Han K, Li Z, Peng R, Zhu L, Zhou T, Wang L, Li S, Zhang X, Hu W, Wu Z et al (2013) Extraordinary expansion of a *Sorangium cellulosum* genome from an alkaline milieu. Sci Rep 3:2101

Hanczyc M, Fujikawa S, Szostak J (2003) Experimental models of primitive cellular compartments: encapsulation, growth, and division. Science 302:618–622

Haroon M, Hu S, Shi Y, Imelfort M, Keller J, Hugenholtz P, Yuan Z, Tyson G (2013) Anaerobic oxidation of methane coupled to nitrate reduction in a novel archaeal lineage. Nature 500:567–567

Hedlund B, Murugapiran S, Alba T, Levy A, Dodsworth J, Goertz G, Ivanova N, Woyke T (2015) Uncultivated thermophiles: current status and spotlight on '*Aigarchaeota*'. Curr Opin Microbiol 25:136–145

Heim S, Lleo M, Bonato B, Guzman C, Canepari P (2002) The viable but nonculturable state and starvation are different stress responses of *Enterococcus faecalis*, as determined by proteome analysis. J Bacteriol 184:6739–6745

Herbert RA (1999) Nitrogen cycling in coastal marine ecosystems. FEMS Microbiol Rev 23:563–590

Hoelher TM, Jørgensen BB (2013) Microbial life under extreme energy conditions. Nature Rev Microbiol 11:83–94

Hoff M (2009) Surviving salt: how do extremophiles do it? PLoS Biol 7:e1000258

Hood M, Macdonell M (1987) Distribution of ultramicrobacteria in a gulf coast estuary and induction of ultramicrobacteria. Microb Ecol 14:113–127

Hoshino T, Toki T, Ijiri A, Morono Y, Machiyama H, Ashi J, Okamura K, Inagaki F (2017) *Atribacteria* from the subseafloor sedimentary biosphere disperse to the hydrosphere through submarine mud volcanoes. Front Microbiol 8:1135

Huber H, Gallenberger M, Jahn U, Eylert E, Berg I, Kockelkorn D, Eisenreich W, Fuchs G (2008) A dicarboxylate/4-hydroxybutyrate autotrophic carbon assimilation cycle in the hyperthermophilic Archaeum *Ignicoccus hospitalis*. Proc Natl Acad Sci U S A 105:7851–7856

Huber H, Thomm M, Konig H, Thies G, Stetter K (1982) *Methanococcus thermolithotrophicus*, a novel thermophilic lithotrophic methanogen. Arch Microbiol 132:47–50

Huber R, Wilharm T, Huber D, Trincone A, Burggraf S, König H, Rachel R, Rockinger I, Fricke H, Stetter K (1992) *Aquifex pyrophilus* gen. nov., sp. nov., represents a novel group of marine hyperthermophilic hydrogen-oxidizing bacteria. System Appl Microbiol 15:340–351

Huggett MJ, Hayakawa DH, Rappé MS (2012) Genome sequence of strain HIMB624, a cultured representative from the OM43 clade of marine *Betaproteobacteria*. Stand Genomic Sci 6:11–20

Hügler M, Sievert S (2011) Beyond the Calvin cycle: autotrophic carbon fixation in the ocean. Ann Rev Mar Sci 3:261–289

Hutchison C 3rd., Chuang R, Noskov V, Assad-Garcia N, Deerinck T, Ellisman M, Gill J, Kannan K, Karas B, Ma L et al (2016) Design and synthesis of a minimal bacterial genome. Science 351:aad6253

Huxley J (1948) Evolution: the modern synthesis. MIT Press, Cambridge, MA

Inagaki F, Hinrichs K, Kubo Y, Bowles M, Heuer V, Hong W, Hoshino T, Ijiri A, Imachi H, Ito M et al (2015) DEEP BIOSPHERE. Exploring deep microbial life in coal-bearing sediment down to ~2.5 km below the ocean floor. Science 349:420–442

Ioannidis P, Johnston K, Riley D, Kumar N, White J, Olarte K, Ott S, Tallon L, Foster J, Taylor M et al (2013) Extensively duplicated and transcriptionally active recent lateral gene transfer from a bacterial *Wolbachia* endosymbiont to its host filarial nematode *Brugia malayi*. BMC Genomics 14:639

Itoh T, Onishi M, Kato S, Iino T, Sakamoto M, Kudo T, Takashina T, Ohkuma M (2016) *Athalassotoga saccharophila* gen. nov., sp. nov., isolated from an acidic terrestrial hot spring, and proposal of *Mesoaciditogales* ord. nov. and *Mesoaciditogaceae* fam. nov. in the phylum *Thermotogae*. Int J Syst Evol Microbiol 66:1045–1051

Jacotot R, Virat B (1954) Longévité des spores de *B. anthrasis* (premier vaccin de Pasteur). Ann Inst Pasteur 87:215–217

Jannasch H (1967) Enrichments of aquatic bacteria in continous culture. Arch Mikrobiol 59:165–173

Jannasch HW (1969) Estimations of bacterial growth rates in natural waters. J Bacteriol 99:156–160

Janto B, Ahmed A, Ito M, Liu J, Hicks DB, Pagni S, Fackelmayer OJ, Smith TA, Earl J, Elbourne LD et al (2011) Genome of alkaliphilic *Bacillus pseudofirmus* OF4 reveals adaptations that support the ability to grow in an external pH range from 7.5 to 11.4. Environ Microbiol 13:3289–3309

Jayarama R (2009) Mutators and hypermutability in bacteria : the *Escherichia coli* paradigm. J Genet 88:379–391

Johnsborg O, Eldholm V, Håvarstein L (2007) Natural genetic transformation: prevalence, mechanisms and function. Res Microbiol 158:767–778

Johnson D (1998) Biodiversity and ecology of acidophilic microorganisms. FEMS Microbiol Ecol 27:307–317

Johnson D (2012) Geomicrobiology of extremely acidic subsurface environments. FEMS Microbiol Ecol 81:2–12

Jones E, Grant W, Duckworth A, Owenson G (1998) Microbial diversity of soda lakes. Extremophiles 2:191–200

Jones S, Lennon J (2010) Dormancy contributes to the maintenance of microbial diversity. Proc Natl Acad Sci U S A 107:5881–5886

Jones W, Leigh J, Mayer F, Woese C, Wolfe R (1983) *Methanococcus jannaschii* sp. nov., an extremely thermophilic methanogen from a submarine hydrothermal vent. Arch Microbiol 136:254–261

Jørgensen B, Boetius A (2007) Feast and famine – microbial life in the deep-sea bed. Nat Rev Microbiol 5:770–781

Jørgensen BB (2011) Deep subseafloor microbial cells on physiological standby. Proc Natl Acad Sci U S A 108:18193–18194

Jørgensen BB, D'Hondt S (2006) Ecology. A starving majority deep beneath the seafloor. Science 314:932–934

Joux F, Lebaron P (1997) Ecological implications of an improved direct viable count method for aquatic bacteria. Appl Environ Microbiol 63:3643–3647

Joye SB (2012) Microbiology: a piece of the methane puzzle. Nature 491:538–539

Juhas M (2016) On the road to synthetic life: the minimal cell and genome-scale engineering. Crit Rev Biotechnol 36:416–423

Juhas M, van der Meer J, Gaillard M, Harding R, Hood D, Crook DW (2009) Genomic islands: tools of bacterial horizontal gene transfer and evolution. FEMS Microbiol Rev 33:376–393

Kallmeyer J, Pockalny R, Adhikari RR, Smith DC, D'Hondt S (2012) Global distribution of microbial abundance and biomass in subseafloor sediment. Proc Natl Acad Sci U S A 109:16213–16236

Kaneko R, Hayashi T, Tanahashi M, Naganuma T (2007) Phylogenetic diversity and distribution of dissimilatory sulfite reductase genes from deep-sea sediments cores. Mar Biotechnol (NY) 9:429–436

Kaneko T, Nakamura Y, Sato S, Asamizu E, Kato T, Sasamoto S, Watanabe A, Idesawa K, Ishikawa A, Kawashima K et al (2000) Complete genome structure of the nitrogen-fixing symbiotic bacterium *Mesorhizobium loti*. DNA Res 7:331–338

Kantor RS, Wrighton KC, Handley KM, Sharon I, Hug LA, Castelle CJ, Thomas BC, Banfield JF (2013) Small genomes and sparse metabolisms of sediment-associated bacteria from four candidate phyla. MBio 4:e00708–e00713

Kaprelyants A, Gottschal J, Kell D (1993) Dormancy in non-sporulating bacteria. FEMS Microbiol Rev 104:271–286

Kaprelyants A, Mukamolova G, Davey H, Kell D (1996) Quantitative analysis of the physiological heterogeneity within starved cultures of Micrococcus luteus by flow cytometry and cell sorting. Appl Environ Microbiol 62:1311–1316

Kashefi K, Holmes DE, Reysenbach AL, Lovley DR (2002) Use of Fe(III) as an electron acceptor to recover previously uncultured hyperthermophiles: isolation and characterization of *Geothermobacterium ferrireducens* gen. nov., sp. nov. Appl Environ Microbiol 68:1735–1742

Kasting JF (1982) Stability of ammonia in the primitive terrestrial atmosphere. J Geophys Res: Oceans 87:3091–3098

Kasting JF (1993) Earth's early atmosphere. Science 259:920–926

Kell D, Kaprelyants A, Weichart D, Harwood C, Barer M (1998) Viability and activity in readily culturable bacteria: a review and discussion of the practical issues. Antonie Van Leeuwenhoek 73:169–187

Kelley D, Karson J, Früh-Green G, Yoerger D, Shank T, Butterfield D, Hayes J, Schrenk M, Olson E, Proskurowski G et al (2005) A serpentinite-hosted ecosystem: the Lost City hydrothermal field. Science 307:1428–1434

Kennedy M, Reader S, Swierczynski L (1994) Preservation records of microorganisms : evidence of the tenacity of life. Microbiology 140:2514–2529

Keyser HH, van Berkum P, Weber DF (1982) A comparative study of the physiology of symbioses formed by *Rhizobium japonicum* with *Glycine max, Vigna unguiculata*, and *Macroptilium atropurpurem*. Plant Physiol 70:1626–1630

Khan M, Pyle B, Camper A (2010) Specific and rapid enumeration of viable but nonculturable and viable-culturable Gram-negative bacteria by using flow cytometry. Appl Environ Microbiol 76:5088–5096

Kimura M (1986) DNA and the neutral theory. Philos Trans R Soc Lond B 312:343–354

Kimura M, Ohta T (1974) On some principles governing molecular evolution. Proc Natl Acad Sci U S A 71:2848–2852

Kjelleberg S, Albertson N, Flärdh K, Holmquist L, Jouper-Jaan A, Marouga R, Östling J, Svenblad B, Weichart D (1993) How do non-differentiating bacteria adapt to starvation? Antonie Van Leeuwenhoek 63:333–341

Knittel K, Boetius A (2009) Anaerobic oxidation of methane: progress with an unknown process. Annu Rev Microbiol 63:311–334

Kobayashi H, Takaki Y, Kobata K, Takami H, Inoue A (1998) Characterization of α-maltotetraohydrolase produced by *Pseudomonas* sp. MS300 isolated from the deepest site of the Mariana Trench. Extremophiles 2:401–407

Koch A (1971) The adaptative responses of *Escherichia coli* to feast existence. Adv Microb Physiol 6:147–217

Kogure K, Simidu U, Taga N (1979) A tentative direct microscopic method for counting living marine bacteria. Can J Microbiol 3:415–420

Konstantinidis KT, Tiedje JM (2004) Trends between gene content and genome size in prokaryotic species with larger genomes. Proc Natl Acad Sci U S A 101:3160–3165

Koonin E (2016) Horizontal gene transfer: essentiality and evolvability in prokaryotes, and roles in evolutionary transitions. F1000Res 5: F1000 Faculty Rev-1805

Koonin E, Makarova K, Aravind L (2001) Horizontal gene transfer in Prokaryotes : quantification and classification. Annu Rev Microbiol 55:709–742

Koonin EV, Wolf YI (2012) Evolution of microbes and viruses: a paradigm shift in evolutionary biology? Front Cell Infect Microbiol 2:119

Kourmentza C, Plácido J, Venetsaneas N, Burniol-Figols A, Varrone C, Gavala H, Reis M (2017) Recent advances and challenges towards sustainable polyhydroxyalkanoate (PHA) production. Bioengineering (Basel) 4:55

Kozak M (1999) Initiation of translation in prokaryotes and eukaryotes. Gene 234:187–208

Krimi Z, Petit A, Mougel C, Dessaux Y, Nesme X (2002) Seasonal fluctuations and long-term persistence of pathogenic populations of *Agrobacterium* spp. in soils. Appl Environ Microbiol 68:3358–3365

Krulwich T, Liu J, Morino M, Fujisawa M, Ito M, Hicks D (2011a) Adaptative mechanisms of extreme alkaliphiles. In: Horikoshi K (ed) Extremophiles handbook. Springer, New York, pp 120–139

Krulwich T, Sachs G, Padan E (2011b) Molecular aspects of bacterial pH sensing and homeostasis. Nat Rev Microbiol 9:330–343

Kurland C, Canback B, Berg O (2003) Horizontal gene transfer: a critical view. Proc Natl Acad Sci U S A 100:9658–9662

Kusumoto A, Asakura H, Kawamoto K (2012) General stress sigma factor RpoS influences time required to enter the viable but non-culturable state in *Salmonella enterica*. Microbiol Immunol 56:228–237

Kuypers M, Lavik G, Woebken D, Schmid M, Fuchs B, Amann R, Jørgensen B, Jetten M (2005) Massive nitrogen loss from the Benguela upwelling system through anaerobic ammonium oxidation. Proc Natl Acad Sci U S A 102:6478–6483

Lafond RE, Lukehart SA (2006) Biological basis for syphilis. Clin Microbiol Rev 19:29–49

Land M, Hauser L, Jun S-R, Nookaew I, Leuze MR, Ahn T-H, Karpinets T, Lund O, Kora G, Wassenaar T et al (2015) Insights from 20 years of bacterial genome sequencing. Funct Integr Genomics 15:141–161

Langille M, Hsiao W, Brinkman F (2010) Detecting genomic islands using bioinformatics approaches. Nat Rev Microbiol 8:373–382

Lanyi J (1974) Salt-dependent properties of proteins from extremely halophilic bacteria. Bacteriol Rev 38:272–290

Lavire C, Normand P, Alekhina I, Bulat S, Prieur D, Birrien JL, Fournier P, Hanni C, Petit JR (2006) Presence of *Hydrogenophilus thermoluteolus* DNA in accretion ice in the subglacial Lake Vostok, Antarctica, assessed using *rrs, cbb* and *hox*. Environ Microbiol 8:2106–2114

Le Bihan T, Rayner J, Roy M, Spagnolo L (2013) *Photobacterium profundum* under pressure: a MS-based label-free quantitative proteomics study. PLoS ONE 8:e60897

Lederberg J, Tatum EL (1946) Gene recombination in *Escherichia coli.* Nature 158:558

Lennon J, Jones S (2011) Microbial seed banks: the ecological and evolutionary implications of dormancy. Nat Rev Microbiol 9:119–130

Léonard L, Bouarab Chibane L, Ouled Bouhedda B, Degraeve P, Oulahal N (2016) Recent advances on multi-parameter flow cytometry to characterize antimicrobial treatments. Front Microbiol 7:1225

Lepper BT, Frolking TA, Fisher DC, Goldstein G, Sanger JE, Wymer DA, Ogden JGI, Hooge PE (1991) Intestinal contents of a late Pleistocene mastodont from midcontinental North America. Quatern Res 36:120–125

Lerner A, Valverde A, Castro-Sowinski S, , Lerner H, Okon, Y, Burdman S (2010) Phenotypic variation in *Azospirillum brasilense* exposed to starvation. Environ Microbiol Rep 2: 577–586

Li L, Mendis N, Trigui H, Oliver J, Faucher S (2014) The importance of the viable but non-culturable state in human bacterial pathogens. Front Microbiol 5:258

Licking E, Gorski L, Kaiser D (2000) A common step for changing cell shape in fruiting body and starvation-independent sporulation of *Myxococcus xanthus*. J Bacteriol 182:3553–3558

Lindgren P, Higgins P, Seifert H, Cars O (2016) Prevalence of hypermutators among clinical *Acinetobacter baumannii* isolates. J Antimicrob Chemother 71:661–665

Lipscomb G, Hahn E, Crowley A, MWW A (2017) Reverse gyrase is essential for microbial growth at 95 °C. Extremophiles 21:603–608

Liu Y, Gilchrist A, Zhang J, Li X-F (2008) Detection of viable but nonculturable *Escherichia coli* O157:H7 bacteria in drinking water and river water. Appl Environ Microbiol 74:1502–1507

Liu Y, Wang C, Fung C, Xing-Fang L (2010a) Quantification of viable but nonculturable *Escherichia coli* O157:H7 by targeting the *rpoS* mRNA. Anal Chem 82:2612–2615

Liu Y, Wang C, Tyrrell G, Li X-F (2010b) Production of Shiga-like toxins in viable but nonculturable *Escherichia coli* O157:H7. Water Res 44:711–718

Liu Y, Whitman W (2008) Metabolic, phylogenetic, and ecological diversity of the methanogenic archaea. Ann N Y Acad Sci 1125:171–189

Ljungdahl L (1986) The autotrophic pathway of acetate synthesis in acetogenic bacteria. Annu Rev Microbiol 40:415–450

Lleò M, Pierobon S, Tafi M, Signoretto C, Canepari P (2000) mRNA detection by reverse transcription-PCR for monitoring viability over time in an *Enterococcus faecalis* viable but nonculturable population maintained in a laboratory microcosm. Appl Environ Microbiol 66:4564–4567

Lopez-Garcia P, Moreira D (2006) Selective forces for the origin of the eukaryotic nucleus. Bioessays 28:525–533

Lorenz M, Wackernagel W (1994) Bacterial gene transfer by natural genetic transformation in the environment. Microbiol Rev 58:563–602

Loveland-Curtze J, Miteva V, Brenchley J (2009) *Herminiimonas glaciei* sp. nov., a novel ultramicrobacterium from 3042 m deep Greenland glacial ice. Int J Syst Evol Microbiol 59:1272–1277

Luef B, Frischkorn KR, Wrighton KC, Holman HY, Birarda G, Thomas BC, Singh A, Williams KH, Siegerist CE, Tringe SG et al (2015) Diverse uncultivated ultra-small bacterial cells in groundwater. Nat Commun 6:6372

Lulchev P, Klostermeier D (2014) Reverse gyrase – recent advances and current mechanistic understanding of positive DNA supercoiling. Nucleic Acids Res 42:8200–8213

Luria SE, Delbruck M (1943) Mutations of bacteria from virus sensitivity to virus resistance. Genetics 28:491–511

Lyons TW, Reinhard CT, Planavsky NJ (2014) The rise of oxygen in Earth's early ocean and atmosphere. Nature 506:307–315

Ma Y, Galinski E, Grant W, Oren A, Ventosa A (2010) Halophiles 2010: life in saline environments. Appl Environ Microbiol 76:6971–6981

Madigan M, Bender K, Buckley D, Sattley W, Stahl D (2015) Brock biology of microorganisms. Pearson, New York

Mailloux BJ, Fuller ME (2003) Determination of in situ bacterial growth rates in aquifers and aquifer sediments. Appl Environ Microbiol 69:3798–3808

Maki JS (2013) Bacterial intracellular sulfur globules: structure and function. J Mol Microbiol Biotechnol 23:270–280

Manina G, McKinney J (2013) A single-cell perspective on non-growing but metabolically active (NGMA) bacteria. Curr Top Microbiol Immunol 374:135–161

Margesin R, Miteva V (2011) Diversity and ecology of psychrophilic microorganisms. Res Microbiol 162:346–361

Margulis L (1970) Origin of eukaryotic cells. Yale University Press, New York

Maron D, Ames B (1983) Revised methods for the Salmonella mutagenicity test. Mutat Res 113:173–215

Marshall BJ, Barrett LJ, Prakash C, McCallum RW, Guerrant RL (1990) Urea protects *Helicobacter (Campylobacter) pylori* from the bactericidal effect of acid. Gastroenterology 99:697–702

Martin D, Bartlett D, Roberts M (2002) Solute accumulation in the deep-sea bacterium *Photobacterium profundum*. Extremophile 6:507–514

Martin W, Müller M (1998) The hydrogen hypothesis for the first eukaryote. Nature 392:37–41

Martínez-Cano D, Reyes-Prieto M, Martínez-Romero E, Partida-Martínez L, Latorre A, Moya A, Delaye L (2015) Evolution of small prokaryotic genomes. Front Microbiol 5:742

Martínez-García E, de Lorenzo V (2016) The quest for the minimal bacterial genome. Curr Opin Biotechnol 42:216–224

Mayr E (1942) Systematics and the origin of species, from the viewpoint of a zoologist. Harvard University Press, Cambridge

McCutcheon JP, Moran NA (2010) Functional convergence in reduced genomes of bacterial symbionts spanning 200 my of evolution. Genome Biol Evol 2:708–718

McCutcheon JP, Moran NA (2011) Extreme genome reduction in symbiotic bacteria. Nat Rev Microbiol 10:13–26

McCutcheon JP, von Dohlen CD (2011) An interdependent metabolic patchwork in the nested symbiosis of mealybugs. Curr Biol 21:1366–1372

McGenity T, Gemmell R, Grant W, Stan-Lotter H (2000) Origins of halophilic microorganisms in ancient salt deposits. Environ Microbiol 2:243–250

McInerney MJ, Bryant MP, Hespell RB, Costerton JW (1981) *Syntrophomonas wolfei* gen. nov. sp. nov., an Anaerobic, Syntrophic, Fatty Acid-Oxidizing Bacterium. Appl Environ Microbiol 41:1029–1039

McKenney P, Driks A, Eichenberger P (2013) The *Bacillus subtilis* endospore: assembly and functions of the multilayered coat. Nature Rev Microbiol 11:33–44

McLeod MP et al (2006) The complete genome of *Rhodococcus* sp. RHA1 provides insights into a catabolic powerhouse. Proc Natl Acad Sci U S A 103:15582–15587

Mei N, Zergane N, Postec A, Erauso G, Ollier A, Payri C, Pelletier B, Fardeau M, Ollivier B, Quéméneur M (2014) Fermentative hydrogen production by a new alkaliphilic *Clostridium* sp. (strain PROH2) isolated from a shallow submarine hydrothermal chimney in Prony Bay, New Caledonia. Int J Hydrogen Energ 39:19465–19473

Mendell JE, Clements KD, Choat JH, Angert ER (2008) Extreme polyploidy in a large bacterium. Proc Natl Acad Sci U S A 105:6730–6734

Mesbah N, Wiegel J (2008) Life at extreme limits: The anaerobic halophilic alkalithermophiles. Ann NY Acad Sci 1125:44–57

Mezzina M, Pettinari M (2016) Phasins, multifaceted polyhydroxyalkanoate granule associated proteins. Appl Environ Microbiol 82:5060–5067

Michoud G, Jebbar M (2016) High hydrostatic pressure adaptive strategies in an obligate piezophile *Pyrococcus yayanosii*. Sci Rep 6:27289

Miller SL, Schlesinger G (1983) The atmosphere of the primitive earth and the prebiotic synthesis of organic compounds. Adv Space Res 3:47–53

Milucka J, Ferdelman T, Polerecky L, Franzke D, Wegener G, Schmi M, Kuypers MM (2012) Zero-valent sulphur is a key intermediate in marine methane oxidation. Nature 491:541

Moeller R, Schuerger A, Reitz G, Nicholson W (2012) Protective role of spore structural components in determining *Bacillus subtilis* spore resistance to simulated Mars surface conditions. Appl Environ Microbiol 78:8849–8853

Moënne-Loccoz Y, Mavingui P, Combes C, Normand P, Steinberg C (2015) Micro-organisms and biotic interactions. In: Bertrand J, Caumette P, Lebaron P, Matheron R, Normand P, Sime-Ngando T (eds) Environmental microbiology: fundamentals and applications. Springer Verlag, Dordrecht, pp 395–444

Mohanta TK, Bae H (2015) The diversity of fungal genome. Biol Proced Online 17:8

Moran NA (1996) Accelerated evolution and Muller's rachet in endosymbiotic bacteria. Proc Natl Acad Sci U S A 93:2873–2878

Mori K, Kim H, Kakegawa T, Hanada S (2003) A novel lineage of sulfate-reducing microorganisms: *Thermodesulfobiaceae* fam. nov., *Thermodesulfobium narugense*, gen. nov., sp. nov., a new thermophilic isolate from a hot spring. Extremophiles 7:283–290

Morita R (1990) The starvation-survival state of microorganisms in nature and its relationship to bioavailable energy. Experientia 46:813–817

Moriyama M, Nikoh N, Hosokawa T, Fukatsu T (2015) Riboflavin Provisioning Underlies *Wolbachia*'s Fitness Contribution to Its Insect Host. MBio 6:e01732–e01715

Morono Y, Terada T, Nishizawa M, Ito M, Hillion F, Takahata N, Sano Y, Inagaki F (2011) Carbon and nitrogen assimilation in deep subseafloor microbial cells. Proc Natl Acad Sci U S A 108:18295–18300

Morris C, Conen F, Alex Huffman J, Phillips V, Pöschl U, Sands D (2014) Bioprecipitation: a feedback cycle linking earth history, ecosystem dynamics and land use through biological ice nucleators in the atmosphere. Glob Chang Biol 20:341–351

Morris RM, Rappé MS, Connon SA, Vergin KL, Siebold WA, Carlson C, Giovannoni S (2002) SAR11 clade dominates ocean surface bacterioplankton communities. Nature 420:806–810

Moya A, Gil R, Latorre A, Peretó J, Pilar Garcillán-Barcia M, de la Cruz F (2009) Toward minimal bacterial cells: evolution vs. design. FEMS Microbiol Rev 33:225–235

Mukamolova G, Kaprelyants A, Kell D, Young M (2003) Adoption of the transiently non-culturable state – a bacterial survival strategy? Adv Microbiol Physiol 47:65–129

Muller HJ (1964) The relation of recombination to mutational advance. Mutat Res 106:2–9

Mushegian A, Koonin E (1996) A minimal gene set for cellular life derived by comparison of complete bacterial genomes. Proc Natl Acad Sci U S A 93:10268–10273

Nakamura Y, Itoh T, Matsuda H, Gojobori T (2004) Biased biological functions of horizontally transferred genes in prokaryotic genomes. Nat Genet 36:760–766

Nicholson W, Schuerger A, Race M (2009) Migrating microbes and planetary protection. Trends Microbiol 17:389–392

Nilsson L, Oliver J, Kjelleberg S (1991) Resuscitation of *Vibrio vulnificus* from the viable but nonculturable state. J Bacteriol 173:5054–5059

Nissen H (1987) Long term starvation of a marine bacterium, *Alteromonas denitrificans*, isolated from a Norwegian fjord. FEMS Microbiol Ecol 45:173–183

Nogva HK, Dromtorp S, Nissen H, Rudi K (2003) Ethidium monoazide for DNA-based differentiation of viable and dead bacteria by 5'-nuclease. PCR BioTechniques 810:812–813

Nohmi T (2006) Environmental stress and lesion-bypass DNA polymerases. Annu Rev Microbiol 60:231–253

Normand P, Lapierre P, Tisa LS, Gogarten JP, Alloisio N, Bagnarol E, Bassi CA, Berry AM, Bickhart DM, Choisne N et al (2007) Genome characteristics of facultatively symbiotic *Frankia* sp. strains reflect host range and host plant biogeography. Genome Res 17:7–15
Novitsky J, Morita R (1977) Survival of a psychrophilic marine *Vibrio* under long-term nutrient starvation. Appl Environ Microbiol 33:635–641
Ochman H, Lawrence JG, Groisman EA (2000) Lateral gene transfer and the nature of bacterial innovation. Nature 405:299–304
Olendzenski L, Gogarten JP (2009) Evolution of genes and organisms: the tree/web of life in light of horizontal gene transfer. Ann N Y Acad Sci 1178:137–145
Oliver A (2005) Hypermutation in natural bacterial populations: consequences for medical microbiology. Rev Med Microbiol 16:25–32
Oliver A, Mena A (2010) Bacterial hypermutation in cystic fibrosis, not only for antibiotic resistance. Eur Soc Clin Microbiol Infect Dis 16:798–808
Oliver J (2010) Recent findings on the viable but nonculturable state in pathogenic bacteria. FEMS Microbiol Rev 34:415–425
Oliver J, Bockian R (1995) In vivo resuscitation, and virulence towards mice, of viable but nonculturable cells of *Vibrio vulnificus*. App Environ Microbiol 61:2620–2623
Ollivier B, Caumette P, Garcia J, Mah R (1994) Anaerobic bacteria from hypersaline ecosystems. Microbiol Rev 58:27–38
Oren A (1999) Bioenergetic aspects of halophilism. Microbiol Mol Biol Rev 63:334–348
Oren A (2008) Microbial life at high salt concentration: phylogenetic and metabolic diversity. Saline Syst 4:2
Oren A (2013) Life at high salt concentrations. The prokaryotes-prokaryotic communities and ecophysiology. Springer, Berlin/Heidelburg
Orfaniotou F, Tzamalis P, Thanassoulas A, Stefanidi E, Zees A, Boutou E, Vlassi M, Nounesis G, Vorgias C (2009) The stability of the archaeal HU histone-like DNA-binding protein from *Thermoplasma volcanium*. Extremophiles 13:1–10
Padan E, Bibi E, Ito M, Krulwich T (2005) Alkaline pH homeostasis in bacteria: new insights. Biochim Biophys Acta 1717:67–88
Paget E, Simonet P (1994) On the track of natural transformation in soil. FEMS Microbiol Ecol 15:109–117
Pais T, Lamosa P, Garcia-Moreno B, Turner D, Santos H (2009) Relationship between protein stabilization and protein rigidification induced by mannosylglycerate. J Mol Biol 394:237–250
Pál C, Papp B, Lercher M (2005) Adaptive evolution of bacterial metabolic networks by horizontal gene transfer. Nat Genet 37:1372–1375
Pande S, Shitut S, Freund L, Westermann M, Bertels F, Colesie C, Bischofs I, Kost C (2015) Metabolic cross-feeding via intercellular nanotubes among bacteria. Nat Commun 6:6238
Parkes J (2000) A case of bacterial immortality? Nature 407:844–845
Parte A (2014) LPSN-list of prokaryotic names with standing in nomenclature. Nucl Acids Res 42:D613–D616
Paul K, Nonoh J, Mikulski L, Brune A (2012) "Methanoplasmatales," *Thermoplasmatales*-related Archaea in termite guts and other environments, are the seventh order of methanogens. Appl Environ Microbiol 78:8245–8253
Pernthaler J, Sattler B, Simek K, Schwarzenbacher A, Psenner R (1996) Top-down effects on the size-biomass distribution of a freshwater bacterioplankton community. Aquat Microb Ecol 10:255–263
Peters V, Rehm B (2005) In vivo monitoring of PHA granule formation using GFP-labeled PHA synthases. FEMS Microbiol Lett 248:93–100
Peterson C, Mandel M, Silhavy T (2005) *Escherichia coli* starvation diets: essential nutrients weigh in distinctly. J Bacteriol 187:7549–7553
Phadtare S, Inouye M (2004) Genome-wide transcriptional analysis of the cold shock response in wild-type and cold-sensitive, quadruple-*csp*-deletion strains of *Escherichia coli*. J Bacteriol 186:7007–7014

Philippe N, Legendre M, Doutre G, Couté Y, Poirot O, Lescot M, Arslan D, Seltzer V, Bertaux L, Bruley C et al (2013) Pandoraviruses: amoeba viruses with genomes up to 2.5 Mb reaching that of parasitic eukaryotes. Science 341:281–286

Pinto D, Santos M, Chambel L (2015) Thirty years of viable but nonculturable state research: unsolved molecular mechanisms. Crit Rev Microbiol 41:61–76

Podar M, Anderson I, Makarova K, Elkins J, Ivanova N, Wall M, Lykidis A, Mavromatis K, Sun H, Hudson M et al (2008) A genomic analysis of the archaeal system *Ignicoccus hospitalis-Nanoarchaeum equitans*. Genome Biol 9:R158

Pöschl U, Shiraiwa M (2015) Multiphase chemistry at the atmosphere-biosphere interface influencing climate and public health in the anthropocene. Chem Rev 115:4440–4475

Postec A, Quéméneur M, Bes M, Mei N, Benaïssa F, Payri C, Pelletier B, Monnin C, Guentas-Dombrowsky L, Ollivier B et al (2015) Microbial diversity in a submarine carbonate edifice from the serpentinizing hydrothermal system of the Prony Bay (New Caledonia) over a 6-year period. Front Microbiol 6:857. https://doi.org/10.3389/fmicb.2015.00857

Pötter M, Steinbüchel A (2005) Poly(3-hydroxybutyrate) granule-associated proteins: Impacts on poly(3-hydroxybutyrate) synthesis and degradation. Biomacromolecules 6:552–560

Poulsen M, Schwab C, Jensen B, Engberg R, Spang A (2013) Methylotrophic methanogenic *Thermoplasmata* implicated in reduced methane emissions from bovine rumen. Nat Commun 4:1428

Pradel N, Ji B, Gimenez G, Talla E, Lenoble P, Garel M, Tamburini C, Fourquet P, Lebrun R, Bertin P et al (2013) The first genomic and proteomic characterization of a deep-sea sulfate reducer: insights into the piezophilic lifestyle of *Desulfovibrio piezophilus*. PLosOne 8:e55130

Preiss L, Hicks DB, Suziki S, Meier T, Krulwich TA (2015) Alkaliphilic bacteria with impact on industrial applications, concepts of early life forms, and bioenergetics of ATP synthesis. Front Bioeng Biotechnol 3:75

Price P (2007) Microbial life in glacial ice and implications for a cold origin of life. FEMS Microbiol Ecol 59:217–231

Pucci F, Kwasigroch J, Rooman M (2017) SCooP: an accurate and fast predictor of protein stability curves as a function of temperature. Bioinformatics 33:3415–3422

Quayle J, Fuller R, Benson A, Calvin M (1957) Enzymatic carboxylation of ribulose diphosphate. J Am Chem Soc 76:3610–3612

Quéméneur M, Bes M, Postec A, Mei N, Hamelin J, Monnin C, Chavagnac V, Payri C, Pelletier B, Guentas-Dombrowsky L et al (2014) Spatial distribution of microbial communities in the shallow submarine alkaline hydrothermal field of the Prony Bay, New Caledonia. Environ Microbiol Rep 6:665–674

Radman M (1975) SOS repair hypothesis: phenomenology of an inducible DNA repair which is accompanied by mutagenesis. Basic Life Sci 5A:355–367

Radman M (1999) Enzymes of evolutionary change. Nature 401:866–867

Raghoebarsing AA, Pol A, van de Pas-Schoonen KT, Smolders AJP, Ettwig KF, Rijpstra WIC (2006) A microbial consortium couples anaerobic methane oxidation to denitrification. Nature 440:918–921

Rakhuba D, Kolomiets E, Dey E, Novik G (2010) Bacteriophage receptors, mechanisms of phage adsorption and penetration into host cell. Pol J Microbiol 59:145–155

Ramamurthy T, Ghosh A, Pazhani G, Shinoda S (2014) Current perspectives on viable but nonculturable (VBNC) pathogenic bacteria. Front Public Health 2:103

Ramírez-Arcos S, Fernández-Herrero L, Marín I, Berenguer J (1998) Anaerobic growth, a property horizontally transferred by an Hfr-like mechanism among extreme thermophiles. J Bacteriol 180:3137–3143

Rappe MS, Connon S, Vergin K, Giovannoni S (2002) Cultivation of the ubiquitous SAR11 marine bacterioplankton clade. Nature 418:630–633

Rasko D, Rosovitz M, Myers G, Mongodin E, Fricke W, Gajer P, Crabtree J, Sebaihia M, Thomson N, Chaudhuri R et al (2008) The pangenome structure of *Escherichia coli*: comparative genomic analysis of *E. coli* commensal and pathogenic isolates. J Bacteriol 190:6881–6893

Rasmussen S, Chen L, Deamer D, Krakauer D, Packard N, Stadler P, Bedau M (2004) Evolution. Transitions from nonliving to living matter. Science 303:963–965

Ravagnani A, Finan CL, Young M (2005) A novel firmicute protein family related to the actinobacterial resuscitation-promoting factors by non-orthologous domain displacement. BMC Genomics 6:39

Raven J (2009) Contributions of anoxygenic and oxygenic phototrophy and chemolithotrophy to carbon and oxygen fluxes in aquatic environments. Aquat Microb Ecol 56:177–192

Redfield R (2001) Do bacteria have sex ? Nat Rev Genet 2:634–639

Reeburgh W (2007) Oceanic methane biogeochemistry. Chem Rev 107:486–513

Rehm B (2003) Polyester synthases: natural catalysts for plastics. Biochem J 376:15–33

Reysenbach A, Liu Y, Banta A, Beveridge T, Kirshtein J, Schouten S, Tivey M, Von Damm K, Voytek M (2006) A ubiquitous thermoacidophilic archaeon from deep-sea hydrothermal vents. Nature 442:444–447

Rinke C, Schwientek P, Sczyrba A, Ivanova NN, Anderson IJ, Cheng JF, Darling A, Malfatti S, Swan BK, Gies EA et al (2013) Insights into the phylogeny and coding potential of microbial dark matter. Nature 499:431–437

Rocap G, Larimer F, Lamerdin J, Malfatti S, Chain P, Ahlgren N, Arellano A, Coleman M, Hauser L, Hess W et al (2003) Genome divergence in two *Prochlorococcus* ecotypes reflects oceanic niche differentiation. Nature 424:1042–1047

Roszak D, Colwell R (1987) Survival strategies of bacteria in the natural environment. Microbiol Rev 51:365–379

Roszak D, Grimes D, Colwell R (1984) Viable but nonrecoverable stage of *Salmonella enteritidis* in aquatic systems. Can J Microbiol 30:334–338

Røy H, Kallmeyer J, Adhikari R, Pockalny R, Jørgensen B, D'Hondt S (2012) Aerobic microbial respiration in 86-million-year-old deep-sea red clay. Science 336:922–925

Rudi K, Moen B, Drømtorp S, Holck A (2005) Use of ethidium monoazide and PCR in combination for quantification of viable and dead cells in complex samples. Appl Environ Microbiol 71:1018–1024

Russell M, Barge L, Bhartia R, Bocanegra D, Bracher P, Branscomb E, Kidd R, MacGlynn S, Meier D, Nitscke W et al (2014) The drive to life on wet and, icy worlds. Astrobiol 14:308–343

Sagan L (1967) On the origin of mitosing cells. J Theor Biol 14:225–274

Samson J, Magadán A, Sabri M, Moineau S (2013) Revenge of the phages: defeating bacterial defences. Nat Rev Microbiol 11:675–687

Sanchez-Andrea I, Knittel K, Amann R, Amils R, Sanz J (2012) Quantification of Tinto River sediment microbial communities: Importance of sulfate-reducing bacteria and their role in attenuating acid mine drainage. Appl Environ Microbiol 78:4638–4645

Santander R, Oliver J, Biosca E (2014) Cellular, physiological, and molecular adaptive responses of *Erwinia amylovora* to starvation. FEMS Microbiol Ecol 88:258–271

Santos CL, Tavares F, Thioulouse J, Normand P (2009) A phylogenomic analysis of bacterial helix-turn-helix transcription factors. FEMS Microbiol Rev 33:411–429

Scanlan D, Ostrowski M, Mazard S, Dufresne A, Garczarek L, Hess W, Post A, Hagemann M, Paulsen I, Partensky F (2009) Ecological genomics of marine picocyanobacteria. Microbiol Mol Biol Rev 73:249–299

Schippers A, Neretin L, Kallmeyer J, Ferdelman T, Cragg B, Parkes R, Jørgensen B (2005) Prokaryotic cells of the deep sub-seafloor biosphere identified as living bacteria. Nature 433:861–864

Schmidt O, Hink L, Horn M, Drake H (2016) Peat: home to novel syntrophic species that feed acetate- and hydrogen-scavenging methanogens. ISME J 10:1954–1966. https://doi.org/10.1038/ismej.2015.256

Schneiker S, Perlova O, Kaiser O, Gerth K, Alici A, Altmeyer M, Bartels D, Bekel T, Beyer S, Bode E et al (2007) Complete genome sequence of the myxobacterium *Sorangium cellulosum*. Nat Biotechnol 25:1281–1289

Schrenk M, Huber J, Edwards K (2010) Microbial provinces in the subseafloor. Ann Rev Mar Sci 2:279–304

Schulz H, Brinkhoff T, Ferdelman T, Mariné M, Teske A, Jorgensen B (1999) Dense populations of a giant sulfur bacterium in Namibian shelf sediments. Science 284:493–495

Senoh M, Ghosh-Banerjee J, Ramamurthy T, Hamabata T, Kurakawa T, Takeda M, Colwell R, Nair G, Takeda Y (2010) Conversion of viable but nonculturable *Vibrio cholerae* to the culturable state by co-culture with eukaryotic cells. Microbiol Immunol 54:502–507

Setlow P (2006) Spores of *Bacillus subtilis*: their resistance to and killing by radiation, heat and chemicals. J Appl Microbiol 101:514–525

Setlow P (2013) When the sleepers wake: the germination of spores of *Bacillus* species. J Appl Microbiol 115:1251–1268

Setlow P (2014a) Germination of spores of *Bacillus* species: what we know and do not know. J Bacteriol 196:1297–1305

Setlow P (2014b) Spore resistance properties. Microbiol Spectr 2:1–15

Sghaier H, Hezbri K, Ghodhbane-Gtari F, Pujic P, Sen A, Daffonchio D, Boudabous A, Tisa LS, Klenk HP, Armengaud J et al (2015) Stone-dwelling actinobacteria *Blastococcus saxobsidens, Modestobacter marinus* and *Geodermatophilus obscurus* proteogenomes. Isme J 10:21–29. https://doi.org/10.1038/ismej.2015.108

Shivanand P, Mugeraya G (2011) Halophilic bacteria and their compatible solutes-osmoregulation and potential applications. Curr Sci 100:1516–1521

Shoun H, Kim DH, Uchiyama H, Sugiyama J (1992) Denitrification by fungi. FEMS Microbiol Lett 94:277–281

Sieburth JM, Smetacek V, Lenz J (1978) Pelagic ecosystem structure: heterotrophic compartments of the plankton and their relationship to plankton size fractions. Limnol Oceanogr 23:1256–1263

Siliakus M, van der Oost J, Kengen S (2017) Adaptations of archaeal and bacterial membranes to variations in temperature, pH and pressure. Extremophiles 21:651–670

Silver WS, Postgate JR (1973) Evolution of asymbiotic nitrogen fixation. J Theor Biol 40:1–10

Simonato F, Campanaro S, Lauro F, Vezzi A, D'Angelo M, Vitulo N, Valle G, Bartlett D (2006) Piezophilic adaptation: a genomic point of view. J Biotechnol 126:11–25

Simpson GG (1944) Tempo and mode in evolution. Columbia University Press, New York

Sinha RP, Hader DP (2002) UV-induced DNA damage and repair: a review. Photochem Photobiol Sci 1:225–236

Slepova T, Sokolova T, Lysenko T, Tourova T, Kolganova T, Kamzolkina O, Karpov G, Bonch-Osmolovskaya E (2006) *Carboxydocella sporoproducens* sp. nov., a novel anaerobic CO-utilizing/H_2-producing thermophilic bacterium from a Kamchatka hot spring. Int J Syst Evol Microbiol 56:797–800

Slobodkin A, Reysenbach A, Slobodkina G, Kolganova T, Kostrikina N, Bonch-Osmolovskaya E (2013) *Dissulfuribacter thermophilus* gen. nov., sp. nov., a thermophilic, autotrophic, sulfur-disproportionating, deeply branching deltaproteobacterium from a deep-sea hydrothermal vent. Int J Syst Evol Microbiol 63:1967–1971

Sneath P (1962) Longevity of micro-organisms. Nature 195:643–646

Sorokin D, Makarova K, Abbas B, Golyshin P, Galinski E, Ciordia S, Mena M, Merkel A, Wolf Y, van Loosdrecht M et al (2017) Discovery of extremely halophilic, methyl-reducing euryarchaea provides insights into the evolutionary origin of methanogenesis. Nat Microbiol 2:17081

Soulas G, Martin-Laurent F (2015) Pesticide biodegradation: genetic basis. The examples of hormone herbicides and s-triazines. In: Bertrand J, Caumette P, Lebaron P, Matheron R, Normand P, Sime-Ngando T (eds) Environmental microbiology: fundamentals and applications. Springer, Dordrecht/Heidelberg/New York/London, pp 693–697

Spring S, Merkhoffer B, Weiss N, Kroppenstedt R, Hippe H, Stackebrandt E (2003) Characterization of novel psychrophilic clostridia from an Antarctic microbial mat: description of *Clostridium frigoris* sp. no., *Clostridium lacusfryxellense* sp. nov., *Clostridium bowmanii* sp. nov. and *Clostridium psychrophilum* sp. nov.and reclassification of *Clostridium laramiense* as *Clostridium estertheticum* subsp *laramiense* subsp. nov. Int J Syst Evol Microbiol 53:1019–1029

Stan-Lotter H, Pfaffenhuemer M, Legat A, Busse H-J, Radax C, Gruber C (2002) *Halococcus dombrowskii* sp. nov., an archaeal isolate from a Permian alpine salt deposit. Int J Syst Evol Microbiol 52:1807–1814

Stanier RY, Ingraham JL, Wheelis ML, Painter PR (1986) The microbial world, 5th edn. Prentice-Hall, Englewood Cliffs

Steinbüchel A, Aerts K, Babel W, Follner C, Liebergesell M, Madkour M, Mayer F, Pieper-Furst U, Pries A, Valentin H (1995) Considerations on the structure and biochemistry of bacterial polyhydroxyalkanoic acid inclusions. Can J Microbiol 41:94–105

Steinbüchel A, Valentin H (1995) Diversity of bacterial polyhydroxyalkanoic acids. FEMS Microbiol Lett 128:219–228

Steinert M, Emödy L, Amann R, Hacker J (1997) Resuscitation of viable but nonculturable *Legionella pneumophila* Philadelphia JR32 by *Acanthamoeba castellanii*. Appl Environ Microbiol 63:2047–2053

Stenström J, Svensson K, Johansson M (2001) Reversible transition between active and dormant microbial states in soil. FEMS Microbiol Ecol 36:93–104

Stetter K (1999) Hyperthermophiles: isolation, classification, and properties. In: Horikoshi K, Grant W (eds) Extremophiles microbial life in extreme environments. Wiley, New York, pp 1–24

Stetter K (2006) History of discovery of the first hyperthermophiles. Extremophiles 10:357–362

Stevenson L (1978) A case for bacterial dormancy in aquatic systems. Microbiol Ecol 4:127–133

Suzuki S, Ishii S, Wu A, Tenney A, Wanger G, Kuenen J, Nealson K (2013) Microbial diversity in The Cedars, an ultrabasic, ultrareducing, and low salinity serpentinizing ecosystem. Proc Natl Acad Sci U S A 110:15336–15341

Suzuki S, Kuenen J, Schipper K, Van Der Velde S, Ishii S, Wu A, Sorokin D, Tenney A, Meng X, Morrill P et al (2014) Physiological and genomic features of highly alkaliphilic hydrogen-utilizing Betaproteobacteria from a continental serpentinizing site. Nature Com 5:3900

Swingley W, Chen M, Cheung P, Conrad A, Dejesa L, Hao J, Honchak B, Karbach L, Kurdoglu A, Lahiri S et al (2008) Niche, adaptation, and, genome, expansion, in, the, chlorophyll, d-producing, cyanobacterium, Acaryochloris, marina. Proc Natl Acad Sci U S A 105:2005–2010

Taddei F, Radman M, Maynard-Smith J, Toupance B, Gouyon P, Godelle B (1997) Role of mutator alleles in adaptive evolution. Nature 387:700–702

Takai K, Nakamura K, Toki T, Tsunogai U, Miyazaki M, Miyazaki J, Hirayama H, Nakagawa S, Nunoura T, Horikoshi K (2008) Cell proliferation at 122 °C and isotopically heavy CH4 production by a hyperthermophilic methanogen under high-pressure cultivation. Proc Natl Acad Sci U S A 105:10949–10954

Takami H, Noguchi H, Takaki Y, Uchiyama I, Toyoda A, Nishi S, Chee G, Arai W, Nunoura T, Itoh T et al (2012) A deeply branching thermophilic bacterium with an ancient acetyl-CoA pathway dominates a subsurface ecosystem. PLoS ONE 7:e30559

Takano S, Pawlowska B, Gudelj I, Yomo T, Tsuru S (2017) Density-dependent recycling promotes the long-term survival of bacterial populations during periods of starvation. MBio 8:pii: e02336–02316

Tamegai H, Ota Y, Haga M, Fujimori H, Kato C, Nogi Y, Kawamoto J, Kurihara T, Sambongi Y (2011) Piezotolerance of the respiratory terminal oxidase activity of the piezophilic *Shewanella violacea* DSS12 as compared with non-piezophilic *Shewanella* species. Biosci Biotechnol Biochem 75:919–924

Tan I, Ramamurthi K (2013) Spore formation in *Bacillus subtilis*. Environ Microbiol Rep 6:212–225

Tenaillon O, Toupance B, Le Nagard H, Taddei F, Godelle B (1999) Mutators, population size, adaptive landscape and the adaptation of asexual populations of bacteria. Genetics 152:485–493

Thao M, Gullan P, Baumann P (2002) Secondary (gamma-Proteobacteria) endosymbionts infect the primary (beta-Proteobacteria) endosymbionts of mealybugs multiple times and coevolve with their hosts. Appl Environ Microbiol 68:3190–3197

Thauer RK, Jungermann K, Deker K (1977) Energy conservation in chemotrophic anaerobic bacteria. Microbiol Mol Biol Rev 41:100–180
Thauer R, Kaster A-K, Seedorf H, Buckel W, Hedderich R (2008) Methanogenic archaea: ecologically relevant differences in energy conservation. Nat Rev Microbiol 6:579–591
Thiel T, Pratte B, Zhong J, Goodwin L, Copeland A, Lucas S, Han C, Pitluck S, Land M, Kyrpides N et al (2014) Complete genome sequence of *Anabaena variabilis* ATCC 29413. Stand Genomic Sci 9:562–573
Thomas C, Nielsen K (2005) Mechanisms of, and barriers to, horizontal gene transfer between bacteria. Nat Rev Microbiol 3:711–721
Tiago I, Chung A, Verissimo A (2004) Bacterial diversity in a non saline alkaline environment: heterotrophic aerobic populations. Appl Environ Microbiol 70:7378–7387
Tiago I, Verissimo A (2013) Microbial and functional diversity of a subterrestrial high pH groundwater associated to serpentinization. Environ Microbiol 15:1687–1706
Tian J, Sinskey AJ, Stubbe J (2005) Kinetic studies of polyhydroxybutyrate granule formationin wautersia eutropha H16 by transmission electron microscopy. J Bacteriol 187:3814–3824
Torrella F, Morlta RY (1981) Microcultural study of bacterial size changes and microcolony and ultramicrocolony formation by heterotrophic bacteria in seawater. Appl Environ Microbiol 41:518–527
Unsworth B, Cross T, Seaward M, Sims R (1977) The longevity of thermoactinomycete endospores in natural substrates. J Appl Bacteriol 42:45–52
Vaïtilingom M, Deguillaume L, Vinatier V, Sancelme M, Amato P, Chaumerliac N, Delort A-M (2013) Potential impact of microbial activity on the oxidant capacity and organic carbon budget in clouds. Proc Natl Acad Sci U S A 110:559–564
van Ditmarsch D, Xavier J (2014) Seeing is believing: what experiments with microbes reveal about evolution. Trends Microbiol 22:2–4
van Niftrik L, Jetten MS (2012) Anaerobic ammonium-oxidizing bacteria: unique microorganisms with exceptional properties. Microbiol Molec Biol Rev 76:585–596
Vanwonterghem I, Evans P, Parks D, Jensen P, Woodcroft B, Hugenholtz P, Tyson G (2016) Methylotrophic methanogenesis discovered in the archaeal phylum *Verstraetearchaeota*. Nat Microbiol 1:16170
Venturi V, Weisbeek P, Koster M (1995) Gene regulation of siderophore-mediated iron acquisition in *Pseudomonas*: not only the Fur repressor. Mol Microbiol 17:603–610
Vreeland R, Rosenzweig W, Powers D (2000) Isolation of a 250 million-year-old halotolerant bacterium from a primary salt crystal. Nature 407:897–900
Waite D, Vanwonterghem I, Rinke C, Parks D, Zhang Y, Takai K, Sievert S, Simon J, Campbell B, Hanson T et al (2017) Comparative genomic analysis of the class Epsilonproteobacteria and proposed reclassification to Epsilonbacteraeota (phyl. nov.). Front Microbiol 8:682
Wallace AR (1870) Contributions to the theory of natural selection. Macmillan and Company, New York/London
Wang G, Mayes M, Gu L, Schadt C (2014a) Representation of dormant and active microbial dynamics for ecosystem modeling. PLoS ONE 9:e89252
Wang HF, Zhang YG, Chen JY, Hozzein WN, Li L, Wadaan MA, Zhang YM, Li WJ (2014b) *Nesterenkonia rhizosphaerae* sp. nov., a novel alkaliphilic actinobacterium isolated from rhizosphere soil of *Reaumuria soongorica* in saline-alkaline desert. Int J Syst Evol Microbiol. https://doi.org/10.1099/ijs.0.066894-0
Wang Q, Cen Z, Zhao J (2015) The survival mechanisms of thermophiles at high temperatures: an angle of omics. Physiology 30:97–106
Waters E, Hohn M, Ahel I, Graham D, Adams M, Barnstead M, Beeson K, Bibbs L, Bolanos R, Keller M et al (2003) The genome of *Nanoarchaeum equitans*: insights into early archaeal evolution and derived parasitism. Proc Natl Acad Sci U S A 100:12984–12988
Werts C, Michel V, Hofnung M, Charbit A (1994) Adsorption of bacteriophage lambda on the LamB protein of *Escherichia coli* K-12: point mutations in gene J of lambda responsible for extended host range. J Bacteriol 176:941–947

Westall F, Hickman-Lewis K, Hinman N, Gautret P, Campbell K, Bréhéret J, Foucher F, Hubert A, Sorieul S, Dass A et al (2018) A hydrothermal-sedimentary context for the origin of life. Astrobiology 18:1–35

Westberry T, Siegel DA (2006) Spatial and temporal distribution of *Trichodesmium* blooms in the world's oceans. Glob Biogeochem Cycles 20:4

Whitesides M, Oliver J (1997) Resuscitation of *Vibrio vulnificus* from the viable but nonculturable state. App Environ Microbiol 63:1002–1005

Whitman WB, Coleman DC, Wiebe WJ (1998) Prokaryotes: the unseen majority. Proc Natl Acad Sci U S A 95:6578–6583

Wiegel J (2011) Anaerobic alkaliphiles and alkaliphilic poly-extremophiles. In: Horikoshi K (ed) Extremophiles handbook. Springer, New York, pp 82–97

Wildman S (2002) Along the trail from Fraction I protein to Rubisco (ribulose bisphosphate carboxylase-oxygenase). Photosynth Res 73:243–250

Winogradsky SN (1949) Microbiologie du Sol. In: Problemes et Methodes: cinquante de recherches. Masson et Cie Eds, Paris

Winters Y, Lowenstein T, Timofeeff M (2015) Starvation-survival in Haloarchaea. Life (Basel) 5:1587–1609

Wirsen CO, Molyneaux SJ (1999) A study of deep-sea natural microbial populations and barophilic pure cultures using a high-pressure chemostat. Appl Environ Microbiol 65:5314–5321

Wolk CP (2000) Heterocyst formation in *Anabaena*. In: Brun YVS, Shimkets LJ (eds) Prokaryotic development. ASM Press, Washington, DC, pp 83–104

Woodford N, Ellinton M (2006) The emergence of antibiotic resistance by mutation. Clin Microbiol Infect 13:5–18

Wright S (1931) Evolution in Mendelian populations. Genetics 16:97–159

Wu B, Liang W, Kan B (2016) Growth phase, oxygen, temperature, and starvation affect the development of viable but non-culturable state of *Vibrio cholerae*. Front Microbiol 7:404

Xu H, Roberts N, Singleton F, Attwell R, Grimes D, Colwell R (1982) Survival and viability of nonculturable *Escherichia coli* and *Vibrio cholerae* in the estuarine and marine environment. Microb Ecol 8:313–323

Yamamoto H (2000) Viable but nonculturable state as a general phenomenon of non-spore- forming bacteria, and its modeling. J Infect Chemother 6:112–114

Yaron S, Matthews K (2002) A reverse transcriptase-polymerase chain reaction assay for detection of viable *Escherichia coli* O157:H7: investigation of specific target genes. J Appl Microbiol 92:633–640

Yayanos AA, Dietz AS, Van Boxtel R (1981) Obligately barophilic bacterium from the Mariana trench. Proc Natl Acad Sci U S A 78:5212–5215

Ye Y, Ma B, Dong C, Zhang H, Chen L, Guo FB (2016) A novel proposal of a simplified bacterial gene set and the neo-construction of a general minimized metabolic network. Sci Rep 6:35082

Young K (2006) The selective value of bacterial shape. Microbiol Mol Biol Rev 70:660–703

Youssef N, Savage-Ashlock K, McCully A, Luedtke B, Shaw E, Hoff W, Elshahed M (2014) Trehalose/2-sulfotrehalose biosynthesis and glycine-betaine uptake are widely spread mechanisms for osmoadaptation in the Halobacteriales. ISME J 8:636–649

Yu JS, Vargas M, Mityas C, Noll KM (2001) Liposome-mediated DNA uptake and transient expression in *Thermotoga*. Extremophiles 5:53–60

Yumoto I, Hirota K, Yoshimune K (2011) Environmental distribution and taxonomic diversity of alkaliphiles. In: Horikoshi K (ed) Extremophiles handbook. Springer, New York, pp 56–79

Zahnle KJ (1986) Photochemistry of methan and the formation of hydrocyanide (HCN) in the earth's early atmosphere. J Geophys Res 91:2819–2834

Zarzycki J, Brecht V, Müller M, Fuchs G (2009) Identifying the missing steps of the autotrophic 3-hydroxypropionate CO_2 fixation cycle in *Chloroflexus aurantiacus*. Proc Natl Acad Sci U S A 106:21317–21322

Žgur-Bertok D (2013) DNA damage repair and bacterial pathogens. PLoS Pathog 9:e1003711

Zhang H, Fu H, Wang J, Sun L, Jiang Y, Zhang L, Gao H (2013) Impacts of nitrate and nitrite on physiology of *Shewanella oneidensis*. PLoS ONE 8:e62629

Zhao X, Schwartz CL, Pierson J, Giovannoni SJ, McIntosh JR, Nicastro D (2017a) Three-dimensional structure of the ultraoligotrophic marine bacterium "*Candidatus* Pelagibacter ubique". Appl Environ Microbiol:83, pii: e02807–02816. https://doi.org/10.01128/AEM.02807-02816

Zhao X, Zhong J, Wei C, Lin C-W, Ding T (2017b) Current perspectives on viable but non-culturable state in foodborne pathogens. Front Microbiol 8:580

Zhaxybayeva O, Swithers K, Lapierre P, Fournier G, Bickhart D, DeBoy R, Nelson K, Nesbø C, Doolittle W, Gogarten J et al (2009) On the chimeric nature, thermophilic origin, and phylogenetic placement of the *Thermotogales*. Proc Natl Acad Sci U S A 106:5865–5870

Zillig W, Gierl A, Schreiber G, Wunderl S, Janekovic D, Stetter KO, Klenk HP (1983) The archaebacterium *Thermofilum pendens* represents, a novel genus of the thermophilic, anaerobic sulfur respiring *Thermoproteales*. Syst Appl Microbiol 4:79–87

Chapter 5
The Evolution of Living Beings Started with Prokaryotes and in Interaction with Prokaryotes

Télesphore Sime-Ngando, Jean-Claude Bertrand, Didier Bogusz, Jean-François Brugère, Claudine Franche, Marie-Laure Fardeau, Emilie Froussart, Anne Geiger, Maria Soledad Goñi-Urriza, Bernard Ollivier, and Paul W. O'Toole

Abstract In natural world, no organism exists in absolute isolation, and thus every organism must interact with the environment and other organisms. Next-generation sequencing technologies are increasingly revealing that most of the cells in the environment resist cultivation in the laboratory and several prokaryotic divisions have no known cultivated representatives. Based on this, we hypothesize that species that

T. Sime-Ngando (✉)
Laboratoire "Microorganismes: Génome et Environnement" (LMGE), CNRS UMR 6023, Université Clermont Auvergne, Clermont-Ferrand, France
e-mail: telesphore.sime-ngando@uca.fr

J.-C. Bertrand
Unité Mixte de Service, UMS 3470, OSU Pythéas, Aix Marseille Université, Marseille Cedex, France

D. Bogusz · C. Franche · E. Froussart
Équipe Rhizogenèse, UMR DIADE (IRD-UM), Institut de Recherche pour le Développement (IRD), Montpellier, Cedex 5, France

J.-F. Brugère
Université Clermont Auvergne, 63000 Clermont-Ferrand, France

M.-L. Fardeau · B. Ollivier
Aix Marseille Université, Université de Toulon, CNRS, IRD, MIO UM 110, Marseille, France

A. Geiger
INTERTRYP, Institut de Recherche pour le Développement, University of Montpellier, Montpellier, France

M. S. Goñi-Urriza
Environmental Microbiology CNRS, Institut des Sciences Analytiques et de Physico-Chimie pour l'Environnemnet et les Matérieaux, UMR5254, Univ. Pau & Pays Adour, Pau, France

P. W. O'Toole
School of Microbiology and APC Microbiome Institute, University College Cork, Co., Cork, Ireland

J.-C. Bertrand et al. (eds.), *Prokaryotes and Evolution*,
https://doi.org/10.1007/978-3-319-99784-1_5

live together in the same ecosystem are more or less dependent upon each other and are very large in diversity and number, outnumbering those that can be isolated in single-strain laboratory culture. In natural environments, bacteria and archaea interact with other organisms (viruses, protists, fungi, animals, plants, and human) in complex ecological networks, resulting in positive, negative, or no effect on one or another of the interacting partners. These interactions are sources of ecological forces such as competitive exclusion, niche partitioning, ecological adaptation, or horizontal gene transfers, which shape the biological evolution. In this chapter, we review the biological interactions involving prokaryotes in natural ecosystems, including plant, animal, and human microbiota, and give an overview of the insights into the evolution of living beings. We conclude that studies of biological interactions, including multipartite interactions, are sources of novel knowledge related to the biodiversity of living things, the functioning of ecosystems, the evolution of the cellular world, and the ecosystem services to the living beings.

Keywords Prokaryotes · Viruses · Protists · Fungi · Plants · Animals · Human · Microbial mats · Biotic interactions · Evolution

5.1 Biological Interactions in the Nature

The current model of the evolution within the cellular world is that the first living organisms were some form of prokaryotes which may have evolved out of the so-called protocells or protobionts, i.e., self-organized, endogenously ordered spherical collection of lipids proposed as a stepping-stone of the origin of life. Because a functional protocell has not yet been discovered nor achieved in a laboratory setting, we considered that the evolution of cellular living beings started with prokaryotes and in interaction with prokaryotes. Recent analyses of new data from molecular phylogenetics, paleontology, bioenergetics, and modern cell biology and biochemistry enable to safely increase support to the "prokaryotes-early" hypothesis, i.e., the two prokaryotic groups are older and are the root of eukaryogenesis (McInerney et al. 2014). This implies that the early involvement of prokaryotes in biological interactions is perhaps the main source of evolutionary processes and genomic innovations which are the mother of the complex and generalized biological interactions in the modern world.

Biological interactions are complex, with at times controversy among biologists concerning the definition of some of them. Herein, we agree with Lidicker (1979) and Faust and Raes (2012) and choose to center biological interactions on their effects, combining positive (i.e., a win), negative (i.e., a loss), and no effect (i.e., neutral) for the involved species (Fig. 5.1).

Different species may corporate to resist adverse conditions, such as a biofilm resistance to antibiotic, i.e., a win-win interaction known as *mutualism*. Certain cases of cross-feeding known as *syntrophy* (i.e., species exchange of metabolic products to the benefit of both) also are mutualism. Competition, i.e., loss-loss interaction, refers to an interaction between antagonist organisms through competitive exclusion, which states that two species with similar *niche* exclude each other. In loss-win interactions,

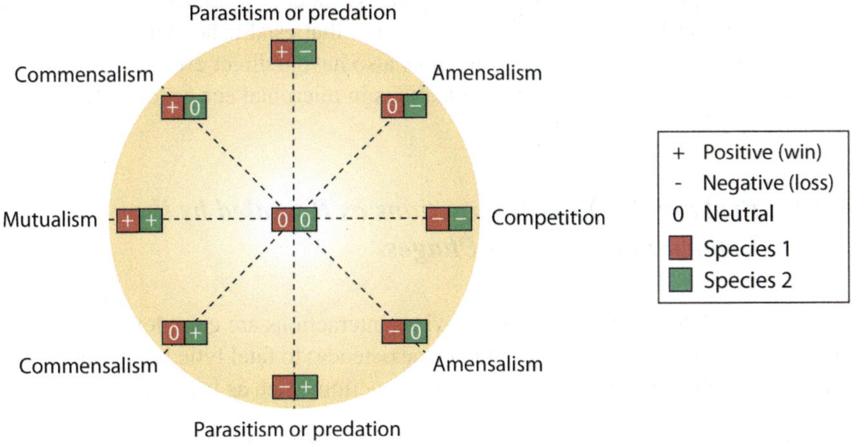

Fig. 5.1 Adopted ecological interactions between members of different species living in the same ecosystem. All possible pairwise interactions are represented based on their outcomes for each of the species partner: positive (+), negative (-) and neutral (0). Modified from Lidicker (1979) by Faust and Raes (2012). Agreement obtained from Springer Nature and Copyright Clearance Center, licence number 4358101124054.

such as predation or parasitism, the interaction between two organisms is positive for one and negative for the other. Predator-prey (interaction between organisms in which one organism captures biomass from another, i.e., grazing) and host-parasite (one organism, the parasite, usually benefits at the expense of the other, the host) relationships are typical examples of loss-win interactions. *Commensalism* is also a win-loss interaction that benefits one organism and the other organism is neither benefiting nor harmed. *Amensalism* is an interaction where an organism inflicts harm to another organism without any costs or benefits received by the latter (i.e., loss-neutral interaction).

In this chapter, the different types of interactions correspond to the definitions provided above that include all possible pairwise interactions between positive, negative, and neutral effects, otherwise clearly defined in particular cases. Indeed, in some cases, there is a very thin line that separates some of these interactions, e.g., mutualism, commensalism, and parasitism, i.e., a continuum concept. Mechanism or factors behind an apparent shift from mutualism to parasitism and vice versa via commensalism remain just a theory.

5.2 Interactions Between Prokaryotes and Viruses

As cellular models, prokaryotes offer an excellent ecological niche for viruses which are well-known as obligatory cellular symbionts because they have no intrinsic metabolism and need the intracellular machinery of a living and sensitive host cell for all processes requiring energy (Sime-Ngando 2014). The various modes of replication of prokaryotic viruses (i.e., phages) indeed intimately depend on the deep-cellular mechanisms (Fig. 5.2), a situation that establishes viruses as obligatory cell partners. These direct partnerships also have indirect effects for both partners, with tremendous ecological implications in microbial ecosystems (Fig. 5.3).

5.2.1 Prokaryote-Virus Interactions as Revealed by the Replication Modes of Phages

Based on viral "lifestyles," prokaryote-virus interactions are complex, ranging in a gradient from true nonlethal (i.e., stable coexistence) to fatal lytic interactions (host cell lysis), with intermediate mutualistic interactions such as lysogeny.

Lytic Interactions All the development cycles of viruses start with diffusive-passive fixation on specific receptors (often transporter proteins) present at the surface of a host cell, followed by injection of the viral genome into the host cell. In the lytic cycle, the viral genome induces the synthesis of viral constituents, including the replication of the viral genetic material. A number of progeny viruses are then pro-

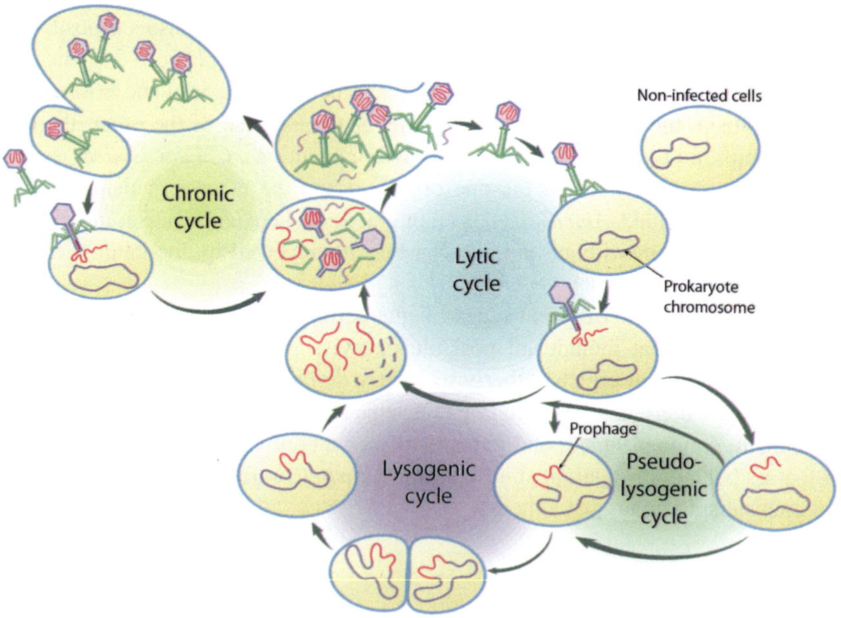

Fig. 5.2 Summary of biotic interactions between viruses and their prokaryotic hosts deduced from the development cycles of bacteriophages: chronic, lytic, lysogenic and pseudo lysogenic cycles. These interactions range from true non-lethal parasitism (i.e. stable coexistence in the chronic cycle) to fatal lytic infection (lytic cycle), with intermediate mutualistic lifestyles (lysogenic and pseudolysogenic cycles). (From Palesse 2014)

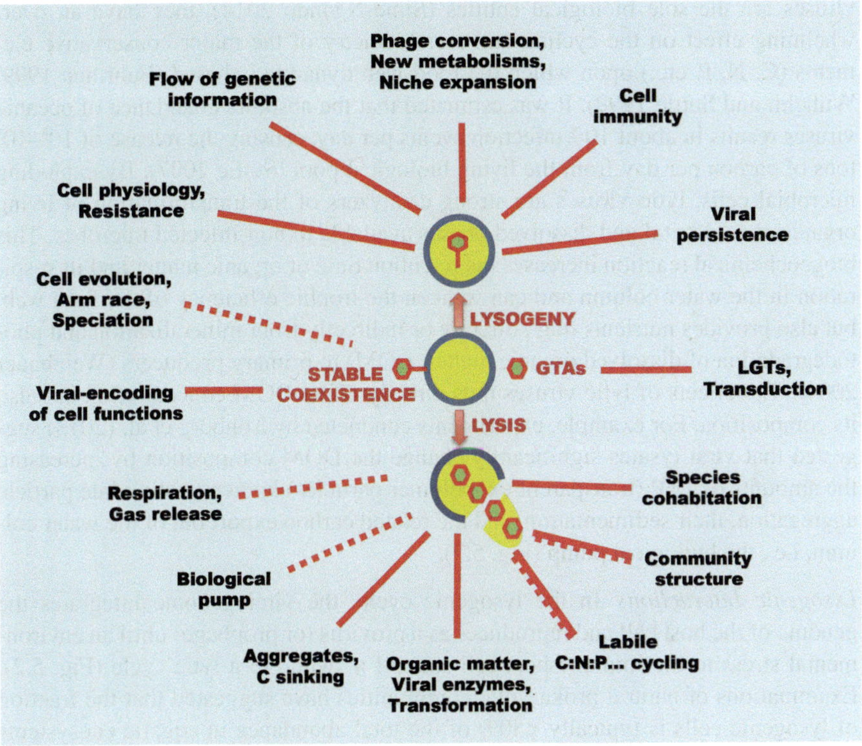

Fig. 5.3 Ecological effects of the biotic interactions between viruses and their microbial hosts on ecosystems processes. Because of the existence of a large panel of lifestyles (cf. Fig. 5.2) and in conjunction with the fact that all types of cells are sensitive to unique viruses, these biological entities are considered the most diverse, abundant, and ubiquitous biological entities in the biosphere where they have tremendous effects on the diversity of living things, the functioning of microbial ecosystems, and the evolution of the cellular world. Some of these direct (solid lines) and indirect (dashed lines) effects on ecologic processes are highlighted. Please refer to the main text for abbreviations

duced and released into the environment by the fatal rupture of the host cell (Fig. 5.2). It is now well accepted that the direct effects of lytic viruses represent one of the main causes of prokaryote mortality in ecosystems. Based on the direct observation of infected cells, viral-mediated mortality averages 10–50% of the daily production of heterotrophic prokaryotes and approximately equals the bacterivory from grazers in aquatic ecosystems (Fuhrman and Noble 1995; Pradeep Ram et al. 2005). The populations of lytic viruses ultimately depend on the availability of specific prokaryotic hosts and could thus respond to the growth rate of the most active hosts. This pattern has the strong feedback effect of preventing species dominance and enhanced species cohabitation within microbial communities, i.e., the so-called "phage kills the winner" hypothesis (Thingstad and Lignell 1997).

Because viruses kill microbial hosts which often dominate the biological biomass in most natural ecosystems, including the most extreme ones where prokaryotes and viruses are the sole biological entities (Sime-Ngando 2014), they have an overwhelming effect on the cycling and stoichiometry of the major conservative elements (C, N, P, etc.) upon which the food-web dynamic is based (Fuhrman 1999; Wilhelm and Suttle 1999). It was estimated that the absolute abundance of oceanic viruses results in about 10^{29} infection events per day, causing the release of 10^8–10^9 tons of carbon per day from the living biological pool (Suttle 2007). By exploding microbial cells, lytic viruses are strong catalyzers of the transformation of living organisms to detrital and dissolved phases available to non-infected microbes. This biogeochemical reaction increases the retention time of organic matter and its respiration in the water column and can weaken the trophic efficiency of the food web, but also provides nutrients (e.g., directly or indirectly from mineralization and photodegradation of dissolved organic matter, DOM) to primary producers (Weinbauer 2004). The effects of lytic viruses thus directly affect DOM concentration but also its composition. For example, experiments conducted by Lønborg et al. (2013) suggested that viral lysates significantly change the DOM composition by increasing the amounts of TEP (transparent exopolymer particles) known to stimulate particle aggregation, their sedimentation, and the related carbon export out of the water column, i.e., the biological pump (Fig. 5.3).

Lysogenic Interactions In the lysogenic cycle, the viral genome integrates the genome of the host cell and reproduces as a provirus (or prophage) until an environmental stress to the immune host cell sets off a switch to a lytic cycle (Fig. 5.2). Examinations of natural prokaryotic communities have suggested that the fraction of lysogenic cells is typically <50% of the total abundance in aquatic ecosystems (Sime-Ngando and Colombet 2009). Both the provirus and the host cell benefit from lysogeny. Lysogenic conversion has been described as a means of survival for viral populations that are threatened by poor host cell abundance and therefore cannot sustain population numbers through lytic infection alone (Stewart and Levin 1984). Lysogeny thus provides a means of persistence for viruses in the nutrient-depleted environments where the abundance of the host cells is very low (Pradeep Ram and Sime-Ngando 2010). Prophages may affect the metabolic properties of host cells which can acquire immunity to superinfections and new phenotypic characteristics such as antibiotic resistance, antigenic changes, and virulence factors, resulting in niche expansion for viral hosts (Fig. 5.3). A spectacular case for such a phage conversion is the finding that cholera infection is due to a lysogenic strain of the *Vibrio cholerae* bacterium. Since the toxin is encoded in the genome of the prophage and is not part of the host genome, non-lysogenic cells do not cause cholera (Weinbauer 2004). More generally, it is likely that all living cells could contain active prophages in their genome. On average, 2.6 prophages have been detected per free-living prokaryotic species (Lawrence et al. 2002), and a number of prokaryotic genomes contain between 3 and 10% of DNA prophages (Brüssow and Hendrix 2002). This highlights the potential of prokaryote-virus interactions as key food-web players in the evolution and maintenance of living things.

A variant to the lysogenic cycle is the so-called carrier state or pseudolysogenic cycle, where the viral genome is not integrated within the host genome but rather remains in an "inactive state" within the host cell (Fig. 5.2). There is no replication of the viral genome which is segregated unequally into progeny cells, most likely for a few generations. Pseudolysogenic viruses probably occur in very poor nutrient conditions where host cells are undergoing starvation and cannot offer the energy necessary for viral gene expression.

Chronic Interactions In the chronic cycle, the progeny viruses are episodically or constantly released from the host cell by budding or extrusion, without immediate lethal events (Fig. 5.2). This cycle is common in eukaryotic viruses such as herpes and hepatitis viruses or rhabdoviruses. Chroniç viral infection is a dynamic and metastable equilibrium process which ends with the lysis of the host cell after serial budding of lipid membrane-coated viruses, as seen in hosts of the marine protist *Emiliania huxleyi* (Mackinder et al. 2009). Chronic infection without host lysis has been reported for the first time in the marine primary producer *Ostreococcus tauri*, where the low rate of viral release through budding (one to three viruses cell^{-1} day^{-1}) allows cell recovery and the stable coexistence of viruses and their hosts (Thomas et al. 2011; Clerissi et al. 2012). This mode of coexistence seems to be widespread in viruses of *Archaea*. The discovery and characterization of unique viruses of *Archaea* are growing increasingly, influencing the field of prokaryotic virology and prokaryote-virus interactions (Prangishvili et al. 2006).

Phage Therapy Gets Revitalized The rise of antibiotic resistance is now increasingly bringing medical research back on a century-old research theme, the use of bacteriophages – viruses that kill bacteria – to fight human diseases. This medical exploitation of prokaryote-virus interactions is still widely used in the East (Russia, Georgia, and Poland) but never took off elsewhere. Now, because of increasing resistance of bacterial pathogens to antibiotics, Western researchers and governments are seriously considered bacteriophages as a resource for fighting drug-resistant bacterial infections. In March 2014, the US National Institute of Allergy and Infectious Diseases listed phage therapy as one of seven prongs in its plan to combat antibiotic resistance. In June 2013, the European Community launched the project *Phagoburn*, the first large, multicenter clinical trial of phage therapy for human infections (http://www.phagoburn.eu/). It is critical to examine questions regarding the phagotherapy issue, such as: (i) Are attempts to use phages for clinical and environmental applications more likely to succeed now than in the past? (ii) Will phage therapy and prophylaxis suffer the same fates as antibiotics – treatment failure due to acquired resistance, and ever-increasing frequencies of resistant pathogens? (Levin and Bull 2004). Recent laboratory and animal studies, exploiting current understandings of phage biology, indeed suggest that phages may be useful as antibacterial agents in certain conditions (Summers 2001).

5.2.2 *Prokaryote-Virus Interactions Drive Critical Ecological and Evolutionary Processes*

Prokaryote-virus interactions can impact ecological and evolutionary processes in two major ways (Sime-Ngando 2014). The first major way includes the direct effects of the intrinsic activities of viruses: (i) keep in check competitive dominant hosts (i.e., lytic viruses), (ii) affect the metabolic properties of host cells which can acquire immunity to superinfections and new phenotypic and genotypic traits such as production of toxins (i.e., temperate phage conversion), and (iii) transfer both viral and host genes between species, thereby influencing speciation (Fig. 5.3). Large-scale metagenomics has shown that viruses contain diverse genes of interest, including virulence genes such as the cholera toxin genes, respiration, nucleic acid, carbohydrate, and protein metabolism genes, as well as genes involved in vitamin and cofactor synthesis, in stress response, and in motility and chemotaxis, which are more common in viromes (metagenomes of viruses) than in their corresponding microbiomes (metagenomes of microbes) (Rohwer and Thurber 2009). Microbes that take up these genes increase their competitive ability and extend their ecological niches. Given the prevalence of phage-encoded biological functions and the occurrence of recombination between phage and host genes, phage populations are thus expected to serve as gene reservoirs that contribute to niche partitioning of microbial species in ecosystems. Gene transfers by transduction may also represent an important mechanism for gene evolution in the environment, and bacteriophage transduction could play an important role in contributing to the genetic diversity and the evolution of biological populations (Sime-Ngando 2014).

Indeed, one of the most surprising findings of whole-genome sequencing is the enormous extent of lateral gene transfers (LGTs). LGTs refer to the gene material exchanges between organisms that happen independently of reproduction (i.e., vertical gene transfers). General mechanisms include transformation (e.g., gene transfer by uptake of free genetic materials, from viral lysis), conjugation (direct gene transfer from cell-to-cell contact), the activity of gene transfer agents (GTAs), and transduction, where viruses are the main vectors that move nucleic materials from one cell to another (Fig. 5.3). The transduction frequency in natural waters ranges from 10^{-8} to 10^{-5} per virus, and they might be up to 100 transductants l^{-1} day^{-1} (Jiang and Paul 1998). An extrapolation exercise suggests that as many as 10^{24} genes are moved by transduction from viruses to hosts each year in the world's ocean (Rohwer and Thurber 2009). Although transduction is a random process, viruses can genetically alter microbial populations through lysogeny and transduction and affect the flow of genetic information in microbial ecosystems.

The second major way that prokaryote-virus interactions impact ecological and evolutionary processes comprises the indirect effects of viral activities, such as (i) the structuring effects of lysis products on species composition and richness, (ii) the sustenance of the amount of information encoded in genomes that may favor horizontal gene transfer mechanisms, and (iii) the effects of physiological mechanisms involved in the resistance of host against viruses, through the host-pathogen arms race (Fig. 5.3). Together with (i) the high abundance and broad geographical distribution of viruses and viral sequences within microbial fractions and (ii) the prevalence of genes among typical viral sequences that encode microbial physiological functions, the above mentioned effects establish environmental prokaryote-virus interactions as a strong vector that generates genetic variability in microbial ecosystems and drive both ecological functions and evolutionary changes (Weinbauer and Rassoulzadegan 2004). Some viral groups such as the *Caudovirales*, the tailed double-stranded DNA phages, are probably older than the separation of life into the three now recognized domains of life (Ackermann 1999; Hendrix 1999). This suggests that, before the occurrence of eukaryotic grazers such as flagellates and ciliates, viruses were probably the main predators of cells in the prokaryotic world and played a major role in the sophisticated forces (dispersal, competition, adaptive radiation, etc.) that shape biogeography and evolution.

5.3 Interactions Between Prokaryotes

5.3.1 Overview of Prokaryote: Prokaryote Associations

During the course of evolution, prokaryotes have entered into numerous interactions, and the associations between prokaryotes have been in place from the earliest stages of life on earth. Stromatolites are the most developed example of these associations of prokaryotic complementary metabolisms (cf. "Stromatolites" in Chap. 3).

However, the life in associations of prokaryotes remained long ignored because microbiology developed from studies of pure cultures of bacteria isolated from their environment (Koch's works). It has been demonstrated by molecular methodologies that most prokaryotes present in the environment are not cultivated in the laboratory; 0.1–1% of the total cells observed under microscope are cultivables (Amann et al. 1995; Torsvik and Øvreås 2002). Thus, only a minor fraction of the tremendous prokaryote diversity is known (Rappé and Giovannoni 2003; Rinke et al. 2013; Eloe-Fadrosh et al. 2016).

Since the development of the works undertaken on the ecology of prokaryotes, primarily those of S. Winograsky and M. Beijerinck, it is known that in nature prokaryotes do not live isolated from each other (*No microbe is an Island existing independently*). They associate transiently or permanently, optionally or mandatorily. The union between partners can be so narrow that it can lead to a particular cell organization, referred to as "parakaryote," where bacteria are living inside another bacterium, e.g., *Parakaryon myojinensis* (Yamaguchi et al. 2012).

Moreover, in nature very few species live in pure culture except those involved in some diseases or those living in extreme environments. An example of a completely isolated lifestyle is the *Candidatus Desulforudis audaxviator*, which is capable of fixing carbon dioxide and nitrogen dioxide and composes >99.9% of the microorganisms inhabiting water collected at 2.8-kilometer depth (Chivian et al. 2008).

Within prokaryotic associations, frequently, the cells signal each other and coordinate their growth, their activities, and movements. Information exchanges between the cells are diverse and numerous. They have been formally shown by the highlighting of conjugation (physical interactions between neighboring bacterial cells) and reinforced by the other works on horizontal gene transfer (HGT) (cf. Chap. 4). Cell-to-cell exchange of informations can also take place in other ways:

1. A process known as "quorum sensing" (QS): Production and detection of low molecular weight sensor molecules ("sensing") used to coordinate gene expression when a certain cell density is reached ("quorum") (cf. biofilms).
2. In many Gram-negative bacteria, the exchange of informations is ensured by vesicles called "outer-membrane vesicles (OMVs)" which convey various molecules (QS molecules, antimicrobial factors, toxins, DNA). These vesicles, liberated from the outer membrane (50–250 nm in diameter), mediate cell communication and have diverse functions (pathogenesis, bacterial survival during stress conditions, induce change as a function of environmental conditions) (Schwechheimer and Kuehn 2015; Mashburn-Warren and Whiteley 2006).
3. Exchange by intimate cytoplasmic connection such as the formation of narrow junctions between nitrogen-fixing heterocysts and adjacent vegetative cell (Fay 1992).
4. Another inter-bacterial communication is the transfer of cytoplasmic constituents (proteins, mRNA, plasmid DNA) through a network of intercellular membranous nanotubes. These nanotubes that connect the bacterium belonging to the same species or between different species unrelated to evolutionary point of view are widespread among bacteria growing in biofilms (Dubey and ben-Yehuda 2011).
5. By extracellular electron transfer without direct cell contact. These so-called cable bacteria (long filamentous bacteria, belonging to the family *Desulfobulbaceae*) were shown to induce electrical currents over centimeter distances in the surface layer of marine sediments. They transfer electron from free sulfide in deeper sediment horizons to oxygen present near the sediment-water interface (Nielsen et al. 2010; Malkin and Meysman 2015).
6. Syntrophy also provides examples of interspecies contact-dependent communication.

In most environments, the association between prokaryotes led to the formation of complex and highly organized structures: biofilms and microbial mats, consisting of organisms of the same species or different species (cf. Sect. 5.3.6 of this chapter).

All kinds of interactions are present in the prokaryotic world. They can be classified into three main types depending on their effects: neutral, conflictual, and beneficial interactions.

5.3.2 *Neutral Interactions*

In neutral interactions two species do not interact with each other; they have no effect on one another. It is difficult to demonstrate this type of interaction especially in natural conditions. Such relationships are likely to exist in various environmental conditions: **(i)** for populations that are spatially distant, **(ii)** when nutrient concentrations are extremely low (oligotrophic conditions) and where population densities are very low, and **(iii)** when the nutritional needs of populations are complete and extremely different.

5.3.3 Conflictual Interactions

In ecosystems conflicting interactions between prokaryotes are numerous and can take various forms.

5.3.3.1 Competition (Loss-Loss Interactions)

In natural environments, most of the time, prokaryotes live in a dietary deficiency state. Thus, populations and/or communities that have the same nutritional needs compete for space and resources (availability of nitrogen source, carbon source, electron donors, electron acceptors, vitamins, etc.). The competition between two populations for the same limiting resource is sometimes so intense that it led to the extinction of one of the two species. This is what is called the competitive exclusion principle which has been demonstrated experimentally the first time by Gause (1935). It states that the populations of two species involved in a single limiting resource cannot coexist indefinitely in a stable and homogeneous medium, the most competitive of the two species ending at more or less long term by eliminating the other. Jacques Monod (1949) was the first to demonstrate the relationship between limiting nutrient concentrations and bacterial growth: in defined medium the bacterial growth yield is linearly dependent on the initial concentration of the limiting nutrient. Hansen and Hubbel (1980) confirm the survival of only one species when several species compete for the same limiting nutrient.

5.3.3.2 Amensalism (Loss-Neutral Interactions)

In this relationship a partner inflicts negative effect on another organism. For example, a strain will release into the environment toxic compounds for its partners and against which it is immune; the production of these toxic compounds is a competitive advantage. Typical examples of amensalism are the production of antibiotics by some prokaryotes that can inhibit or kill other prokaryotes that are sensitive to these compounds (Madigan et al. 2015). The bacteriocins are another example of compounds produced by some prokaryotes (Cotter et al. 2013). These antimicrobial agents are active against other bacteria and act on cell envelope, gene expression, and protein production. They represent a viable potential alternative to fight against antibiotic-resistant prokaryotes (e.g., antibiotics are bacteriocins which prevent and control the growth of staphylococci and/or enterococci in and on catheter tubing).

5.3.3.3 Predation and Parasitism (Loss-Win Interactions)

In this kind of conflictual relationship – predation, parasitism – the interaction between two organisms is positive for one and negative for the other. The border between parasitism and predation is sometimes difficult to establish, and distinction between these two types of relationships varies according to the authors. For some authors, the term "predation" is used for any relationship that is positive for one partner and negative for the other. Other authors separate between two types of interactions: (i) with contact between the two partners at least for part of their life, then they speak of "parasitism," and (ii) without direct contact between the two partners, which corresponds to "predation."

In the framework of this chapter, we distinguish four types of "loss-win" strategies: social predation ("wolfpack" group predation), epibiotic attachment, periplasmic invasion, and cytoplasmic invasion (Martin 2002; Pasternak et al. 2014; Pérez et al. 2016) (Fig. 5.4).

Wolfpack Strategy A number of cells produce hydrolytic enzymes that degrade nearby bacteria and consume the released compounds. *Myxococcus* and *Lysobacter*, common in soil, are examples of this attack strategy of a prey. *Myxococcus xanthus* has a complex life cycle, which is facultatively multicellular; it is a model system for social and cooperative behavior in bacteria (Fig. 5.5). Indeed, these gliding bacteria hunt in group, and cells must work together to lyse other bacteria and grow on the nutrient release. This hunting strategy is generally referred as "wolfpack" attack,

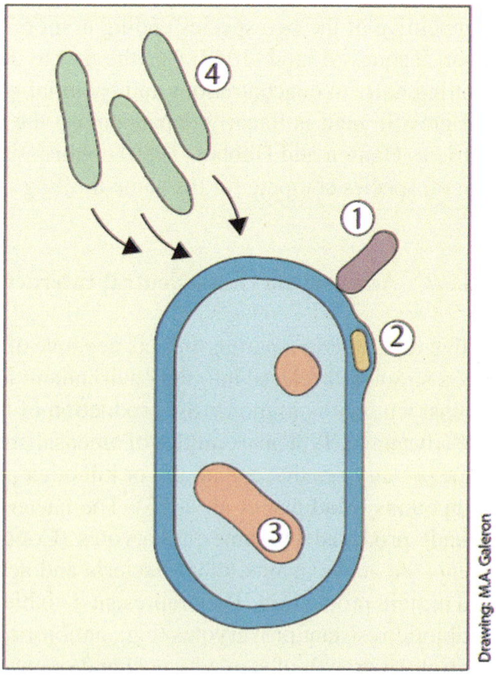

Fig. 5.4 Different strategies of « loss-win » interactions between prokaryotes: epibiotic attachment (1) ; periplasmic invasion (2) ; cytoplasmic invasion (3) ; « wolfpack » predation (4). (Modified and redrawn from Guerrero et al. 1986; Martin 2002)

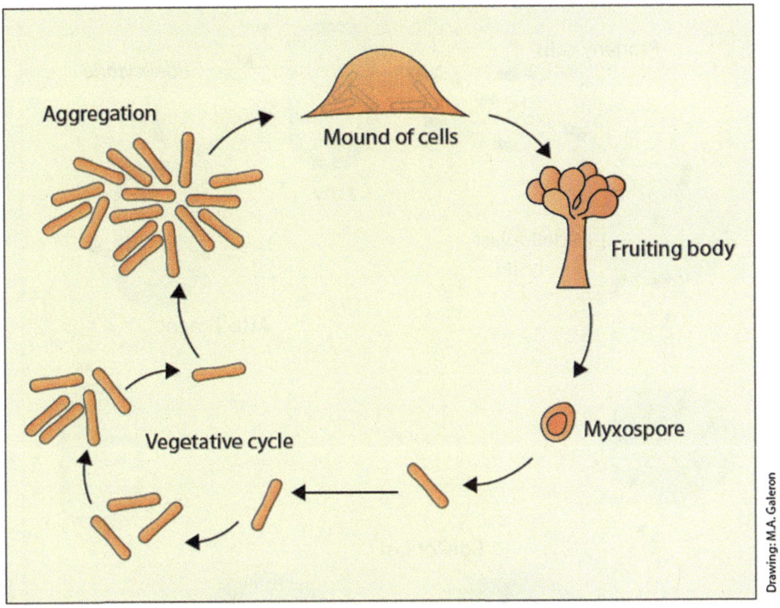

Fig. 5.5 Schematic life cycle of a gliding myxobacteria : *Myxococcus xanthus*. Under non-limiting nutrient conditions *M. xanthus* has a vegetative cycle. The vegetative cells (Gram-negative rods and nonflagellated cells) glide together across surface releasing compounds that lyse other bacteria. When nutrients decrease, vegetative cells aggregate, forming multicellular structures called mounds of cells and fruiting body. In fruiting body, the majority of cells are converted to myxospore more resistant to drying, UV radiation, heat, than vegetative cells. Under favorable nutritional and physical conditions, myxospores germinate. (Modified and redrawn from Madigan et al. 2015)

because a high cell density seems to be required to lyse the prey; the term "wolf-pack" is questioned by some authors (Pérez et al. 2016). To lyse the prey, *M. xanthus* use antibiotics (myxovirescin), hydrolytic enzymes (protease MepA), and extracellular outer-membrane vesicles (OMVs), produced in large quantities, which may facilitate delivery of lytic compounds (Berleman et al. 2014; Keane and Berleman 2016; Pérez et al. 2016).

Epibiotic Attachment In the cell-to-cell contact with attachment with the prey, the predator remains attached at the outer surface of the prey where it degrades and assimilates the prey molecules (Fig. 5.4). This is the strategy adopted by the species *Vampirococcus* sp., *Micavibrio aeruginosavorus*, and *Bdellovibrio exovorus*. For example, *Vampirococcus*, motile by polar flagella, attaches to the surface of the phototrophic bacterium *Chromatium* spp. and secretes degradative enzymes that lead to the release of cytoplasmic content of the prey, which is "sucked" by *Vampirococcus*.

Invasion of the Periplasm Predator (or the parasite according to the terminology adopted by the author), propelled by a polar flagellum, swims to encounter the prey and fixes on it (Fig. 5.6). At the point of attachment, it secretes hydrolytic enzymes

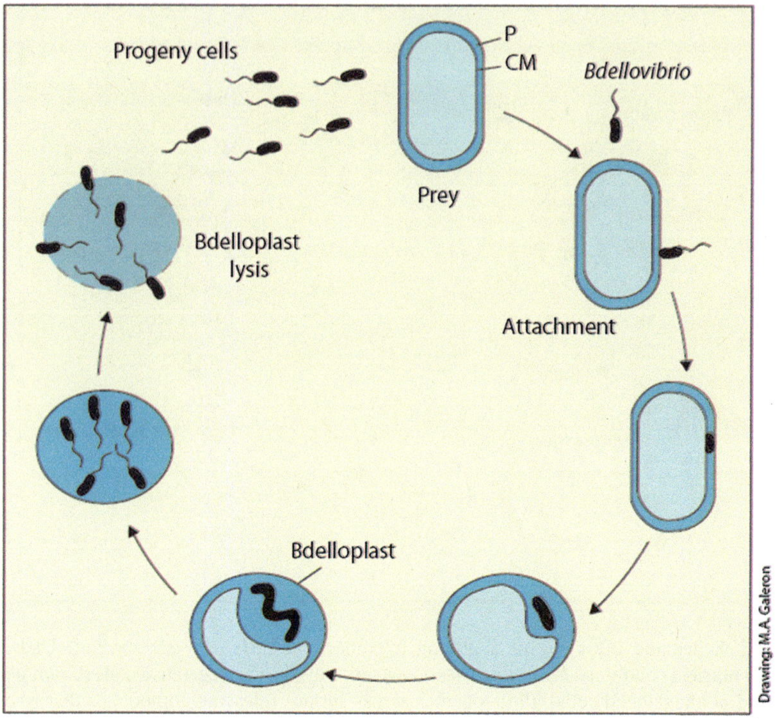

Fig. 5.6 The predatory cycle of *Bdellovibrio bacteriovorus*. After attachment (attack phase), the predator penetrates the prey and replicates in the periplasmic space (P). The prey becomes spherical, forming the bdelloplast, and *Bdellovibrio* takes the form of a filament. After *Bdellovibrio* septation, the bdelloplast lyses releases progeny cells. (Modified and redrawn from Davidov and Jurkevitch 2009)

that perfore the cell wall of the prey and then moved into its periplasmic space, where it grows and replicates. The predator kills the prey. The best known example of prokaryotic that has adopted this strategy is that of *Bdellovibrio bacteriovorus* (Sockett 2009).

Invasion of the Cytoplasm In this strategy, predator penetrates the prey until the cytoplasm where it grows and divides. The only known species which adopted this strategy is the genus *Daptobacter*.

Predator as Therapeutic Agents It is potentially possible to use predatory bacteria as "living antibiotics" to fight against certain pathogenic bacteria. The ability of predator of Gram-negative bacteria could be regarded as an alternative approach to fight against multiresistant Gram-negative pathogens. It has been demonstrated that *Bdellovibrio* and *Micavibrio* have the potential to attack a wide range of human pathogens (as planktonic cell or as biofilm). However, further investigations are needed to definitively demonstrate the use of these "living antibiotics" in medicine, agriculture, veterinary science, and food industry (van Essche et al. 2011; Kadouri, et al. 2013; Lebba et al. 2014).

Origin of Eukaryotic Cell: A Hypothesis Different hypotheses have been proposed to account to the passage from a cellular organization of the prokaryotic type to a cellular organization of the eukaryotic type. The final explanation is still pending. Some authors suggested that the eukaryotic cell is the result of the merger of two prokaryotes by the establishment of a syntrophic association between bacteria and archaea. Davidov and Jurkevitch (2009) propose that the eukaryotic cell is the result of a predatory or parasitic interaction between prokaryotes: *The predator was a small (facultative) aerobic alpha-proteobacterium, which penetrated and replicated within the host periplasm, and later became the mitochondria.*

The "Intimate Association" of Nanoarchaeum Equitans and Ignicoccus Hospitalis, A Unique Interaction Between Two Archaea: Commensalism or Parasitism? Ignicoccus hospitalis is an anaerobic and obligate chemolithoautotrophic organism: it fixes CO_2 via a novel pathway and grows by reduction of elemental sulfur with molecular hydrogen as electron donor. *I. hospitalis* is one of the smallest genomes of free-living organisms (a single circular chromosome with less than 1500 genes) (Huber et al. 2012; Hamerly et al. 2015). Cell envelope of this archaeon is very specific: absence of a rigid cell wall component (like pseudoprotein or S-layer protein) and presence of a double-membrane system separated by intermembrane space, an "intermembrane compartment (IMC), width from 20 to 1000 nm, containing vesicles or tubes budding from cytoplasmic membrane. *Nanobacterium equitans* cells are cocci (350–500 nm in diameter), with a very small genome (about 490 Kbp), growing on surface of *Ignicoccus*, but the contact between the two organisms is very limited (Huber et al. 2012, Moissl-Eichinger and Hubert 2011) (Fig. 5.7a).

The relationship between the two partners of this "intimate" association is difficult to define and can belong to two types:

1. Commensalism – *N. equitans* benefited and is even dependent for its survival on the presence of *I. hospitalis*, while *I. hospitalis* is unaffected by the presence of *N. equitans*.
2. Parasitism – in this case, *I. hospitalis* is affected by the presence of *N. equitans*.

In the framework of this chapter, the association will be considered as a parasitic association where *N. equitans* is "a nutritional parasite on *I. hospitalis*" (Giannone et al. 2015) which "exhibits parasitic function" (Hamerly et al. 2015).

I. hospitalis and *N. equitans* can be cultivated in a stable coculture. *I. hospitalis* can grow alone and together with *N. equitans*. In contrast, *N. equitans* does not grow alone, neither autotrophically nor heterotrophically (Huber et al. 2012): *I. hospitalis* is a mandatory host for *N. equitans* which behaves like a nutritional and energetic parasite (Giannone et al. 2015). Indeed, from multi-"omics" data (Hamerly et al. 2015), it has been demonstrated, under laboratory conditions, that *N. equitans* was nutritionally and energetically dependent on *I. hospitalis* for growth and survival: it uses a large fraction of the molecule pool produce by *I. hospitalis* to satisfy its own metabolic needs and to build its membrane lipids (amino acids, lipids are synthesized by *I. hospitalis* and transferred to *N. equitans*). Moreover, the presence of *N. equitans* seems to have a significant metabolic cost (Hamerly et al. 2015) for its host. The combined relative

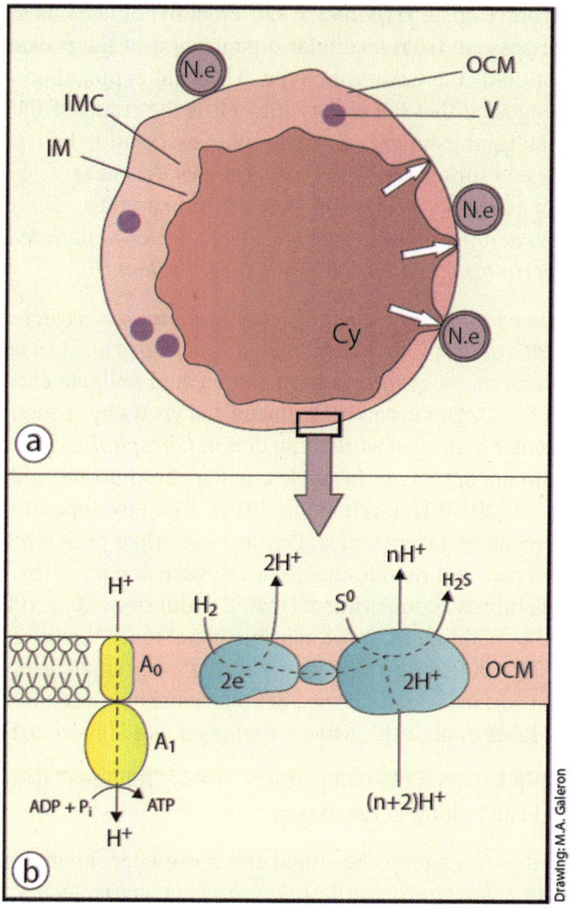

Fig. 5.7 Interaction between *Nanobacterium equitans* (N.e) and *Ignicoccus hospitalis*. (**a**) Drawing illustring *N. equitans* cells attachment to an *I. hospitalis* host cell; OCM : outer cellular membrane; IMC : intermembrane compartment; IM : inner membrane; V : vacuoles; Cy : cytoplasm. (**b**) Scheme of ATP synthase and H_2 : sulfur oxidoreductase. White arrows : contact site between OCM and IM. (Modified and redrawn from Jahn et al. 2008; Moissl-Eichinger and Hubert 2011; Huber et al. 2012; Giannone et al. 2015)

abundance of membrane proteins increased by 50% in coculture compared when *I. hospitalis* was grown alone (e.g., the ATP synthase, H2/sulfur oxidoreductase) (Fig. 5.7b) (Giannone et al. 2015). Finally, if *I. hospitalis* continued to be metabolically active in the presence of *N. equitans*, its growth is slowed down by the presence of its host (Giannone et al. 2015). These data, obtained from laboratory culture conditions, are in favor of parasitic relationship; *N. equitans* "acts much more like a parasite than a commensal" (Giannone et al. 2011).

5.3.4 Beneficial Interactions

The evolution and life of prokaryotes are not based only on competition and struggle; prokaryotes can coexist and cooperate for survival.

5.3.4.1 Commensalism (Win-Neutral)

Commensalism describes an interaction in which one organism benefits in the relationship (commensal) and the other organism (host) is neither benefited nor harmed. This interaction is therefore the counterpart of amensalism. Commensal relationships can be established in different physical and chemical conditions.

Habitat Modification For example, when a facultative anaerobic population consumes oxygen entirely in a biotope, changing the physical environment creates favorable conditions for the development of strict anaerobic bacteria. The commensal, the anaerobic population, benefits from the activity of facultative anaerobes which remain unaffected by the relationship if the two populations do not compete for the same substrates. Such a situation settles in the intestinal tract where a bacterium such as *E. coli*, facultative anaerobe, consumes the oxygen, allowing the growth of anaerobic microflora.

Production of Growth Factors Frequently, prokaryotes secrete growth factors (vitamins, amino acids) which are used by other prokaryotes for their own growth.

Transformation of Insoluble Compounds to Soluble Compounds During the biodegradation of hydrocarbons, some strains produce biosurfactants, which increase the bioavailability of poorly soluble hydrocarbons, thereby promoting uptake by hydrocarbonoclastic microorganisms (Bertrand et al. 2015b).

Conversion of Soluble Compound to Gaseous Compound The change of state from soluble compound to gas in a biotope generates the appearance of a substrate, which can move to another biotope and can be used by another population. For instance, the methane produced by methanogenic populations in sediment can benefit to methane-oxidizing populations in the overlying water column (Bertrand et al. 2015a).

Desorption Processes Certain populations have the property to adsorb without transforming a compound, which thus becomes available for another population. The proposed example is populations that produce biosurfactants, aforementioned in the transformation of insoluble to soluble compounds. Some biosurfactants (e.g., rhamnolipid) are also capable of causing desorption of polycyclic aromatic hydrocarbons (PAHs) from contaminated soils (Mao et al. 2015). After release of these compounds, they may be biodegraded by hydrocarbonoclastic populations.

Conversion of Organic Molecules by One Population into Substrates for Other Population For instance, in anoxic ecosystem (marine and lake sediments, soils, digesters), fermentative prokaryotes are the only microorganisms able to degrade complex polymers (enzymatic hydrolysis) into monomers (sugars, amino acids, long-chain fatty acids, etc.). In anoxic conditions, these monomers will be substrates for different respiratory metabolic pathways: denitrification, manganese and iron reduction, sulfate reduction, or methanogenesis (Bertrand et al. 2015a).

Removal of a Toxic Compound A population is able to remove a compound, which is toxic for another population. In marine sediments, the degradation of organic matter by sulfate-reducing bacteria results in the formation of H_2S, a toxic compound for the biotope. The so-formed H_2S can be removed by chemolithotrophic sulfur-oxidizing prokaryotes (CSOP) and anoxygenic phototrophic sulfur bacteria (APB). Four kinds of APB are able to perform this detoxification: among the purple bacteria, *Chromatiaceae* (e.g., *Chromatium okenii, Thiospirillum jenense, Thiococcus pfennigii, etc.*), *Ectothiorhodospiraceae* (*Ectothiorhodospira mobilis, Halorhodospira halophila, Thiorhodospira sibirica,* etc.), some green nonsulfur bacteria (*Chloroflexus,* multicellular, filamentous), and green sulfur bacteria (Bertrand et al. 2015a).

It is sometimes difficult to distinguish between commensalism and cooperation. For example, in the case of nitrification, the *ammonium-oxidizing bacteria* (AOB) oxidize ammonium to nitrite. The nitrite formed by these populations is used by *nitrite-oxidizing bacteria* (NOB). At first glance, this interaction is a commensal interaction, but nitrite is often toxic to ammonium oxidizers. In this case the removal of nitrite is beneficial to ammonia oxidizers. In the latter case, the interaction AOB-NOB is rather a cooperative interaction.

Cometabolism The cometabolism can be considered as a variation of commensalism. In this process a microorganism transformed a compound on which the microorganism is unable to grow. The microorganism derives no energetic or nutritional benefits of this transformation. In cometabolism, the biotransformed compound is used as a growth substrate by another microorganism which benefits from the action of the first. The process was initially described by Leadbetter and Foster (1958), with the strain *Pseudomonas methanica* growing on methane used as a carbon and energy source. This strain could oxidize ethane but was unable to use this substrate as a carbon source. Similarly, the strain *Mycobacterium vaccae* is able to grow on propane and simultaneously cometabilize cyclohexane which is oxidized to cyclohexanol and which can be utilized by other bacterial populations (Beam and Perry 1974). Since, it has been demonstrated that a wide range of microorganisms can degrade many compounds by cometabolisms, especially xenobiotic compounds (Nzila 2013) (Table 5.1).

Table 5.1 Examples of cometabolism of polycyclic aromatic hydrocarbons and chlorinated compounds

Non-growth substrate	Growth substrate	Bacteria
Benzo [a] pyrene[a]	Fluoranthene	*Sphingomonas paucimobilis*
Pyrene[a]	Phenanthrene	*Cycloclasticus* sp. *Sphingomonas* sp.
Fluoranthene[a]	Fluorene	*Sphingomonas* sp.
Phenanthrene[a]	Fluorene	*Sphingomonas* sp.
Fluorene[a]	Sucrose	*Rhodococcus rhodochrous*
Trichloroethylene[b]	Methane	Methane degrading bacteria *Methylosinus trichosporium*
Chloroform[b]	Methane	Methane degrading bacteria *Methylosinus trichosporium*
Chlorobenzene[c]	Glucose	*Pseudomonas* sp. *Staphylococcus xylosus*
PCB (1-6 chlorine atoms)[d]	Biphenyl	*Nocardia* sp. *Pseudomonas* sp.

Nzila (2013)
[a]Polyclic aromatic hydorcarbons
[b]Chlorinated aliphatics
[c]Chorinated mono-aromatics
[d]Polychlorinated biphenyl

5.3.4.2 Mutualism (Win-Win Interactions)

Mutualism is an interaction – obligatory and nonobligatory – where both partners benefit with the interaction.

5.3.4.2.1 Metabolic Cooperation

Metabolic cooperation is very common among prokaryotes and concerns prokaryotes that carry out complementary metabolisms. This interaction allows the use of substrates that a strain alone cannot use. Such cooperation occurs frequently in the case of the degradation of complex compounds, natural or anthropogenic. Such biodegradation requires the intervention of several populations. Grouped, these populations synthesize all the enzymes necessary for the biodegradation of these complex compounds. Below are three examples of organisms that carry out complementary metabolisms.

Organic Matter Mineralization in Anoxic Sediment In anoxic marine sediments, the mineralization of organic matter cannot be achieved without the joint action of several prokaryotic populations with different and complementary metabolisms: fermentative bacteria are the only organisms able of attacking complex polymers by action of their hydrolytic enzymes. Fermentation-produced monomers are then used by various respiratory metabolisms and by methanogens (Bertrand et al. 2015a).

Petroleum Biodegradation Petroleum biodegradation can be made only by different prokaryotic populations with different and complementary enzymatic activities. No pure strain has all the enzymes necessary for the biodegradation of a compound containing thousands of chemically different molecules (Bertrand et al. 1983, 2015b; McGenity et al. 2012).

Polychlorobiphenyls (PCBs) Mineralization Mineralization of PCBs can only be performed by the joint and complementary actions of anaerobic and aerobic populations; generally, a single strain is unable to mineralize PCBs. In a first step, anaerobic bacteria remove the chlorine atoms of highly chlorinated PCBs. In a second step, the weakly chlorinated congeners produced by anaerobic bacteria are substrates for aerobic bacteria that are able to open the related benzene cycle, leading to the mineralization of PCBs (Abramowicz 1990, Bertrand et al. 2015b).

5.3.4.2.2 Obligatory Relationship

During evolution some species have adopted a lifestyle in tight association to each other. These close interspecies associations are highly structured, maintaining a permanent cell-to-cell contact termed "consortia" (Schink 2002) that are mandatory for one or both partners. Such associations of prokaryotes exist in various biotopes. These consortia are commonly found in photosynthetic bacteria (Overmann and Schubert 2002).

The activity of the phototrophic consortium "*Chlorochromatium aggregatum*" is one example of an organism that cannot be grown in the absence of a partner (Overmann and Schubert 2002; Müller and Overmann 2011; Liu et al. 2013; Morris et al. 2013; Cerqueda-García et al. 2014). This highly structured phototrophic consortium consists of colored green sulfur bacteria epibionts (*Chlorobium chlorochromatii* strain CaD) surrounding a flagellated rod-shaped central bacterium, "*Candidatus* Symbiobacter mobilis," which is a chemoheterotrophic colorless β-proteobacteria; the two species are phylogenetically distant (Fig. 5.8a, b).

This consortium can grow in laboratory conditions and cell division of all epibionts, and central bacteria are highly synchronized. The epibiont grows photolithoautotrophically utilizing exogenous sulfide as photosynthetic electron donor, and its bacteriochlorophylls function as light sensors. It has been successfully cultivated in pure culture, but all attempts to grow the central bacterium alone have failed so far. Exchange of metabolites between the two partners of consortium was observed, and it is likely that organic compounds (sugars, amino acids) are transferred from epibionts to heterotroph "*Candidatus* Symbiobacter mobilis," and this explains its inability to develop independently. The inability of "*Candidatus* Symbiobacter mobilis" to develop independently could also be explained by a massive loss of genes during evolution (when compared to eight close relative strains), and the growth of this *Candidatus* now depends on its epibiont for several key metabolites.

Intact consortium exhibits a strong chemotaxis toward sulfide and a scotophobic response: the central bacterium provides mobility to reach location in water column

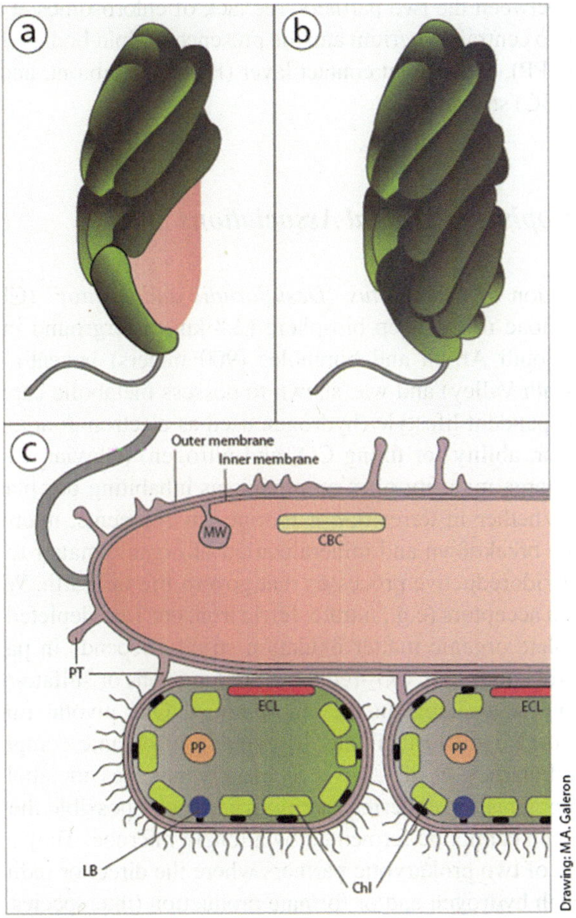

Fig. 5.8 Schematic representation of the photosynthetic consortium *Chlorochromatium aggregatum*. (**a**): Consortium at an early stage of the division when epibiont (green) cover partially the flagelated central bacterium *Candidatus* Symbiobacter mobilis (pink); (**b**) consortium after division into two daughter consortia. (**c**): Schematic representation of the ultrastructure and morphological relations between the two consortium partners. Chlorosomes (Chl); lipid bodies (LB); polyphosphate globules (PP); epibiont contact layer (ECL); central bacterium crystal (CBC); complex cytoplasmic membrane invaginations (MW); periplasmic tubules (PT) formed by the outer membrane of central bacterium which are in contact with the outer membrane of the epibionte. Wanner et al. (2008) postulated a "common periplasmic space". (Modified and redrawn from Wanner et al. 2008; Liu et al. 2013)

of stratified lakes toward sulfate and light, ensuring optimal conditions for epibiont photosynthesis which is adapted to the low light intensities (≤ 7 mol quanta·m^{-2}·s^{-1}).

A detailed ultrastructural study was conducted by Wanner et al. (2008) and led the authors to propose an ultrastructural model of the phototrophic consortia (Fig. 5.8c). The rod-shaped epibionts were connected by polysaccharide chains, and each consortium harbored, on average, 16 ± 3 epibionts. Authors highlighted a common periplasmic space between the two partners, the lack of chlorosomes at the attachment site of epibiont to central bacterium and the presence of lipid bodies (LB), polyphosphate globules (PP), an epibiont contact layer (ECL) in epibiont, and central bacterium crystal (CBC) structures.

5.3.5 Syntrophic Microbial Associations

With the exception of *Candidatus* "*Desulforudis audaxviator*" (CDa) which has been detected alone in the deep biosphere (2.8 km underground in the Mponeng gold mine in South Africa and boreholes (900 meters) beneath the surface of California's Death Valley) and was shown to possess metabolic capabilities necessary for an independent lifestyle (hydrogen used as electron donor, sulfate used as electron acceptor, ability for fixing CO_2 and nitrogen) (Chivian et al. 2008), it is reported that a large majority of microorganisms inhabiting our planet live within communities. Whether in terrestrial or marine environments, microorganisms are essential for the breakdown and mineralization of organic matter together with all the microbial oxidoreductive processes that govern life on Earth. When oxygen or mineral electron acceptors (e.g., nitrate, ferric iron, etc.) are depleted from the environment, complete organic matter oxidation strictly depends in particular on the activity of hydrogenotrophic CO_2-reducing methanogens or sulfate-reducing bacteria when sulfate is available in the environment. One pivotal function existing between microorganisms to facilitate degradation of organic compounds is called syntrophy (McInnerney et al. 2009). Syntrophy involves metabolic interactions between microorganisms rendering thermodynamically possible the use of organic compounds which cannot be fermented by a lonely microbe. They mainly result in the cooperation of two prokaryotic partners where the direct or indirect removal of electrons through hydrogen and/or formate production (interspecies electron transfer: IET) is essential to succeed in oxidizing organic material (syntrophic propionate, butyrate, or benzoate oxidation ....) under anaerobic conditions (Stams and Plugge 2009; Morris et al. 2013). In this respect, IET is an important mechanism for energy exchange in a range of anaerobic microbial communities (Smith et al. 2015). Recent works have contributed to better understand the functioning of syntrophic cultures and have demonstrated that besides the transfer of reducing agents such as hydrogen and/or formate, microbial interactions may also involve the exchange of organic, sulfurous, and nitrogenous compounds or the removal of toxic compounds (Morris et al. 2013). One of the most studied strategies related to IET is H_2 interspecies transfer (HIT) where protons are reduced by the electrons released by an organic compound-oxidizing microorganism into H_2 to be used by hydrogenotrophs (methanogens within the domain *Archaea* or sulfate reducers within the domain *Bacteria*), thus favoring the oxidative activity of the former bacterium. Hydrogenotrophs are thus recognized as primordial partners in syntrophic associations. This is not only true for obligatory syntrophic communities but also for facultative ones where the removal of H_2 resulted in the production of more oxidized volatile fatty acids (e.g., acetate), thus leading to more energy delivered through substrate level phosphorylation and to the increase in the use of substrate (Stams and Plugge 2009). HIT was first reported by Bryant et al. (1967) who described that *Methanobacillus omelianski* maintained in ethanol medium was in fact an association of two species, the first one (the S-organism) fermented ethanol with production of hydrogen, acetate, and CO_2 (reaction 1) and the second one which was later described as *Methanobacterium bryantii* used hydrogen for growth and produced methane by reducing CO_2 (reaction 2), thus creating thermodynamically favorable conditions allowing the S-organism to oxidize ethanol.

$$\text{Ethanol} + \text{H}_2\text{O} \rightarrow \text{Acetate}^- + \text{H}^+ + 2\text{H}_2 \quad (\text{reaction 1}) \qquad \Delta G^{0'} = +9.6\,\text{kJ / reaction}$$

$$4\text{H}_2 + \text{HCO}_3^- + \text{H}^+ \rightarrow \text{methane} + 3\text{H}_2\text{O} \quad (\text{reaction 2}) \qquad \Delta G^{0'} = -135{,}6\,\text{kJ / reaction}$$

Despite, in the case of *M. omelianski*, ethanol oxidation required the presence of the S-organism, it has been clearly established that some methanogens may also use directly ethanol to produce methane (Frimmer and Widdel 1989; Garcia et al. 2000).

In contrast to ethanol, propionate, butyrate, and also benzoate are only degraded by a coculture involving the organic compound oxidizer and hydrogenotroph partners (reaction 2) following the reactions 3, 4, and 5.

$$\text{Propionate}^- + 3\text{H}_2\text{O} \rightarrow \text{acetate}^- + \text{HCO}_3^- + \text{H}^+ + 3\text{H}_2 \quad (\text{reaction 3}) \quad \Delta G^{0'} = +76.1\,\text{kJ / reaction}$$

$$\text{Butyrate}^- + 2\text{H}_2\text{O} \rightarrow 2\,\text{acetate}^- + \text{H}^+ + 2\text{H}_2 \quad (\text{reaction 4}) \quad \Delta G^{0'} = +48.1\,\text{kJ / reaction}$$

$$\text{Benzoate}^- + 7\text{H}_2\text{O} \rightarrow 3\,\text{acetate}^- + 3\text{H}^+ + \text{HCO}_3^- + 3\text{H}_2 \quad (\text{reaction 5}) \quad \Delta G^{0'} = +70.1\,\text{kJ / reaction}$$

(reactions 1, 2, 3, and 4 from Thauer et al. 1977; reaction 5 from McInnerney et al. 2009)

Many mesophilic and thermophilic propionate-degrading bacteria have been described. They include Gram-negative (*Syntrophobacter* and *Smithella* spp.) and Gram-positive bacteria (*Pelotomaculum* and *Desulfotomaculum* spp.) (Stams et al. 2012) (Fig. 5.9). Syntrophic butyrate-degrading bacteria comprise members of the genera *Syntrophomonas, Syntrophus,* and *Thermosyntropha* (Stams et al. 2012). The oxidation of benzoate has been reported for the syntrophic bacterium, *Syntrophus aciditrophicus* (Jackson et al. 1999). When producing hydrogen, the syntrophic bacteria required an energy input for reoxidizing reduced flavoproteins involved in electron transport together with NADH. This is made possible by the ability of these bacteria to perform a reverse electron transport where peculiar Rnf-typr and Qrc membrane complexes, but also oxidoreductases and electron transfer flavoproteins, play a significant role (Stams and Plugge 2009; Morris et al. 2013).

Formate to be used by methanogens (reaction 6) or sulfate-reducing bacteria may substitute for H_2 as the electron carrier causing a formate interspecies transfer (FIT).

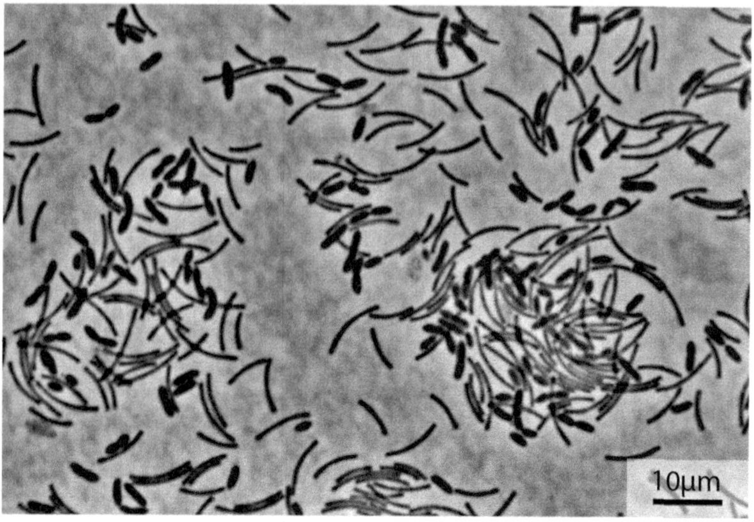

Fig. 5.9 Phase contrast of syntrophic coculture between strain MGP affiliated with *Desulfotomaculum* subcluster Ih (short thick rods) and *Methanospirillum hungatei* (long curved rods) grown on propionate. Bar, 10 µm. From Imachi et al. (2006). Agreement obtained from American Society for Microbiology and Copyright Clearance Center, order number 4347631076628

$$4\,\text{Formate}^- + H^+ \rightarrow 3\,HCO_3^- + 2\,H_2O + CH_4 \left(\text{reaction}\,6\right) \quad \Delta G^{0'} = -130\,\text{kJ / reaction}$$

(reaction 6 from Stams et al. 2012)

Both FIT and HIT may be involved in the electron transfer between the syntrophic microorganisms and hydrogenotrophs (Stams and Plugge 2009). FIT has been noticed in cocultures using propionate, butyrate, but also proteins (Shrestha and Rotaru 2014). Molecular approaches provided evidence that multiple hydrogenase and formate dehydrogenase genes were expressed in coculture experiments with *Syntrophobacter fumaroxidans* grown on propionate in the presence of *Methanospirillum hungatei*, thus suggesting that hydrogen and formate were involved simultaneously as electron carriers in these syntrophic associations (Stams and Plugge 2009). Hamilton et al. (2015) demonstrated that formate was even preferred over hydrogen for electron transfer due to thermodynamic considerations. In this respect, we may expect the use of both FIT and HIT by syntrophic bacteria to be essential in maintaining the adequate redox potential equilibrium during exchange of electrons from substrate to hydrogenotrophic methanogen resulting in its optimal oxidation. The importance of syntrophic associations has been highlighted in the degradation of amino acids with a peculiar emphasis for alanine, valine, leucine, and isoleucine. The deamination of these amino acids to the corresponding alpha-keto acids was thermodynamically favorable only after removal of hydrogen by methanogens or SRB (Baena et al. 1998, 1999, 2000; Morris et al. 2013). Syntrophic communities are also known to be involved in the degradation of aliphatic and aromatic hydrocarbons. They comprise *Syntrophus* and *Smithella* spp. (Morris et al. 2013). It is quite intriguing to note that some sugar-utilizing bacteria require the presence of hydrogenotrophic microorganisms to consume their substrate. Despite the term "syntrophy" is not appropriate for such association as sugars (e.g., glucose and fructose) are easily fermentable, the latter bacteria should be nonetheless considered as obligate syntrophs as recommended by Stams and Plugge (2009) since they cannot grow alone on sugars and do not therefore ferment sugars. In this respect, the use of the adjective "fermenting" for such microorganisms should not be applied as they exhibit an oxidative metabolism of sugars which is most probably only possible, thanks to removal of H_2 allowing oxidation of NADH (Stams and Plugge 2009). Microbiologists should pay much more attention to these syntrophs as they have been found of ecological significance in the environments that they inhabit even surpassing the number of sugar-fermenting bacteria existing in situ. This is the case of *Bacillus* spp. originated from a deep freshwater sample in Germany with growth being dependent of the presence of hydrogenotrophic methanogens (e.g., *Methanospirillum hungatei)* to use carbohydrates (Mueller et al. 2008). In order to enumerate bacteria which demethoxylate benzoids, Krumholz and Bryant (1986) isolated from the rumen *Syntrophococcus sucromutans* which was demonstrated to oxidize sugars in the presence of the hydrogenotrophic methanogen, *Methanobrevibacter smithii*, or other organic molecules (e.g., formate, methoxymonobenzoids) possibly acting as terminal electron acceptors. This oxidative process was considered by these authors as a major catabolic function by this bacterium in the rumen (Krumholz and Bryant 1986). It is only recently that mesophilic members (*Mesotoga* spp.) of the order *Thermotogales* have been also recognized to actively oxidize sugars in the presence of hydrogenotrophic sulfate-reducing bacteria (Fadhlaoui et al. 2018) (Fig. 5.10). Based on these observations, we may expect such types of bacteria to have been underestimated so far in natural environments

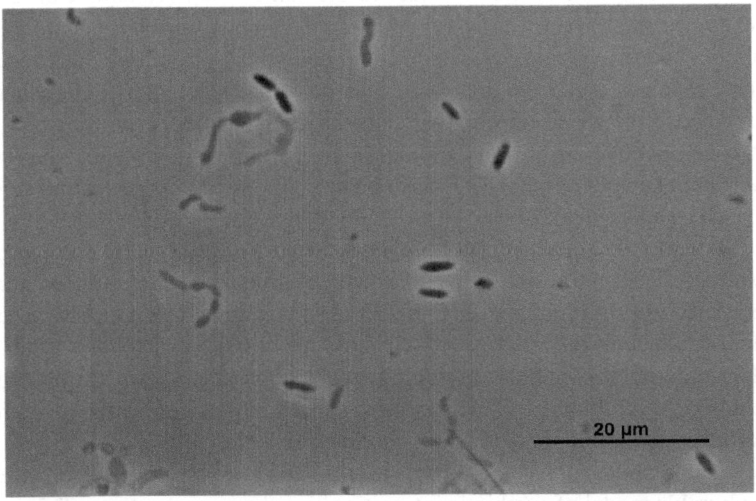

Fig. 5.10 Phase contrast of syntrophic coculture between *Mesotoga prima* strain PhosAc3 (grey rods and filaments) and *Desulfotomaculum gibsoniae* (black rods) grown on glucose

whether they are poor or rich in sugar content. Further experiments are therefore clearly needed to understand their ecological role in nature as compared to classical fermenting microbes, which are believed to be the main sugar users in any environment on Earth.

Direct interspecies electron transfer (DIET) is an alternative to HIT and FIT in which syntrophic partners exchange directly electrons (Stams and Plugge 2009; Morris et al. 2013; Shrestha et al. 2013). DIET has been first demonstrated via microbial nanowires in the ferric iron-reducing cultures of *Geobacter sulfurreducens* and *Shewanella oneidensis* when grown on acetate and lactate, respectively (Gorby et al. 2006; Reguera et al. 2005). In that case, both bacteria were shown to be directly in contact, thanks to pili formation, with Fe(III) oxides to be used as electron acceptors. However extracellular electron transfer between two prokaryotes has been established only in 2010 (Summers et al. 2010). At this occasion, two *Geobacter* species were associated to oxidize ethanol. While *G. metallireducens* was involved in the oxidation of ethanol to acetate, this oxidative process was rendered thermodynamically possible by the use of electrons by *G. sulfurreducens* having the ability to use fumarate as terminal electron acceptor and to possibly consume acetate as electron donor (Summers et al. 2010). The ecological significance of *Geobacter* spp. in DIET has been emphasized when they were cocultured with methanogens. This was highlighted by the syntrophic coculture *Geobacter metallireducens-Methanosaeta harundinacea* which stoichiometrically converted ethanol to methane. Transcriptomic, radiotracer, and genetic analyses demonstrated that *M. harundinacea* used electrons via DIET for the reduction of carbon dioxide to methane (Rotaru et al. 2014a). While the production of pili by *G. metallireducens* was required for transferring electrons to the methanogen (Rotaru et al. 2014a), a conductive mediator was not necessary when *G. metallireducens* was associated with *Methanosarcina barkeri* to perform the same reaction, thus demonstrating that direct cell-to-cell contact may conduct to the same result (Rotaru et al. 2014b). Interestingly, members of *Methanosaeta* and *Methanosarcina* genera possess membrane-bound cytochromes, which could potentially play a role in extracellular electron exchange, thus giving ecological advantages to them to compete for electron use in the anaerobic environments that they inhabit. As an example, DIET offers a competitive advantage in some methanogenic environments, and there is circumstantial evidence that DIET may enhance the function of anaerobic digesters (Malvankar and Lovley 2014).

A less explorated mechanism of IET is electron exchange via quinone-mediated interspecies electron transfer (QUIET) in which compounds with quinone moieties serve as electron shuttles between the electron-donating and the electron-accepting partner (Smith et al. 2015).

Many microbial communities exhibit interactions that do not fit the classical definition of syntrophy (e.g., bacteria from sulfureous environments, phototrophic consortia, hyperthermophilic archaea, etc.) (Morris et al. 2013).

Regarding phototrophic consortia, "*Chlorochromatium aggregatum*" was the first culturable model of symbiotic interactions involving prokaryotes. This consortium was enriched from Lake Dagow in Germany (Fröstl and Overmann 1998) and comprised 12–20 green sulfur bacteria belonging to the species *Chlorobium chlorochromatii* (epibionts) that surround a central, colorless, and motile β-proteobacterium, phylogenetically related to *Comamonadaceae* for which the metabolical involvement in the consortium has not been established so far (Kanzler et al. 2005). Indeed, while *Chlorobia* can be cultured axenically (Cerqueda-García et al. 2014), isolation of the obligate-symbiotic β-proteobacterium has never been successful (Müller and Overmann 2011). A more simple phototrophic consortium involved *Desulfuromonas acetoxidans* (Pfennig and Biebl 1976) which was shown to grow with photosynthetic green sulfur bacteria. Their relationship should be considered as mutualistic in that the *Chlorobium* spp. oxidize sulfide to elemental sulfur which is used as a terminal electron acceptor by *Desulfuromonas* during acetate oxidation with subsequent production of sulfide to be reused as electron donor by the former bacterium (Warthmann et al. 1992). In this case, sulfide as a strong reducing agent also helps in keeping highly reducing conditions as required for growth of these phototrophs (Müller and Overmann 2011).

Under anaerobic conditions, interactions between only the members of the domain *Archaea* are scarce. The hyperthermophile *Nanoarchaeum equitans* (Huber et al. 2002) was shown to grow symbiotically with the crenarchaeon *Ignicoccus hospitalis* (Fig. 5.7). Genome sequence analysis of *N. equitans* revealed that it lacks metabolic capacity to synthesize many cell components (e.g., biosynthesis of amino acids, nucleotides, and lipids) (Waters et al. 2003). This intimate association represents to date the only natural, cultivated community of two archaeal species (Morris et al. 2013).

The demonstration of anaerobic methane oxidation (AOM) coupled to sulfate reduction was first established to occur in marine sediments (Reeburgh 1976; Stams and Plugge 2009; Morris et al. 2013). The process was found to be catalyzed by communities of anaerobic methanotrophic archaea (ANME) and sulfate-reducing bacteria (SRB) of the *Deltaproteobacteria* rendering the methane oxidative process thermodynamically favorable by most probably removing electrons (Stams and Plugge 2009; Morris et al. 2013). However, neither the methanogenic nor the sulfate-reducing microorganisms associated in this consortium have been isolated in pure culture with metabolism of each partner being not elucidated so far. Timmers et al. (2016) found that AOM activity was only performed at low-sulfate concentrations in marine and low-salinity environments. It is only recently that AOM has been demonstrated to be performed in the presence of nitrate and nitrite (Haroon et al. 2013), iron and/or manganese (Riedinger et al. 2014), and humic acids (Gupta et al. 2013) as terminal electron acceptors, but also through a new pathway for dissimilatory sulfate reduction by the methanotrophic archaea (Milucka et al. 2012).

As discussed above, the large majority of syntrophic relationships takes place in anoxic environments. However, there are some examples of syntrophy which exist under aerobic conditions. Association of *Hyphomicrobium* sp. with aerobic methanotrophic bacteria was found to be helpful to oxidize methane as the sole energy and carbon source (Hanson and Hanson 1996). The importance of *Hyphomicrobium* sp. in this coculture consisted in its ability to remove methanol and/or formaldehyde, known as toxic compounds, resulting from methane oxidation by the methanotroph (Wilkinson et al. 1974, Moore 1981, Schink 2002).

5.3.6 *Life in Community, from Biofilms to Microbial Mats*

Microorganisms predominantly live in aggregates or biofilms, i.e., structured microbial communities embedded in an extracellular matrix. Biofilm's lifestyle provides numerous advantages, including the access to resources, the protection from predators, viruses, antibiotics, and other chemical toxins and environmental stress factors. The biofilms lifestyle leads microbial communities to develop in hostile environmental conditions in which planktonic cells would not develop. Combining unfavorable environments with proximity of cells promotes genetic mobility and modifications, and thus, accelerates evolution of microorganisms.

5.3.6.1 Biofilms

Biofilms formation has been largely investigated because they can provoke deleterious effects in human health by reducing the effectiveness of antibiotics or the cleaning up of medical material such as catheters. Biofilms can also have major economic repercussions as a consequence of biofouling or biocorrosion. Because harmful health biofilms are generally monospecific and because the development and regulation of each biofilm are dependent on its bacterial composition, most of the studies focused on one species-specific biofilm.

The transition from the bacterial free-living lifestyle to the biofilm lifestyle is dependent on the production of extracellular matrix by the bacteria. Despite the striking differences in lifestyles, only about 1% of genes showed differential expression in biofilm-attached versus free-living *Pseudomonas aeruginosa*; half of the genes were activated and half were repressed in biofilms (Whiteley et al. 2001). The microbial cells only constitute a minor part of the biofilm; the majority is composed of the biofilm matrix. This extracellular matrix consists of polysaccharides, proteins, and extracellular DNA (Pamp et al. 2007). It also contains functional components such as extracellular enzymes and vesicles, signaling molecules (intra- or interspecific), and toxins. The specific composition of the biofilm matrix may vary greatly depending on the microorganism(s) that compose the biofilm. The composition of the extracellular matrix will also depend on the environmental conditions in which the biofilm develops. The mechanism of biofilm formation is, however, reversible by the induction of extracellular matrix degradation (Gjermansen et al. 2010).

The biofilm development and dispersion are regulated processes. Although submitted to quorum sensing (QS) regulation in most of the cases, the bis-(3′-5′)-cyclic dimeric guanosine monophosphate (c-di-GMP) and regulatory small RNAs (sRNAs) appear like central regulators for biofilm development. But the regulation of biofilm formation varies depending on species composition of the biofilm. For example, in *Salmonella*, in association with the previous regulatory systems, the curli subunit gene D (*csgD*) is involved in the activation of the extracellular polymeric matrix biosynthesis, leading to the transition from the planktonic stage of bacterial cells to the multicellular state (Liu et al. 2014). The two former regulatory systems associated with QS are also involved in biofilm formation in various *Pseudomonas* species and in *Burkholderia cenocepacia*. Nevertheless, the molecular mechanisms involved in transducing the signals into expression of the biofilm matrix components differ between species, since the composition of the extracellular matrix differs: exopolysaccharides are the most important biofilm matrix components for *P. aeruginosa*, whereas large surface proteins appear to be the most important biofilm matrix components for *P. putida*, *P. fluorescens*, and *B. cenocepacia* (Fazli et al. 2014).

The availability to form biofilms can be lost very quickly; a single expontaneous nucleotide change in a protein involved in an ABC transporter of a type I secretion system was sufficient to eliminate biofilm formation in *D. vulgaris* Hildenborough (De León et al. 2017). Moreover, biofilms are hotspot structures for genetic evolution. In a biofilm, isogenic populations will metabolically and genetically diversify into distinct variants in order to occupy biofilm niches and thus improve biofilm productivity by reducing competition between cells. New types of interactions are developed. A recent study working with *B. subtilis* biofilms demonstrated a dramatically different availability to biofilm forming after isolation of colonies from a biofilm. This variability was related to changes in gene expression of biofilm-related genes and to subtle genetic differences between the new strain variants (Dragoš et al. 2018). The diversification of niche occupancy in the biofilm and the interactions between cells allow the evolutionary diversification of cells in the biofilm.

In nature, biofilms are rarely monospecific but composed of different microorganisms, and biofilms are as diverse as their constituent microbes. For example, the fluffy acidophilic biofilms formed on acid mine drainage rocks are strikingly different from the stiff photosynthetic biofilms developing on salt marsh or the dental plaque formed on teeth. Except the fact that they are embedded in a matrix, all these different communities should be considered as specific communities, with specific structure and functioning. In those natural biofilms, genetic adaptations are enhanced because of the genetic diversity of microorganisms composing the biofilm. Laboratory studies using simplified communities demonstrated that simple mutations in the genome of one partner lead it to adapt to other partners, modifying the interactions between them in a specialized association (Hansen et al. 2007). Biofilms, as spatially structured environment, promote evolution in interactions between species, provoke changes in their relationships, and affect overall community functioning.

5.3.6.2 Microbial Mats, a Self-Sustaining Community?

Microbial mats are complex communities in which microorganisms are structured along biogeochemical gradients. In turn, microbial metabolisms modify those biogeochemical gradients, maintaining the ecosystem. The interactions between microorganisms determine the functioning of the whole system and favor the biomass and energy production as well as elements' cycling. Microbial mats can be view as metaorganisms based on a nonpermanent functional symbiosis, optimizing their development. They are autonomous, dependent only on external energy sources.

In this chapter, two types of microbial mats will be described, the chemolithoautotrophic mats developing on the hydrothermal vents and seeping areas and the photosynthetic mats developing on coastal sediments.

Chemolithoautotrophic Mats Chemolithotrophs include microbes that can oxidize inorganic compounds such as H_2S, H_2, and reduced metals to keep energy for growth. Life in deep-sea vent environments is supported by these microorganisms. The chemical and thermal gradients in vent chimneys and seepage areas provide a wide range of microhabitats leading to the development of chemolithoautotrophic microbial mats.

The mats are composed of chemolithoautotrophs and chemolithoheterotrophs. The colorless sulfur-oxidizing bacteria, which include filamentous gliding bacteria such as *Beggiatoa*, *Thiotrix*, and *Thioploca* species or nonmotile such as *Thiomargarita* sp., are key members of the mats. Filaments of these bacteria can reach up to 10 cm long and 200 μm of diameter, allowing the physical cohesion of the mat at the water-sediment interface. Some *Beggiatoa* mats are composed of strictly autotrophic microorganisms, but others include also mixotrophs. Other microorganisms such as neutrophilic iron oxidizing and chemosynthetic microorganisms are known to be able to form mats.

Heterotrophs are associated with those chemolithoautotrophs. In deep seawater, the association of methane seeps with high concentrations of sulfate favors the development of sulfate-reducing bacteria and archaea from the ANME 1, 2, and 3 groups, able to anaerobic oxidation of methane (Teske et al. 2002). This oxidation involves a metabolic association between both microorganisms, resulting in the production of bicarbonate and sulfide, by this equation:

$$CH_4 + \left(SO^{2-}\right)_4 > HCO_3^- + HS^- + H_2O$$

The increase of alkalinity induces the precipitation of carbonate chimneys. The chimneys associated with colorless sulfur-oxidizing bacteria form mats such as those observed in the Lucky Strike and Lost City vent fields (Mid-Atlantic Ridge), in the Guaymas Basin hydrothermal vent (Gulf of California), or in the Juan de Fuca and Hydrate Ridge (North Pacific Ridge).

Finally, other microorganisms with metabolisms based on ammonium, hydrogen, or metal oxidation are involved in chemolithotrophic mats. Among them, the ammonium oxidizers have a key role in mat's nitrogen cycle. The ammonium and oxygen form an opposite gradient in the sediments. The *Thaumarchaeota* growing chemolithoautotrophically can participate to the aerobic oxidation of ammonium to nitrate, while the *Beggiatoa* couple the nitrate respiration with the sulfur oxidation. A close spatial coupling of those *Thaumarchaeota* ammonium oxidizing and the *Beggiatoa* filaments has been observed in the Guaymas Basin hydrothermal vent (Winkel et al. 2014), showing the nitrification and nitrate respiration coupling in the mats, contributing to the nitrogen cycling.

However, the microbial diversity varies from vent to vent and in the time and space within the same vent field. The hydrothermal vent chimneys have steep chemical and thermal gradients, selecting specific communities. Changes in archaeal diversity were related to temperature in the Guaymas Basin chimney mat and to the maturity level, with the immature chimneys harboring methanogens, replaced by methylotrophs or acetoclastic methanogens in mature chimneys (Pagé et al. 2008). The maturity level of the chimneys and their physical chemical characteristic affect also the overall metabolism of the mats (Wang et al. 2009a). Besides the sulfur oxidation metabolism associated with nitrate reduction that supported inorganic carbon fixation through the Calvin-Benson-Bassham cycle, the mats developing on the smoker chimney 4143-1 in the Mothra hydrothermal vent field (temperature > 300 °C, pH 2–3) were enriched in genes involved in DNA repair, transposition, and recombination, compared to other mats developing at less extreme conditions (lower temperature, higher pH). This enrichment allows probably the mats to resist to extreme environmental conditions operating in the smoker chimney 4143-1 (Xie et al. 2011).

Photosynthetic Mats Photosynthetic mats develop all through the world, including coastal sediments, hyperacidic, hypersaline or hyperalkaline environments, polar ponds, or hot springs. Among the photosynthetic mats, stromatoliths are cemented mats (see Chap. 3), showing particular interest for living being evolution understanding, since fossilized stromatolites have been dated to 3.7 billion years. Modern microbial mats colonize the water-sediment interface, and microorganisms are distributed along a vertical gradient of light, oxygen, and sulfide (Fig. 5.11). The main energy driving photosynthetic mats is solar light. The colonization of sediments starts with the

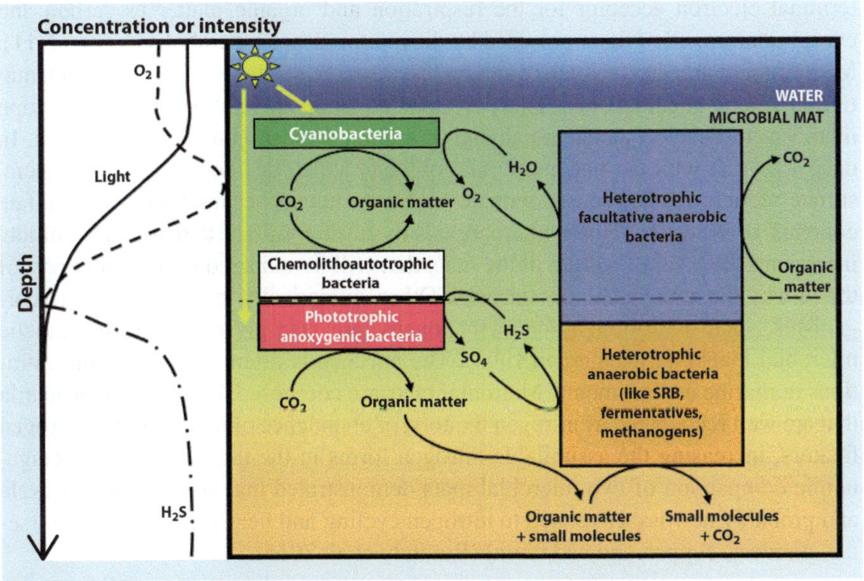

Fig. 5.11 Stratification of bacterial populations in a photosynthetic microbial mat related to physical chemical gradients. The main exchanges of carbon sources and electron donors and acceptors between the autotrophic component of the mats (in the left side of the scheme) and the heterotrophic members (right side) are shown

installation of cyanobacteria at the water-sediment interface, performing photosynthesis and fixing N_2; they exudate organic matter and, thus, enrich the sediments. Those organic compounds would be consumed by heterotrophic bacteria, respiring oxygen and, thus, creating an anoxic environment. In such conditions, sulfate-reducing prokaryotes develop, respiring sulfate and producing sulfur and degrading the organic compounds secreted by upper layer's microorganisms. The concentration of sulfur increases in the mat, allowing the development of sulfur-oxidizing bacteria. Mats develop on water-sediment interface with limited grazing organisms; otherwise, they would consume the cyanobacterial top layer. Once the mat built, the day/night shifts in oxygen concentration and the sulfide toxicity prevent grazers to overdevelop.

Oxygenic photosynthetic microorganisms (including microalgae and cyanobacteria) have key role in the support of the mats. They produce oxygen and fix N_2 while producing biomass. The sturdiness of the mats depend mainly on this component of the mat; mats dominated by *Microcoleus* or other filamentous cyanobacteria will be more solid than mats dominated by diatoms or nonfilamentous cyanobacteria. Another important group of (para)-primary producers in photosynthetic mats are anoxygenic phototrophs, developing below the oxygenic phototrophs in a thin anoxic layer, rich in sulfide and that receives light. They use sulfide as electron donor for the photosynthesis and produce sulfate. They include purple sulfur bacteria such as those belonging to the *Chromatiaceae* and the *Ectothiorhodospiraceae* families and the green sulfur bacteria, i.e., *Chlorobiaceae*.

Between both photosynthetic layers heterotrophs develop using use oxygen as terminal electron acceptor for the respiration and organic matter as carbon and energy sources, all of them provided by the photosynthetic metabolisms (Fig. 5.11). Most of them are also facultative anaerobes, through nitrate respiration or fermentative energetic metabolisms. This versatile metabolism leads to the heterotroph members to develop under the changing diurnal oxygen contents in the mats. In deeper layers, with permanent anoxia, obligate anaerobes develop. Among them, sulfate-reducing prokaryotes dominate sea environments because of the high sulfate contents in seawater. Those microorganisms respire sulfate and produce sulfide, increasing the level of sulfide in the mats that will be oxidized in the upper part of the mat by the anoxygenic phototrophs. Other, numerically less abundant microorganisms, such as acetogens or methanogens, are also present in photosynthetic microbial mats (van Gemerden 1993). The nitrogen is often at limiting concentrations in marine environments. Microbial mats are composed, besides cyanobacteria that are well recognized as nitrogen fixators, of abundance of heterotrophic nitrogen fixators, increasing the assimilable nitrogen forms in the mats. Recently, metagenomic comparison of two microbial mats demonstrated mat-specific carbon cycle and probably new issues related to nitrogen cycling and heavy metal cycling (arsenic, mercury, copper, and cadmium) (Ruvindy et al. 2016).

One of the main characteristics of the photosynthetic mats is the extremely changing physical and chemical parameters during a day/night cycle. The oxygenic photosynthesis during the day induces the penetration of oxygen in the upper layers of the mats, favoring oxygenic respiration. Sulfate reduction is highest also during the day, probably because substrates are more available or due to interactions between sulfate reducers and cyanobacteria that favor this metabolism. At night, sulfate reducers are the main mineralization actors; sulfide would participate to the nighttime O_2 consumption (Wieland et al. 2005). Vertical migration and metabolic switches are the main response of microorganisms to oxygen and sulfide fluctuations. Among sulfate reducers, some groups are able to migrate downward when oxygen concentration increase in the upper layers of the mats, but other form aggregates, allowing them to deal with oxygen (Fourçans et al. 2008).

Microbial mats are robust systems, allowing their resistance to extreme environmental conditions. Probably one of most famous microbial mats develops on the acidic hot spring of the Yellowstone Park. Their microbial community has been largely studied, and their relation with the extreme conditions (mainly temperature and pH) has been described. The coevolution of microbial and viral communities has been proved through the CRISPR spacers sequences study (Heidelberg et al. 2009), and this coevolution was considered key on the structuring of Yellowstone microbial mat communities. Besides, microbial mats are known to be robust enough to resist to environmental stresses, such as increasing UV irradiation and temperature, or contamination. The capacity to adapt to such contamination and to be resilient after an oil contamination have proved that, although there is a transient modification of the community structure, the mat acquired a stable structure, similar to that of a not contaminated mat (Bordenave et al. 2007).

Overall, microbial mats are constituted by primary producers associated with heterotrophs. In photosynthetic mats, the main primary producers are phototrophs, but chemolithotrophs are usually present, although numerically in less abundance. For chemolithotrophic mats, the main primary producers are chemoautotrophic bacteria, but some questions remain about the possible role of phototrophs. An obligate photosynthetic bacterial anaerobe have been isolated from a deep-sea hydrothermal vent at 2391 m in depth (Beatty et al. 2005) able to grow probably using geothermal radiations as energy source. In recent metagenomic studies, *cbbL* genes (involved in carbon fixation) related to photosynthetic bacteria are frequently detected in deep-sea hydrothermal vents, suggesting that the presence of obligate phototrophs in deep sea is not anecdotal. The metabolisms of microbial mats maintain the whole system active. In this case, should the microbial mats be considered as a community in which different microbial species are in equilibrium? Or should it be considered as a meta-organism in which the species composing it do not matter but the metabolism they are able to perform? In this later scenario, the presence of key metabolisms (photosynthesis, sulfate reduction, as example) is necessary for the functioning of the mat, but it does not matter which species of cyanobacteria or sulfate reducer is bearing it.

5.4 Interaction Between Prokaryotes and Unicellular Eukaryotes

Unicellular eukaryotes, often named protists, i.e., nucleated cell organisms, form a complex polyphyletic group of microbes where about all trophic modes along the continuum between autotrophy and heterotrophy, as strict or transient mixotrophic behavior, are represented (Sime-Ngando and Niquil 2011). On an evolutionary point of view, one of the most significant cell innovations in protists is their ability to engulf and internalize particles and other cells, primarily prokaryotes, a process called endocytosis or phagocytosis (literally meaning "cell eating"). This mode of nutrition opened up many new niches that ultimately facilitated the formation of permanent associations between very different life forms via endosymbiosis (Stanier 1970). Endosymbiotic associations with prokaryotes have provided eukaryotes with much of their central metabolism, which has remained relatively conserved throughout the group's history.

5.4.1 Fundamental Roles of Prokaryotic Endosymbionts in Eukaryotic Evolution

Endosymbiosis is the process by which one cell is taken up by another and retained internally, such that the two cells live together and integrate at some level, sometimes permanently. This typical protistan innovation forms the roots of one of the famous theories on the origin of eukaryotic cells with the advent of endocytosis of prokaryotes in an ancestor of living eukaryotes. Endosymbiotic interactions have been common in eukaryotic evolution, and many such partnerships persist today (Margulis 1981). In two cases, however, endosymbiotic events had far-reaching effects on the evolution of life: these are the origins of eukaryotic mitochondria and plastids (chloroplasts) which are considered as prokaryotic-like endosymbiotic remains of protobacterium and cyanobacterium, respectively (Fig. 5.12).

Mitochondria are generally known as the energy-generating powerhouses of eukaryotic cells, where oxidative phosphorylation and electron transport metabolism takes place (Reichert and Neupert 2004). They are also involved in several other metabolisms such as oxidation of fatty acids, amino acid metabolism, and assembly of iron-sulfur clusters (Lill and Kispal 2000). They are bounded by two membranes, the innermost of which is generally highly infolded to form "cristae" that take characteristic shapes, either flat, tubes, or paddle shapes (Fig. 5.12) (Taylor 1978). The presence of mitochondria is an ancestral trait in eukaryotes, although in certain anaerobes and microaerophiles they have radically reduced or transformed functions: in some cases they are not involved in energy production at all (e.g., the "mitosomes" of microsporidia, diplomonads, and archaemoebae or "hydrogenosomes" of parabasalia, some ciliates, and some chytrid fungi) (van der Giezen et al. 2005). Mitochondria can be traced back to a single endosymbiosis of an alpha-proteobacterium (Gray et al. 2004).

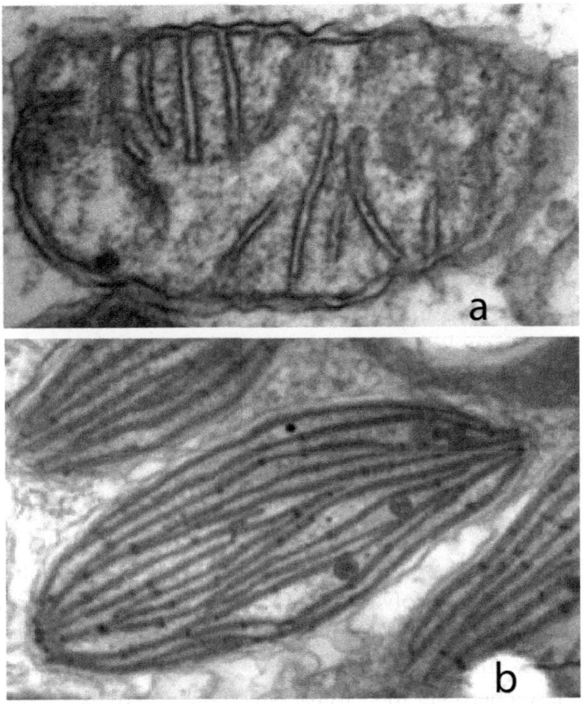

Fig. 5.12 Eukaryotic mitochondria (**a**, ©Kevin Carpenter) and plastids (chloroplasts) (**b**, ©Brian S. Leander) are considered as prokaryotic-like endosymbiotic remnants of protobacterium and cyanobacterium, respectively. From the tree of life web project (http://tolweb.org/Eukaryotes/3)

Plastids are the photosynthetic organelles of plants and algae, including the most abundant planktonic unicellular eukaryotes, i.e., phytoplankton (Fig. 5.12). "Plastid" is a general term for all such organelles, including chloroplasts (in the green lineage), rhodoplasts (in the red lineage), leucoplasts (colorless plastids), etc. Plastids have diverse functions in addition to photosynthesis, including the biosynthesis of amino acids and fatty acids. As in the case of mitochondria, plastids in many lineages have been radically reduced or transformed, primarily through the loss of photosynthesis (e.g., the "apicoplast" of Apicomplexa and the relict plastids of many parasitic algae and plants (Gould et al. 2008)). Plastids can also be traced back to a single endosymbiosis event involving a cyanobacterium and the ancestor of the Archaeplastida (Reyes-Prieto et al. 2007). However, unlike mitochondria, plastids then spread to other eukaryotic lineages by secondary and tertiary endosymbiotic events where one eukaryotic cell took up another eukaryote that already contained a plastid (an alga), and this second, endosymbiotic eukaryote was then reduced and integrated. In most cases all that remains of this alga is the plastid surrounded by the remains of the endosymbiont's plasma membrane (Gould et al. 2008). Other endosymbiotic relationships based on photosynthesis are also known, but typically these are not integrated to the extent that they are generally accepted to be "organelles" rather than "endosymbionts." One possible exception is the euglyphid amoeba *Paulinella chromatophora*, i.e., a unicellular eukaryote, where a cyanobacterium similar to *Synechococcus* or *Prochlorococcus* has been integrated to an extent approaching that of canonical plastids (Nowack et al. 2008).

5.4.2 Protist-Prokaryote Phagotrophic Interactions

Bacterivorous protists and their prokaryotic resources form the most ancient predator-prey interaction in the nature. Phagotrophy of prokaryotes is considered the origin of eukaryotes (Cavalier-Smith 2002) and can thus be regarded as an evolutionary precursor of many other interactions between prokaryotic and eukaryotic organisms. It is thus likely that the complexity and the sophistication of phagocytosis in unicellular eukaryotes, which is probably more economical than exoenzyme secretion into the environment, is the basis of the evolutionary steps in digestive processes. Some prokaryotes have evolved means to take control of the protistan food vacuole by circumventing digestion, giving rise to fundamental ecological phenomena such as phototrophy in eukaryotes and other forms of endosymbiosis such as the typical bacterial endosymbionts of amoeba (Pernthaler 2005). It is thus conceivable that the invention of bacterivory by unicellular eukaryote has shaped microbial evolution and perhaps is the origin of oxygenic photosynthesis.

In natural environments, this predator-prey system form the root of the microbial food webs composed of viruses, chemoorganotrophic prokaryotes (including cyanobacteria), as well as autotrophic and heterotrophic unicellular eukaryotes. In these food webs, dissolved organic matter exudates from primary producers are taken up by prokaryotes which then serve as food sources for bacterivorous ciliates and flagellates. In pelagic ecosystems, most organic carbon is produced and consumed in these microbial food webs, which then serve as a relay in the transfer of dissolved organic matter up to metazoan grazers and the classical food chain (e.g., algae, zooplankton, fish), via the grazing of bacteriovorus protists by metazoan grazers (Sime-Ngando et al. 2011). It is now considered that lysis by viruses but mainly predation by unicellular eukaryotes are the main causes of bacterial mortality in microbial ecosystems, resulting in the consumption of prokaryotic biomass at approximately the same rate as it is produced (Sime-Ngando 2014). Mechanisms of particle uptake and handling are diverse in the protistan world (filter feeding, interception feeding, etc.) and protists cannot feed on all prokaryotes with equal efficiency. In addition, protistan feeding activity is constraint by the sizes of prey, with preferential prey size range between 1 and 3 μm (González et al. 1990). Protistan feeding also depends on prokaryote species, e.g., Gram-positive bacteria are more easily digested than Gram-negative bacteria. The above observations highlight the existence of uptake selectivity in omnivorous protists, a mechanism that can structure the community composition and diversity in natural ecosystems (Pernthaler 2005).

Grazing on prokaryotes by unicellular eukaryotes is also considered to have a key role in nutrient regeneration in natural systems, primarily of phosphorus and nitrogen in aquatic ecosystems. These conservative elements, often limiting for primary producers, are more concentrated in prokaryotic than in eukaryotic cells. This is because prokaryotic cell has higher ratio of proteins and nucleic acids to total cell mass compared to eukaryotes, implying that protists that graze on prokaryotes release excess nutrients in the environment, thereby stimulating primary producers and perhaps other bacteria with high affinity for nutrient uptake (Caron et al. 1988).

5.4.3 Protist-Prokaryote Mixotrophic Interactions

Mixotrophy refers to a broad range of trophic strategies, from predominantly phototrophic to almost exclusively heterotrophic modes. Mixotrophy is a common feature in unicellular eukaryotes, most of which are able to form and maintain both phagotrophic and photosynthetic apparatus. Many pigmented protists eat prokaryotes, and it has been shown that the related metabolic cost is rather low (< 10% of total metabolic cost) compared to the potential benefits (Raven 1997). This is indeed an efficient way of acquiring both limiting nutrients and organic carbon in oligotrophic light-limiting environments. Furthermore, considering that the prokaryotic affinity for nutrients is higher than that of eukaryotes, phagotrophy in phototrophs favors mixotrophic primary producers, because they are able to eat their competitors (Thingstad et al. 1996). In some particular conditions, such as in nutrient-poor, high mountain lakes, photosynthetic unicellular eukaryotes that eat prokaryotes can form the dominant primary producer species, creating a scenario for prokaryotes-eukaryote interactions of combined commensalism and predation (Medina-Sanchez et al. 2004).

5.4.4 Other Interactions Between Prokaryotes and Unicellular Eukaryotes

As provided previously, prokaryotes and unicellular eukaryotes have coexisted ever since the early stages of evolution. This coevolution has revolutionized life on earth in many aspects. Prokaryotes and unicellular eukaryotes together influence ecosystems and represent all conceivable modes of interactions, from mutualism to parasitism (Rishiram et al. 2016). In the environment, algae are closely associated with other microorganisms, especially heterotrophic prokaryotes, forming, for example, the phycosphere in aquatic systems. Phycosphere is a microscale region that is rich in organic matter immediately surrounding and influenced phytoplankton cells. This area is high in nutrients due to extracellular waste from the phytoplankton cell, and it has been suggested that bacteria inhabiting this area feed on these nutrients. This high nutrient environment, also known as oasis or resource hotspots in oligotrophic environments, creates a microbiome and a diverse food web for microbes such as viruses, bacteria, and protists that are distinct from the surrounding water. Mutualistic exchanges of organic matter, energy, oxygen, and nutrients are intense in the phycosphere (Paerl and Pinckney 1996), in which prokaryotes can be free living, attached to the surface of algae, or occur as intracellular algal symbionts (Jasti et al. 2005). A well-known case for bacteria-algae mutualism is the bacterial supply of micronutrients such as vitamin B12 in exchange of fixed carbon. In general, algae supply fixed organic carbon to mutualistic bacteria, and bacteria in return supply dissolved inorganic nutrients and low molecular organic carbon for algal consumption. Such interactions can also be facultative or opportunistic (Rishiram et al. 2016).

However, some prokaryotes, far from being involved in a mutualistic relationship, lyse algae as a food source. Many bacteria are known to negatively affect algae and hence very encouraging for scientists studying microalgae and cyanobacterial bloom control. Several studies have suggested that bacterium-phytoplankton interactions have the potential to dramatically influence harmful algal bloom dynamics, and some evidence of specificity in these interactions was provided (Jasti et al. 2005). The influence of bacteria on toxic algae has been of particular interest, and bacteria have been implicated in the production and/or modification of algal toxins (Anderson et al. 1998). However, very few studies have been conducted on bacterial parasites of algae, their mechanism of elimination, and its ecological reasoning. These studies demonstrate that the algal cell lysis is achieved through a mechanism similar to plant-pathogen interaction. Glucosidases, chitinases, cellulases, and other enzymes that help degrade plant cell wall are also involved in the lysis of algal cells (review in Rishiram et al. 2016).

Overall, bacterium-phytoplankton interactions play a key role in ecological processes, ranging from biogeochemical cycling within the microbial food webs to biochemically mediated interactions that influence phytoplankton growth, reproduction, cyst formation, and mortality. A few recent studies have shown that bacteria not only affect algal growth but also help in flocculation, both essential processes in algal biotechnology (Rishiram et al. 2016).

5.5 Interactions Between Prokaryotes and Multicellular Eukaryotes

5.5.1 Interactions with Animals

5.5.1.1 Symbiosis and Evolution

Prokaryotic microorganisms are widely distributed throughout all environments, such that the earth's biosphere can be described as a network where different organisms establish various complex interactions with each other. One such interaction is symbiosis, which is a major factor involved in the evolution of species (along with predation or competition). Symbiosis is the sustainable coexistence of two (or more) organisms and can occur over a whole lifespan or part of it, irrespective of the nature of the exchanges between them. Symbiosis is omnipresent and concerns all living organisms, either as a host or as a symbiont.

Since the 1960s, the establishment of a symbiotic theory explaining cell evolution was put in advance (Sagan 1967). Today, there is a large consensus on the crucial role of symbioses in the origin and evolution of eukaryotic cells. Furthermore, on the basis of its wide distribution across major phylogenetic taxa, symbiosis could have an important role in the evolution of the species that are involved in such partnerships. In all major biological phenomena classified as symbiotic, novel cellular structures and/or metabolic capabilities emerge as a result of evolutionary forces, favoring the maintenance of this association (Margulis 1993). The interplay of both partners, or between a single host and more than one symbiont in some cases, forms a biological community that evolves with time.

Depending on whether the symbiont has a deleterious, beneficial, or neutral effect on the reproductive success (i.e., a selective value or fitness) of its host, three types of symbiotic associations can be distinguished: (i) parasitism, in which only the symbiont takes advantage of the association by imposing a selective cost (fitness cost) on the host; (ii) mutualism, where both the host and the symbiont benefit from their association; and (iii) commensalism, where the two partners coexist without any cost or profit (Cheng 1991). It is the balance between benefit and cost of the association at the host level that makes it possible to define the nature of the relationship between the host and its symbiont. Thus, the selective value of the host does not result from the simple expression of its genome, but also from that of the symbiont, and finally from the interaction between the two genomes.

Mutualism and parasitism are not fixed characters. Rather, there is a continuum in the interactions between these two extremes, and transitions from one stage to another may occur depending on the physiological conditions of the partners, which in turn may evolve with the environmental conditions (Bronstein 1994). This transition from mutualism to parasitism is well illustrated by the interaction between the bacterium *Hamiltonella defensa* and the aphid *Acyrthosiphon pisum*. *Hamiltonella* promotes the resistance of the aphid against one of its natural enemies, the parasitoid *Aphidius ervi*. In the presence of the parasitoid, infected aphids show increased survival, as compared to uninfected aphids. This benefit disappears in the absence of the parasitoid: *Hamiltonella* becomes parasitic, and infected individuals become less competitive than uninfected aphids (Oliver et al. 2003, 2008).

One important factor driving the evolution of symbiotic interactions is the transmission mode of the symbiont (Ebert and Herre 1996; Toft and Karter 1990). When symbionts are transmitted horizontally (i.e., between unrelated hosts), the cost imposed by the symbiont may be high if it is correlated with the success of its transmission. In cases such as with bacteriophages, the death of the host (lysis of the bacterial cell) is essential to complete the symbiont life cycle. In contrast, the reproductive success of the symbiont (when transmitted vertically) depends entirely on the reproduction of its host. In this situation, natural selection will tend to reduce the selective cost induced by symbionts in order to increase the reproductive success of the host. Thus, there is a tendency for vertically transmitted symbionts to evolve toward mutualism (Frank 1996). The classic example of this type of evolution is the integration of mitochondria and chloroplasts into cell organelles in eukaryotic cells (Margulis 1993). For example, the virulence of *Barley stripe mosaic virus* (BSMV) in barley increases in cases of horizontal transmission and decreases significantly when transmission is vertical (Stewart et al. 2005). However, not all vertically transmitted symbionts evolve toward mutualism; some, such as the *Wolbachia* bacterium, alter the nature or quantity of the host offspring and are therefore referred to as reproductive parasites (Werren et al. 2008).

In the past 20 years, the accumulation of information on molecular evolution and symbiont ecology has increased our understanding of the evolution of symbiotic interactions and the role of symbionts in the evolution of host species. This chapter focuses on endosymbiosis, a specific type of interaction that concerns microorganisms (viral, prokaryotic, or eukaryotic) living within their host cells. Due to their intracellular biology, most of them are transmitted vertically via the cytoplasm of oocytes, especially in arthropod hosts.

5.5.1.2 Endosymbiotic Bacteria of Arthropods

The endosymbiotic bacteria of arthropods have a diverse phylogenetic origin. Endosymbionts are found particularly within the *Proteobacteria* (e.g., *Wolbachia*, *Buchnera*, *Hamiltonella*, *Arsenophonus*, and *Sodalis*), *Mollicutes* (including *Spiroplasma*), and *Bacteroidetes* (e.g., *Cardinium*, *Flavobacterium*, *Sulcia*, and *Blattabacterium*) phyla (Duron et al. 2008; Moran et al. 2008). Depending on the nature of the interactions with their hosts, these bacteria fall into two main categories: the primary endosymbionts (obligate mutualists) and the secondary endosymbionts, which include the "facultative mutualist" and "reproductive parasite" subcategories (Moran et al. 2008).

5.5.1.2.1 The Primary Endosymbionts

Primary endosymbionts infect different hosts belonging to the class Insecta (Clark et al. 2010). These endosymbionts are strictly located within giant cells known as bacteriocytes and are exclusively transmitted maternally. Their primary role is to provide nutrients that supplement the diet of their insect hosts (Douglas 1998). Importantly, primary endosymbionts allow insects to specialize in and occupy ecological niches that otherwise would not be possible without their nutritional assistance. For instance, several hemipterans including aphids and leafhoppers can exist on nutrient-poor diets such as sap plants, thanks to the primary endosymbionts that provide them with the synthesis of essential amino acids. Without the presence of their symbionts, these insects would be unable to grow and reproduce normally. Conversely, these symbionts cannot live outside of their bacteriocytes, since they have lost the genes necessary to overcome the host immune defenses and the ability to biosynthesize surface-active compounds such as lipopolysaccharides and phospholipids (Shigenobu et al. 2000).

The associations between primary endosymbionts and insects are very ancient, as demonstrated by the familiar example between the bacterium *Buchnera aphidicola* and aphids dating back 160–280 million years (Moran et al. 1993). The genomes of primary endosymbionts are characterized by a nucleotide bias toward adenine and thymine (Moran 1996) and a highly reduced size, as compared to the genomes of free but phylogenetically related bacteria. Within the *Enterobacteriaceae*, for example, the *Buchnera* genome is 640 kb (Shigenobu et al. 2000), whereas the confamilial *Escherichia coli* genome is 4.5–5.5 Mb (Bergthorsson and Ochman 1998). Indeed, endosymbionts display the smallest known bacterial genomes, as exemplified by the bacterium *Carsonella* (159 kb), which is an endosymbiont of psyllids (Nakabachi et al. 2006), and *Sulcia* (246 kb), an endosymbiont in leafhoppers (McCutcheon and Moran 2007). The reduced size of primary endosymbiont genomes may be linked to a rapid evolution in DNA sequences favored by the absence of recombination (asexuality). This could result from several factors, including their sequestration in bacteriocytes, the absence of mobile genetic elements, the absence of bacteriophages (or phages) and transposons, and the loss of certain genes involved in recombination events (Wernegreen 2002). Primary endosymbionts and their hosts display congruent phylogenies, demonstrating that their association has a close specificity (Moran et al. 1993). Here, the symbiotic association functions as a single evolutionary unit and therefore will involve both partners in the case of diversification. The absence of horizontal transmission explains this strict cospeciation and highlights a chronological correspondence between the speciation of the hosts and their symbionts.

5.5.1.2.2 Secondary Endosymbionts

Secondary endosymbionts are most often found in different hosts belonging to the phylum Arthropoda (Engelstadter and Hurst 2009). In comparison to primary symbionts, secondary symbionts have a broader host spectrum with more diverse interactions situated along a continuum from parasitism to mutualism and a wider distribution within host organs (e.g., reproductive organs, hemolymph, and digestive tract) (Dobson et al. 1999; Fukatsu et al. 2000). Finally, they are generally not essential to the survival of their hosts, although they have developed strategies to maintain themselves in the host populations.

The evolutionary success of secondary symbionts can be explained by three main strategies. The first strategy, reproductive parasitism, is utilized by several secondary endosymbionts to manipulate the reproduction of their hosts, in order to increase the success of their own transmission (Duron et al. 2008; Werren et al. 2008; Engelstadter and Hurst 2009). The second strategy, which involves horizontal transfers within the same host species and between different host species, has been documented in different secondary endosymbionts and contributes to broadening their host spectrum in arthropods (Sandstrom et al. 2001; Russell et al. 2003). Regarding the third strategy, facultative mutualism (in which the symbiont is advantageous only under certain environmental conditions) may explain the persistence of secondary endosymbionts in host populations (Jaenike and Brekke 2011).

Theoretical studies predict that the selection process should favor "heritable" symbionts that serve as "indirect" mutualists, by allowing their hosts to become resistant to other parasites (Lively et al. 2005; Jones et al. 2007). These studies are supported by the increasing evidence that secondary endosymbionts protect hosts against their natural enemies. Examples of this type of protection include *Spiroplasma*, which protects *Drosophila neotestacea* against the parasitic nematode *Howardula aoronymphium* (Jaenike et al. 2010); *Wolbachia*, which protects *Drosophila melanogaster* against RNA viruses (Hedges et al. 2008; Teixeira et al. 2008); *Hamiltonella defensa*, which protects *Acyrthosiphon pisum* against *Aphidius ervi* (Oliver et al. 2009); and

Regiella insecticola, which provides aphids with protection against pathogenic fungi (Scarborough et al. 2005). However, one of the most spectacular examples is the bacterium *Rickettsiella*, which allows aphids to change their color, making them less visible to their coccinellid predators (Tsuchida et al. 2010).

The diversity of interactions between secondary endosymbionts and their hosts at the phenotypic and genomic levels is illustrated in the next section by *Wolbachia*.

5.5.1.2.3 The Secondary Endosymbiont *Wolbachia*

The bacterium *Wolbachia* belongs to the *Alphaproteobacteria* class, which includes *Rickettsia* and *Ehrlichia*. The latter two bacteria are well-known due to their role in various mammalian pathologies, such as typhus in humans (caused by *Rickettsia prowazekii*) or cowdriosis, a bovine pathology caused by *Ehrlichia ruminantium* (Valbuena and Walker 2009). *Wolbachia* is essentially transmitted vertically through the cytoplasm of oocytes. This endosymbiont is almost totally eliminated during spermatogenesis; thus it cannot be transmitted by males (Bressac and Rousset 1993).

Wolbachia pipientis (Hertig 1936) was described for the first time in the mosquito *Culex pipiens* (Hertig and Wolbach 1924). Since then, it has been observed in a wide range of species belonging to a variety of classes. Among the arthropods, *Wolbachia* has been found in hosts belonging to Arachnida (particularly mites, spiders, and pseudoscorpions) and Insecta, where it infects hosts belonging to different orders such as Diptera, Lepidoptera, Hemiptera, Hymenoptera, Thysanoptera, Coleoptera, and Orthoptera (see Werren et al. 2008 for review). It has also been shown to infect filarial nematodes (Bandi et al. 2001; Casiraghi et al. 2001; Bordenstein et al. 2003). Among all of these host groups, the current data indicate that *Wolbachia* prevalence is highest in the class Insecta, where it is estimated to vary between 20 and 66% (Werren et al. 1995; Werren and Windsor 2000; Jeyaprakash and Hoy 2000; Duron et al. 2008; Hilgenboecker et al. 2008). The evolutionary success of *Wolbachia* is essentially related to its ability to manipulate the reproduction of its hosts. *Wolbachia* is an obligate mutualist in nematodes and is essential for their juvenile growth and reproduction (Bandi et al. 2001). In contrast, *Wolbachia* generally acts as a reproductive parasite in arthropods, where it employs four reproductive manipulation strategies, all of which favor females.

Phenotypes Induced by Wolbachia in Arthropods The feminization phenotype is typically observed in woodlice such as *Armadillidium vulgare* (Rigaud et al. 1997), although it has been described in the lepidopterans *Eurema hecabe* (Hiroki et al. 2002) and *Ostrinia furnacalis* (Kageyama et al. 2002), as well as the leafhopper *Zyginidia pullula* (Negri et al. 2006). *Wolbachia* inhibits the androgen gland in these hosts, preventing the expression of any male characters. *Wolbachia* is also responsible for the male-killing phenomenon in many groups, particularly *Drosophila innubila* (Dyer et al. 2005), the butterflies *Acraea encedon* (Jiggins et al. 2001) and *Hypolimnas bolina* (Charlat et al. 2005), the scarab beetle *Tribolium madens* (Fialho and Stevens 2000), and the pseudoscorpion *Cordylochernes scorpioides* (Zeh et al. 2005). The male-killing phenomenon causes the death of infected male offspring of infected females. This mortality usually takes place at the embryonic stage, leading to an approximately 50% reduction in the egg hatching rate of infected vs uninfected females (Dyer and Jaenike 2004; Hornett et al. 2006). The benefit conferred by *Wolbachia* is less obvious in this case, since the males are not transformed into females but are purely eliminated. However, this process can be interpreted as an increase in the selective value of females (Engelstadter and Hurst 2009). Thelytokous parthenogenesis, which exclusively produces females, is known only in haplodiploid species such as *Bryobia praetiosa* mites (Weeks and Breeuwer 2001), *Franklinothrips vespiformis* thrips (Arakaki et al. 2001), and *Trichogramma* wasps (Stouthamer et al. 1990). Parthenogenesis in these species is typically arrhenotokous, meaning that males are haploid and develop from unfertilized eggs, whereas females are diploid and develop from fertilized eggs. In these species, *Wolbachia* transforms non-fertilized eggs into females by generating diploidization. The strategies of feminization, male-killing, and parthenogenesis may limit the invasive capacity of *Wolbachia*, since the scarcity of males induces an obvious reproductive cost for the host (Rigaud and Moreau 2004).

The most common phenotype induced by *Wolbachia* is cytoplasmic incompatibility (CI) (Werren 1997; Werren et al. 2008). CI has been reported in terrestrial isopods (*Cylisticus convexus*; Moret et al. 2001) and mites (*Tetranychus urticae*; Gotoh et al. 2007), as well as in several insects, including the mosquitoes *Culex pipiens* (Yen and Barr 1971), *Aedes albopictus* (Sinkins et al. 1995), and *Armigeres subalbatus*, the flies *Drosophila simulans* (Hoffmann et al. 1986) and *Rhagoletis Cerasi* (Boller et al. 1976), and the beetle *Chelymorpha alternans* (Keller et al. 2004). In general, CI occurs when crossing *Wolbachia*-infected males with uninfected females, resulting in abnormally high embryonic mortality. CI thus reduces the reproductive success of uninfected females and allows *Wolbachia* to invade host populations (Engelstadter and Telschow 2009).

Host Response to the Reproductive Manipulation Induced by Wolbachia To escape extinction, species must adapt to the selective pressures exerted by other species with which they interact (Van Valen 1973). This principle is also valid for the symbiosis between *Wolbachia* and arthropods. In some hosts, two types of genes that suppress the reproductive manipulation by the symbiont can be selected: male-killing suppressors (Hornett et al. 2006; Charlat et al. 2007; Jaenike and Dyer 2008) and nuclear restorers of compatibility (Rousset et al. 1991; Sinkins et al. 2005). Another way for hosts to escape reproductive manipulation is by homogamy, which may favor the maintenance of incompatible *Wolbachia* in host populations (Rousset et al. 1991).

Metabolic Assessment of Infection: Costs and Benefits Conferred by Wolbachia The cost incurred by *Wolbachia* depends on the host and symbiont genotypes, as well as the environmental conditions. In *D. simulans*, for example, the *Wolbachia* strain wRi reduces the fertility of laboratory-raised females, but has no effect in natural populations (Hoffmann et al. 1990). Similarly, *D. melanogaster* adults infected with wMelPop display normal survival at 19 °C, but die prematurely when the temperature is raised to 25 °C (Reynolds et al. 2003). Bacterial density is also an essential factor in the expression of cost. The rapid mortality induced by the wMelPop strain in *D. melanogaster* is linked to an exponential growth of bacteria in nerve and muscle tissues (Min and Benzer 1997). Bacterial density also influences the *Wolbachia*-induced cost in *Asobara tabida* wasps (Mouton et al. 2004). It is generally accepted that a high cost produced by the symbiont represents a transitory state and that the host and symbiont should rapidly coevolve toward a decrease in this cost (Turelli 1994). Several examples of this have been described in the literature (Hoffmann et al. 1990; Weeks et al. 2007; Echaubard et al. 2010).

Mutualism between *Wolbachia* and its arthropod hosts has also been reported. Here, *Wolbachia* has been reported to increase the fertility of females in *Aedes albopictus* (Dobson et al. 2002) and *D. melanogaster* (Fry et al. 2004; Brownlie et al. 2009). Recent studies have also reported that some *Wolbachia* strains protect their hosts against pathogenic RNA viruses as a result of specific interactions between *Wolbachia* and the virus. In other instances, *Wolbachia* can become a mandatory mutualist, as seen with nematodes. For example, when *Asobara tabida* wasps were coinfected with the three *Wolbachia* strains WAtab1, wAtab2, and wAtab3, the latter strain was necessary for the oogenesis of its host, whereas the other two strains were facultative and responsible for CI (Dedeine et al. 2001, 2004). Additional mutualism examples have been described in the literature (Hosokawa et al. 2010; Starr and Cline 2002). Finally, other phenomena deserve to be mentioned, including the indirect effect of *Wolbachia* on host mitochondrial diversity, as well as the effect of the interaction type between symbiont and host on the evolution of *Wolbachia* genomes.

5.5.1.3 Tsetse Flies and Sleeping Sickness: A Case Study

5.5.1.3.1 Brief Summary

Certain protozoa in the genus *Trypanosoma* are the causative agents of either human African trypanosomiasis (HAT, or sleeping sickness) or animal African trypanosomiasis (AAT, or nagana). These are digenetic parasites as they successively infect a *Glossina* spp. tsetse fly, a prerequisite for their transmission to a mammal. After a fly bites a trypanosome-infected mammal and takes an infected blood meal, the ingested trypanosomes reach the fly midgut, where they differentiate from the bloodstream form into the procyclic form (Sbicego et al. 1999). The parasites then differentiate into several forms during their migration from the gut to the salivary glands or the proboscis (depending on the trypanosome species), before finally generating the nonproliferating metacyclic form (Van den Abbeele et al. 1999). This last form is the only one that is infective for mammals and is transmitted from the fly's saliva to a subsequent mammalian host's bloodstream during ingestion of a new blood meal. The ability of the fly to acquire the parasite, favor its maturation, and transmit it to a mammalian host is called "vector competence" and depends on both the *Glossina* and trypanosome species, among other factors. In natural fly populations, and even after experimental infections, the "infected tsetse" prevalence is low. Indeed, most individuals escape infection because they have developed mechanisms capable of eliminating their ingested parasites. The latter individuals are defined as "refractory" to infection. Thus, elucidating the mechanisms that control fly susceptibility and refractoriness to trypanosome infection are of equal importance. In addition to the trypanosomes, tsetse flies harbor one obligate symbiotic microorganism, *Wigglesworthia glossinidia*, and possibly two facultative species, *Wolbachia* and *Sodalis glossinidius* (Aksoy 2000).

5.5.1.3.2 Characteristics of Tsetse Fly/Trypanosome/Bacteriome Tripartite Interactions

In most symbioses between prokaryotes and eukaryotes, the relationship between a host and its symbionts is so close that the microorganisms cannot be cultured, making them difficult to study. However, in the past few years, genome sequencing, transcriptomics, and the recently emergent field of metagenomics (which bypasses the need to culture microbial cells) have offered new opportunities in symbiosis research. This new ability to study cohabiting bacterial endosymbionts within the same host has enabled the discovery of complex and stable associations and has helped determine the contribution of each partner to the association. Nevertheless, many questions remain unresolved about the nature of the genome-reduction process, the type of genes maintained by the endosymbiont, the molecular pathways used by the host to control the endosymbiont population, and the complex set of factors that determine whether the final outcome of the association is parasitic or mutualistic.

Bacteria-Tsetse Fly Symbiosis

Wigglesworthia Interactions All *Glossina* flies harbor the obligate symbiont *Wigglesworthia* (Wgm) (Aksoy et al. 1995) in a relationship that is very ancient (50–80 million years old) (Chen et al. 1999). The symbiont genome carries genes encoding enzymes for vitamin biosynthesis pathways, and the main function of the Wgm population within bacteriocytes (constituting the "bacteriome") is to produce vitamins absent from the tsetse fly diet (Nogge 1982; Akman et al. 2002; Pais et al. 2008; Rio et al. 2012).

Recent studies have shown that Wgm has an immunomodulatory effect in tsetse flies. For instance, *G. morsitans morsitans* (Gmm) that lack their symbiont (GmmWgm) are immunodepressed as compared to wild-type flies (GmmWT) (Weiss et al. 2011; Pais et al. 2008; Wang et al. 2009a, b, c). In fact, Wgm must be present during the tsetse larval stages so that its immune system develops and becomes functional in adulthood (Weiss et al. 2011). The tsetse immune system as it relates to the symbiosis with Wgm could affect the ability of the fly to transmit the trypanosome. The peptidoglycan recognition protein LB (PGRP-LB) of the host, which blocks the peptidoglycans of pathogens, prevents stimulation of the Imd pathway in the fly and is closely associated with maintaining its symbiosis with Wgm (Wang et al. 2009a, b, c; Wang and Aksoy 2012). The PGRP-LB protein is transmitted by the female fly through milk gland secretions to the offspring. Because this protein also has trypanocidal activity, increased PGRP-LB concentrations could increase the refractory status of flies to infection by trypanosomes, in non-teneral tsetse flies.

Sodalis Glossinidius Interactions *S. glossinidius* (Cheng and Aksoy 1999) is a commensal symbiont harbored by all insectary-raised tsetse flies. This bacterium is vertically transmitted through the milk gland secretions by the female tsetse to its offspring. *Sodalis* displays a wide tissue tropism and is found both intracellularly and extracellularly in the intestine, hemolymph, and salivary glands of the tsetse fly. Bacteria belonging to this genus are closely genetically related (even when they are harbored by different *Glossina* species), indicating that the association of *Sodalis* with its host is relatively recent (Weiss et al. 2006). *Sodalis*, which coinhabits the intestine of the infected tsetse with the trypanosomes, can be isolated, cultivated in vitro, and genetically manipulated.

Epidemiological surveys in several HAT foci have shown that *Sodalis* promotes tsetse infection by trypanosomes (Farikou et al. 2010). One previous report demonstrated that *Sodalis* populations are not genetically homogeneous and that the genetic structure of these populations depends on the hosting tsetse species (Geiger et al. 2005). It has also been demonstrated that this genetic structure can differ between HAT foci and that a low but significant flow of genes could exist between different foci (Farikou et al. 2011). However, given the biological modalities of heritability for the symbiont, this is most likely due to a flow of flies hosting different symbiont genotypes. A final result to highlight here is the relationship between a given tsetse symbiont genotype and infection of the tsetse by a specific trypanosome species (Geiger et al. 2007). This is an important finding because it demonstrates that the effect of the symbiont on trypanosome infection is not based on a simple "presence-absence" of the symbiont in the tsetse fly. Indeed, the relationship appears to be far more complex and will require the deciphering of the molecular cross talk between the partners to better understand the involved infection mechanisms.

One recent study from Western Zambia conducted on *G. m. centralis* showed that the prevalence of microbiota may vary within the same tsetse species depending on the geographical location of the sampling site, as it is the case for *Wolbachia*. The study also demonstrated that *Sodalis* infection could affect vector competence (Mbewe et al. 2015), thus confirming previous studies. Consequently, *Sodalis* may become a target in vector control strategies, due to its inheritance modalities that could make it an ideal candidate for paratransgenic experiments in tsetse flies (Weiss and Aksoy 2011). Another factor that could promote this is that the symbiont is present throughout the entire life cycle of the fly (from the larval stage to the adult stage). This means that it has a persistent role throughout the tsetse life cycle and that the fly will be able to transmit *S. glossinidius* to its offspring throughout its adult life, from one generation to the next (Hamidou Soumana et al. 2013).

Wolbachia Interactions The α-proteobacterium *Wolbachia* infects approximately 40% of arthropods (O'Neill et al. 1993; Hilgenboecker et al. 2008; Werren et al. 2008). *Wolbachia* is intracellularly localized in the reproductive organs (Cheng and Aksoy 1999) and is mainly transmitted vertically by the female to its offspring. As reviewed above, *Wolbachia* generates certain reproductive abnormalities in infected arthropods, such as cytoplasmic incompatibility (CI), feminization, and parthenogenesis, and may physiologically intervene with the host at the following levels: "fitness," fertility, immunity, longevity, and host development (Kambris et al. 2009; Moreira et al. 2009; Glaser and Meola 2010; Kambris et al. 2010). *Wolbachia* strains are hosted by many tsetse fly species and populations that are closely genetically related but nevertheless distinct (Cheng et al. 2000; Doudoumis et al. 2012). Cross-breeding studies between *Wolbachia*-infected and *Wolbachia*-uninfected *Glossina morsitans morsitans* individuals have shown that *Wolbachia* is responsible for the induction of a strong cytoplasmic incompatibility in this tsetse fly species (Alam et al. 2011).

Other Bacteria-Tsetse Fly Interactions

Identification and Diversity The establishment of trypanosomes in the tsetse intestine is modulated by the immune barrier of the host itself regulated by the intestinal microflora (Weiss et al. 2013). This illustrates the need for studies on the tsetse fly microbiome composition. Along these lines, studies have been performed on insectary-raised *G. palpalis gambiensis* (Gpg) flies, as well as different species sampled from HAT foci in Angola (epidemic areas) and Cameroon (endemic areas), in order to investigate the possible presence of other bacteria species (besides the three main symbionts) in the tsetse gut (Geiger et al. 2009, 2010, 2011). In these studies, bacteria were isolated from more than 50% of tsetse flies, revealing a large variety of bacteria species as well as a great diversity in the composition of microflora, according to the tsetse species or subspecies and/or their geographical origin. The identification of these bacteria could greatly assist vector control, since some of them are known for their ability to affect insect survival and/or its vector competence (Hertle et al. 1999; Moss 2002; Azambuja et al. 2005). Similar studies performed on *G. fuscipes* flies from Kenya (Lindh and Lehane 2011) confirmed the higher prevalence of nonsymbiotic bacteria in the digestive tract of *Glossina* belonging in particular to the genera *Enterobacter*, *Acinetobacter*, and *Enterococcus*.

Recent work has investigated bacterial diversity in the gut of *G. pallidipes* sampled in Tanzania (Malele et al. n.d.). Most of the identified bacteria from this study belong to the *Firmicutes*, *Actinobacteria*, and *Proteobacteria* phyla. Additional phylogenetic analyses from this study allowed the classification of these isolates into the genera *Bacillus*, *Acinetobacter*, *Mesorhizobium*, *Paracoccus*, *Microbacterium*, *Micrococcus*, *Arthrobacter*, *Corynebacterium*, *Curtobacterium*, *Vagococcus*, and *Dietzia* spp., indicating that the *G. pallidipes* gut represents a well-adapted environmental reservoir for a vast number of bacterial species. The presence of these bacteria in the gut of field-collected tsetse flies has also been assessed by bacterial isolation and culture approaches (Geiger et al. 2009, 2011; Lindh and Lehane 2011; Malele et al. n.d.). In contrast to this abundance of species, direct molecular identification methods (such as deep sequencing of the V4 hypervariable region of the *16S rRNA* gene) revealed a limited diversity of gut-specific microbiota in wild tsetse flies from Uganda (Aksoy et al. 2014). However, a similar approach (using barcoded Illumina sequencing of the hypervariable V3-V4 region of the *16S rRNA* gene) performed on *G. p. palpalis* flies infected or not with trypanosomes and sampled in different HAT foci revealed a large diversity in gut bacteria, in which more than 80 genera (mostly belonging to the *Actinobacteria*, *Bacteroidetes*, *Firmicutes*, and *Proteobacteria* phyla) were identified (Anne Geiger, personal communication). Although molecular approaches associated with biostatistics have the ability to process a large number of samples in a short time, they do not make it possible to ensure that the identified microorganisms are actually alive and not the residues of organisms with intact targeted DNA regions. This illustrates the need to isolate and cultivate bacteria that are considered to be important within the context of the research project objectives (Fig. 5.13).

Production of Antiparasitic Molecules Several mechanisms may be involved in the modulation of parasite infection by bacteria from the tsetse intestinal microflora. Synergy between the coinhabiting bacteria Wgm and *Sodalis* has been observed, in which elimination of Wgm resulted in the loss of *Sodalis* over several host generations (Wang et al. 2013). In addition to supplying the tsetse with vitamins, Wgm can provide thiamine to *Sodalis*, which does not possess the proper biosynthetic pathway, although it has preserved the transporter that allows the fly to absorb the exogenously produced vitamin (Snyder et al. 2010). The existence of functional complementation between the genomes of both symbionts has been suggested, which would reduce their competition. Furthermore, comparing the Wgm genomes, respectively, hosted by Gmm and *G. brevipalpis* has revealed potential differences in their metabolomes (Rio et al. 2012). For example, chorismate, phenylalanine, and folate biosyntheses are

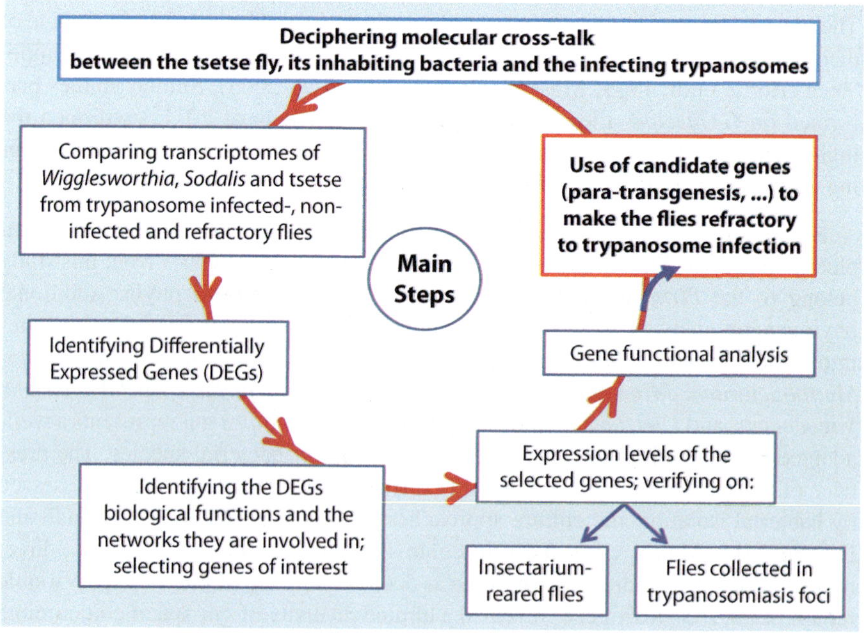

Fig. 5.13 Steps involved in the production of trypanosomes refractory tsetse flies. Transcriptomic analyses, identification of differentially expressed genes, identification of gut bacteria from field flies have been performed

only present in Wgm strains from Gmm. Since African trypanosomes are auxotrophic for folate and phenylalanine, they are able to absorb these two compounds from their immediate environment. Therefore, it is possible in Gmm that the presence of Wgm (which is capable of synthesizing these two compounds) could promote infection of the host by trypanosomes (Rio et al. 2012). In contrast, many antiparasitic molecules are produced by a variety of bacteria species that have been identified in the tsetse gut. These molecules include cytotoxic metalloproteases produced by *S. marcescens* and *Pseudomonas aeruginosa* (Maeda and Morihara 1995); hemolysins secreted by *Enterobacter* spp., *E. coli*, *S. marcescens*, and *Enterococcus* spp. (Hertle et al. 1999; Coburn and Gilmore 2003); antibiotics produced by *Serratia* spp. (Thomson et al. 2000); and hemagglutinins (Gilboa-Garber 1972) and siderophores secreted by *P. aeruginosa* (Schalk et al. 2002). Similarly, *P. fluorescens* produces an antitrypanosome factor (Mercado and Colon-Whitt 1982), and pigments such as prodigiosin are secreted by Gram-negative bacteria including *Serratia* spp. and *Enterobacter* spp. (Moss 2002). Prodigiosin is known to be toxic to *P. falciparum* (Lazaro et al. 2002) and *T. cruzi* (Azambuja et al. 2005). All of these species have been identified in the tsetse fly intestine (Geiger et al. 2009, 2010, 2011; Lindh and Lehane 2011).

It has been suggested that the anti-plasmodium effect mediated by bacteria is due to the antibacterial immune response of the mosquito, potentially through the activation of immunity (Dong et al. 2009). In Zambia, *Enterobacter* spp. have been isolated from a mosquito that is resistant to *P. falciparum* infection; this anti-plasmodium effect is suggested to be due to the production of active oxygen by this bacterium (Cirimotich et al. 2011).

Aphids harbor the facultative endosymbiont *Candidatus Hamiltonella defensa*, which can increase survival of a host attacked by a parasitoid wasp (Degnan et al. 2009). Strains of *H. defensa* that differ in the level of host protection that they confer have been characterized. In some strains of the symbiont, the protective character mediated by the symbiont depends on the presence of a lysogenic bacteriophage, APSE (Duron 2014). Aphids that host the *H. defensa* symbiont infected with APSE are significantly more resistant to parasitoid wasps than aphids that host the uninfected symbiont. Protection is then ensured when the bacteriophage expresses a toxin (encoded by its genome) directed against eukaryotes (Degnan and Moran 2008a, b).

5.5.1.4 Recent Transcriptomic Approaches for Investigating Molecular Interactions

Many reports highlight the complex interactions between symbionts (such as *S. glossinidius*) and tsetse flies that act to control fly infection by trypanosomes (and consequently their vector competence). This situation is potentially complicated by the recently revealed presence of diverse indigenous bacteria that compose the fly intestinal microflora. These bacteria likely interact with each other and with the fly at different levels that include host immunity and production of metabolites that are favorable or toxic to the host and/or other coinhabitants (i.e., symbionts and trypanosomes).

In order to decipher the complex interactions between several partners and identify the genes that are involved in fly susceptibility or refractoriness, the transcriptome of each partner in the interaction must be analyzed at the time point at which all partners interact (including the fly and its immune system). This requires a technique that can compare the different transcriptomes recorded from non-infected flies, trypanosome-infected (susceptible) flies, and refractory flies (i.e., flies fed on trypanosome-infected mice but that have eliminated the ingested parasites) during the crucial steps of the infection time course (ideally at days 3, 10, and 20 post-infected or non-infected meal uptake) (Van den Abbeele et al. 1999; Ravel et al. 2003). Comparing the transcriptomes in this manner can provide the identification of differentially expressed genes (DEGs) with reference to the evolution in fly susceptibility/refractoriness and/or the infection time course. These DEGs must then be annotated in order to determine their biological functions and metabolic pathways (i.e., gene ontology) and canonical paths (KEEG pathways) in which they are involved. Today, these types of comparative transcriptomic studies are facilitated by the availability of sequences from tsetse genomes (International Glossina Genome Initiative 2014), trypanosomes genomes (El-Sayed et al. 2005; Jackson et al. 2010; Kolev et al. 2010), and symbiont genomes (Akman et al. 2002; Darby et al. 2005; Toh et al. 2006).

In the following subsections, we review the literature on the transcriptomic changes occurring in *Sodalis* and *Wigglesworthia* underlying their interactions in tsetse flies, as well as the changes in tsetse gene expression in response to trypanosome infection.

5.5.1.4.1 The *Sodalis* Transcriptome

The *Sodalis* chromosome and its 4 plasmids represent 2523 genes (Darby et al. 2005; Toh et al. 2006), from which 176 DEGs have been identified. Transcriptomic experiments have been further designed using protocol of Fig. 5.14. Interestingly, overexpression of bacterial cell wall lysis proteins encoded by phages has been shown to enrich the cytolysis, lysozyme activity, peptidoglycan catabolism, and bacteriolytic enzyme functions in *Sodalis* from tsetse flies refractory to trypanosome infection (Hamidou Soumana et al. 2014a). This finding suggests that these cell wall hydrolysis products may be involved in activating the tsetse immune system (Fig. 5.15). The expression of phage genes indicates that the analyzed symbionts were lysogenic and that a prophage activation mechanism was triggered. Similarly, enrichment of the KEGG pathway "bacterial secretion systems" was observed due to the overexpression of transcripts encoding proteins from type III secretion systems (Hamidou Soumana et al. 2014a). The secretion of toxic proteins in the cytosol of target cells, via the type III secretion system (i.e., the injectisome), has been demonstrated in certain pathogenic Gram-negative bacteria such as *Pseudomonas aeruginosa* (Cornelis 2000). In contrast, the "chitinase hypothesis" (Welburn et al. 1993) could not be assessed since no chitinase-encoding gene was found to be expressed (Hamidou Soumana et al. 2014a, b).

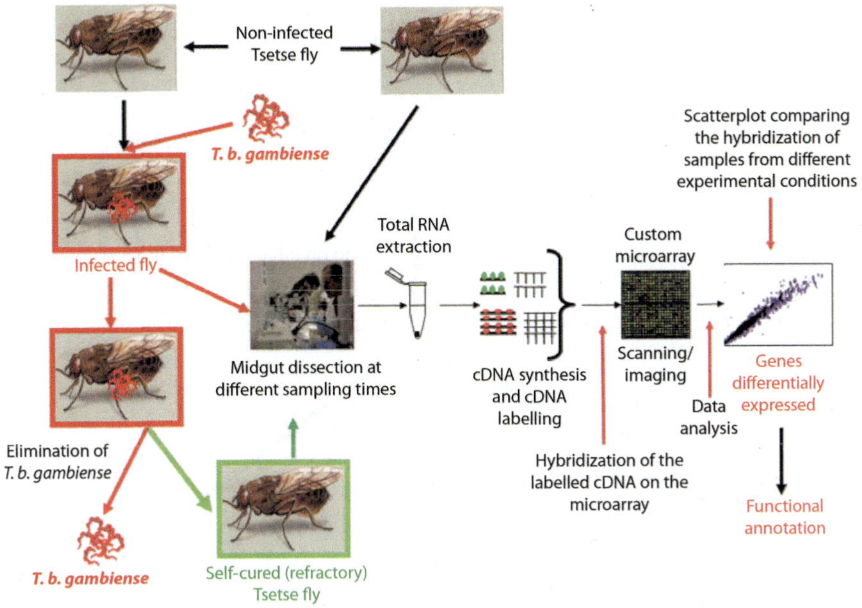

Fig. 5.14 Experiment general protocol. *G. p. gambiensis* midgut was sampled at three times post-*T. b. gambiense* infected bloodmeal. Biological replicates of several midguts were constituted for each time points and analyzed for *Sodalis* or *Wigglesworthia* transcriptome by using different steps described on the picture

5.5.1.4.2 The *Wigglesworthia* Transcriptome

There are 673 genes on the *Wigglesworthia* chromosome, more than 200 of which are differentially expressed when *Wigglesworthia* is hosted in infected tsetse flies versus refractory flies. Protocol used was the same as for transcriptomic analyses of *Sodalis* (Fig. 5.14). However, the DEGs differ depending on the infection time course. At day 3, genes involved in "metabolic and binding processes" were overexpressed, whereas genes related to "processes of development, morphogenesis, and cellular networks" were overexpressed at day 10. In addition, transcripts encoding GroEL and GroES chaperones, noncoding RNAs, proteins involved in the transport of bacterial toxins, and proteins involved in the synthesis of thiamine were under-expressed in symbionts from stimulated tsetse flies, 3 days after ingestion of an infected meal. In contrast, the transcripts corresponding to GroEL chaperones, GroES chaperones, and noncoding RNAs were overexpressed in infected tsetse flies at day 10 (Hamidou Soumana et al. 2014c). These *Wigglesworthia* chaperones could function as toxins against trypanosomes, similar to what has been reported for *Enterobacter aerogenes* (an antlion symbiont), which produces a toxic chaperone that paralyzes the antlion's prey (Yoshida et al. 2001).

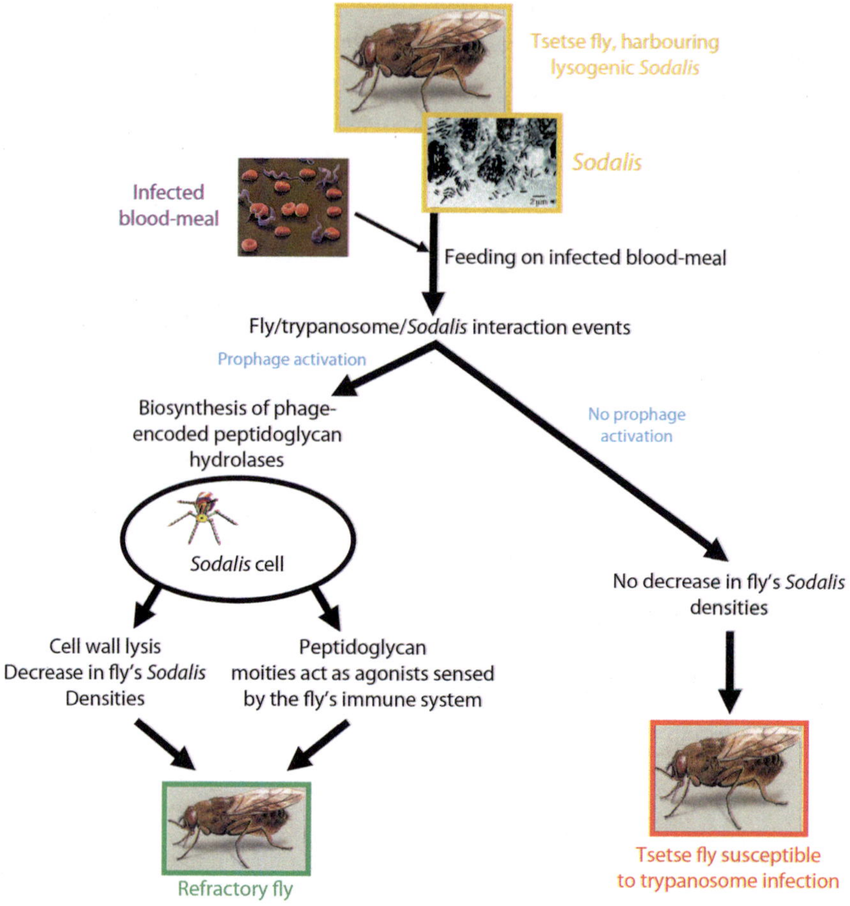

Fig. 5.15 The Sodalis'phage strategy. Picture representing the possible effects of phage of Sodalis in the refractoriness of tsetse fly to trypanosomes

5.5.1.4.3 G*lossina palpalis gambiensis* Transcriptomes and Identification of Orthologous Genes in *G. morsitans morsitans*

The *Glossina palpalis gambiensis* transcriptomes were previously analyzed from RNAseq libraries produced from stimulated/infected flies as well as non-stimulated/uninfected flies, at three selected time points after ingesting an infected or non-infected meal (3, 10, or 20 days). Subsequently, de novo assembly of RNAseq reads resulted in the identification and functional annotation of 16,936 contigs (Hamidou Soumana et al. 2015). At 3 days post-feeding, 1373 genes differentially expressed between stimulated and non-stimulated flies were identified. Fifty-two DEGs were identified between infected and self-cured flies at 10 days post-feeding, and 1025 DEGs were similarly identified at 20 days post-feeding. The most significant biological functions and canonical pathways associated with the infection were also identified (Hamidou

Soumana et al. 2015). In this study, *Glossina* candidate genes were identified that encode a broad selection of proteases, chitin-binding proteins, and antimicrobial peptides such as Pro3 protein, transferrin, mucin, attacin, and cecropin. These candidate genes would be excellent choices for further functional analyses.

Sleeping sickness is caused by either *Trypanosoma brucei gambiense*, which is transmitted by the vector *Glossina palpalis gambiensis* (Gpg) in West and Central Africa and causes the chronic form of the disease, or *T. b. rhodesiense*, which is transmitted by *G. m. morsitans* (Gmm) in East Africa and causes the acute form of the disease. In the context of a common anti-vector strategy, the Gmm genome has been examined for genes orthologous to Gpg DEGs by mapping the RNAseq sequences on the Gmm genome, revealing that approximately 50% of Gpg DEGs have orthologs in the Gmm genome (Hamidou Soumana et al. 2017). This result assesses the possibility of developing common strategies and tools to fight sleeping sickness, whether the causative parasites are transmitted by *Glossina palpalis gambiensis* or by *Glossina morsitans morsitans*. This approach should also be considered for the fight against AAT, which is caused by a different *Trypanosoma* species and transmitted by a different *Glossina* species.

5.5.1.5 Potential Uses of Bacteria in a *Glossina* Vector Control Strategy

The use of *Wolbachia* for anti-vector control represents an alternative to the methods currently used to reduce vector density and/or create conditions that lead to the reduction of pathogen transmission. In this context, four potential strategies are considered in this section.

a) *Wolbachia* could be used as a vector to transfer genes of interest to host populations (Beard et al. 1998; Dobson 2003; Sinkins and Gould 2006). This method relies on the introduction of pathogen resistance genes into the *Wolbachia* genome. The transformed bacterium that will express the molecules of interest is then introduced into the target vector through microinjection. When released into natural vector populations, the individuals infected with transgenic *Wolbachia* will gradually invade populations through their induced cytoplasmic incompatibility (CI), enabling the spread of the gene of interest. However, this technique remains difficult to implement due to the intracellular biology of *Wolbachia*, which complicates its genetic transformation.
b) *Wolbachia* could also be used to reduce the lifespan of adult female *Glossina* flies, which would prevent the trypanosome from completing its parasitic cycle within the host. The fact that *Wolbachia* induces CI also represents an advantage, since it favors the conquest of natural fly populations by these short-lived hosts. The main challenge of this strategy will be the artificial transfer of the symbiont into the fly.
c) *Wolbachia*-induced CI could also be used to reduce the size of host populations. This is similar to the sterile male technique and consists in releasing males infected with incompatible *Wolbachia*. Mating with females that are not infected with *Wolbachia* will then induce CI and in turn embryo degeneration, which will reduce the size of targeted populations. *Wolbachia* presents an additional interest

in terms of vector and pathogen control, since some strains have been reported to stimulate the immune system of the host, consequently inhibiting infections by viruses and parasites. For example, in *Aedes aegypti*, the wMelPop *Wolbachia* strain has been shown to inhibit the multiplication of dengue virus, chikungunya virus, and parasites such as *Plasmodium gallinaceum* (responsible for malaria; Moreira et al. 2009), *Plasmodium berghei* (Kambris et al. 2010), and the filarial parasite *Brugia pahangi* (a parasite of rodents; Kambris et al. 2009).

d) A final possibility is to exploit the association between *Sodalis* and *Wolbachia*. *Sodalis glossinidius* appears to be well-adapted for a "paratransgenesis" (Rio et al. 2004) approach (Fig. 5.16) on several grounds: (1) it is able to inhabit different tissues of the tsetse fly (including the middle intestine and hemolymph) (Cheng and Aksoy 1999) that are in close proximity to the fly-infecting trypanosomes; (2) it can be cultivated and genetically modified in vitro (Dale and Maudlin 1999; Cheng and Aksoy 1999; De Vooght et al. 2012, 2014); (3) it can be transferred into the tsetse fly (Weiss et al. 2006); (4) it is maternally transmitted to the offspring (Balmand et al. 2013; Wang et al. 2013), while paternal transmission has recently been demonstrated during mating (De Vooght et al. 2015); and (5) due to large-scale erosion of its genome (Akman et al. 2001; Rio et al. 2003), *Sodalis* is metabolically dependent on its tsetse host. This last point is important because it means that *Sodalis* will not be transferred to any non-*Glossina* species, suggesting that this symbiont is an "environmentally" safe candidate for use in paratransgenesis strategies. A key step will be the identification of antitrypanosome targets as effector transgenes (Fig. 5.13). The transcriptomic approaches under development to decipher the multiple tsetse fly/microbiome/trypanosome interactions (Fig. 5.13) are expected to provide several possible candidates as antitrypanosome effector molecules for in vivo and in situ "delivery" by the *Sodalis* bacterium.

The modalities of an effective dissemination in HAT foci of tsetse flies hosting the transformed symbiont remain to be determined. As previously shown, a strong cytoplasmic incompatibility occurs in female flies when a (W-) female that does not host the symbiont mates with a (W+) male fly. In contrast, crossing a (W +) female with a (W +) or (W-) male results in fertile progeny that are more numerous than the cross from a (W-) female with a (W-) male (which is the classical event occurring in the field). Therefore, based on their reproductive advantage, dissemination in HAT foci of (W+) female tsetse flies that host the modified *Sodalis* symbiont will result in the progressive replacement (over generations) of the natural tsetse population. Accordingly, one study suggests that the dissemination of a number of flies corresponding to 10% of the natural population in a focus should lead to the replacement of 90% of this population within 2 years (Alam et al. 2011).

In conclusion, just as we recognize an irrefutably crucial role for *E. coli* and other intestinal bacteria in human health, there is an extreme diversity of bacterial species present in the intestines of insects that is essential to their biology. This review brings into focus this diversity and the complexity of the relationships (whether beneficial or harmful) between these bacteria, their hosts, and the pathogens that can invade the insect host's "intestinal niche." All of these factors make it

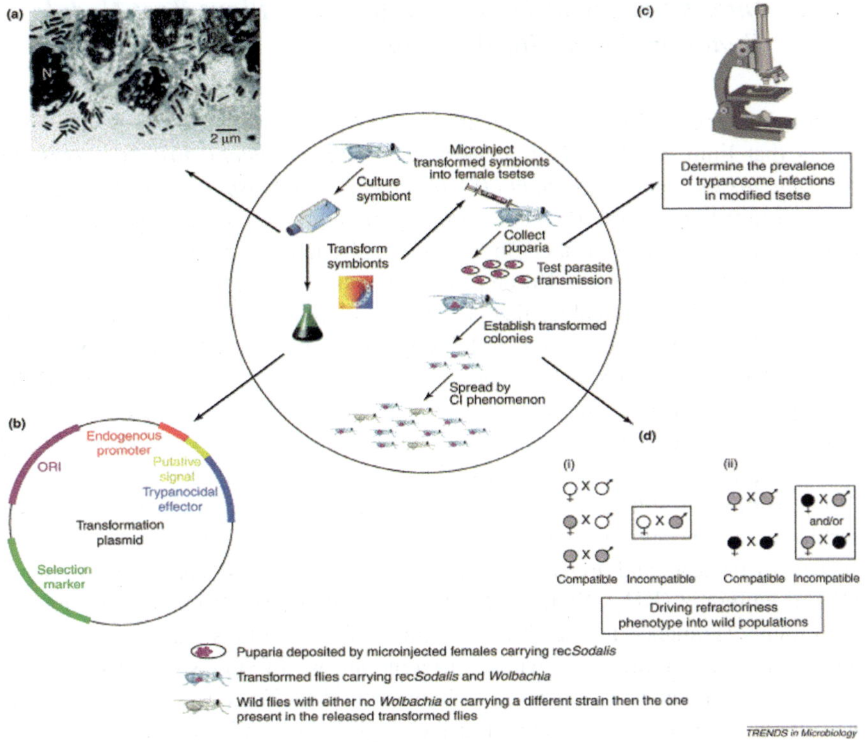

Fig. 5.16 Paratransgenesis approach. (**a**) *Sodalis* is isolated from tsetse fly hemolymph and cultured first on a feeder *Aedes albopictus* C6/36 cell layer. After what *Sodalis* is grown in cell-free medium, under microaerophilic conditions. (**b**) A plasmid containing the trypanocidal effector expressed by an endogenous control element with spatial and temporal specificity is used to transform *Sodalis*. Recombinant *Sodalis* (rec*Sodalis*) is microinjected into pregnant mothers. (**c**) The offspring harboring rec*Sodalis* are then challenged with trypanosomes. Prevalence of trypanosomes in tsetse fly tissues is examined by microscopy to evaluate the extent of the refractoriness due to the transgene product. (**d**) Colonies harboring the rec*Sodalis* are then established and can replace their susceptible counterparts in field by cytoplasmic incompatibility (CI) conferred by *Wolbachia* symbionts. This approach requires that either the established colony should have a *Wolbachia* infection with the native population being uninfected, or that the colony carries a different *Wolbachia* strain to that found in the field population. (i), CI is expressed when uninfected females (open circles) mate with infected males (hatched circles), resulting in loss of progeny; (ii), CI is expressed when infected females mate with males that are infected with a different *Wolbachia* strain or with multiple *Wolbachia* strains. Using CI, *Wolbachia*-infected insects can replace naive populations while simultaneously driving other maternally inherited elements such as rec*Sodalis* into the tsetse fly populations

difficult to inventory these interactions and their induced effects. However, thanks to rapid advances in genetics, DNA and protein sequencing, identification of biological molecules, bioinformatics, and biostatistics, deciphering these interactions in tsetse flies is making astonishing progress. Future studies will need to focus on how to efficiently sort all of the recorded data in order to maximize the best way to use this information to combat such diseases as sleeping sickness.

5.5.2 *Interactions with Plants: Evolution of Plant Root Nodule Endosymbioses with Bacteria*

Although nitrogen is the most abundant element in the air, it is one of the main elements that limit plant production worldwide. *Plants cannot directly access the nitrogen gas that makes up about 80% of the earth's atmosphere.* Most plants rely for their nutrition on soil inorganic nitrogen (ammonium, nitrate) derived from the atmosphere through the Haber-Bosch industrial chemical process and biological nitrogen fixation. The Haber-Bosch process, which turns the nitrogen in the air into ammonia to make chemical fertilizers, consumes large quantities of fossil fuels and results in CO_2 and N_2O emissions. Additional nitrogen pollution is caused by chemical fertilizers in the runoff from agricultural fields, much of which is lost to the atmosphere as gaseous emissions. Excessive use of nitrogen fertilizers increases *global warming* and climate change, can have serious impacts on biodiversity and on water quality, and is a major threat to human and animal health (Good and Beatty 2011). Biological nitrogen fixation is a possible alternative to chemical nitrogen fertilizers, but, because of the stability of the triple bond of dinitrogen, only a few species of bacteria and archaea with the enzyme *nitrogenase* can reduce nitrogen (N_2) to ammonia (NH_3). Some of these bacteria can form mutualistic symbioses with plants (Franche et al. 2009). Two kinds of associations lead to root nodule endosymbiotic symbiosis in which different nitrogen-fixing soil eubacteria are hosted in root nodules: Gram-positive filamentous *Frankia* actinobacteria associate with about 260 plant species belonging to eight different families called actinorhizal plants (Dawson 2007), whereas Gram-negative rhizobia associate only with legumes (*Fabaceae*) (Sprent 2001) and *Parasponia* (*Cannabaceae*) (Trinick 1973, 1979). The high efficiency of these root nodule symbioses contrasts with endophytic associations of plant roots with other diazotrophic bacteria that contribute only slightly to the *nitrogen nutrition* of their host plants.

Molecular phylogenetic analyses *have* greatly advanced our knowledge of the relationships on both sides of the *plant*-bacteria *nodulation* symbiosis. In recent years, molecular tools have made it possible to characterize the molecular mechanisms that control the formation of root nodules in legumes and in non-legumes, thereby revealing some of the genetic constraints underlying the evolution of these root nodule symbioses (Downie 2014; Froussart et al. 2016). This chapter examines how molecular data has provided new information on the evolution of nodulation.

5.5.2.1 Phylogeny of Nodulating Plants

Legume and the non-legume nodules differ in their ontogeny and structure. Legume nodules have a stem-like structure with peripheral vascular bundles and infected cells in the central tissues and nodule primordia that originate in the cortex. In contrast, actinorhizal nodules are composed of multiple lobes, each of which represents a modified lateral root with a central vascular bundle and infected cells in the cortex.

Ontogenesis and the final structure of the *Parasponia* nodule lobe is similar to that observed in actinorhizal plants and resembles a lateral root (Fig. 5.17). Despite the diversity of nodule ontogenesis and structure, molecular phylogenetic analyses revealed that angiosperm families of nodulating plants, together with 18 families of non-nodulating plants, *belong to a single "nitrogen-fixing" clade* within Rosid I clade (Fabidae) (Soltis et al. 1995; Wang et al. 2009a, b, c) (Fig. 5.18). The Fabidae clade, which is monophyletic, is composed of the orders Rosales, Cucurbitales, Fagales, and Fabales and had a single origin for symbiotic nitrogen fixation (Fig. 5.18). Root nodule symbiosis is restricted in some but not all species belonging to four related orders of flowering plants: Fabales, Fagales, Curcubitales, and Rosales. Altogether, plants that are able to develop these symbioses account for only about 2.5% of angiosperm families (Soltis et al. 1995). In Fabales, only one of the four families contains nodulators; this family, Leguminosae (or Fabaceae), comprises three subfamilies, Caesalpinioideae, Mimosoideae, and Papilionoideae representing approximately 19,500 *species* in about *750 genera* (Lewis et al. 2005). Due to the ability of more than 85% of the species to form nodules with rhizobia, they account for important grain, pasture, and agroforestry species that are sources of protein for human and animal nutrition and of nitrogen for many cropping systems (Graham and Vance 2000). Around 90% of the species in the papilionoid and mimosoid genera can

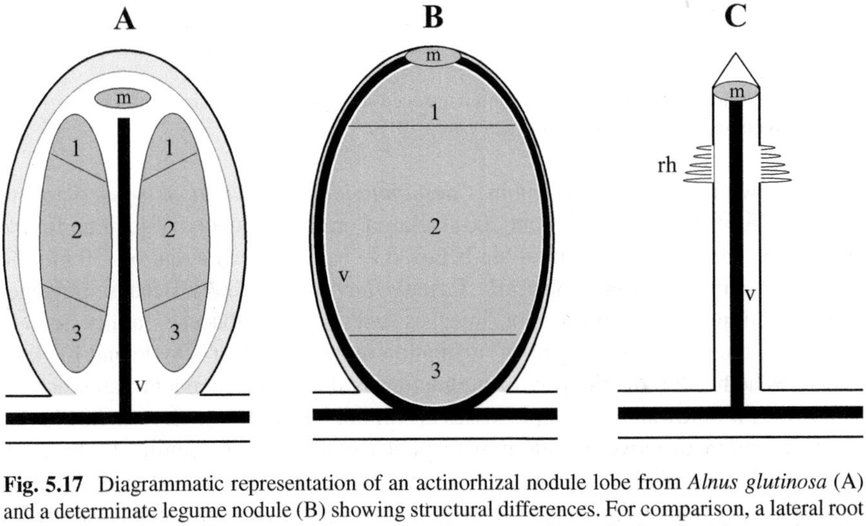

Fig. 5.17 Diagrammatic representation of an actinorhizal nodule lobe from *Alnus glutinosa* (A) and a determinate legume nodule (B) showing structural differences. For comparison, a lateral root is shown (C). (**a**)- Actinorhizal nodule lobe contains a central vascular bundle (v), a meristem at the apex (m). Developmental zonation with specific patterns of gene expression can be defined in the cortex: (1) infection zone where cells originated from the apex become infected by *Frankia*; (2) nitrogen fixation zone where *Frankia nif* genes are expressed; (3) senescence zone where plant cells and *Frankia* are being degraded. (**b**)- Indeterminate legume nodule contains a peripheral vascular system (v) and infected cells in the central tissue. Due to the activity of the meristem (m) the cells in the central tissue are arranged in a developmental gradient ranging from prefixation zone (1) to senescence zone (3). (**c**)- Lateral root contains a root meristem (m) covered by the cap, root hairs (rh) and a central vascular bundle (v). (Adapted from Pawlowski and Bisseling 1996)

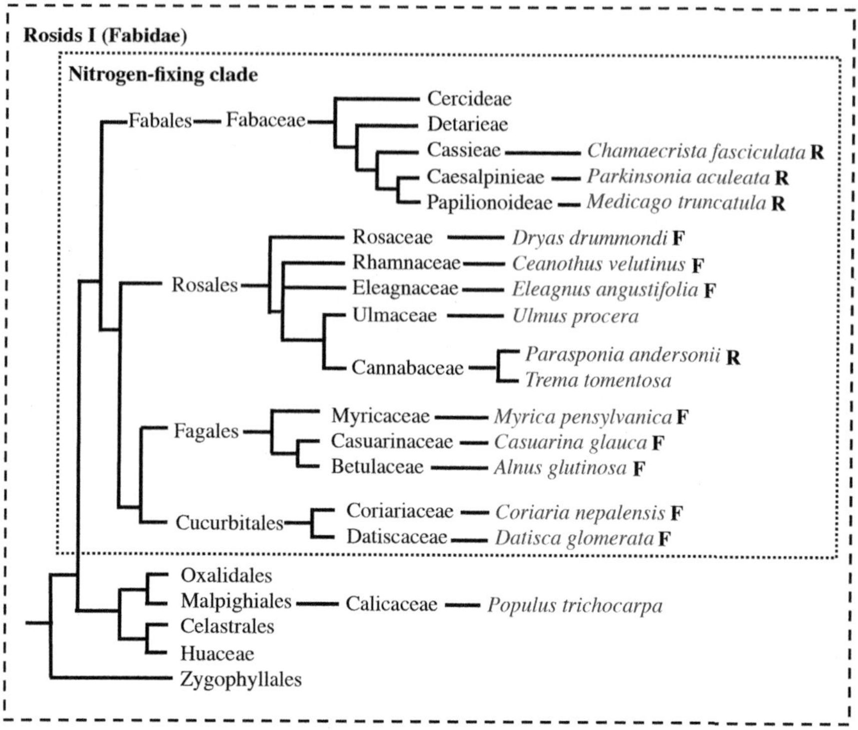

Fig. 5.18 Phylogenetic representation of the nitrogen-fixing clade. R : Rhizobium symbiosis; F : *Frankia* symbiosis. (Adapted from Op den Camp 2012)

nodulate with rhizobia, but within *Caesalpinioideae, nodulation* is restricted to around 5% of the species (Sprent 2001). Based on the discovery of legume fossils and molecular analyses, legumes are believed to have evolved about 65–50 million years ago in the late Cretaceous/early Tertiary (Lavin et al. 2005). Because phylogenetic relationships among major lineages within the Fabaceae family remain unknown, the number of origins of nodulation remains unclear (Doyle and Luckow 2003). Whether the distribution of nodulation is due to independent gains or to a single origin followed by multiple losses is still not known (Cannon et al. 2014).

As mentioned above, in addition to plants of the legume family (Fabaceae), *Parasponia* (Rosales) (originally classified as Trema), a tropical genus in the Cannabaceae, is also able to interact with rhizobia (Trinick 1973). The *Parasponia*-rhizobia symbiosis includes only five species of medium-sized tropical and pioneer trees (up to 15 m in height growing on nitrogen-poor and disturbed soils). These trees originated from the Malay Archipelago. Among the five nodulated species identified (Becking 1992), *P. andersonii/Rhizobium* is the most widely studied symbiotic association. Several characteristics of the *Parasponia*-rhizobia symbiosis, including a basal mode of infection and less sophisticated control of symbiotic association than with legumes, suggest the recent emergence of the ability of the host plants to be nodulated by rhizobia (Op den Camp et al. 2012). As mentioned previously, despite being nodulated by rhizobia, *Parasponia* nodules resemble actinorhizal nodules, and phylogenetically, *Parasponia* is more closely related to actinorhizal plants than to legumes (Soltis et al. 1995) (Fig. 5.18).

Actinorhizal plants are distributed in the three orders: Rosales, Fagales, and Cucurbitales (Fig. 5.18); they have the ability to develop endosymbiosis with the nitrogen-fixing soil actinobacteria *Frankia.* Actinorhizal plants are woody shrubs and trees, except for the genus *Datisca*, which is herbaceous. Ecologically speaking, the majority of actinorhizal plants are pioneer species that colonize nitrogen-poor sites and disturbed areas where they play important ecological roles. For example, in Africa and Asia, *Casuarina* are planted as wind belts, to stabilize coastal and desert dunes, and in agroforestry (used in inter-cropping) (Diagne et al. 2013; Zhong et al. 2010). They are distributed worldwide from cold regions (except Antarctica) to warm latitudes. Actinorhizal plants represent a diverse group of about 220 species belonging to 25 genera from Betulaceae, Casuarinaceae, Coriariaceae, Datiscaceae, Elaeagnaceae, Myricaceae, Rhamnaceae, and Rosaceae (Swensen 1996) (Table 5.2).

Table 5.2 Currently identified actinorhizal families and genera. Actinorhizal plants represent a diverse group of 220 dicotyledons species belonging to 8 families and 25 genera

Plant orders	Family	Genus
Fagales	Betulaceae	*Alnus*
	Casuarinaceae	*Allocasuarina*
		Allocasuarina
		Ceuthostoma
		Gymnostoma
	Myricaceae	*Comptonia*
		Myrica
Rosales	Eleagnaceae	*Eleagnus*
		Hippophae
		Shepherdia
	Rhamnaceae	*Adolphia*
		Ceanothus
		Colletia
		Discaria
		Kentrothamnus
		Retanilla
		Talguenea
		Trevoa
	Rosaceae	*Cercocarpus*
		Chamaebatia
		Cowania
		Dryas
		Purshia
Cucurbitales	Coriariaceae	*Coriaria*
	Datiscaceae	*Datisca*

Adapted from Dawson (2007)

In symbioses with actinorhizal lineages, root nodulation has evolved via a combination of three or four independent gains or losses (Swensen 1996). More recently, using large-scale phylogenetic analyses, Li et al. (2015) hypothesized that several actinorhizal nitrogen-fixing lineages originated during the late Cretaceous and Eocene periods of harsh conditions with high levels of atmospheric carbon dioxide and high temperature and high light conditions.

Thus, the fact that all nodule-forming plants *belong* to a single *clade* suggests that all nodulating flowering plants are more closely related than was previously thought and that the predisposition to nodulate might have arisen only once in the common ancestor of the Rosid 1 clade about 100 million years ago upon which root nodule symbiosis could develop (Soltis et al. 1995; Doyle 1998, 2011). The molecular bases of this predisposition to establish nodule symbiosis are so far unknown. Since non-nodulating plants belong to the "nitrogen-fixing clade," this implies a loss of nodulation capacity in the plants concerned. However, although there is evidence for a single origin of the predisposition for nodulation, as mentioned for legume and actinorhizal plants, subsequent analyses based on the molecular and morphological evidence suggested that innovations appeared independently in the different nitrogen-fixing lineages. Recent work using a N_2 fixation database containing 9156 angiosperm species and evolutionary models supports the hypothesis of predisposition, suggesting that a single evolutionary innovation occurred over 100 million years ago (MYA) and was followed by several evolutionary events leading to the emergence of the different kinds of symbiotic associations (Werner et al. 2014).

5.5.2.2 Origin of Bacteria Symbiotic Signaling Molecules

Legume and non-legume symbioses start with a molecular dialogue between the host plants and the symbionts. A major advance in our understanding of nitrogen-fixing bacteria plant interactions was the finding, in the early 1990s, that plant-derived flavonoids induce rhizobia nodulation genes (*nod*) to produce lipochitooligosaccharides (LCOs) known as nod factors (NFs). Rhizobial NF molecules consist of an acylated chitin oligomeric backbone with various functional group substitutions at the terminal or nonterminal residues (Long 1996). The synthesis of the chitin oligomer backbone is controlled by three specific enzymes, encoded by canonical *nod* genes *nodABC* (NodA, acyl transferase; NodB, chitin deacetylase; NodC, chitin synthase) (Oldroyd et al. 2011). The NF core is then modified by species-specific proteins resulting in various substitutions, including glycosylation and sulfation (Long 1996). These substitutions are specific to each host legume and are major determinants of host specificity. The perception of NFs by plant LysM-receptor-like kinase (LysM-RLKs) induces a signal transduction cascade that is required for infection and for all symbiotic responses that lead to the development of fully differentiated nodules on legume and *Parasponia* roots (Long 1996; Oldroyd et al. 2011).

Rhizobia are Gram-negative bacteria distributed in *Alphaproteobacteria* and *Betaproteobacteria* subclasses and include hundreds of species in only 14 genera. Recent data suggest that, during evolution, rhizobia arose from very rare horizontal transfer of genes essential for nodulation and nitrogen fixation between these two subclasses, from infrequent transfer between genera, and from frequent transfer within genera (Remigi et al. 2016). *Frankia* is a Gram-positive bacterium belonging to the high-GC subgroup of the phylum Actinobacteria. Phylogenetically, *Frankia* and rhizobia are quite distant, suggesting that they acquired their ability to enter root nodule symbiosis from different directions. Several phylogenetic chronometers including RAPD PCR, 16S and 23S-rRNA, gene sequences (ITS region) between 16S and 23S-rRNA, and *nif* gene sequences have been used to identify the genetic variability and relationships among the *Frankia* strains isolated from different host plants (reviewed in Hahn 2008). Based on the 16S database, Normand et al. (1996) grouped the *Frankia* genus in four clusters: cluster 1, strains effective in *Alnus* (Betulaceae), *Casuarina*, *Allocasuarina* (Casuarinaceae), and *Comptonia*, *Myrica*, and *Morella* (Myricaceae); cluster 2, uncultured strains present in nodules of *Dryas* (Rosaceae), *Coriaria* (Coriariaceae), and *Datisca* (Datiscaceae); cluster 3, strains effective in *Elaeagnaceae* and *Gymnostoma* (Casuarinaceae); and cluster 4, non-infective or ineffective strains isolated from a range of host plants.

The ubiquitous presence of *nod* genes in all rhizobia led to a proposal of a universal paradigm for specific recognition between nitrogen-fixing bacteria and plants. However, the universality of this model was called into question by the finding that canonical *nod* genes are absent in photosynthetic bradyrhizobia (Giraud et al. 2007) as well as in several *Frankia* strains (Normand et al. 2007). The candidate molecules for *Frankia* signals were found to be hydrophilic and chitinase-resistant (Cérémonie et al. 1999; Chabaud et al. 2016). These diffusible biologically active molecules have the ability to induce calcium spiking in *Casuarina glauca*, thereby strengthening the hypothesis of their role as signaling molecules (Chabaud et al. 2016). These data suggest that *Frankia* and *Rhizobium* NFs may not be related. Purification and complete characterization of *Frankia* signaling molecules involved in symbiotic process will provide further clues to the relationship and to the origin of actinorhizal and legume symbioses. Recently, however, genome sequencing of the non-culturable cluster 2 *Frankia datiscae* Dg1 and Dg2 revealed the presence of canonical *nod* genes *nodABC* (Persson et al. 2011; Van Nguyen et al. 2016). Since actinorhizal symbioses involving cluster 2 *Frankia* strains are thought to be the oldest actinorhizal symbioses, it was hypothesized that canonical *nod* genes were probably lost during the evolution of *Frankia* symbionts (Persson et al. 2015). Interestingly, phylogenetic analyses have recently shown that *Frankia* Dg1 (cluster 2) nod proteins are related to those of nodulator proteobacteria, suggesting that transfer between actinobacteria and proteobacteria is possible (Persson et al. 2015).

In the nod-independent photosynthetic *Bradyrhizobium-Aeschynomene* symbiotic signaling process, the nature of the symbiotic signal remains unknown. One striking feature of photosynthetic *Bradyrhizobium* is our inability to isolate strict nodulation minus mutant (Giraud et al. 2007; Bonaldi et al. 2010). Interestingly, Okazaki et al. (2015) recently found that non-photosynthetic bradyrhizobia can interact symbiotically through two distinct processes, one involving the type III secretion system (T3SS), which is known to be connected with the plant immune system. These observations led Okazaki et al. (2015) to conclude that direct contact between the plant and the bacterium play a crucial role in early symbiotic signaling probably through the perception of a bacterial surface component by a plant receptor leading to suppression of defense responses. Taken together, these data suggest that photosynthetic bradyrhizobia and most *Frankia* strains use an alternative nod-independent pathway to initiate nitrogen-fixing symbioses. The signaling molecules secreted by bacteria and plant host symbiotic components for this *alternative* nod-independent *pathway remain to be identified.*

Mycorrhizal fungi secrete a mixture of lipochitooligosaccharides (LCOs) and chitooligosaccharides (COs), called Myc factors (Maillet et al. 2011; Genre et al. 2013). These secreted molecules are assumed to play a role in fungal signaling by regulating symbiotic-related genes (Kosuta et al. 2003). Interestingly, rhizobia NFs have a striking structural similarity to Myc factors LCOs, including the same N-acetylglucosamine backbone (Fig. 5.19). In contrast to rhizobia, AM fungi have a broad *range* of more than 200,000 species of *host* plants, suggesting these fungi have the ability to produce a broad spectrum of LCOs and the existence of a conserved communication process (Oldroyd 2013). The origin of bacterial *nod* genes required for LCO synthesis is not yet clear. Bacterial *nod* genes could have been acquired by lateral transfer to the ancestral *Rhizobium* species from AM fungi about 60 million years ago and subsequently spread to a variety of soil bacteria (Streng et al. 2011). It has also been hypothesized that at least part of the *nod* operon, the LCO transport system, originates from ancestral *Burkholderia* (Aoki et al. 2013). The common *nodABC* genes that determine the synthesis of the LCO backbone

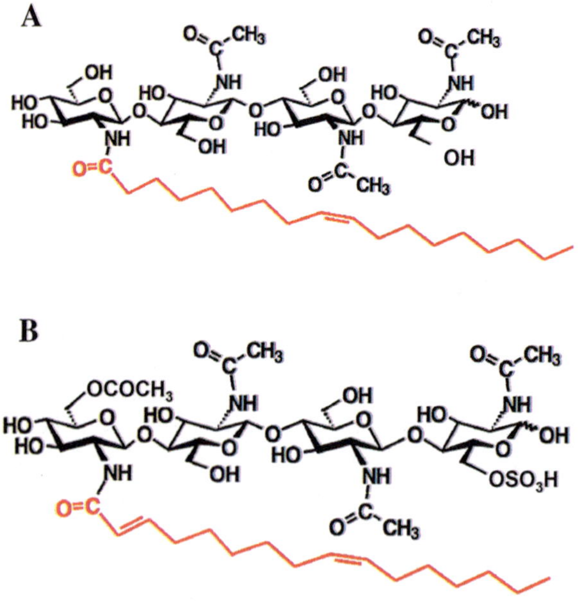

Fig. 5.19 Chemical structure Myc-LCOs and Nod factors. (**a**) Major Myc-LCOs produced by *Rhizophagus irregularis* involved in mycorrhization. (**b**) Major Nod factor produced by *Sinorhizobium meliloti* involved in symbiosis with *M. truncatula*. (Adapted from Maillet et al. 2011)

display a greater degree of diversity in *Alphaproteobacteria* than in *Burkholderia*, thus suggesting that these genes evolved first in an ancestral alphaproteobacterial species (Aoki et al. 2013; Bontemps et al. 2010).

5.5.2.3 Genes that Are Important for Legume, *Parasponia*, and Actinorhizal Root Nodulation Derived from Arbuscular-Mycorrhizal Symbiosis

Studies of the model legumes, *Medicago truncatula* and *Lotus japonicus*, provided insights into the molecular components of the symbiotic signaling pathway. Several of the corresponding genes have been identified (Oldroyd et al. 2011). Interestingly, genetic analyses demonstrated that rhizobia legumes and AM associations share a single signaling pathway or "common symbiosis pathway" (CSP) (Oldroyd et al. 2011) (Fig. 5.20). This pathway contains a receptor-like kinase SYMRK/DMI2, nuclear pores, and the potassium channel proteins required for the induction of calcium oscillations (Gutjahr and Parniske 2013). A nuclear calcium- and calmodulin-dependent kinase (CCaMK/DMI3) interacting with transcription factors (Cyclops/IPD3) is also part of this common pathway. The nuclear Ca^{2+}-spiking induced by AM fungi and rhizobia is likely decoded by CCaMK, thus triggering infection and organogenesis programs (Mitra et al. 2004; Miwa et al. 2006). As mentioned above, the *Rhizobium*/legume symbiosis evolved more recently (about 65 million years ago) than the much older AM symbiosis (at least 400 million years ago) (Remy et al. 1994). Thus, the identification of a common symbiotic signaling pathway governing legume nodule and AM formation, together with the ability of the fungi to synthesize molecules similar to NFs, suggests that legume nodulation evolved by recycling at least part of the ancestral and widespread AM program (Markmann and Parniske 2009). Interestingly, the CSP has been shown to be essential in three other nodulating lineages including two actinorhizal species: *C. glauca* (Fagales), *Datisca glomerata* (Curcubitales) (Gherbi et al. 2008; Markmann et al. 2008; Svistoonoff et al. 2013), and *Parasponia andersonii* (Rosales) nodulated by rhizobia (Op den Camp et al. 2011). Beyond the CSP, several genes including several transcription factors of the GRAS and NF-Y families have been shown to play a critical role in rhizobia and AM symbioses (Rípodas et al. 2014). These genes are involved in the developmental program of nodule and AM organogenesis and in the infection process. In legume, the transcription factor NODULE INCEPTION, NIN, is essential for the coordination of the symbiotic organogenesis development program resulting in nitrogen-fixing nodules (Marsh et al. 2007; Vernié et al. 2015). Recently, Clavijo et al. (2015) demonstrated that *CgNIN*, an actinorhizal *Casuarina* homolog to the legume *NIN* gene, plays a central role in the control of nodulation by *Frankia* (Clavijo et al. 2015). Using transcriptional and phylogenetic analyses, Diédhiou et al. (2014) demonstrated that several potential transcription factors homologous to those involved in legume *Rhizobium* and AM symbioses are present in the two actinorhizal plants *C. glauca* and *Alnus glutinosa*. Therefore it seems that the conservation of the genetic components of the symbiosis signaling pathway controlling legume-rhizobia, actinorhizal, and AM symbioses extends far beyond the CSP.

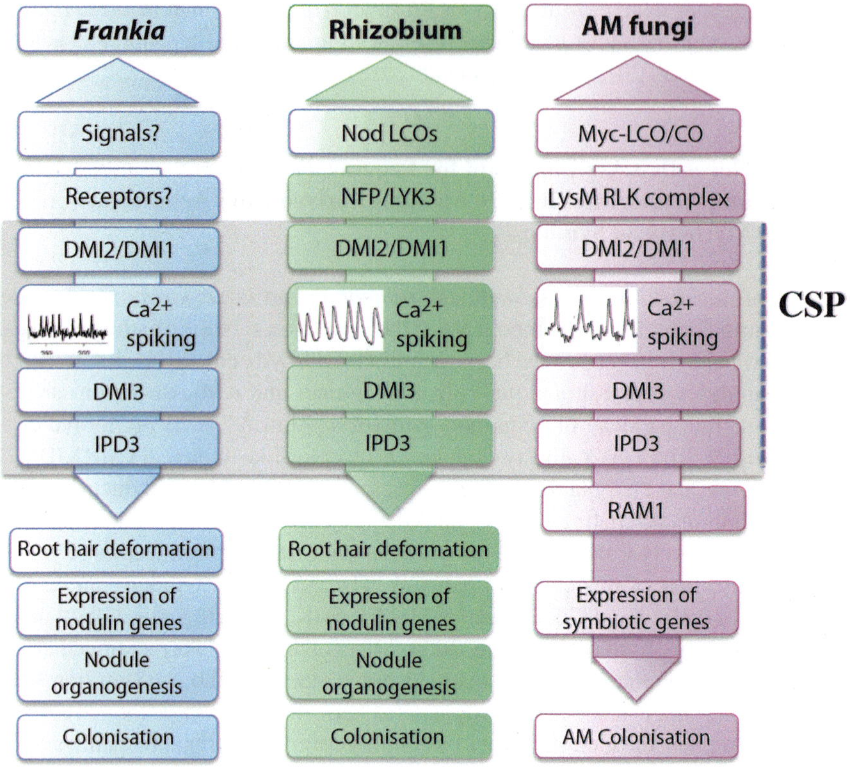

Fig. 5.20 Simplified model of the symbiotic signaling pathway in plant root endosymbioses with *Frankia*, rhizobia bacteria and arbuscular mycorrhiza fungi. *Frankia* factor(s) and actinorhizal plant root receptor(s) are yet unknown

5.5.2.4 Implication of the Evolution of Symbiotic N_2 Fixation in Transferring the Ability to Nodulate Cereals

To cope with the expanding world population, it is indispensable that cereal crop productivity increases in the near future. Using current agricultural practices, this increased productivity will require nitrogen inputs, resulting in higher production costs and in irreversible ecological damage. Thus, in the context of global food security and environmental issues such as water pollution and global warming, there is renewed interest in the feasibly of engineering the ability of cereals to fix nitrogen from the atmosphere (Mus et al. 2016). This is a challenge that includes recognition, infection, intracellular accommodation of the nitrogen-fixing endosymbionts, and nodule organogenesis in which plants provide an appropriate niche and carbohydrate nutrition to bacteria in exchange for fixed nitrogen.

The findings that CSP genes in legume, *Parasponia*, and actinorhizal plants may have evolved in part from a pre-existing pathway that regulates the more ancient AM-symbiosis suggest that most of the genes required for root nodule symbiosis are already *present* in cereals. Indeed, several rice orthologs of symbiotic legume genes have been identified; these include *NFR1*, *NFR5*, *SYMRK/DMI2*, *CASTOR*, *POLLUX/DMI1*, *DMI3*, *MtNSP1*, and *MtNSP2* (Zhu et al. 2006). Moreover, it has been shown that most of the rice orthologs are able to restore AM symbiosis in legume mutants that fail to develop AM (reviewed in Nakagawa and Imaizumi-Anraku 2015). These data indicate functional conservation of these genes between rice and legumes. In addition, several rice orthologs of legume CSP genes such as rice *CCaMK* were shown to be able to restore nodulation in a legume mutant (Godfroy et al. 2006). Although most nodules did not contain rhizobia, and bacteria were not released from the infection threads, these data show that elements of the rice genetic program can trigger the appropriate downstream signaling pathway leading to legume nodulation. In contrast to other rice CSP genes, the rice *SYMRK* that encodes a symbiotic leucine-rich repeat (LRR) receptor kinase rescued the AM legume mutant but failed to rescue the nodulation-defective mutant (Markmann et al. 2008). Interestingly, it was shown that the rice SYMRK was shorter than the legume orthologs, suggesting that acquisition of SYMRK contributed to the evolution of root nodule symbiosis from a pre-existing signaling network from AM (Markmann et al. 2008).

In conclusion, legumes and a number of non-legume plants have developed a number of strategies for NO_3 acquisition to cope with nitrogen deficiency. The most efficient biological nitrogen fixation systems known are root nodule endosymbioses between *Rhizobium* legume and *Parasponia* sp. and *Frankia*-actinorhizal plants. Exploring basic mechanisms of nodulation in model legumes and non-legumes provides insights into the evolution and diversity of symbioses. Interactions between plants and nitrogen-fixing bacteria have an ancient history that led to several evolutionary patterns for both endosymbionts and host plants and resulted in specific symbiotic mechanisms. For example, on the bacteria side, in contrast to most rhizobia, bradyrhizobia and *Frankia* are able to induce nod factor-independent symbiosis using an as yet unknown mechanism. On the plant side, several genes that are important for root nodulation are derived from the AM symbiosis, suggesting that root nodule symbioses have been partially recruited from the more ancient AM process. Thus, although microorganisms feature a wide range of symbiotic determinants, N2-fixing plants appear to share a similar symbiotic genetic program. So far, most basic research has been conducted on the model legumes *M. truncatula* and *L. japonicus* and a few non-legume symbiotic plants such as *Parasponia* and *Casuarina*. Next-generation genomics will include *non-model* nodulating plant species and their non-nodulating close relatives and should make it possible to identify the evolutionary inventions that led to nodulation.

5.6 Interactions Between Prokaryotes and Man

Microorganisms preceded on earth all more complex life forms including insects and metazoans, such that outer and inner surfaces of the latter could constitute ecological niches immediately colonized by the former. This is true for man and the simian ancestors of *Homo* spp., which coevolved with a complex consortium of microbes that in turn became tightly adapted. Therefore, humans are a metaorganism where microbes are found on/in skin and hair, all the gastrointestinal tract (GIT) from the mouth to the anus, the lungs, the external auditory canal, the nostril, the eye, or the vagina. Considering the large variations in the physical-chemical of these "locations, encompassing also availability of O_2 and types/quantities of nutrients, it is not surprising that they are of very different compositions and densities, differing also in their intrinsic diversity. Advances have been recently made in our knowledge of this microbial world within and on us, mainly due to benefits taken from technological progresses and decreases in their cost (Claesson et al. 2017; Mondot and Lepage 2016): more specifically, next-generation sequencing and large data analyses have circumvented some limitations of usual microbiological methods (time-consuming, non-cultivability of a significant part of these host-adapted microbes, especially for gut members where anaerobic conditions prevail), and coordinated efforts of large international research programs have been deployed, like, for example, the US Human Microbiome Project (HMP) (Human Microbiome Project 2012) or the European Metagenomics of the Human Intestinal Tract (MetaHIT) (Qin et al. 2010). Figure 5.21a illustrates the large microbial diversity encountered at various body locations, as represented here for bacteria (the main constituents) at the phylum level. The term "microbiome" has two different definitions: the first one, which will be the one used here, comes from ecology and refers to the different microorganisms which live in a particular environment. A second more recent use of the word "microbiome" refers to the genomes of all microorganisms present in a particular ecosystem and considered as a whole, as deduced mainly from metagenomic studies.

5.6.1 *From a Sterile Human to a Metaorganism*

It is generally accepted that the dense and diverse microbial community associated with humans is acquired starting from the very moment of birth as the fetus is kept sterile throughout pregnancy due to the barrier property of the placenta. An inoculation by maternal transmission in utero, i.e., before birth, is however suggested (e.g., Jimenez et al. 2008). Initial colonization is likely a strong determinant of the microbial composition, at least in the GIT, that will imprint the host and help to educate the immune system (Gomez de Aguero et al. 2016). In fact, the host has to develop the mucosal immune system to achieve the required equilibrium between a tolerance toward the gut microbiota, i.e., to not induce excessive/deleterious immune responses, but sufficient to control an overgrowth/a translocation from the GIT onto the human body. This initial colonization will later lead on to these hundred trillions of microbes in adult and depends to which microbes humans are first exposed and therefore depends notably on the location, the delivery mode (caesarean vs vaginally delivered), and the feeding of the newborn (breast-fed vs formula-fed). While vaginally delivered newborns show bacterial communities related to their mother's vaginal ones (*Lactobacillus* and *Prevotella* spp.), caesarean delivery leads to bacterial communities typically found on the skin (e.g., *Corynebacterium*, *Propionibacterium*, and *Staphylococcus* spp.) (Dominguez-Bello et al. 2016). Concerning the GIT, a mother-to-GIT infant transfer

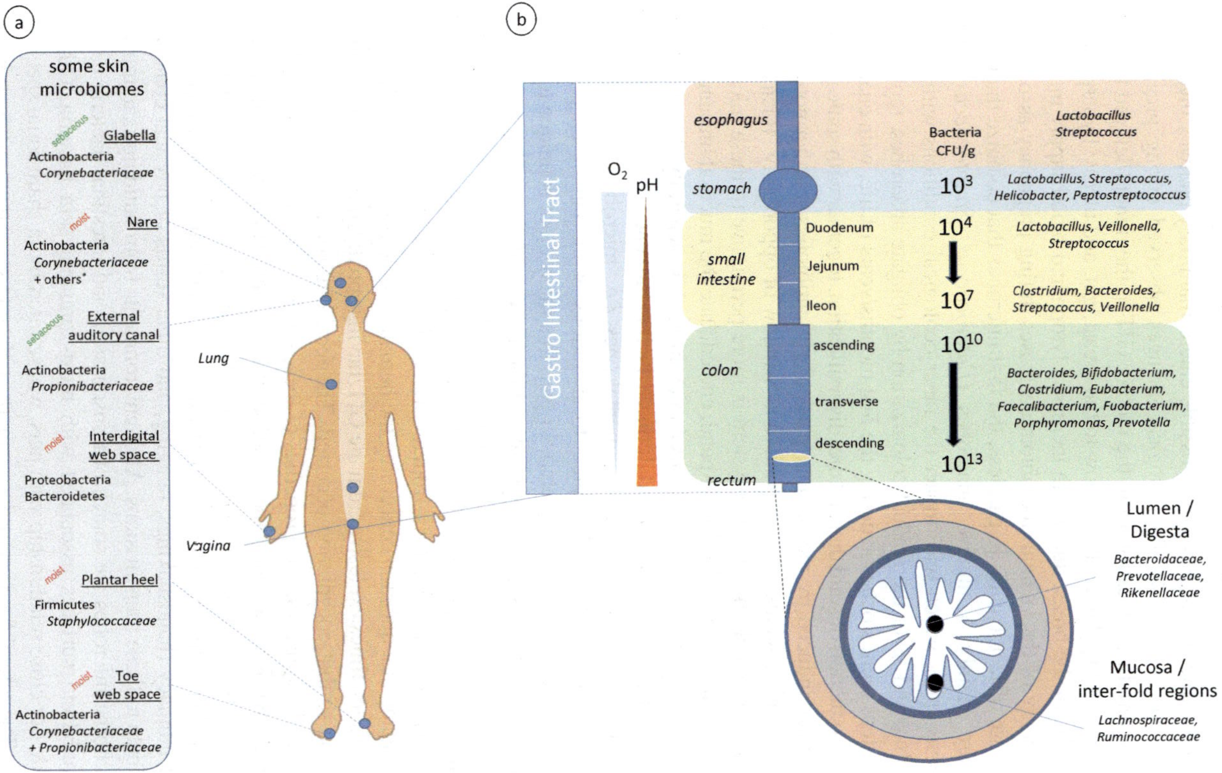

Fig. 5.21 Some examples of the biogeography of microbiotas in humans. (**a**)- at left, the main phylum and family of skin microbiotas are exemplified based on different areas of skin. (**b**)- at center, quantities (in colony forming unit CFU per gram of content) and main genera of bacteria of the gastro-intestinal tract are indicated in the different anatomical parts. (**c**)- at bottom right, the differences in microbial composition in a section of the colon are shown. (Adapted from Costello et al. 2009; Spor et al. 2011; Sanford and Gallo 2013; Donaldson et al. 2016; Mondot and Lepage 2016; *indicate *Actinobacteria* which are not either *Corynebacteriaceae*, *Micrococcaceae* or *Propionibacteriaceae*)

of microbes happens through breast-feeding, since milk and colostrum in addition to sebaceous skin harbor bacteria (mainly staphylococci and streptococci in addition to bifidobacteria, corynebacteria, lactic acid bacteria, and propionibacteria): some of these milk bacteria could in fact originate not only from the mother sebaceous skin but also from her gut microbiota, through a possible entero-mammary pathway involving phagocytes (Rodriguez 2014). Physiological changes observed during pregnancy could favor bacterial translocation from the mother GIT to the infant (delayed transit time/decreased motility; higher blood flow to the breast, gut, and uterus; hyperplasia; and hypertrophy of the breast ducts with increased lymph and blood supply mainly). Also, the diversity of the vaginal microbiota decreases during pregnancy associated with an increase of lactobacilli. Regardless, the first microbes that colonize the newborn GIT are mainly facultative anaerobes (*Enterococcus*, *Streptococcus*, and *Staphylococcus* spp.), which will impact the inner ecological conditions within the GIT (anaerobia) and favor the gradual colonization by obligate anaerobes, replacing them. These steps seem stochastic during the first months/years of life, with changing affected by introduction of solid diet and diet diversification, together with some potent dramatic changes induced by medication, particularly antibiotics. Furthermore, this early life colonization seems to correspond to a "window of opportunity" (Gensollen et al. 2016) during which the immune system is more permissive to these microbes, which, in turn, would help to develop, expand, and educate the mucosal immune system, leading to "learn" to tolerate commensal microbiota. It is generally assumed that the gut microbiota evolved next to an adult one, which is influenced by many factors (environmental ones, but also non-environmental ones, i.e., microbial and host factors), supposed mainly stable and resilient to alterations, while largely adapting to changes (diet, medication, infections, etc.). In fact, this microbiota is unique to each human host, with an increase diversity observed throughout lifespan. The way these new microbes are added and selected remains hypothetical and mainly extrapolated from the route of transmission of pathogens (Browne et al. 2017), supposing a likely commonality. This transmission (exit from a donor and colonization of a recipient) is however a fundamental question concerning the human commensals, mainly obligate anaerobes, and sometimes exclusive from the GIT, having spread and stably colonized hosts over time so that coevolution occurred. The fecal-oral route is very likely the principal one and implies various properties of these commensals and among them a survival in the environment and/or the presence of reservoirs. Humans, especially at the level of the family unit, are of course the main reservoir as previously mentioned through the mother skin, vagina, and breast milk, but social interactions, their frequency, and intimacy (handshakings, hugging, and other forms of physical contact) likely contribute to these transmissions. Also, environments from buildings and transport with airborne microbes are source of transmission, and intestinal bacteria have been retrieved from surfaces associated with most-dominant skin bacteria. Other routes include our close relationship with pets, whose microbiota share some species with humans. The role of food seems less important, despite some dairy products host some bacteria also retrieved from the human gut, while the role of water (important as an environmental reservoir of pathogens) remains currently unknown (Browne et al. 2017). However, considering obligate anaerobes, the survival in the environment (as well as in the upper part of the GIT, encompassing the stomach with is highly acidic pH) is primordial for efficient spreading among humans. This can be achieved by three main strategies developed by microbes, encompassing sporulation (several gut-associated microbes, up to 30%, are spore-forming bacteria: *Clostridiaceae*, *Lachnospiraceae*, *Peptostreptococcaceae*, *Ruminococcaceae*), viable but non-culturable dormancy (decrease in metabolic activity coupled to a strengthened cell wall), and aerotolerance (mechanisms developed to counterpart the damaging effects of atmospheric oxygen on DNA and proteins). This last strategy is however limited for obligate anaerobes, some being classified as EOS (extremely oxygen-sensitive) which can survive only a few minutes when exposed to atmospheric oxygen, like, for example, *Roseburia* spp. (Browne et al. 2017).

5.6.2 The Human Gut Microbiome

Of all, the intestinal microbiota is far the most studied community: the results from a PubMed search covering recent years (from 2011) gave more than 10, 000 records referring to microbiota (and synonyms) and this body site (Lloyd-Price et al. 2016). This represents two thirds of all microbiota studies, while the other third encompasses mainly the oral microbiota (around 15%), as much as collectively the cutaneous, lung, and urogenital microbiota. The intestinal microbiota will therefore be used throughout this chapter to illustrate our recent knowledge about the mechanisms underlying the interactions between humans and prokaryotes. Possible reasons of such an effort and attraction on the gut rely on the fact that it is known for a long time to be one of the ecosystems with the highest microbial density, whose presence/activity participates to some physiological traits of humans, like nutrition. Moreover, the microbiota in this location is involved or is linked to pathologies encompassing non-digestive ones. It has to be noticed that what we know from the gut microbiota results mainly from fecal studies, for evident ease of sampling (technically and ethically), or from animal studies, basically rodents. These last are sometimes "humanized" by the use of gnotobiotic animals (initially germ-free animals seeded by some microbes or a human fecal microbiota). However, the fecal microbiota differs from the gut microbiota, as it encompasses microbes from other location in the GIT, with very different physical-chemical and trophic conditions, between, for example, the stomach or the colon and its different anatomical and functional parts (e.g., right vs left large intestine), but also within the lumen of the cavity, the diet components in digestion, and on/within more or less deeply the mucus layer that surrounds the epithelium (Fig. 5.21b and c). All of this contributes to many different conditions to which in turn microbes contribute also.

The gut microbiota participates and is necessary for the correct development and functioning of the human organism. It is likely that its functions, initially discovered through germ-free mice almost half a century ago, are still incompletely understood. The GIT mucosa forms the largest human surface physically exposed to the external environment, i.e., microbes. By their presence, commensals prevent the colonization and host tissue dissemination of harmful microbes through the forming of a physical barrier encompassing also the blocking of adherence of some pathogen (e.g., *Listeria*) and through the consumption of nutrients necessary for pathogen growth (Kamada et al. 2013). Furthermore, some of these commensals stimulate the production of antimicrobial peptides by the host or produce them (bacteriocins), as well as some growth inhibition factors resulting from their metabolism (e.g., L-lactate). They actively participate also in the immune development of the host. In fact, morphology and ultrastructural development of the gut is impaired in germ-free animals, especially intestinal epithelial cells which show a lower turnover and altered microvilli formation, while lymphocytes population is decreased: In fact, the gut-associated lymphoid tissues, isolated lymphoid follicles, Peyer's patches, and mesenteric lymph nodes all show altered developments (Round and Mazmanian 2009). Interestingly, some of these impaired development can be restored by adding some bacterial components like polysaccharide A from *Bacteroides fragilis* (Mazmanian et al. 2005).

The gut microbiota has also a nutritional role for the host at various levels. It forms a typical (anaerobic) ecosystem where microbial populations co-participate to the hydrolysis and fermentations of nutrients, resulting from diet constituents nondigested by the host, mainly due to an enzymatic defects of the host for such metabolisms: this is especially the case of some complex sugars (namely, nondigestible carbohydrates like cellulose, resistant starch, or xylans) which will be metabolized by some specialized gut microbiota members like more specifically some *Ruminococcus*, *Roseburia*, *Faecalibacterium*, *Bacteroides*, or *Prevotella* spp.. Also partially nondigested and nonabsorbed nutrients by the host in the upper part of the GIT (encompassing proteins), dead host cells, and mucins from the mucus layer surrounding the intestinal epithelial cells contribute to feed these microbes, which, in turn, will provide some important nutrients for the host through fermentations: especially, the gut microbiota metabolism provides lactate, succinate, and importantly short-chain fatty acids (SCFAs). These lasts are mainly acetate, propionate, and butyrate (roughly in the proportion of 2:1:1) and minor ones, encompassing branched SCFA resulting from amino acid metabolism. Besides important cross-feeding implications in between gut microbes, they have important functions for the host, notably by providing 5 to 10% of the energy retrieval from diet: butyrate feeds the colonocytes (70% of colonocyte need), exerts trophic activity on the epithelia, and helps to inhibit colorectal cancer cells (Wong et al. 2006). Acetate and propionate are absorbed and used as energy source for various organs. Importantly, SCFAs can activate G protein-coupled receptors GPR41 and GPR43, expressed in colon epithelial cells, but also adipocytes, spleen, lymph nodes, and lymphocytes among some other peripheral blood mononuclear cells: SCFAs participate in and/or modulate energy, lipid, and glucose metabolisms, regulate gut hormones that control appetite and satiety (e.g., PYY peptide), and modulate the immune system through notably differentiation of regulatory T cells (T-reg cells). Besides this gut microbial metabolisms/fermentations, the metabolic activity of the gut microbiota can lead to biotransformation of several dietary (e.g., dietary phenolic compounds) or xenobiotic/drug compounds (e.g., inactivation of digoxin). At last, gut microbes provide also vitamins to the host, and particularly vit K (menaquinone (MK) form), together with quite all water-soluble vitamin B (biotin, cobalamin, folate, niacin, pantothenate, pyridoxine, riboflavin, and thiamine) (LeBlanc et al. 2013).

All of this highlight the tight association established through evolution between prokaryotes and man, however originating from a sterile organism at birth.

5.6.3 A Shared Prokaryotic Composition Among Humans?

Using the human gut microbiota as a model, referring to a healthy or normal microbiota across humans is based upon the hypothesis that some microbial taxa would be universally shared among healthy individuals (absence of diagnosed disease); it does not obligatorily imply that the absence of one or more of these members of the core taxa indicate a disease state. It is a challenging task to determine such a healthy microbiota because many factors contribute to the very high variability among humans; it should therefore rely on a large dataset reflecting the human populations (gender, age, geographic area, ethnicity, diets, etc.). Large-scale amplicon sequencing of markers and metagenomics have made such studies possible due to "next-generation sequencing" technologies and have revolutionized our vision of these microbiomes. Such designed studies have to be standardized enough from the sample and the sampling to the bioinformatics and database analyses, to allow strength cross comparisons between studies.

Considering the gut, large datasets are now available, encompassing large cohort studies. The standard composition of the gut microbiota has therefore been evaluated, at least for a specific human population, but the question of the constitution of a standard human "healthy" microbiota remains very hard to answer. Falony et al. (2016) and Zhernakova et al. (2016) analyzed the microbiota composition of a large Western European population in a large-scale cross-sectional study on fecal samples from a confined geographic region (respectively, Flanders, Belgium, Flemish Gut Flora Project and the Netherlands, LifeLines-DEEP project). More than 1100 subjects were included in each studies, giving a solid framework to a description of typical "healthy" gut microbiota, in addition to previous studies on other populations (USA, Europe, China, Papua New Guinea, Tanzania, Amazonian, etc.): in mean, metagenomics indicates that bacteria prevail in the Dutch population (97.6% of the reads) also with archaea (2.2%), while viruses and eukaryotic cells were less abundant (respectively, 0.2% and less than 0.01%) (Zhernakova et al. 2016). Major phyla are *Firmicutes*, *Bacteroidetes*, and *Actinobacteria*, which form typically more than 90% of typical gut microbiota (Fig. 5.21b), with some discrepancies due to populations and methodologies (16S vs metagenomics). Interestingly, when combined with other data from Western countries, Falony et al. (2016) identified a total richness of 664 genera for a total of nearly 4000 subjects, which is however suspected to be still under-sampled: an extrapolation estimated a richness at the genus level of 784 ± 40 genera in the Western population that would require around 45,000 samples for being fully determined. Around 630 different species are predicted in each individual (using the Dutch data). Interestingly, some gut microbes are shared among humans, defining a "core" microbiota, which is however heavily influenced by geographical sampling collections. When defined as genera found in 95% of individuals, this core microbiota was determined to be composed of only 35 genera in one confined geographic area (the Flemish cohort) and represented a mean abundance in each individual of around 90% of total gut microbes. This core microbiota was drastically decreased to 17 genera (which account in mean for less than 75% of the abundance) when the Dutch and Flemish cohorts were combined, despite they correspond to near geographical areas. If combined with other published data from other geographical/ethnic origin populations (UK, USA, Papua New Guinea, Peru, and Tanzania), only 14 genera were found shared in 95% of individuals (Falony et al. 2016).

In short, while being heterogeneous across different geographic/ethnic areas and showing an important interindividual diversity, the gut microbiota evolves rapidly and apparently in a stochastically manner from the birth and less rapidly from the age of 2–3. The genus richness correlates next positively with age (Falony et al. 2016), while a decrease of the abundance of the core microbiota is concomitantly observed. A cross-sectional analysis of gut microbiota of healthy laboratory rats under controlled dietary and environmental conditions give consistency to a model of constant evolution of the gut microbiota during lifespan, with an increased diversity and a decreased abundance of initial microbes. These lasts would be present for most of them all the lifespan and progressively completed by other various ones selected through their functions and metabolic behavior rather than their taxonomy (Flemer et al. 2017). At the last step, aging process also affects this composition notably through host physiological state (O'Toole and Jeffery 2015). Therefore, a healthy microbiota is most likely to rely on its functional potential, which can be addressed through metagenomics, rather than on its strict taxa composition.

5.6.4 *Factors Affecting Colonization*

Environmental, microbial, and host factors contribute to the colonization of humans by prokaryotes and control the evolution of microbiota over the lifespan. Considering the gut microbiome, these factors act of course due to intrinsic properties, as well of the host and of the microbe (see examples below), but the positive or negative selection of one organism is multifactorial and opportunistic. This leads to an important interindividual variation (each individual being considered as a unique metaorganism, apart from his own genetics), but most of differences are partly attributable to abundance variations in members of the core or dominant species (Falony et al. 2016). As mentioned above, initial colonization is likely a strong determinant of the microbial composition, at least in the GIT, that will imprint the host and help to educate the immune system (Gomez de Aguero et al. 2016). In fact, the host has to develop the mucosa immune system to find the right equilibrium between a tolerance toward the gut microbiota, i.e., to not induce excessive/deleterious immune responses, but sufficient to control an overgrowth/a translocation from the GIT onto the human body. This is achieved by secreted antimicrobial peptides (AMPs) like defensins or cathelicidins and specific host receptors, notably the *toll*-like receptors

(TLR), and among them TLR5 (bacterial flagellin sensing) or TLR4 (LPS sensing) (Cario 2010; Ostaff et al. 2013). Geography is a strong distinctive element among human gut microbiota (see above and Yatsunenko et al. 2012), but common different traits have been identified among distinct populations, giving clues that common rules govern the establishment of the gut microbiota all over the world not linked to ethnicity: initially, three so-called enterotypes were identified in Europeans (Danish, French, and Italian), North Americans, and Japanese allowing a categorization into three classes of humans (Arumugam et al. 2011) based on a predominance of some bacteria (*Bacteroides*, *Prevotella*, *Ruminococcus*). This classification is however not stringent as studies indicate gradient with no marked discontinuity across a population (Jeffery et al. 2012) and also the predominance of some other types in large cohort studies (Falony et al. 2016). This indicated however the strong influence of diet, confirmed in various studies (e.g., De Filippo et al. 2010; Claesson et al. 2012). This makes sense as nutrient availability within the gut depends at least partly on the diet. For example, the presence and quantity of nondigestible fiber greatly favors some bacteria, among others, giving a rationale to the use of prebiotics. This is of importance in the early colonization of the gut when the breast-fed newborn has to deal with the human milk composed of a complex mixture of oligosaccharides, some containing *N*-acetylglucosamine and fucose in addition to glucose and galactose. Besides the fact that these sugars share common structural motifs with host intestinal receptors known to bind pathogens and therefore that they likely prevent adhesion of pathogens to epithelium, these modifications render these sugars inaccessible to most intestinal bacteria, except in case of presence of sialidases and fucosidases: bifidobacteria (especially *Bifidobacterium infantis* and *Bifidobacterium bifidum*) are more specifically known to be able to grow on these human milk oligosaccharides (Barile and Rastall 2013), and indeed, these bacteria are the dominant ones in breast-fed infants' intestine. This effect of human milk oligosaccharides (HMO) is described as a bifidogenic activity, one example of the role of diet as a major driving force of the gut microbial composition. Stool consistency (reflecting transit time of dietary compounds in the gut) is another major explanatory factors of the gut microbiota composition, likely through different nutrient availability for gut microbes associated with transit time. Besides drugs (of various nature), another important factor is the red cell count in blood (Falony et al. 2016). Apparently hard to link, this fact likely indicates the influence of molecular oxygen (diffusion into the gut from the epithelium). They are interesting clues that commensal bacteria themselves also control the luminal bioavailability of oxygen (that favor many gut pathogens, as being facultative anaerobes like *Salmonella enterica* or *Escherichia coli*) and how it could work: gut butyrate-producing bacteria sense host epithelial metabolism by PPAR-γ (peroxisome proliferator-activated receptor γ), leading to higher consumption of oxygen through β-oxidation and conversely to lower oxygen in the lumen (Byndloss et al. 2017). Moreover, loss of these butyrate-producing bacteria/lower activation of PPAR-γ induces nitrate formation in the lumen (through activation of host-inducible nitric oxide synthase 2, Nos2), which in turn favors the development of these kinds of pathogens which use nitrate as a respiratory electron acceptor. Therefore, a dialogue and fine tuning exist within the gut between commensals and the host.

The functioning of the gut microbiota relies on various favorable microbial factors and on interactions (encompassing metabolic ones) among microbes, therefore affecting the colonization behavior of each microbe: in addition to already mentioned bifidobacteria and α-fucosases, many other examples highlight the importance of microbial factors for colonization. One can cite the interesting case of the main methanogenic archaeon retrieved from human GIT *Methanobrevibacter smithii*. It shows also a close adaptation to the human GIT, with several genomic and metabolic adaptations helping it to live in the colon like expression of adhesin-like proteins or surface glycans similar to those of the host mucosa (Samuel et al. 2007) and a mutualistic relationship with fiber-degrading/saccharolytic bacteria like *Bacteroides thetaiotaomicron* (Samuel and Gordon 2006). This close coevolution of some gut microbes with their host and specialization to the GIT environment is also illustrated by the environmental distribution of members of the archaeal order *Methanomassiliicoccales*. Members of this methanogenic lineage show two different clades, one being exclusively retrieved from GIT (from insects to humans), while the second thrives in various anaerobic environments (Borrel et al. 2017). It is unknown however if different archaeal species or strains are specialized to one specific host. One of this peculiar GIT-associated archaeal strains, found in humans, possesses also a bile salt hydrolase gene likely acquired through a lateral gene transfer from gut bacteria (Borrel et al. 2014). Host bile salts have antimicrobial properties, and many gut bacteria have specific hydrolases which confer them with resistance to host bile salts, besides also allowing regulation of cholesterol metabolism (Joyce et al. 2014). Also, bacteria thriving in the gut synthesize antimicrobials called bacteriocins which usually target closely related species. These antimicrobial peptides (to which the producer is immune) have a broad diversity of structure and mechanism of action (Cotter et al. 2013).

5.6.5 *Considerations About Gut Microbiota in Health and Disease*

While being known for a long time as inhabitants of human, microbes were seen mostly as commensals. More than one century ago, two new roles for microbes were considered, either beneficial or deleterious. Elie Metchnikoff imagined that there were host-friendly bacteria (in Bulgarian yogurt) that could be used to delay senility, deduced from the observation of healthy longer life in Bulgarians: this opened door to the probiotics concept (Mackowiak 2013). Oppositely, some microbes are known to be the direct causal link with one defined disease, based on the postulates of Koch: these lasts mention that, if the microbe is the causative agent, it must be present in all organisms having the disease and not in healthy organisms; it can be grown in pure culture from an isolate of the diseased organism, and it causes the disease when inoculated into a healthy organism, from which it must be re-isolated, being identical, when the inoculated organism has declared the disease. Many pathogenic bacteria, which our microbiota helps us to fight, obey to these rules, and these postulates formed and form a solid framework for the identification of such pathogens and next the identification of targeted fight against them. These postulates seem however of limited interest considering our current view: some organisms are now recognized as pathobionts, meaning that they live as symbionts and become pathogens under altered circumstances.

Differences in gut microbiota composition are currently observed in several disease states, encompassing both GIT disorders (inflammatory bowel disease, irritable bowel disease, colorectal cancer, coeliac disease) and disorders not located in the GIT: these lasts are very diverse (type 2 diabetes, chronic renal disease, obesity, rheumatoid arthritis, cardiovascular disease, autism spectrum disorder, depression, anxiety, allergy, etc.). For some of these diseases, a causal link or, at least, some mechanisms relying on the gut microbe activities themselves have been proposed (e.g., low-grade inflammation and type 2 diabetes, gut-derived metabolite TMAO, and cardiovascular diseases). In fact, the trillions of microbes living within/on us form organized structures interacting with us, and they have useful and even essential roles for the host. This forms a normobiosis, meaning a supposed stable and right balance of these commensal microbes in the gut associated with a balanced and beneficial response of the host. This normobiosis does not rely on a unique and shared microbial composition among all humans, at least at a taxonomical point of view. Inversely, temporal or permanent disruption of this normobiosis, known as dysbiosis, may generate various alterations leading to a considered disease state.

Overall, the microbial community is now recognized as part of our physiology, mentioning the gut microbiota as "a forgotten organ" (O'Hara and Shanahan 2006). It is constituted of a complex network of prokaryotes highly adapted and interacting together and with us, with host benefits taken from this long-term coevolution. However, its role in health and disease is becoming more and more evident. This questions us about the rapid evolution in our modern way of life, which has likely deeply affected this close relationship elaborated through ages: among these, use of medication (e.g., antibiotics), delivery mode, and feeding of babies, diet, hygiene, etc. are examples of factors that affect our microbiota, especially in its early development steps; this would impact the host comportment toward microbes in the gut for his lifespan and forms a hypothetical origin of improper host reactions later in life. C-section delivery (about one third of births in the USA) is known to be associated with asthma, allergies, and metabolic diseases, leading to experiment the inoculation of newborns immediately after delivery with their mothers' vaginal microbes (Dominguez-Bello et al. 2016): in this pilot study, the first 30-day survey indicated a partial restoration of the gut, oral, and skin microbiota compared to naturally delivered or untreated C-delivered infants: long-term consequences remain to be determined; also, heterologous fecal microbiota transplantation (FMT) is now more and more used with success as treatment of recurrent *Clostridium difficile* infections (CDI). All of this illustrates the importance that our commensals microbes gained collectively in our current view and that we now recognize as being a constitutive and active part of ourselves.

5.7 General Conclusion

Large-scale genomic surveys (e.g., Tara Oceans Expedition, Malaspina Expedition, Human Microbiome Project, etc.) have clearly demonstrated that uncultivated viruses and microbial symbionts largely dominate the diversity and the network of ecological interactions in natural ecosystems, including living organisms (plant, animal, and human) which are now considered as metaorganisms when considering their associated microbes. Prokaryotes form one of the earliest forms of life and likely are the root of eukaryogenesis. As long as this "prokaryote-early" hypothesis is safely supported, it can be considered that the evolution of living beings started with prokaryotes and in interaction with prokaryotes. In this chapter, we have reviewed some of the crucial biological interactions involving prokaryotes and have given an overview of the insights into the evolution of living beings. These interactions are ancient and highly complex, concerning different scales of biological organization, from gene to ecosystem. They can be ranged in three categories depending on whether the prokaryotic symbiont has deleterious, beneficial, or neutral effects on the reproductive success (i.e., a selective value or fitness) of its host. Although these effects can be more complex and sophisticated, shaping ecosystems functioning through major evolutionary forcing processes (Red Queen dynamics, inter-actomics, molecular dialogue, host manipulation, coevolution, effects on food webs, and biogeochemical cycles), it has also been realized, relatively recently, that human, animal, and plant health and the response of these hosts to therapeutics and environmental changes can be significantly modulated by microorganisms living in or on these hosts (i.e., microbiota, microbiome).

However, the bulk of our knowledge of biology and evolution is still based on the minority of free-living cultivable cells (Windsor 1998), which is a pity but a situation that also opens promising avenues for biology and ecology in the future. This should (must) not only be based on the studies of associations between one host and one symbiont as it is widely the case now, but also includes multipartite biological interactions. Theses interactions are not only common but undoubtedly have important organismal, ecological, and evolutionary impacts. We are at the frontiers of background research, scholarship, and engineering in the domain of biological sciences. The insights gained from studying biological interactions, including multipartite interactions, will indeed have far-reaching benefits, ranging from basic understanding of their ecological impacts to the manipulation of these interactions in biotechnology, e.g., improving bioenergetic aquaculture of algae (which has been developing exponentially worldwide over the last decades) or biocontrol strategies against unwanted microorganisms (antibiotic-resistant pathogens, biofilm formations, biofouling microbes, etc.). Overall, we believe that studies of biological interactions are sources of novel knowledge related to the biodiversity of living things, the functioning of ecosystems, the evolution of the cellular world, and the related ecosystem services.

Acknowledgments The authors thank the "Institut de Recherche pour le Développement," the International Atomic Energy Agency (IAEA), and "France Génomique" for their support. This project has received funding from the French ANR under grant agreement ANR-12-BSV7-0019. The initial figures were improved by M. A. Galeron; thanks to our co-author B.O. for the payment of the associated costs.

References

Abramowicz DA (1990) Aerobic and anaerobic biodegradation of PCBs: a review. Crit Rev Biotechnol 10:241–151

Ackermann HW (1999) Tailed bacteriophages : the order Caudovirales. AdvVirus Res 51(135):201. https://doi.org/10.1016/S0065-3527(08)60785-X

Akman L, Rio RV, Beard CB, Aksoy S (2001) Genome size determination and coding capacity of *Sodalis glossinidius*, an enteric symbiont of tsetse flies, as revealed by hybridization to *Escherichia coli* gene arrays. J Bacteriol 183:4517–4525

Akman L, Yamashita A, Watanabe H, Oshima K, Shiba T, Hattori M, Aksoy S (2002) Genome sequence of the endocellular obligate symbiont of tsetse flies, Wigglesworthia glossinidia. Nat Genet 32:402–407

Aksoy S (2000) Tsetse - a haven for microorganisms. Parasitol Today 16:114–118

Aksoy S, Pourhosseini AA, Chow A (1995) Mycetome endosymbionts of tsetse flies constitute a distinct lineage related to *Enterobacteriaceae*. Insect Mol Biol 4:15–22

Aksoy E, Telleria EL, Echodu R, Wu Y, Okedi LM, Weiss BL, Aksoy S, Caccone A (2014) Analysis of multiple tsetse fly populations in Uganda reveals limited diversity and species-specific gut microbiota. Appl Environ Microbiol 80:4301–4312

Alam U, Medlock J, Brelsfoard C, Pais R, Lohs C, Balmand S, Carnogursky J, Heddi A, Takac P, Galvani A, Aksoy S (2011) *Wolbachia* symbiont infections induce strong cytoplasmic incompatibility in the tsetse fly *Glossina morsitans*. PLoS Pathog 7:e1002415

Amann RI, Ludwig W, Schleifer K-H (1995) Phylogenetic identification and in situ detection of individual microbial cells without cultivation. Microbiol Rev 59(1):143–169

Anderson D.M., Cembella A.D., and G. M. Hallegraeff (ed.) (1998). Physiological ecology of harmful algal blooms, vol. G 41. Springer, Berlin, Germany

Aoki S, Ito M, Iwasaki W (2013) From β- to α-Proteobacteria: the origin and evolution of Rhizobial nodulation genes *nodIJ*. Mol Biol Evol 30:2494–2508. https://doi.org/10.1093/molbev/mst153

Arakaki N, Miyoshi T, Noda H (2001) *Wolbachia*-mediated parthenogenesis in the predatory thrips *Franklinothrips vespiformis* (*Thysanoptera*: *Insecta*). Proc R Soc London Ser B 268:1011–1016

Arumugam M, Raes J, Pelletier E, Le Paslier D, Yamada T, Mende DR, Fernandes GR, Tap J, Bruls T, Batto JM, Bertalan M, Borruel N, Casellas F, Fernandez L, Gautier L, Hansen T, Hattori M, Hayashi T, Kleerebezem M, Kurokawa K, Leclerc M, Levenez F, Manichanh C, Nielsen HB, Nielsen T, Pons N, Poulain J, Qin J, Sicheritz Ponten T, Tims S, Torrents D, Ugarte E, Zoetendal EG, Wang J, Guarner F, Pedersen O, de Vos WM, Brunak S, Dore J, Antolin M, Artiguenave F, Blottiere HM, Almeida M, Brechot C, Cara C, Chervaux C, Cultrone A, Delorme C, Denariaz G, Dervyn R, Foerstner KU, Friss C, van de Guchte M, Guedon E, Haimet F, Huber W, van Hylckama-Vlieg J, Jamet A, Juste C, Kaci G, Knol J, Lakhdari O, Layec S, Le Roux K, Maguin E, Merieux A, Melo Minardi R, M'Rini C, Muller J, Oozeer R, Parkhill J, Renault P, Rescigno M, Sanchez N, Sunagawa S, Torrejon A, Turner K, Vandemeulebrouck G, Varela E, Winogradsky Y, Zeller G, Weissenbach J, Ehrlich SD, Bork P (2011) Enterotypes of the human gut microbiome. Nature 473(7346):174–180. https://doi.org/10.1038/nature09944

Azambuja P, Garcia ES, Ratcliffe NA (2005) Gut microbiota and parasite transmission by insect vectors. Trends Parasitol 21:568–572

Baena S, Fardeau ML, Labat M, Ollivier B, Thomas P, Garcia JL, Patel BK (1998) *Aminobacterium colombiense* gen. Nov., sp. nov., an amino acid-degrading anaerobe isolated from anaerobic sludge. Anaerobe 4:241–250

Baena S, Fardeau ML, Ollivier B, Labat M, Thomas P, Garcia JL, Patel BK (1999) *Aminomonas paucivorans* gen. Nov., sp. nov., a mesophilic, anaerobic, amino-acid-utilizing bacterium. Int J Syst Bacteriol 49:975–982

Baena S, Fardeau ML, Labat M, Ollivier B, Garcia JL, Patel BK (2000) *Aminobacterium mobile* sp. nov., a new anaerobic amino-acid-degrading bacterium. Int J Syst Evol Microbiol 50:259–264

Balmand S, Lohs C, Aksoy S, Heddi A (2013) Tissue distribution and transmission routes for the tsetse fly endosymbionts. J Invertebr Pathol 112(Suppl):S116–S122

Bandi C, Trees AJ, Brattig NW (2001) *Wolbachia* in filarial nematodes: evolutionary aspects and implications for the pathogenesis and treatment of filarial diseases. Vet Parasitol 98:215–238

Barile D, Rastall RA (2013) Human milk and related oligosaccharides as prebiotics. Curr Opin Biotechnol 24(2):214–219. https://doi.org/10.1016/j.copbio.2013.01.008

Beam HW, Perry JJ (1974) Microbial degradation of cycloparaffinic hydrocarbons via co-metabolism and commensalism. J Gen Microbiol 82:163–169

Beard CB, Durvasula RV, Richards FF (1998) Bacterial symbiosis in arthropods and the control of disease transmission. Emerg Infect Dis 4:581–591

Beatty JT, Overmann J, Lince MT, Manske AK, Lang AS, Blankenship RE, Van Dover CL, Martinson TA, Plumley FG (2005) An obligately photosynthetic bacterial anaerobe from a deep-sea hydrothermal vent. Proc Natl Acad Sci U S A 102(26):9306–9310

Becking JH (1992) The rhizobium symbiosis of the non-legume Parasponia. In: Stacey G, Burris RH, Evans HJ (eds) Biological nitrogen fixation: achievements and objectives. Chapman & Hall, New York, pp 497–559

Bergthorsson U, Ochman H (1998) Distribution of chromosome length variation in natural isolates of Escherichia coli. Mol Biol Evol 15:6–16

Berleman JE et al (2014) The lethal cargo of *Myxococcus xanthus* outer membrane vesicles. Front Microbiol 5:474. https://doi.org/10.3389/fmicb.2014.00474

Bertrand J-C, Rambeloarisoa E, Rontani J-F, Guisti G, Mattei G (1983) Microbial degradation of crude oil in sea water and continuous culture. Biotechnol Lett 5:567–572

Bertrand J-C, Bonin P, Caumette P, Gattuso J-P, Grégori G, Guyoneaud LRX, Matheron R, Poly F (2015a) Biogeochemical cycles. In: Bertrand JC, Caumette P, Lebaron P, Matheron R, Normand P, Sime-Ngando T (eds) Environmental microbiology: fundamentals and aplications. Springer, Dordrecht/Heidelberg/New York/London, pp 511–617

Bertrand J-C, Doumenq P, Guyoneaud R, Marrot B, Martin-Laurent F, Matheron R, Moulin P, Soulas G (2015b) Applied microbial ecology and bioremediation. Microorganisms as major actors of pollution elimination in the environment. In: Bertrand JC, Caumette P, Lebaron P, Matheron R, Normand P, Sime-Ngando T (eds) Environmental microbiology: fundamentals and aplications. Springer, Dordrecht/Heidelberg/New York/London, pp 659–753

Boller EF, Russ K, Vallo V, Bush GL (1976) Incompatible races of European cherry fruit fly, *Rhagoletis cerasi* (*Diptera, Tephritidae*), their origin and potential use in biological control. Entomol Exp Appl 20:237–247

Bonaldi K, Gourion B, Fardoux J, Hannibal L, Cartieaux F, Boursot M, Vallenet D, Chaintreuil C, Prin Y, Nouwen N, Giraud E (2010) Large-scale transposon mutagenesis of photosynthetic Bradyrhizobium sp. strain ORS278 reveals new genetic loci putatively important for nod-independent symbiosis with *Aeschynomene indica*. Mol Plant-Microbe Interact 23:760–770. https://doi.org/10.1094/MPMI-23-6-0760

Bontemps C, Elliott GN, Simon MF, Dos Reis Júnior FB, Gross E, Lawton RC, Neto NE, De Fatima Loureiro M, De Faria SM, Sprent JI, James EK, Young JP (2010) Burkholderia species are ancient symbionts of legumes. Mol Ecol 19:44–52. https://doi.org/10.1111/j.1365-294X.2009.04458.x

Bordenave S, Goñi-Urriza MS, Caumette P, Duran R. 2007. Effects of heavy fuel oil on the bacterial community structure of a pristine microbial mat. Appl Environ Microbiol Oct;73(19):6089–6097. Epub 2007 Aug 17

Bordenstein SR, Fitch DHA, Werren JH (2003) Absence of *Wolbachia* in nonfilariid nematodes. J Nematol 35:266–270

Borrel G, Parisot N, Harris HM, Peyretaillade E, Gaci N, Tottey W, Bardot O, Raymann K, Gribaldo S, Peyret P, O'Toole PW, Brugere JF (2014) Comparative genomics highlights the unique biology of Methanomassiliicoccales, a Thermoplasmatales-related seventh order of methanogenic archaea that encodes pyrrolysine. BMC Genomics 15:679. https://doi.org/10.1186/1471-2164-15-679

Borrel G, McCann A, Deane J, Neto MC, Lynch DB, Brugere JF, O'Toole PW (2017) Genomics and metagenomics of trimethylamine-utilizing archaea in the human gut microbiome. ISME J 11(9):2059–2074. https://doi.org/10.1038/ismej.2017.72

Bressac C, Rousset F (1993) The reproductive incompatibility system in *Drosophila simulans*: DAPI-staining analysis of the *Wolbachia* symbionts in sperm cysts. J Invertebr Pathol 61:226–230

Bronstein JL (1994) Conditional outcomes in mutualistic interactions. Trends Ecol Evol 9:214–217

Browne HP, Neville BA, Forster SC, Lawley TD (2017) Transmission of the gut microbiota: spreading of health. Nat Rev Microbiol 15(9):531–543. https://doi.org/10.1038/nrmicro.2017.50

Brownlie JC, Cass BN, Riegler M, Witsenburg JJ, Iturbe-Ormaetxe I, McGraw EA, O'Neill SL (2009) Evidence for metabolic provisioning by a common invertebrate endosymbiont, *Wolbachia pipientis*, during periods of nutritional stress. PLoS Pathog 5:e1000368

Brüssow H, Hendrix RW (2002) Phagegenomics : small is beautiful. Cell 108:13–16. https://doi.org/10.1016/S0092-8674(01)00637-7

Bryant MP, Wolin EA, Wolin MJ, Wolfe RS (1967) *Methanobacillus omelianskii,* a symbiotic association of two species of bacteria. Arch Microbiol 59:20–31

Byndloss MX, Olsan EE, Rivera-Chavez F, Tiffany CR, Cevallos SA, Lokken KL, Torres TP, Byndloss AJ, Faber F, Gao Y, Litvak Y, Lopez CA, Xu G, Napoli E, Giulivi C, Tsolis RM, Revzin A, Lebrilla CB, Baumler AJ (2017) Microbiota-activated PPAR-gamma signaling inhibits dysbiotic Enterobacteriaceae expansion. Science 357(6351):570–575. https://doi.org/10.1126/science.aam9949

Cannon SB, McKain MR, Harkess A, Nelson MN, Dash S, Deyholos MK, Peng Y, Joyce B, Stewart CN, Rolf M et al (2014) Multiple polyploidy events in the early radiation of nodulating and non-nodulating legumes. Mol Biol Evol 32:193–210. https://doi.org/10.1093/molbev/msu296

Cario E (2010) Toll-like receptors in inflammatory bowel diseases: a decade later. Inflamm Bowel Dis 16(9):1583–1597. https://doi.org/10.1002/ibd.21282

Caron DA, Goldman JC, Dennett MR (1988) Experimental demonstration of the roles of bacteria and bacterivorous protozoa in plankton nutrient cycles. Hydrobiologia 159:27–40

Casiraghi M, Anderson TJ, Bandi C, Bazzocchi C, Genchi C (2001) A phylogenetic analysis of filarial nematodes: comparison with the phylogeny of *Wolbachia* endosymbionts. Parasitology 122:93–103

Cavalier-Smith T (2002) The phagotrophic origin of eukaryotes and phylogenetic classification of protozoa. Int J Syst Evol Microbiol 52:297–354

Cérémonie H, Debellé F, Fernandez MP (1999) Structural and functional comparison of *Frankia* root hair deforming factor and rhizobia nod factor. Can J Bot 77:1293–1301. https://doi.org/10.1139/b99 060

Cerqueda-García D, Martínez-Castilla LP, Falcón LI, Delaye L (2014) Metabolic analysis of *Chlorobium chlorochromatii* CaD3 reveals clues of the symbiosis in '*Chlorochromatium aggregatum*. ISME J 8:991–998

Chabaud M, Gherbi H, Pirolles E, Vaissayre V, Fournier J, Moukouanga D, Franche C, Bogusz D, Tisa LS, Barker DG, Svistoonoff S (2016) Chitinase-resistant hydrophilic symbiotic factors secreted by *Frankia* activate both Ca^{2+} spiking and *NIN* gene expression in the actinorhizal plant *Casuarina glauca*. New Phytol. https://doi.org/10.1111/nph.13732

Charlat S, Hornett EA, Dyson EA, Ho PP, Loc NT, Schilthuizen M, Davies N, Roderick GK, Hurst GD (2005) Prevalence and penetrance variation of male-killing *Wolbachia* across indo-Pacific populations of the butterfly *Hypolimnas bolina*. Mol Ecol 14:3525–3530

Charlat S, Hornett EA, Fullard JH, Davies N, Roderick GK, Wedell N, Hurst GD (2007) Extraordinary flux in sex ratio. Science 317:214

Chen X, Li S, Aksoy S (1999) Concordant evolution of a symbiont with its host insect species: molecular phylogeny of genus *Glossina* and its bacteriome-associated endosymbiont, *Wigglesworthia glossinidia*. J Mol Evol 48:49–58

Cheng TC (1991) Is parasitism symbiosis ? A definition of terms and the evolution of concepts. In: Toft CA, Aeschlimann A, Bolis L (eds) Parasite-host associations, coexistence or conflict? Oxford University Press, New York, pp 15–36

Cheng Q, Aksoy S (1999) Tissue tropism, transmission and expression of foreign genes in vivo in midgut symbionts of tsetse flies. Insect Mol Biol 8:125–132

Cheng Q, Ruel TD, Zhou W, Moloo SK, Majiwa P, O'Neill SL, Aksoy S (2000) Tissue distribution and prevalence of *Wolbachia* infections in tsetse flies, *Glossina* spp. Med Vet Entomol 14:44–50

Chivian D, Brodie EL, Alm EJ, Culler DE, Dehal PS, Desantis TZ, GihringI TM, Lapidus A, Lin LH, Lowry SR, Moser DP, Richardson PM, Southam G, Wanger G, Pratt LM, Andersen GL, Hazen TC, Brockman FJ, Adam P, Arkin AP, Onstott TC (2008) Environmental genomics reveals a single-species ecosystem deep within earth. Science 322(5899):275–278. https://doi.org/10.1126/science.1155495

Cirimotich CM, Dong Y, Clayton AM, Sandiford SL, Souza-Neto JA, Mulenga M, Dimopoulos G (2011) Natural microbe-mediated refractoriness to *Plasmodium* infection in *Anopheles gambiae*. Science 332:855–858

Claesson MJ, Jeffery IB, Conde S, Power SE, O'Connor EM, Cusack S, Harris HM, Coakley M, Lakshminarayanan B, O'Sullivan O, Fitzgerald GF, Deane J, O'Connor M, Harnedy N, O'Connor K, O'Mahony D, van Sinderen D, Wallace M, Brennan L, Stanton C, Marchesi JR, Fitzgerald AP, Shanahan F, Hill C, Ross RP, O'Toole PW (2012) Gut microbiota composition correlates with diet and health in the elderly. Nature 488(7410):178–184. https://doi.org/10.1038/nature11319

Claesson MJ, Clooney AG, O'Toole PW (2017) A clinician's guide to microbiome analysis. Nat Rev Gastroenterol Hepatol 14(10):585–595. https://doi.org/10.1038/nrgastro.2017.97

Clark EL, Karley AJ, Hubbard SF (2010) Insect endosymbionts: manipulators of insect herbivore trophic interactions. Protoplasma 244:25–51

Clavijo F, Diedhou I, Vaissayre V, Brottier L, Ascolate J, Moukouanga D, Crabos A, Auguy F, Franche C, Gherbi H, Champion A, Hocher V, Barker D, Bogusz D, Tisa L, Svistoonoff S (2015) The Casuarina *NIN* gene is transcriptionally activated throughout *Frankia* root infection as well as in response to bacterial diffusible signals. New Phytol 208:887–903. https://doi.org/10.1111/nph.13506

Clerissi C, Desdevises Y, Grimsley N (2012) Prasinoviruses of the marine green alga *Ostreococcus tauri* : are mainly species specific. J Virology 86:4611–4619

Coburn PS, Gilmore MS (2003) The *Enterococcus faecalis* cytolysin: a novel toxin active against eukaryotic and prokaryotic cells. Cell Microbiol 5:661–669

Cornelis GR (2000) Type III secretion: a bacterial device for close combat with cells of their eukaryotic host. Philos Trans R Soc Lond Ser B Biol Sci 355:681–693

Costello EK, Lauber CL, Hamady M, Fierer N, Gordon JI, Knight R (2009) Bacterial community variation in human body habitats across space and time. Science 326(5960):1694–1697. https://doi.org/10.1126/science.1177486

Cotter PD, Ross RP, Hill C (2013) Bacteriocins - a viable alternative to antibiotics? Nat Rev Microbiol 11(2):95–105. https://doi.org/10.1038/nrmicro2937

Dale C, Maudlin I (1999) *Sodalis* gen. Nov. and *Sodalis glossinidius* sp. nov., a microaerophilic secondary endosymbiont of the tsetse fly *Glossina morsitans morsitans*. Int J Syst Bacteriol 1:267–275

Darby AC, Lagnel J, Matthew CZ, Bourtzis K, Maudlin I, Welburn SC (2005) Extrachromosomal DNA of the symbiont *Sodalis glossinidius*. J Bacteriol 187:5003–5007

Davidov Y, Jurkevitch E (2009) Predation between prokaryotes and the origin of eukaryotes. BioEssays 31:748–757. https://doi.org/10.1002/bies.200900018

Dawson JO (2007) Ecology of Actinorhizal plants. In: Pawlowski K, Newton WE (eds) Nitrogen-fixing Actinorhizal symbioses. Nitrogen fixation: origins, applications, and research Progress, vol 6. Springer, Dordrecht

De Filippo C, Cavalieri D, Di Paola M, Ramazzotti M, Poullet JB, Massart S, Collini S, Pieraccini G, Lionetti P (2010) Impact of diet in shaping gut microbiota revealed by a comparative study in children from Europe and rural Africa. Proc Natl Acad Sci U S A 107(33):14691–14696. https://doi.org/10.1073/pnas.1005963107

De León, K.B., Zane, G.M., Trotter, V.V., Krantz, G.P., Arkin, A.P., Butland, G.P., Walian, P.J., Fields, M.W., Wall, J.D. (2017). Unintended laboratory-driven evolution reveals genetic requirements for biofilm formation by *desulfovibrio vulgaris* Hildenborough (2017) MBio, 8 (5), art. no. e01696–17

De Vooght L, Caljon G, Stijlemans B, De Baetselier P, Coosemans M, Van den Abbeele J (2012) Expression and extracellular release of a functional anti-trypanosome Nanobody® in *Sodalis glossinidius*, a bacterial symbiont of the tsetse fly. Microb Cell Factories 11:23

De Vooght L, Caljon G, De Ridder K, Van Den Abbeele J. 2014. Delivery of a functional anti-trypanosome Nanobody in different tsetse fly tissues via a bacterial symbiont, *Sodalis glossinidius*. Microb Cell Fact. 13:156

De Vooght L, Caljon G, Van Hees J, Van Den Abbeele J (2015) Paternal transmission of a secondary Symbiont during mating in the viviparous tsetse Fly. Mol Biol Evol 32:1977–1980

Dedeine F, Vavre F, Fleury F, Loppin B, Hochberg ME, Bouletreau M (2001) Removing symbiotic *Wolbachia* bacteria specifically inhibits oogenesis in a parasitic wasp. Proc Natl Acad Sci U S A 98:6247–6252

Dedeine F, Vavre F, Shoemaker DD, Bouletreau M (2004) Intra-individual coexistence of a *Wolbachia* strain required for host oogenesis with two strains inducing cytoplasmic incompatibility in the wasp *Asobara tabida*. Evolution 58:2167–2174

Degnan PH, Moran NA (2008a) Evolutionary genetics of a defensive facultative symbiont of insects: exchange of toxin-encoding bacteriophage. Mol Ecol 17:916–929

Degnan PH, Moran NA (2008b) Diverse phage-encoded toxins in a protective insect endosymbiont. Appl Environ Microbiol 74:6782–6791

Degnan PH, Yu Y, Sisneros N, Wing RA, Moran NA (2009) *Hamiltonella defensa*, genome evolution of protective bacterial endosymbiont from pathogenic ancestors. Proc Natl Acad Sci U S A 106:9063–9068

Diagne N, Diouf D, Svistoonoff S, Kane A, Noba K, Franche C et al (2013) *Casuarina* in Africa: distribution, role and importance of arbuscular mycorrhizal, ectomycorrhizal fungi and Frankia on plant development. J Environ Manag 128:204–209. https://doi.org/10.1016/j.jenvman.2013.05.009

Diédhiou I, Tromas A, Cissoko M, Gray K, Parizot B, Crabos A, Alloisio N, Fournier P, Carro L, Svistoonoff S, Gherbi H, Hocher V, Diouf D, Laplaze L, Champion A (2014) Identification of potential transcriptional regulators of actinorhizal symbioses in *Casuarina glauca* and *Alnus glutinosa*. BMC Plant Biol 14:342. https://doi.org/10.1186/s12870-014-0342-z

Dobson SL (2003) Reversing *Wolbachia* based population replacement. Trends Parasitol 19:128–133

Dobson SL, Bourtzis K, Braig HR, Jones BF, Zhou W, Rousset F, O'Neill SL (1999) *Wolbachia* infections are distributed throughout insect somatic and germ line tissues. Insect Biochem Mol Biol 29:153–160

Dobson SL, Marsland EJ, Rattanadechakul W (2002) Mutualistic *Wolbachia* infection in *Aedes albopictus*: accelerating cytoplasmic drive. Genetics 160:1087–1094

Dominguez-Bello MG, De Jesus-Laboy KM, Shen N, Cox LM, Amir A, Gonzalez A, Bokulich NA, Song SJ, Hoashi M, Rivera-Vinas JI, Mendez K, Knight R, Clemente JC (2016) Partial restoration of the microbiota of cesarean-born infants via vaginal microbial transfer. Nat Med 22(3):250–253. https://doi.org/10.1038/nm.4039

Donaldson GP, Lee SM, Mazmanian SK (2016) Gut biogeography of the bacterial microbiota. Nat Rev Microbiol 14(1):20–32. https://doi.org/10.1038/nrmicro3552

Dong Y, Manfredini F, Dimopoulos G (2009) Implication of the mosquito midgut microbiota in the defense against malaria parasites. PLoS Pathog 5:e1000423

Doudoumis V, Tsiamis G, Wamwiri F, Brelsfoard C, Alam U, Aksoy E, Dalaperas S, Abd-Alla A, Ouma J, Takac P, Aksoy S, Bourtzis K (2012) Detection and characterization of *Wolbachia* infections in laboratory and natural populations of different species of tsetse flies (genus *Glossina*). BMC Microbiol 12 Suppl 1:S3

Douglas AE (1998) Nutritional interactions in insect-microbial symbioses: aphids and their symbiotic bacteria *Buchnera*. Annu Rev Entomol 43:17–37

Downie JA (2014) Legume nodulation. Curr Biol 24:184–190. https://doi.org/10.1016/j.cub.2014.01.028

Doyle JJ (1998) Phylogenetic perspectives on nodulation: an evolving views of plants and symbiotic bacteria. Trends Plant Sci 3:473–478. https://doi.org/10.1016/S1360-1385(98)01340-5

Doyle JJ (2011) Phylogenetic perspectives on the origins of nodulation. Mol Plant-Microbe Interact 24:1289–1295

Doyle JJ, Luckow MA (2003) The rest of the iceberg. Legume diversity and evolution in a phylogenetic context. Plant Physiol 131:900–910. https://doi.org/10.1104/pp.102.018150

Dragoš A, Lakshmanan N, Martin M, Horváth B, Maróti G, Falcón García C, Lieleg O, Kovács ÁT (2018) Evolution of exploitative interactions during diversification in Bacillus subtilis biofilms. FEMS Microbiol Ecol 94(1)

Dubey GP, Ben-Yehuda S (2011) Intercellular nanotubes mediate bacterial communication. Cell 144:590–600

Duron O (2014) *Arsenophonus* insect symbionts are commonly infected with APSE, a bacteriophage involved in protective symbiosis. FEMS Microbiol Ecol 90:184–194

Duron O, Bouchon D, Boutin S, Bellamy L, Zhou L, Engelstädter J, Hurst GD (2008) The diversity of reproductive parasites among arthropods: *Wolbachia* do not walk alone. BMC Biol 6:27

Dyer KA, Jaenike J (2004) Evolutionarily stable infection by a male-killing endosymbiont in *Drosophila innubila*: molecular evidence from the host and parasite genomes. Genetics 168:1443–1455

Dyer KA, Minhas MS, Jaenike J (2005) Expression and modulation of embryonic male killing in *Drosophila innubila*: opportunities for multilevel selection. Evolution 59:838–848

Ebert D, Herre EA (1996) The evolution of parasitic diseases. Parasitol Today 12:96–101

Echaubard P, Duron O, Agnew P, Sidobre C, Noël V, Weill M, Michalakis Y (2010) Rapid evolution of *Wolbachia* density in insecticide resistant *Culex pipiens*. Heredity 104:15–19

Eloe-Fadrosh EA, Ivanova NN, Woyke T, Kyrpides NC (2016) Metagenomics uncovers gaps in amplicon-based detection of microbial diversity. Nature Microbiol, 15032, doi:https://doi.org/10.1038/nmicrobiol.2015.32

El-Sayed NM, Myler PJ, Blandin G, Berriman M, Crabtree J, Aggarwal G, Caler E, Renauld H, Worthey EA, Hertz-Fowler C, Ghedin E, Peacock C, Bartholomeu DC, Haas BJ, Tran AN, Wortman JR, Alsmark UC, Angiuoli S, Anupama A, Badger J, Bringaud F, Cadag E, Carlton JM, Cerqueira GC, Creasy T, Delcher AL, Djikeng A, Embley TM, Hauser C, Ivens AC, Kummerfeld SK, Pereira-Leal JB, Nilsson D, Peterson J, Salzberg SL, Shallom J, Silva JC, Sundaram J, Westenberger S, White O, Melville SE, Donelson JE, Andersson B, Stuart KD, Hall N (2005) Comparative genomics of trypanosomatid parasitic protozoa. Science 309:404–409

Engelstadter J, Hurst GDD (2009) The ecology and evolution of microbes that manipulate host reproduction. Annual Review of Ecology. Evol Syst 40:127–149

Engelstadter J, Telschow A (2009) Cytoplasmic incompatibility and host population structure. Heredity 103:196–207

Fadhlaoui K, Ben-Hania W, Armougom F, Bartoli M, Fardeau ML, Erauso G, Brasseur G, Aubert C, Hamdi M, Brochier-Armanet C, Dolla A, Ollivier B (2018) Obligate sugar oxidation in Mesotoga spp., phylum Thermotogae, in the presence of either elemental sulfur or hydrogenotrophic sulfate-reducers as electron acceptor. Environ Microbiol 20:281–292

Falony G, Joossens M, Vieira-Silva S, Wang J, Darzi Y, Faust K, Kurilshikov A, Bonder MJ, Valles-Colomer M, Vandeputte D, Tito RY, Chaffron S, Rymenans L, Verspecht C, De Sutter L, Lima-Mendez G, D'Hoe K, Jonckheere K, Homola D, Garcia R, Tigchelaar EF, Eeckhaudt L, Fu J, Henckaerts L, Zhernakova A, Wijmenga C, Raes J (2016) Population-level analysis of gut microbiome variation. Science 352(6285):560–564. https://doi.org/10.1126/science.aad3503

Farikou O, Njiokou F, Mbida Mbida JA, Njitchouang GR, Djeunga HN, Asonganyi T, Simarro PP, Cuny G, Geiger A (2010) Tripartite interactions between tsetse flies, *Sodalis glossinidius* and trypanosomes--an epidemiological approach in two historical human African trypanosomiasis foci in Cameroon. Infect Genet Evol 10:115–121

Farikou O, Thevenon S, Njiokou F, Allal F, Cuny G, Geiger A (2011) Genetic diversity and population structure of the secondary symbiont of tsetse flies, *Sodalis glossinidius*, in sleeping sickness foci in Cameroon. PLoS Negl Trop Dis 5:e1281

Faust K, Raes J (2012) Microbial interactions: from networks to models. Nat Rev Microbiol 10:538–550

Fay P (1992) Oxygen relations of nitrogen fixation in cyanobacteria. Microbiol Rev 56:340–373

Fazli M, Almblad H, Rybtke ML, Givskov M, Eberl L, Tolker-Nielsen T (2014) Regulation of biofilm formation in Pseudomonas and Burkholderia species. Environ Microbiol 16(7):1961–1981

Fialho RF, Stevens L (2000) Male-killing *Wolbachia* in a flour beetle. Proc R Soc B Biol Sci 267:1469–1473

Flemer B, Gaci N, Borrel G, Sanderson IR, Chaudhary PP, Tottey W, O'Toole PW, Brugere JF (2017) Fecal microbiota variation across the lifespan of the healthy laboratory rat. Gut Microbes 8(5):428–439. https://doi.org/10.1080/19490976.2017.1334033

Fourçans A, Ranchou-Peyruse A, Caumette P, Duran R (2008) Molecular analysis of the spatio-temporal distribution of sulfate-reducing bacteria (SRB) in Camargue (France) hypersaline microbial mat. Microb Ecol 56(1):90–100

Franche C, Lindstrom K, Elmerich C (2009) Nitrogen-fixing bacteria associated with leguminous and non-leguminous plants. Plant Soil 321:35–59. https://doi.org/10.1007/s11104-008-9833-8

Frank SA (1996) Host control of symbiont transmission. The separation of symbionts into germ and soma. Am Nat 148:1113–1124

Frimmer U, Widdel F (1989) Oxidation of ethanol by methanogenic bacteria. Arch Microbiol 152:479–483

Fröstl MJ, Overmann J (1998) Physiology and tactic response of the phototrophic consortium *chlorochromatium aggregatum*. Arch Microbiol 169:129–135

Froussart E, Bonneau J, Franche C, Bogusz D (2016) Recent advances in actinorhizal symbiosis signaling. Plant Mol Biol. https://doi.org/10.1007/s11104-016-0450-2

Fry AJ, Palmer MR, Rand DM (2004) Variable fitness effects of *Wolbachia* infection in *Drosophila melanogaster*. Heredity 93:379–389

Fuhrman JA (1999) Marine viruses and their biogeochemical and ecological effects. Nature 399:541–548

Fuhrman JA, Noble RT (1995) Viruses and protists cause similar bacterial mortality in coastal water. Limnol Oceanogr 40:1236–1242

Fukatsu T, Nikoh N, Kawai R, Koga R (2000) The secondary endosymbiotic bacterium of the pea aphid *Acyrthosiphon pisum* (*Insecta: Homoptera*). Appl Environ Microbiol 66:2748–2758

Garcia JL, Patel BKC, Ollivier B (2000) Phylogenetic and ecological diversity of methanogenic archaea. Anaerobe 6:205–226

Gause GF (1935) Vérifications expérimentales de la théorie mathématique de la lutte pour la vie. Actual Scient ind 277:1–63

Geiger A, Cuny G, Frutos R (2005) Two tsetse fly species, *Glossina palpalis gambiensis* and *Glossina morsitans morsitans,* carry genetically distinct populations of the secondary symbiont *Sodalis glossinidius*. Appl Environ Microbiol 71:8941–8943

Geiger A, Ravel S, Mateille T, Janelle J, Patrel D, Cuny G, Frutos R (2007) Vector competence of *Glossina palpalis gambiensis* for *Trypanosoma brucei* s.l. and genetic diversity of the symbiont *Sodalis glossinidius*. Mol Biol Evol 24:102–109

Geiger A, Fardeau ML, Grebaut P, Vatunga G, Josénando T, Herder S, Cuny G, Truc P, Ollivier B (2009) First isolation of *Enterobacter, Enterococcus*, and *Acinetobacter* spp. as inhabitants of the tsetse fly (*Glossina palpalis palpalis*) midgut. Infect Genet Evol 9:1364–1370

Geiger A, Fardeau ML, Falsen E, Ollivier B, Cuny G (2010) *Serratia glossinae* sp. nov., isolated from the midgut of the tsetse fly *Glossina palpalis gambiensis*. Int J Syst Evol Microbiol 60:1261–1265

Geiger A, Fardeau ML, Njiokou F, Joseph M, Asonganyi T, Ollivier B, Cuny G (2011) Bacterial diversity associated with populations of *Glossina* spp. from Cameroon and distribution within the campo sleeping sickness focus. Microb Ecol 62:632–643

Gemerden V (1993) Microbial mats: a joint venture. Mar Geol 113:3–25

Genre A, Chabaud M, Balzergue C, Puech-Pagès V, Novero M, Rey T, Fournier J, Rochange S, Bécard G, Bonfante P, Barker DG (2013) Short-chain chitin oligomers from arbuscular mycorrhizal fungi trigger nuclear Ca^{2+} spiking in *Medicago truncatula* roots and their production is enhanced by strigolactone. New Phytol 198:190–202. https://doi.org/10.1111/nph.12146

Gensollen T, Iyer SS, Kasper DL, Blumberg RS (2016) How colonization by microbiota in early life shapes the immune system. Science 352(6285):539–544. https://doi.org/10.1126/science.aad9378

Gherbi H, Markmann K, Svistoonoff S, Estevan J, Autran D, Giczey G, Auguy F, Péret B, Laplaze L, Franche C, Parniske M, Bogusz D (2008) *SymRK* defines a common genetic basis for plant root endosymbioses with arbuscular mycorrhiza fungi, rhizobia, and *Frankia* bacteria. Proc Natl Acad Sci U S A 105:4928–4932. https://doi.org/10.1073/pnas.0710618105

Giannone RJ, Huber H, Karpinets T, Heimerl T, Küper U, Rachel R, Keller M, Hettich RL, Podar M (2011) Proteomic characterization of cellular and molecular processes that enable the *Nanoarchaeum equitans-Ignicoccus hospitalis* relationship. PLoS One 6(8):e22942. https://doi.org/10.1371/journal.pone.0022942

Giannone RJ, Wurch LL, Heimerl T, Martin S, Yang Z, Huber H, Rachel R, Hettich RL, Podar M (2015) Life on the edge: functional genomic response of *Ignicoccus hospitalis* to the presence of *Nanoarchaeum equitans*. ISME J 9:101–114

Gilboa-Garber N (1972) Purification and properties of hemagglutinin from *Pseudomonas aeruginosa* and its reaction with human blood cells. Biochim Biophys Acta 273:165–173

Giraud E, Moulin L, Vallenet D, Barbe V, Cytryn E, Avarre JC et al (2007) Legumes symbioses: absence of *nod* genes in photosynthetic bradyrhizobia. Science 316:1307–1312. https://doi.org/10.1126/science.1139548

Gjermansen M, Nilsson M, Yang L, Tolker-Nielsen T (2010) Characterization of starvation-induced dispersion in Pseudomonas putida biofilms: genetic elements and molecular mechanisms. Mol Microbiol 75:815–826

Glaser RL, Meola MA (2010) The native *Wolbachia* endosymbionts of *Drosophila melanogaster* and *Culex quinquefasciatus* increase host resistance to West Nile virus infection. PLoS One 5:e11977

Godfroy O, Debellé F, Timmers T, Rosenberg C (2006) A rice calcium- and calmodulin-dependent protein kinase restores nodulation to a legume mutant. Mol Plant-Microbe Interact 19:495–501. https://doi.org/10.1094/MPMI-19-0495

Gomez de Aguero M, Ganal-Vonarburg SC, Fuhrer T, Rupp S, Uchimura Y, Li H, Steinert A, Heikenwalder M, Hapfelmeier S, Sauer U, McCoy KD, Macpherson AJ (2016) The maternal microbiota drives early postnatal innate immune development. Science 351(6279):1296–1302. https://doi.org/10.1126/science.aad2571

González JM, Sherr EB, Sherr BF (1990) Size selective grazing on bacteria by natural assemblages of estuarine flagellates and ciliates. Appl Environ Microbiol 56:583–589

Good AG, Beatty PH (2011) Fertilizing nature: a tragedy of excess in the commons. PLoS Biol 9:e1001124. doi:10.1371/journal.pbio.1001124

Gorby YA, Yanina S, McLean JS, Rosso KM, Moyles D, Dohnalkova A, Beveridge TJ, Chang IS, Kim BH, Kim KS, Culley DE, Reed SB, Romine MF, Saffarini DA, Hill EA, Shi L, Elias DA, Kennedy DW, Pinchuk G, Watanabe K, Ishii S, Logan B, Nealson KH, Fredrickson JK (2006)

Electrically conductive bacterial nanowires produced by *Shewanella oneidensis* strain MR-1 and other microorganisms. Proc Natl Acad Sci U S A 103:11358–11363

Gotoh T, Sugasawa J, Noda H, Kitashima Y (2007) *Wolbachia*-induced cytoplasmic incompatibility in Japanese populations of *Tetranychus urticae* (*Acari: Tetranychidae*). Exp Appl Acarol 42:1–16

Gould SB, Waller RF, McFadden GI (2008) Plastid evolution. Annu Rev Plant Biol 59:491–517

Graham PH, Vance CP (2000) Nitrogen fixation in perspective: an overview of research and extension needs. Field Crops Res 65:93–106. https://doi.org/10.1016/S0378-4290(99)00080-5

Gray MW, Lang BF, Burger G (2004) Mitochondria of protists. Annu Rev Genet 38:477–524

Guerrero R, Pedrós-Alió C, Esteves I, Mas J, Chase D, Margulis L (1986) Predatoryprokaryotes: predation and primary consumption evolved in bacteria. Proc Natl Acad Sci U S A 83:2138–2142

Gupta V, Smemo KA, Yavitt J, Fowle DA, Branfireun BA, Basiliko N (2013) Stable isotopes reveal widespread anaerobic methane oxidation across latitude and peatland type. Environ Sci Technol 47:8273–8279

Gutjahr C, Parniske M (2013) Cell and developmental biology of arbuscular mycorrhiza symbiosis. Annu Rev Cell Dev Biol 29:593–617. https://doi.org/10.1146/annurev-cellbio-101512-122413

Hahn D (2008) Polyphasic taxonomy of the genus *Frankia*. In: Pawlowski K, Newton WE (eds) Nitrogen –fixing actinorhizal symbioses. Springer, New York, pp 25–45. https://doi.org/10.1007/978-1-4020-3547-0-2

Hamerly T, Tripet BP, Tigges M, Giannone RJ, Wurch L, Hettich RL, Podar M, Copie V, Bothner B (2015) Untargeted metabolomics studies employing NMR and LC–MS reveal metabolic coupling between *Nanoarcheum equitans* and its archaeal host *Ignicoccus hospitalis*. Metabolomics 11:895–907. https://doi.org/10.1007/s11306-014-0747-6

Hamidou Soumana I, Berthier D, Tchicaya B, Thevenon S, Njiokou F, Cuny G, Geiger A (2013) Population dynamics of *Glossina palpalis gambiensis* symbionts, *Sodalis glossinidius*, and *Wigglesworthia glossinidia*, throughout host-fly development. Infect Genet Evol 13:41–48

Hamidou Soumana I, Loriod B, Ravel S, Tchicaya B, Simo G, Rihet P, Geiger A (2014a) The transcriptional signatures of *Sodalis glossinidius* in the *Glossina palpalis gambiensis* flies negative for *Trypanosoma brucei gambiense* contrast with those of this symbiont in tsetse flies positive for the parasite: possible involvement of a *Sodalis*-hosted prophage in fly *Trypanosoma* refractoriness? Infect Genet Evol 24:41–56

Hamidou Soumana I, Tchicaya B, Loriod B, Rihet P, Geiger A (2014b) Identification of overexpressed genes in *Sodalis glossinidius* inhabiting trypanosome-infected self-cured tsetse flies. Front Microbiol 5:255

Hamidou Soumana I, Tchicaya B, Simo G, Geiger A (2014c) Comparative gene expression of *Wigglesworthia* inhabiting non-infected and *Trypanosoma brucei gambiense*-infected *Glossina palpalis gambiensis* flies. Front Microbiol 5:620

Hamidou Soumana I, Klopp C, Ravel S, Nabihoudine I, Tchicaya B, Parrinello H, Abate L, Rialle S, Geiger A. 2015. RNA-seq de novo assembly reveals differential gene expression in *Glossina palpalis gambiensis* infected with *Trypanosoma brucei gambiense* vs. non-infected and self-cured flies. Front Microbiol. 6:1259

Hamidou Soumana I, Tchicaya B, Rialle S, Parrinello H, Geiger A (2017) Comparative genomics of *Glossina palpalis gambiensis* and *G. morsitans morsitans* to reveal gene orthologs involved in infection by *Trypanosoma brucei gambiense*. Frontiers Microbiol in press

Hamilton JJ, Contreras M, Reed JL (2015) Thermodynamics and H_2 transfer in a methanogenic community. PLoS Comput Biol 11(7):e1004364. https://doi.org/10.1371/journal.pcbi.1004364

Hansen SR, Hubbel SP (1980) Single-nutrient microbial competition : qualitative agreement between experimental and theoretically forecast outcomes. Science 207:1491–1493

Hansen SK, Rainey PB, Haagensen JAJ, Molin S (2007) Evolution of species interactions in a biofilm community. Nature 445. https://doi.org/10.1038/nature05514

Hanson RS, Hanson TE (1996) Methanotrophic bacteria. Microbiol Rev:439–471

Haroon MF, Hu SH, Shi Y, Imelfort M, Keller J, Hugenholtz P, Yuan Z, Tyson GW (2013) Anaerobic oxidation of methane coupled to nitrate reduction in a novel archaeal lineage. Nature 500:567–570

Hedges LM, Brownlie JC, O'Neill SL, Johnson KN (2008) *Wolbachia* and virus protection in insects. Science 322:702–702

Heidelberg JF, Nelson WC, Schoenfeld T, Bhaya D (2009) Germ warfare in a microbial mat Community: CRISPRs provide insights into the co-evolution of host and viral genomes. PLoS One 4(1):e4169

Hendrix R (1999) The long evolutionary reach of viruses. Curr Biol 9:9914–9917

Hertig M (1936) The rickettsia, *Wolbachia pipientis* (gen. Et sp. n.) and associated inclusions of the mosquito *Culex pipiens*. Parasitology 28:454–486

Hertig M, Wolbach SB (1924) Studies on rickettsia-like microorganisms in insects. J Med Res 44:329–374

Hertle R, Hilger M, Weingardt-Kocher S, Walev I (1999) *Cytotoxic action of Serratia marcescens* hemolysin on human epithelial cells. Infect Immun 67:817–825

Hilgenboecker K, Hammerstein P, Schlattmann P, Telschow A, Werren JH (2008) How many species are infected with *Wolbachia*?--a statistical analysis of current data. FEMS Microbiol Lett 281:215–220

Hiroki M, Kato Y, Kamito T, Miura K (2002) Feminization of genetic males by a symbiotic bacterium in a butterfly, *Eurema hecabe* (*Lepidoptera: Pieridae*). Naturwissenschaften 89:167–170

Hoffmann AA, Turelli M, Simmons GM (1986) Unidirectional incompatibility between populations of *Drosophila simulans*. Evolution 40:692–701

Hoffmann AA, Turelli M, Harshman LG (1990) Factors affecting the distribution of cytoplasmic incompatibility in Drosophila simulans. Genetics 126:933–948

Hornett EA, Charlat S, Duplouy AM, Davies N, Roderick GK, Wedell N, Hurst GD (2006) Evolution of male-killer suppression in a natural population. PLoS Biol 4:e283

Hosokawa T, Koga R, Kikuchi Y, Meng XY, Fukatsu T (2010) *Wolbachia* as a bacteriocyte associated nutritional mutualist. Proc Natl Acad Sci U S A 107:769–774

Huber H, Hohn MJ, Rachel R, Fuchs T, Wimmer VC, Stetter KO (2002) A new phylum of *Archaea* represented by a nanosized hyperthermophilic symbiont. Nature 417:63–67

Huber H, Küper U, Daxer S, Rachel R (2012) The unusual cell biology of the hyperthermophilic Crenarchaeon *Ignicoccus hospitalis*. Antonie Van Leeuwenhoek 102:203–219. https://doi.org/10.1007/s10482-012-9748-5

Human Microbiome Project (2012) Structure, function and diversity of the healthy human microbiome. Nature 486(7402):207–214. https://doi.org/10.1038/nature11234

Imachi H, Sekiguchi Y, Kamagata Y, Loy A, Qiu YL, Hugenholtz P, Kimura N, Wagner M, Ohashi A, Harada H (2006) Non-sulfate-reducing, syntrophic bacteria affiliated with *Desulfotomaculum* cluster I are widely distributed in methanogenic environments. Appl Environ Microbiol 72:2080–2091

International Glossina Genome Initiative (2014) Genome sequence of the tsetse fly (*Glossina morsitans*): vector of African trypanosomiasis. Science 344:380–386

Jackson BE, Bhupathiraju VK, Tanner RS, Woese CR, McInerney MJ (1999) *Syntrophus aciditrophicus* sp. nov., a new anaerobic bacterium that degrades fatty acids and benzoate in syntrophic association with hydrogen-using microorganisms. Arch Microbiol 171:107–114

Jackson AP, Sanders M, Berry A, McQuillan J, Aslett MA, Quail MA, Chukualim B, Capewell P, MacLeod A, Melville SE, Gibson W, Barry JD, Berriman M, Hertz-Fowler C (2010) The genome sequence of *Trypanosoma brucei gambiense*, causative agent of chronic human african trypanosomiasis. PLoS Negl Trop Dis 4:e658

Jaenike J, Brekke TD (2011) Defensive endosymbionts: a cryptic trophic level in community ecology. Ecol Lett 14:150–155

Jaenike J, Dyer KA (2008) No resistance to male-killing *Wolbachia* after thousands of years of infection. J Evol Biol 21:1570–1577

Jaenike J, Unckless R, Cockburn SN, Boelio LM, Perlman SJ (2010) Adaptation via symbiosis: recent spread of a *Drosophila* defensive symbiont. Science 329:212–215

Jahn U, Gallenberger M, Paper W, Junglas B, Eisenreich W, Stetter KO, Rachel R, Huber H (2008) *Nanoarchaeum equitans* and *Ignicoccus hospitalis*: new insights into a unique, intimate association of two archaea. J Bacteriol 190(5):1743–1750. https://doi.org/10.1128/JB.01731-07

Jasti S, Sieracki ME, Poulton NJ, Giewat MW, Rooney-Varga JN (2005) Phylogenetic diversity and specificity of bacteria closely associated with Alexandrium spp. and other phytoplankton. Appl Environ Microbiol 71:3483–3494

Jeffery IB, Claesson MJ, O'Toole PW, Shanahan F (2012) Categorization of the gut microbiota: enterotypes or gradients? Nat Rev Microbiol 10(9):591–592

Jeyaprakash A, Hoy MA (2000) Long PCR improves *Wolbachia* DNA amplification: wsp sequences found in 76% of sixty-three arthropod species. Insect Mol Biol 9:393–405

Jiang SC, Paul JH (1998) Significance of lysogeny in the marine environments: studies with isolates and a model of lysogenic phage-production. Microb Ecol 35:235–243

Jiggins FM, Hurst GD, Schulenburg JH, Majerus ME (2001) Two male-killing *Wolbachia* strains coexist within a population of the butterfly *Acraea encedon*. Heredity 86:161–166

Jimenez E, Marin ML, Martin R, Odriozola JM, Olivares M, Xaus J, Fernandez L, Rodriguez JM (2008) Is meconium from healthy newborns actually sterile? Res Microbiol 159(3):187–193. https://doi.org/10.1016/j.resmic.2007.12.007

Jones EO, White A, Boots M (2007) Interference and the persistence of vertically transmitted parasites. J Theor Biol 246:10–17

Joyce SA, MacSharry J, Casey PG, Kinsella M, Murphy EF, Shanahan F, Hill C, Gahan CG (2014) Regulation of host weight gain and lipid metabolism by bacterial bile acid modification in the gut. Proc Natl Acad Sci U S A 111(20):7421–7426. https://doi.org/10.1073/pnas.1323599111

Kadouri DE, To K, Shanks RMQ, Doi Y (2013) Predatory bacteria: a potential ally against multidrug-resistant gram-negative pathogens. PLoS One 8(5):e63397

Kageyama D, Nishimura G, Hoshizaki S, Ishikawa Y (2002) Feminizing *Wolbachia* in an insect, *Ostrinia furnacalis* (*Lepidoptera: Crambidae*). Heredity 88:444–449

Kamada N, Seo SU, Chen GY, Nunez G (2013) Role of the gut microbiota in immunity and inflammatory disease. Nat Rev Immunol 13(5):321–335. https://doi.org/10.1038/nri3430

Kambris Z, Cook PE, Phuc HK, Sinkins SP (2009) Immune activation by life-shortening *Wolbachia* and reduced filarial competence in mosquitoes. Science 326:134–136

Kambris Z, Blagborough AM, Pinto SB, Blagrove MS, Godfray HC, Sinden RE, Sinkins SP (2010) *Wolbachia* stimulates immune gene expression and inhibits plasmodium development in *Anopheles gambiae*. PLoS Pathog 6:e1001143

Kanzler BE, Pfanes KR, Vogl K, Overmann J (2005) Molecular characterization of the nonphotosynthetic partner bacterium in the consortium '*Chlorochromatium aggregatum*. Appl Environ Microb 71:7434–7441

Keane R, Berleman J (2016) The predatory life cycle of *Myxococcus xanthus*. Microbiology 162:1–11. https://doi.org/10.1099/mic.0.00020

Keller GP, Windsor DM, Saucedo JM, Werren JH (2004) Reproductive effects and geographical distributions of two *Wolbachia* strains infecting the Neotropical beetle, *Chelymorpha alternans Boh.* (*Chrysomelidae, Cassidinae*). Mol Ecol 13:2405–2420

Kolev NG, Franklin JB, Carmi S, Shi H, Michaeli S, Tschudi C (2010) The transcriptome of the human pathogen *Trypanosoma brucei* at single-nucleotide resolution. PLoS Pathog 6:e1001090

Kosuta S, Chabaud M, Lougnon G, Gough C, Dénarié J, Barker DG, Bécard G (2003) A diffusible factor from arbuscular mycorrhizal fungi induces symbiosis-specific *MtENOD11* expression in roots of *Medicago truncatula*. Plant Physiol 131:952–962. https://doi.org/10.1104/pp.011882

Krumholz LR, Bryant MP (1986) *Syntrophococcus sucromutans* sp. nov. gen. Nov. uses carbohydrates as electron donors and formate, methoxymonobenzenoids or *Methanobrevibacter* as electron acceptor systems. Arch Microbiol 143:313–318

Lavin M, Herendeen PS, Wojciechowski MF (2005) Evolutionary rates analysis of Leguminosae implicates a rapid diversification of lineages during the tertiary. Syst Biol 54:575–594. https://doi.org/10.1080/10635150590947131

Lawrence JG, Hatfull GF, Hendrix RW (2002) Imbroglios of viral taxonomy: genetic exchange and failings of phenetic approaches. J Bacteriol 184:4891–4905

Lazaro JE, Nitcheu J, Predicala RZ, Mangalindan GC, Nesslany F, Marzin D, Concepcion GP, Diquet B (2002) Heptyl prodigiosin, a bacterial metabolite is antimalarial in vivo and non-mutagenic in vitro. J NatToxins 11:367–377

Leadbetter ER, Foster JW (1958) Studies on some methane-utilizing bacteria. Archiv Mikrobio 30:91–118

Lebba V et al (2014) *Bdellovibrio bacteriovorus* directly attacks *Pseudomonas aeruginosa* and *Staphylococcus aureus* cystic fibrosis isolates. Front Microbiol 5:1–9., article 280. https://doi.org/10.3389/fmicb.2014.00280

LeBlanc JG, Milani C, de Giori GS, Sesma F, van Sinderen D, Ventura M (2013) Bacteria as vitamin suppliers to their host: a gut microbiota perspective. Curr Opin Biotechnol 24(2):160–168. https://doi.org/10.1016/j.copbio.2012.08.005

Levin BR, Bull JJ (2004) Population and evolutionary dynamics of phage therapy. Nat Rev Microb 2:166–173

Lewis G, Schrire B, Mackind B, Lock M (2005) Legumes of the world. Royal Botanic Gardens, Kew, UK

Li H-L, Wang W, Mortimer PE et al (2015) Large-scale phylogenetic analyses reveal multiple gains of actinorhizal nitrogen-fixing symbioses in angiosperms associated with climate change. Sci Rep 5:14023. https://doi.org/10.1038/srep14023

Lidicker WZA (1979) Clarification of interactions in ecological systems. Bioscience 29:475–477

Lill R, Kispal G (2000) Maturation of cellular Fe-S proteins: an essential function of mitochondria. Trends Biochem Sci 25:352–356

Lindh JM, Lehane MJ (2011) The tsetse fly *Glossina fuscipes fuscipes* (*Diptera: Glossina*) harbours a surprising diversity of bacteria other than symbionts. Antonie Van Leeuwenhoek 99:711–720

Liu Z et al (2013) Genomic analysis reveals key aspects of prokaryotic symbiosis in the phototrophic consortium *"Chlorochromatium aggregatum"*. Genome Biol 14:R127. https://doi.org/10.1186/gb-2013-14-11-r127

Liu Z, Niu H, Wu S, Huang R (2014) CsgD regulatory network in a bacterial trait-altering biofilm formation. Emerg Microbes Infect 3:e1. https://doi.org/10.1038/emi.2014.1

Lively CM, Clay K, Wade MJ, Fuqua C (2005) Competitive co-existence of vertically and horizontally transmitted parasites. Evol Ecol Res 7:1183–1190

Lloyd-Price J, Abu-Ali G, Huttenhower C (2016) The healthy human microbiome. Genome Med 8(1):51. https://doi.org/10.1186/s13073-016-0307-y

Lønborg C, Middelboe M, Brussaard CP (2013) Viral lysis of Micromonas pusilla: impacts on dissolved organic matter production and composition. Biogeochemistry 116:231–240

Long SR (1996) Rhizobium symbiosis: nod factors in perspective. Plant Cell 8:1885–1898. https://doi.org/10.1105/tpc.8.10.1885

Mackinder LCM, Worthy C, Biggi G, Hall M, Ryan KP, Varsani A et al (2009) A unicellular algal virus, Emiliania huxleyi virus 86, exploits an animal-like infection strategy. J Gen Virol 90:2306–2316

Mackowiak PA (2013) Recycling metchnikoff: probiotics, the intestinal microbiome and the quest for long life. Front Public Health 1:52. https://doi.org/10.3389/fpubh.2013.00052

Madigan MT, Martinko JM, Bender KS, Buckley DH, Stahl DA, Brock T (2015) Brock Biology of Microorganisms (14th Edition) Pearson Education

Maeda H, Morihara K (1995) Serralysin and related bacterial proteinases. Methods Enzymol 248:395–413

Maillet F, Poinsot V, André O, Puech-Pagés V, Haout A, Gueunier M, Cromer L, Giraudet D, Formey D, Niebel A, Martinez EA, Driguez H, Bécard G, Dénarié J (2011) Fungal lipochitooligosaccharide symbiotic signals in arbuscular mycorrhiza. Nature 469:58–63. https://doi.org/10.1038/nature09622

Malele I, Nyingilili H, Lyaruu E, Tauzin M, Ollivier B, Fardeau M-L, Geiger A Bacterial diversity in the gut of *G. pallidipes* population from a non sleeping sickness focus in Tanzania and its implication for species' vectorial capacity. BMC Microbiol in submission

Malkin SY, Meysman FJR (2015) Rapid redox signal transmission by "cable Bacteria" beneath a photosynthetic biofilm. Appl Environ Microbial 81(3):948–956

Malvankar NS, Lovley DR (2014) Microbial nanowires for bioenergy applications. Curr Opin Biotechnol 27:88–95

Mao X, Jiang R, Xiao W, Yu J (2015) Use of surfactants for the remediation of contaminated soils: a review. J Hazard Mater 285:419–435
Margulis L (1981) Symbiosis in cell evolution. W. H. Freeman and Co, San Francisco
Margulis L (1993) Symbiosis in cell evolution. Microbial communities in the Archaean and Proterozoic eons. W. H. Freeman and Co, New York
Markmann K, Parniske M (2009) Evolution of root endosymbiosis with bacteria: how novel are nodules? Trends Plant Sci 14:77–86. https://doi.org/10.1016/j.tplants.2008.11.009
Markmann K, Giczey G, Parniske M (2008) Functional adaptation of a plant receptor- kinase paved the way for the evolution of intracellular root symbioses with Bacteria. PLoS Biol 6:e68. https://doi.org/10.1371/journal.pbio.0060068
Marsh JF, Rakocevic A, Mitra RM, Brocard L, Eschstruth A, Long SR, Schultze M, Ratet P, Oldroyd GE (2007) Medicago truncatula NIN is essential for rhizobial-independent nodule organogenesis induced by autoactive calcium/calmodulin-dependent protein kinase. Plant Physiol 144:324–335. https://doi.org/10.1104/pp.106.093021
Martin MO (2002) Predatory prokaryotes: an emerging research opportunity. J Mol Microbiol Biotechnol 4(5):467–477
Mashburn-Warren LM, Whiteley M (2006) Special delivery: vesicle trafficking in prokaryotes. Mol Microbiol 61(4):839–846. https://doi.org/10.1111/j.1365-2958.2006.05272.x
Mazmanian SK, Liu CH, Tzianabos AO, Kasper DL (2005) An immunomodulatory molecule of symbiotic bacteria directs maturation of the host immune system. Cell 122(1):107–118. https://doi.org/10.1016/j.cell.2005.05.007
Mbewe NJ, Mweempwa C, Guya S, Wamwiri FN (2015) Microbiome frequency and their association with trypanosome infection in male *Glossina morsitans centralis* of Western Zambia. Vet Parasitol 211:93–98
McCutcheon JP, Moran NA (2007) Parallel genomic evolution and metabolic interdependence in an ancient symbiosis. Proc Natl Acad Sci U S A 104:19392–19397
McGenity TJ, Folwell BD, McKew BA, Sanni GO (2012) Marine crude-oil biodegradation : a central role for interspecies interactions. Aquat Biosyst 8(10). https://doi.org/10.1186/2046-9063-8-10
McInerney JO, O'Connell MJ, Pisani D (2014) The hybrid nature of the *Eukaryota* and a consilient view of life on earth. Nat Rev Microbiol 12:449–455
McInnerney MJ, Sieber JR, Gunsalus RP (2009) Syntrophy in anaerobic global carbon cycles. Curr Opin Biotechnol 20:623–632
Medina-Sanchez JM, Villar-Argaiz M, Carrillo P (2004) Neither with nor without you: a complex algal control on bacterioplankton in a high mountain lake. Limnol Oceanogr 49:1722–1733
Mercado TI, Colon-Whitt A (1982) Lysis of Trypanosoma cruzi by Pseudomonas fluorescens. AntimicrobAgents Chemother 22:1051–1057
Milucka J, Ferdelman TG, Polerecky L, Franzke D, Wegener G et al (2012) Zero-valent Sulphur is a key intermediate in marine methane oxidation. Nature 491:541–546
Min KT, Benzer S (1997) *Wolbachia*, normally a symbiont of *Drosophila*, can be virulent, causing degeneration and early death. Proc Natl Acad Sci U S A 94:10792–10796
Mitra RM, Gleason CA, Edwards A, Hadfield J, Downie JA, Oldroyd GED, Long SR (2004) A Ca^{2+}/calmodulin-dependent protein kinase required for symbiotic nodule development: gene identification by transcript based cloning. Proc Natl Acad Sci U S A 101:4701–4705. https://doi.org/10.1073/pnas.0400595101
Miwa H, Sun J, Oldroyd GED, Downie JA (2006) Analysis of nod-factor-induced calcium signaling in root hairs of symbiotically defective mutants of *Lotus japonicus*. Mol Plant-Microbe Interact 19:914–923. https://doi.org/10.1094/MPMI-19-0914
Moissl-Eichinger C, Hubert H (2011) Archaeal symbionts and parasites. Curr Opin Microbiol 14:364–370
Mondot S, Lepage P (2016) The human gut microbiome and its dysfunctions through the metaomics prism. Ann N Y Acad Sci 1372(1):9–19. https://doi.org/10.1111/nyas.13033
Monod J (1949) The growth of bacterial cultures. Annu Rev Microbiol 3:371–394
Moore RL (1981) The biology of *Hyphomicrobium* and other prosthecate, budding bacteria. Annu Rev Microbiol 35:567–594

Moran NA (1996) Accelerated evolution and Muller's rachet in endosymbiotic bacteria. Proc Natl Acad Sci U S A 93:2873–2878

Moran NA, Munson MA, Baumann P, Ishikawa H (1993) A molecular clock in endosymbiotic bacteria is calibrated using the insect hosts. Proc R Soc London Ser B 253:167–171

Moran NA, McCutcheon JP, Nakabachi A (2008) Genomics and evolution of heritable bacterial symbionts. Annu Rev Genet 42:165–190

Moreira LA, Iturbe-Ormaetxe I, Jeffery JA, Lu G, Pyke AT, Hedges LM, Rocha BC, Hall-Mendelin S, Day A, Riegler M, Hugo LE, Johnson KN, Kay BH, McGraw EA, van den Hurk AF, Ryan PA, O'Neill SL (2009) A *Wolbachia* symbiont in *Aedes aegypti* limits infection with dengue, Chikungunya, and Plasmodium. Cell 139:1268–1278

Moret Y, Juchault P, Rigaud T (2001) *Wolbachia* endosymbiont responsible for cytoplasmic incompatibility in a terrestrial crustacean: effects in natural and foreign hosts. Heredity 86:325–332

Morris BEL, Henneberger R, Huber H, Moissl-Eichinger C (2013) Microbial syntrophy: interaction for the common good. FEMS Microbiol Rev 37:384–406. https://doi.org/10.1111/1574-6976.12019

Moss M (2002) Bacterial pigments. Microbiologist 3:10–12

Mouton L, Dedeine F, Henri H, Boulétreau M, Profizi N, Vavre F (2004) Virulence, multiple infections and regulation of symbiotic population in the *Wolbachia/Asobara tabida* symbiosis. Genetics 168:181–189

Mueller N, Griffin BM, Stingl U, Schink B (2008) Dominant sugar utilizers in sediment of Lake Constance depend on syntrophic cooperation with methanogenic partner organisms. Environ Microbiol 10:1501–1511

Müller J, Overmann J (2011) Close interspecies interactions between prokaryotes from sulfureous environments. Front Microbiol 2:146. https://doi.org/10.3389/fmicb.2011.0014

Mus F, Crook MB, Garcia K, Costas A, Geddes BA, Kouri ED, Paramasivan P, Ryu M-H, Oldroyd GED, Poole PS, Udvardi MK, Voigt CA, Ané J-M, Peters JW (2016) Symbiotic nitrogen fixation and the challenges to its extension to nonlegumes. Appl Environ Microbiol 82:3698–3710. https://doi.org/10.1128/AEM.01055-16

Nakabachi A, Yamashita A, Toh H, Ishikawa H, Dunbar HE, Moran NA, Hattori M (2006) The 160-kilobase genome of the bacterial endosymbiont Carsonella. Science 314:267–267

Nakagawa T, Imaizumi-Anraku H (2015) Rice arbuscular mycorrhiza as a tool to study the molecular mechanisms of fungal symbiosis and a potential target to increase productivity. Rice 8:32. https://doi.org/10.1186/s12284-015-0067-0

Negri I, Pellecchia M, Mazzoglio PJ, Patetta A, Alma A (2006) Feminizing *Wolbachia* in *Zyginidia pullula (Insecta, Hemiptera)*, a leafhopper with an XX/X0 sex-determination system. Proc R Soc London Ser B 273:2409–2416

Nielsen LP, Risgaard-Petersen N, Fossing H, Christensen PB, Sayama M (2010) Electric currents couple spatially separated biogeochemical processes in marine sediment. Nature 463:1071–1074. https://doi.org/10.1038/nature08790

Nogge G (1982) Significance of symbionts for the maintenance of an optional nutritional state for successful reproduction in hematophagous arthropods. Parasitology 82:299–304

Normand P, Orso S, Cournoyer B, Jeannin P, Chapelon C, Dawson J, Evtushenko L, Misra AK (1996) Molecular phylogeny of the genus *Frankia* and related genera and emendation of the family *Frankiaceae*. Int J Syst Bacteriol 46:1–9. https://doi.org/10.1099/00207713-46-1-1

Normand P, Lapierre P, Tisa LS, Gogarten JP, Alloisio N, Bagnarol E, Bassi CA, Berry AM, Bickhart DM, Choisne N, Couloux A, Cournoyer B, Cruveiller S, Daubin V, Demange N, Francino MP, Goltsman E, Huang Y, Kopp OR, Labarre L, Lapidus A, Lavire C, Marechal J, Martinez M, Mastronunzio JE, Mullin BC, Niemann J, Pujic P, Rawnsley T, Rouy Z, Schenowitz C, Sellstedt A, Tavares F, Tomkins JP, Vallenet D, Valverde C, Wall LG, Wang Y, Medigue C, Benson DR (2007) Genome characteristics of facultatively symbiotic *Frankia sp.* strains reflect host range and host plant biogeography. Genome Res 17:7–15. https://doi.org/10.1101/gr.5798407

Nowack EC, Melkonian M, Glockner G (2008) Chromatophore genome sequence of Paulinella sheds light on acquisition of photosynthesis by eukaryotes. Curr Biol 18:410–418

Nzila A (2013) Update on the cometabolism of organic pollutants by bacteria. Environ Poll 178:474–482

O'Hara AM, Shanahan F (2006) The gut flora as a forgotten organ. EMBO Rep 7:688–693

O'Neill SL, Gooding RH, Aksoy S (1993) Phylogenetically distant symbiotic microorganisms reside in *Glossina* midgut and ovary tissues. Med Vet Entomol 7:377–383

O'Toole PW, Jeffery IB (2015) Gut microbiota and aging. Science 350(6265):1214–1215. https://doi.org/10.1126/science.aac8469

Okazaki S, Noisangiam R, Okubo T, Kaneko T, Oshima K, Hattori M, Teamtisong K, Songwattana P, Tittabutr P, Boonkerd N, Saeki K, Sato S, Uchiumi T, Minamisawa K, Teaumroong N (2015) Genome analysis of a novel *Bradyrhizobium* sp. DOA9 carrying a symbiotic plasmid. PLoS One 10:e0117392. doi:10.1371/journal.pone.0117392

Oldroyd GE (2013) Speak, friend, and enter: signalling systems that promote beneficial symbiotic associations in plants. Nature Rev. Microbiol. 11:252–263. https://doi.org/10.1038/nrmicro2990

Oldroyd GE, Murray JD, Poole PS, Downie JA (2011) The rules of engagement in the legume-rhizobial symbiosis. Annu Rev Genet 45:119–144. https://doi.org/10.1146/annurev-genet-110410-132549

Oliver KM, Russell JA, Moran NA, Hunter MS (2003) Facultative bacterial symbionts in aphids confer resistance to parasitic wasps. Proc Natl Acad Sci U S A 100:1803–1807

Oliver KM, Campos J, Moran NA, Hunter MS (2008) Population dynamics of defensive symbionts in aphids. Proc R Soc B Biol Sci 275:293–299

Oliver KM, Degnan PH, Hunter MS, Moran NA (2009) Bacteriophages encode factors required for protection in a symbiotic mutualism. Science 325:992–994

Op den Camp, R.H. (2012). Evolution of rhizobium symbiosis. Thesis, Wageningen University, ISBN 978-94-6137-198-2

Op den Camp R, Streng A, De Mita S, Cao Q, Polone E et al (2011) LysM-type mycorrhizal receptor recruited for rhizobium symbiosis in nonlegume Parasponia. Science 331:909–912. https://doi.org/10.1126/science.1198181

Op den Camp RH, Polone E, Fedorova E et al (2012) Nonlegume *Parasponia andersonii* deploys a broad rhizobium host range strategy resulting in largely variable symbiotic effectiveness. Mol Plant-Microbe Interact 25:954–963

Ostaff MJ, Stange EF, Wehkamp J (2013) Antimicrobial peptides and gut microbiota in homeostasis and pathology. EMBO Mol Med 5(10):1465–1483. https://doi.org/10.1002/emmm.201201773

Overmann J, Schubert K (2002) Phototrophic consortia: model systems for symbiotic interrelations between prokaryotes. Arch Microbiol 177:201–208

Paerl H, Pinckney J (1996) A mini-review of microbial consortia: their roles in aquatic production and biogeochemical cycling. Microb Ecol 31:225–247

Pagé A, Tivey MK, Stakes DS, Reysenbach AL (2008) Temporal and spatial archaeal colonization of hydrothermal vent deposits. Environ Microbiol 10(4):874–884

Pais R, Lohs C, Wu YN, Wang JW, Aksoy S (2008) The obligate mutualist *Wigglesworthia glossinidia* influences reproduction, digestion, and immunity processes of its host, the tsetse fly. Appl Environ Microbiol 74:5965–5974

Palesse S. (2014) Déterminisme de la Decision lysogénique au sein des communautés virales aquatiques: importance des fluctuations physiologiques et métaboliques. PhD Thesis, Université Clermont Auvergne, 165 pp.

Pamp SJ, Gjermansen M, Tolker-Nielsen T (2007) The biofilm matrix: a sticky framework. In the biofilm mode of life: mechanisms and adaptations. Kjelleberg, S., and Givskov, M. (eds). Norfolk, UK: horizon. Bioscience:37–69

Pasternak Z et al (2014) In and out: an analysis of epibiotic vs periplasmic bacterial predators. ISME J 8:625–635

Pawlowski K, Bisseling T (1996) Rhizobial and Actinorhizal symbioses: what are the shared features? Plant Cell 8:1899–1913

Pérez J, Moraleda-Muñoz A, Marcos-Torres FJ, Muñoz-Dorado J (2016) Bacterial predation: 75 years and counting! Environ Microbiol 18(3):766–779

Pernthaler J (2005) Predation on prokaryotes in the water column and its ecological implications. Nature Review Microbiol 3:537–546

Persson T, Benson DR, Normand P, Vanden Heuvel B, Pujic P, Chertkov O, Teshima H, Bruce DC, Detter C, Tapia R, Han S, Han J, Woyke T, Pitluck S, Pennacchio L, Nolan M, Ivanova N, Pati A, Land ML, Pawlowski K, Berry AM (2011) Genome sequence of "*Candidatus Frankia datiscae*" Dg1, the uncultured microsymbiont from nitrogen-fixing root nodules of the dicot *Datisca glomerata*. J Bacteriol 193:7017–7018. https://doi.org/10.1128/JB.06208-11

Persson T, Battenberg K, Demina IV, Vigil-Stenman T, Vanden Heuvel B, Pujic P, Facciotti MT, Wilbanks EG, O'Brien A, Fournier P, Cruz Hernandez MA, Mendoza Herrera A, Médigue C, Normand P, Pawlowski K, Berry AM (2015) *Candidatus Frankia Datiscae Dg1*, the actinobacterial microsymbiont of *Datisca glomerata*, expresses the canonical *nod* genes *nodABC* in symbiosis with its host plant. PLoS One e0127630:10. https://doi.org/10.1371/journal.pone.0127630

Pfennig N, Biebl H (1976) "Desulfuromonas acetoxidans" gen. Nov. and sp. nov., a new anaerobic, sulfur-reducing, acetate-oxidizing bacterium. Arch Microbiol 110:3–12

Pradeep Ram AS, Sime-Ngando T (2010) Resources drive trade-off between viral lifestyles in the plankton: evidence from freshwater microbial microcosms. Environ Microbiol 12:467–479

Pradeep Ram AS, Boucher D, Sime-Ngando T, Debroas D, Romagoux JC (2005) Phage bacteriolysis, protistan bacterivory potential, and bacterial production in a freshwater reservoir: coupling with temperature. Microb Ecol 50:64–72

Prangishvili D, Forterre P, Garrett RA (2006) Viruses of the archaea: a unifying view. Nature Rev. Microbiol. 4:837–848

Qin J, Li R, Raes J, Arumugam M, Burgdorf KS, Manichanh C, Nielsen T, Pons N, Levenez F, Yamada T, Mende DR, Li J, Xu J, Li S, Li D, Cao J, Wang B, Liang H, Zheng H, Xie Y, Tap J, Lepage P, Bertalan M, Batto JM, Hansen T, Le Paslier D, Linneberg A, Nielsen HB, Pelletier E, Renault P, Sicheritz-Ponten T, Turner K, Zhu H, Yu C, Li S, Jian M, Zhou Y, Li Y, Zhang X, Li S, Qin N, Yang H, Wang J, Brunak S, Doré J, Guarner F, Kristiansen K, Pedersen O, Parkhill J, Weissenbach J, MetaHIT Consortium BP, Ehrlich SD, Wang J (2010) A human gut microbial gene catalogue established by metagenomic sequencing. Nature 464:59–65

Rappé MS, Giovannoni SJ (2003) The uncultured microbial majority. Annu Rev Microbiol 57:369–394. https://doi.org/10.1146/annurev.micro.57.030502.090759

Ravel S, Grébaut P, Cuisance D, Cuny G (2003) Monitoring the developmental status of *Trypanosoma brucei gambiense* in the tsetse fly by means of PCR analysis of anal and saliva drops. Acta Trop 88:161–165

Raven JA (1997) Phagotrophy in phototrophs. Limnol Oceanogr 42:198–205

Reeburgh WS (1976) Methane consumption in Cariaco trench waters and sediments. Earth Planet Sci Lett 28:337–344

Reguera G, McCarthy KD, Metha T, Nicoli JS, Teominen MT, Loveley DR (2005) Extracellular electron transfer via microbial nanowires. Nature 435:1098–1101

Reichert AS, Neupert W (2004) Mitochondriomics or what makes us breathe. Trends Genet 20:555–562

Remigi P, Zhu J, Young JPW, Masson-Boivin C (2016) Symbiosis within Symbiosis: evolving nitrogen-fixing legume Symbionts. Trends Microbiol 24:63–75. https://doi.org/10.1016/j.tim.2015.10.007

Remy W, Taylor TN, Hass H, Kerp H (1994) Four hundred-million-year-old vesicular arbuscular mycorrhizae. Proc Natl Acad Sci U S A 91:11841–11843

Reyes-Prieto A, Weber AP, Bhattacharya D (2007) The origin and establishment of the plastid in algae and plants. Annu Rev Genet 41:147–168

Reynolds KT, Thomson LJ, Hoffmann AA (2003) The effects of host age, host nuclear background and temperature on phenotypic effects of the virulent *Wolbachia* strain popcorn in Drosophila melanogaster. Genetics 164:1027–1034

Riedinger N, Formolo MJ, Lyons TW, Henkel S, Beck A, Kasten S (2014) An inorganic geochemical argument for coupled anaerobic oxidation of methane and iron reduction in marine sediments. Geobiology 12:172–181

Rigaud T, Moreau M (2004) A cost of *Wolbachia*-induced sex reversal and female-biased sex ratios: decrease in female fertility after sperm depletion in a terrestrial isopod. Proc R Soc London Ser B 271:1941–1946

Rigaud T, Juchault P, Mocquard JP (1997) The evolution of sex determination in isopod crustaceans. BioEssays 19:409–416

Rinke C et al (2013) Insights into the phylogeny and coding potential of microbial dark matter. Nature 499:431–437. https://doi.org/10.1038/nature12352

Rio RV, Lefevre C, Heddi A, Aksoy S (2003) Comparative genomics of insect-symbiotic bacteria: influence of host environment on microbial genome composition. Appl Environ Microbiol 69:6825–6832

Rio RV, Hu Y, Aksoy S (2004) Strategies of the home-team: symbioses exploited for vector-borne disease control. Trends Microbiol 12:325–336

Rio RVM, Symula RE, Wang J, Lohs C, Wu YN, Snyder AK, Bjornson RD, Oshima K, Biehl BS, Perna NT, Hattori M, Aksoy S (2012) Insight into the transmission biology and species-specific functional capabilities of tsetse (*Diptera: Glossinidae*) obligate symbiont *Wigglesworthia*. MBio 3:e00240–e00211

Rípodas C, Clúa J, Battaglia M, Baudin M, Niebel A, Zanetti ME, Blanco F (2014) Transcriptional regulators of legume-rhizobia symbiosis: nuclear factors Ys and GRAS are two for tango. Plant Signal Behav 9:e28847. https://doi.org/10.4161/psb.28847

Rishiram R, Byung-Hyuk K, Dae-Hyun C, Hee-Mock O, Hee-Sik K (2016) Algae-bacteria interactions: evolution, ecology and emerging applications. Biotechnol Adv 34:14–29

Rodriguez JM (2014) The origin of human milk bacteria: is there a bacterial entero-mammary pathway during late pregnancy and lactation? Adv Nutrit (Bethesda, Md) 5(6):779–784. https://doi.org/10.3945/an.114.007229

Rohwer F, Thurber RV (2009) Viruses manipulate the marine environment. Nature 459:207–212

Rotaru AE, Shrestha PM, Liu F, Shrestha M, Shrestha D, Embree M, Zengler K, Wardman C, Nevin KP, Loveley DR (2014a) A new model for electron flow during anaerobic digestion: direct interspecies electron transfer to *Methanosaeta* for the reduction of carbon dioxide to methane. Ener Environ Sci 7:408–415

Rotaru AE, Shrestha PM, Liu F, Markovaite B, Chen S, Nevin KP, Loveley DR (2014b) Direct interspecies electron transfer between *Geobacter metallireducens* and *Methanosarcina barkeri*. Appl Environ Microbiol 80:4599–4605

Round JL, Mazmanian SK (2009) The gut microbiota shapes intestinal immune responses during health and disease. Nat Rev Immunol 9:313–323

Rousset F, Raymond M, Kjellberg F (1991) Cytoplasmic incompatibilities in the mosquito *Culex pipiens*: how to explain a cytotype polymorphism. J Evol Biol 4:69–81

Russell JA, Latorre A, Sabater-Munoz B, Moya A, Moran NA (2003) Side-stepping secondary symbionts: widespread horizontal transfer across and beyond the *Aphidoidea*. Mol Ecol 12:1061–1075

Ruvindy R, White RA 3rd, Neilan BA, Burns BP. 2016 Unravelling core microbial metabolisms in the hypersaline microbial mats of Shark Bay using high-throughput metagenomics. ISME J 10(1):183–196

Sagan L (1967) On the origin of mitosing cells. J Theor Biol 14:255–274

Samuel BS, Gordon JI (2006) A humanized gnotobiotic mouse model of host-archaeal-bacterial mutualism. Proc Natl Acad Sci U S A 103(26):10011–10016. https://doi.org/10.1073/pnas.0602187103

Samuel BS, Hansen EE, Manchester JK, Coutinho PM, Henrissat B, Fulton R, Latreille P, Kim K, Wilson RK, Gordon JI (2007) Genomic and metabolic adaptations of Methanobrevibacter smithii to the human gut. Proc Natl Acad Sci U S A 104(25):10643–10648. https://doi.org/10.1073/pnas.0704189104

Sandstrom JP, Russell JA, White JP, Moran NA (2001) Independent origins and horizontal transfer of bacterial symbionts of aphids. Mol Ecol 10:217–228

Sanford JA, Gallo RL (2013) Functions of the skin microbiota in health and disease. Semin Immunol 25(5):370–377. https://doi.org/10.1016/j.smim.2013.09.005

Sbicego S, Vassella E, Kurath U, Blum B, Roditi I (1999) The use of transgenic *Trypanosoma brucei* to identify compounds inducing the differentiation of bloodstream forms to procyclic forms. Mol Biochem Parasitol 104:311–322

Scarborough CL, Ferrari J, Godfray HCJ (2005) Aphid protected from pathogen by endosymbiont. Science 310:1781–1781

Schalk IJ, Abdallah MA, Pattus F (2002) Recycling of pyoverdinon the FpvA receptor after ferric pyoverdin uptake and dissociation in *Pseudomonas aeruginosa*. Biochemistry 41:1663–1671

Schink B (2002) Synergistic interactions in the microbial world. Antonie Van Leeuwenhoek 81:257–261

Schwechheimer C, Kuehn MJ (2015) Outer-membrane vesicles from gram-negative bacteria: biogenesis and functions. Nature Rev Microbiol 13:605–619. https://doi.org/10.1038/nrmicro3525

Shigenobu S, Watanabe H, Hattori M, Sakaki Y, Ishikawa H (2000) Genome sequence of the endocellular bacterial symbiont of aphids *Buchnera* sp APS. Nature 407:81–86

Shrestha PM, Rotaru AE (2014) Plugging in or going wirelesss: strategies for interspecies electron transfer. Front Microbiol 5:237

Shrestha PM, Rotaru AE, Aklujkar M, Liu F, Shrestha M, Zarath M, Summers ZM, Malvankar N, Flores DC, Lovley DR (2013) Syntrophic growth with direct interspecies electron transfer as the primary mechanism for energy exchange. Environ Microbiol Rep 5:904–910

Sime-Ngando T (2014) Environmental bacteriophages: viruses of microbes in aquatic ecosystems. Frontiers Microbiol 5:355

Sime-Ngando T, Colombet J (2009) Virus et prophages dans les écosystèmes aquatiques. Can J Microbiol 55:95–109

Sime-Ngando T, Niquil N (eds) (2011) Disregarded microbial diversity and ecological potentials in aquatic systems. Springer, New York

Sime-Ngando T, Lefèvre E, Gleason F (2011) Hidden diversity among aquatic heterotrophic flagellates: ecological potentials of zoosporic fungi. In: Sime-Ngando T, Niquil N (eds) Disregarded microbial diversity and ecological potentials in aquatic systems. Springer, New York

Sinkins SP, Gould F (2006) Gene drive systems for insect disease vectors. Nat Rev Genet 7:427–435

Sinkins SP, Braig HR, O'Neill SL (1995) *Wolbachia pipientis*: bacterial density and unidirectional cytoplasmic incompatibility between infected populations of *Aedes albopictus*. Exp Parasitol 81:284–291

Sinkins SP, Walker T, Lynd AR, Steven AR, Makepeace BL, Godfray HC, Parkhill J (2005) *Wolbachia* variability and host effects on crossing type in *Culex* mosquitoes. Nature 436:257–260

Smith JA, Nevin KP, Lovley DR (2015) Syntrophy growth via quinone-mediated interspecies electron transfer. Front Microbiol 6:121

Snyder AK, Deberry JW, Runyen-Janecky L, Rio RV (2010) Nutrient provisioning facilitates homeostasis between tsetse fly (*Diptera: Glossinidae*) symbionts. Proc Biol Sci 277:2389–2397

Sockett RE (2009) Predatory lifestyle of *Bdellovibrio bacteriovorus*. Annu Rev Microbiol 63:523–539. https://doi.org/10.1146/annurev.micro.091208.073346

Soltis DE, Soltis PS, Morgan DR, Swensen SM, Mullin BC, Dowd JM, Martin PG (1995) Chloroplast gene sequence data suggest a single origin of the predisposition for symbiotic nitrogen fixation in angiosperms. Proc Natl Acad Sci U S A 92:2647–2651

Spor A, Koren O, Ley R (2011) Unravelling the effects of the environment and host genotype on the gut microbiome. Nat Rev Microbiol 9(4):279–290. https://doi.org/10.1038/nrmicro2540

Sprent JI (2001) Nodulation in legumes. Royal Botanical Gardens, Kew, UK

Stams AJM, Plugge CM (2009) Electron transfer in syntrophic communities of anaerobic bacteria and archaea. Nat Rev Microbiol 7:568–577

Stams AJM, Sousa DZ, Kleerebezem R, Plugge CM (2012) Role of syntrophic microbial communities in high-rate methanogenic bioreactors. Water Sci Technol 66:352–362

Stanier RY (1970) Some aspects of the biology of cells and their possible evolutionary significance. Symp Soc Gen Mircrobiol 20:1–38

Starr DJ, Cline TW (2002) A host parasite interaction rescues *Drosophila* oogenesis defects. Nature 418:76–79

Stewart FM, Levin BR (1984) The population biology of bacterial viruses: why be temperate? Theor. Popul Biol 26:93–117

Stewart AD, Logsdon JM, Kelley SE (2005) An empirical study of the evolution of virulence under both horizontal and vertical transmission. Evolution 59:730–739

Stouthamer R, Luck RF, Hamilton WD (1990) Antibiotics cause parthenogenetic *Trichogramma (Hymenoptera, Trichogrammatidae)* to revert to sex. Proc Natl Acad Sci U S A 87:2424–2427

Streng A, Op den Camp R, Bisseling T, Geurts R (2011) Evolutionary origin of rhizobium nod factor signaling. Plant Signal Behav 6:1510–1514. https://doi.org/10.4161/psb.6.10.17444

Summers WC (2001) Bacteriophage therapy. Ann Rev Microbiol 55:437–451

Summers ZM, Fogarty HE, Leang C, Franks AE, Malvankar NS, Lovley DR (2010) Direct exchange of electrons within aggregates of an evolved syntrophic coculture of anaerobic bacteria. Science 330:1413–1415

Suttle CA (2007) Marine viruses: major players in the global ecosystem. Nature Rev Microbiol 5:801–812

Svistoonoff S, Benabdoun FM, Nambiar-Veetil M, Imanishi L, Vaissayre V, Cesari S, Diagne N, Hocher V, de Billy F, Bonneau J, Wall L, Ykhlef N, Rosenberg C, Bogusz D, Franche C, Gherbi H (2013) The independent acquisition of plant root nitrogen-fixing symbiosis in Fabids recruited the same genetic pathway for nodule organogenesis. PLoS One 8:e64515. https://doi.org/10.1371/journal.pone.0064515

Swensen SM (1996) The evolution of actinorhizal symbioses : evidence for multiple origins of the symbiotic association. Am J Bot 83:1503–1512

Taylor FJ (1978) Problems in the development of an explicit hypothetical phylogeny of the lower eukaryotes. Biosystems 10:67–89

Teixeira L, Ferreira A, Ashburner M (2008) The bacterial symbiont *Wolbachia* induces resistance to RNA viral infections in *Drosophila melanogaster*. PLoS Biol 6:2753–2763

Teske M, Hinrichs K-U, Edgcomb V, Gomez A d V, Kysela D, Sylva SP, Sogin ML, Jannasch HW (2002) Microbial diversity of hydrothermal sediments in the Guaymas Basin: evidence for anaerobic Methanotrophic communities. Appl Environ Microbiol 68(4):1994–2007

Thauer RK, Jungermann K, Decker K (1977) Energy conservation in chemotrophic anaerobic bacteria. Bacteriol Rev 41:100–180

Thingstad, T..F, and Lignell, R. (1997). Theoretical models for the control of bacterial growth rate, abundance, diversity and carbon demand, Aquat Microb Ecol 13, 19–27

Thingstad TF, Havskum H, Garde K, Riemann B (1996) On the strategy of 'eating your competitor': a mathematical analysis of algal mixotrophy. Ecology 77:2108–2118

Thomas R, Grimsley N, Escande M-L, Subirana L, Derelle E, Moreau H (2011) Acquisition and maintenance of resistance to viruses in eukaryotic phytoplankton populations. Environ Microbiol 13:1412–1420

Thomson NR, Crow MA, McGowan SJ, Cox A, Salmond GPC (2000) Biosynthesis of carbapenem antibiotic and prodigiosin pigment in *Serratia* is under quorum sensing control. MolMicrobiol 36:539–556

Timmers PHA, Diego A, Suarez-Zuluaga DAS, van Rossem M, Diender M, Stams AJM, Plugge CM (2016) Anaerobic oxidation of methane associated with sulfate reduction in a natural freshwater gas source. ISME J 10:1400–1412

Toft CA, Karter AJ (1990) Parasite-host coevolution. Tren Ecol Evol 5:326–329

Toh H, Weiss BL, Perkin SA, Yamashita A, Oshima K, Hattori M, Aksoy S (2006) Massive genome erosion and functional adaptations provide insights into the symbiotic lifestyle of *Sodalis glossinidius* in the tsetse host. Genome Res 16:149–156

Torsvik V, Øvreås L (2002) Microbial diversity and function in soil: from genes to ecosystems. Curr Opin Microbiol 5:240–245

Trinick MJ (1973) Symbiosis between rhizobium and the non-legume, *Trema aspera*. Nature 244:459–460. https://doi.org/10.1038/244459a0

Trinick MJ (1979) Structure of nitrogen-fixing nodules formed by rhizobium on roots of *Parasponia andersonii*. Planch Can J Microbiol 25:565–578. https://doi.org/10.1139/m79-082

Tsuchida T, Koga R, Horikawa M, Tsunoda T, Maoka T, Matsumoto S, Simon JC, Fukatsu T (2010) Symbiotic bacterium modifies aphid body color. Science 330:1102–1104

Turelli M (1994) Evolution of incompatibility-inducing microbes and their hosts. Evolution 48:1500–1513

Valbuena G, Walker DH (2009) Infection of the endothelium by members of the order Rickettsiales. Thromb Haemost 102:1071–1079

Valen V (1973) A new evolutionary law. Evol Theory 1:1–30

Van den Abbeele J, Claes Y, Bockstaele D, Ray D, Coosemans M (1999) Trypanosoma brucei spp. development in the tsetse fly: characterization of the post-mesocyclic stages in the foregut and proboscis. Parasitology 118:469–478

van der Giezen M, Tovar J, Clark CG (2005) Mitochondrion-derived organelles in protists and fungi. Int Rev Cytol 244:175–225

Van Essche M, Quirynen M, Sliepen I, Loozen G, Boon N, Van Eldere J, Teughels W (2011) Killing of anaerobic pathogens by predatory bacteria. Mol Oral Biol 26:52–61

Van Nguyen T, Wibberg D, Battenberg K, Blom J, Vanden Heuvel B, Berry AM, Kalinowski J, Pawlowski K (2016) An assemblage of Frankia cluster II strains from California contains the canonical *nod* genes and also the sulfotransferase gene *nodH*. BMC Genomics 17:796. https://doi.org/10.1186/s12864-016-3140-1

Vernié T, Kim J, Frances L, Ding Y, Sun J, Guan D, Niebel A, Gifford ML, de Carvalho Niebel F, Oldroyd GE (2015) The NIN transcription factor coordinates diverse nodulation programs in different tissues of the *Medicago truncatula* root. Plant Cell 27:3410–3424. https://doi.org/10.1105/tpc.15.00461

Wang J, Aksoy S (2012) PGRP-LB is a maternally transmitted immune milk protein that influences symbiosis and parasitism in tsetse's offspring. Proc Natl Acad Sci U S A 109:10552–10557

Wang F, Zhou H, Meng J, Peng X, Jiang L, Sun P, Zhang C, Van Nostrand JD, Deng Y, He Z, Wu L, Zhou J, Xiao X. 2009a. GeoChip-based analysis of metabolic diversity of microbial communities at the Juan de Fuca ridge hydrothermal vent. Proc Natl Acad Sci U S A 24;106(12):4840–4845

Wang J, Wu Y, Yang G, Aksoy S (2009b) Interactions between mutualist *Wigglesworthia* and tsetse peptidoglycan recognition protein (PGRP-LB) influence trypanosome transmission. Proc Natl Acad Sci U S A 106:12133–12138

Wang H, Moore MJ, Soltis PS, Bell CD, Brockington SF, Alexandre R, Davis CC, Latvis M, Manchester SR, Soltis DE (2009c) Rosid radiation and the rapid rise of angiosperm-dominated forests. Proc Natl Acad Sci U S A 106:3853–3858. https://doi.org/10.1073/pnas.0813376106

Wang J, Weiss BL, Aksoy S (2013) Tsetse fly microbiota: form and function. Front Cell Infect Microbiol 3:69

Wanner G, Vogl K, Overmann J (2008) Ultrastructural characterization of the prokaryotic Symbiosis in "*Chlorochromatium aggregatum*". J Bacteriol 190(10):3721. https://doi.org/10.1128/JB.00027-08

Warthmann R, Cypionka H, Pfennig N (1992) Photoproduction of H_2 from acetate by syntrophic cocultures of green sulfur bacteria and sulfur-reducing bacteria. Arch Microbiol 157:343–348

Waters E, Hohn MJ, Ahel I, Graham DE, Adams MD, Barnstead M, Beeson KY, Bibbs L, Bolanos R, Keller M, Kretz K, Lin X, Mathur E, Jingwei Ni J, Podar M, Richardson T, Granger GG, Simon M, Söll D, Stetter KO, Short JM, Noordewier M (2003) The genome of *Nanoarchaeum equitans*: insights into early archaeal evolution and derived parasitism. Proc Natl Acad Sci U S A 100:12984–12988

Weeks AR, Breeuwer JA (2001) *Wolbachia*-induced parthenogenesis in a genus of phytophagous mites. Proc R Soc London Ser B 268:2245–2251

Weeks AR, Turelli M, Harcombe WR, Reynolds KT, Hoffmann AA (2007) From parasite to mutualist: rapid evolution of *Wolbachia* in natural populations of Drosophila. PLoS Biol 5:e114
Weinbauer MG (2004) Ecology of prokaryotic viruses. FEMS Microbiol Rev 28:127–181
Weinbauer MG, Rassoulzadegan F (2004) Are viruses driving microbial diversification and diversity? Environ Microbiol 4:1–11
Weiss B, Aksoy S (2011) Microbiome influences on insect host vector competence. Trends Parasitol 27:514–522
Weiss BL, Mouchotte R, Rio RV, Wu YN, Wu Z, Heddi A, Aksoy S (2006) Interspecific transfer of bacterial endosymbionts between tsetse fly species: infection establishment and effect on host fitness. Appl Environ Microbiol 72:7013–7021
Weiss BL, Wang J, Aksoy S (2011) Tsetse immune system maturation requires the presence of obligate symbionts in larvae. PLoS Biol 9:e1000619
Weiss BL, Wang J, Maltz MA, Wu Y, Aksoy S (2013) Trypanosome infection establishment in the tsetse fly gut is influenced by microbiome-regulated host immune barriers. PLoS Pathog 9:e1003318
Welburn SC, Arnold K, Maudlin I, Gooday GW (1993) *Rickettsia*-like organisms and chitinase production in relation to transmission of trypanosomes by tsetse flies. Parasitology 107:141–145
Wernegreen JJ (2002) Genome evolution in bacterial endosymbionts of insects. Nat Rev Genet 3:850–861
Werner GDA, Cornwell WK, Sprent JI, Kattge J, Kiers ET (2014) A single evolutionary innovation drives the deep evolution of symbiotic N2-fixation in angiosperms. Nat Commun 5:4087. https://doi.org/10.1038/ncomms5087
Werren JH (1997) Biology of *Wolbachia*. Annu Rev Entomol 42:587–609
Werren JH, Windsor DM (2000) *Wolbachia* infection frequencies in insects: evidence of a global equilibrium? Proc R Soc London Ser B 267:1277–1285
Werren JH, Windsor D, Guo L (1995) Distribution of *Wolbachia* among neotropical arthropods. Proc R Soc London Ser B 262:197–204
Werren JH, Baldo L, Clark ME (2008) *Wolbachia*: master manipulators of invertebrate biology. Nat Rev Microbiol 6:741–751
Whiteley M, Bangera MG, Bumgarner RE, Parsek MR, Teitzel GM, Lory S, Greenberg EP (2001) Gene expression in Pseudomonas aeruginosa biofilms. Nature 413(6858):860–864
Wieland A, Zopfi J, Benthien M, Kühl M (2005) Biogeochemistry of an iron-rich hypersaline microbial mat (Camargue, France). Microb Ecol 49(1):34–49
Wilhelm SW, Suttle CA (1999) Viruses and nutrient cycles in the sea. Bioscience 49:781–788
Wilkinson TG, Topiwals H, Hamer G (1974) Interactions in a mixed bacterial population growing on methane in continuous culture. Biotechnol Bioeng 6:41–59
Windsor DA (1998) Most of the species on earth are parasites. Int J Parasitol 28:1939–1941
Winkel M, de Beer D, Lavik G, Peplies J and Mußmann M. 2014. Close association of active nitrifiers with Beggiatoa mats covering deep-sea hydrothermal sediments. Environ Microbiol 16(6), 1612–1626
Wong JM, de Souza R, Kendall CW, Emam A, Jenkins DJ (2006) Colonic health: fermentation and short chain fatty acids. J Clin Gastroenterol 40(3):235–243
Xie W, Wang F, Guo L, Chen Z, Sievert SM, Meng J, Huang G, Li Y, Yan Q, Wu S, Wang X, Chen S, He G, Xiao X, Xu A (2011) Comparative metagenomics of microbial communities inhabiting deep-sea hydrothermal vent chimneys with contrasting chemistries. ISME. J Mar 5(3):414–426
Yamaguchi M et al (2012) Prokaryote or eukaryote? A unique microorganism from the deep sea. J Electron Microsc 61(6):423–431. https://doi.org/10.1093/jmicro/dfs062
Yatsunenko T, Rey FE, Manary MJ, Trehan I, Dominguez-Bello MG, Contreras M, Magris M, Hidalgo G, Baldassano RN, Anokhin AP, Heath AC, Warner B, Reeder J, Kuczynski J, Caporaso JG, Lozupone CA, Lauber C, Clemente JC, Knights D, Knight R, Gordon JI (2012) Human gut microbiome viewed across age and geography. Nature 486(7402):222–227. https://doi.org/10.1038/nature11053

Yen JH, Barr AR (1971) New hypothesis of the cause of cytoplasmic incompatibility in *Culex* pipiens L. Nature 232:657–658

Yoshida N, Oeda K, Watanabe E, Mikami T, Fukita Y, Nishimura K, Komai K, Matsuda K (2001) Protein function: chaperonin turned insect toxin. Nature 411:44

Zeh DW, Zeh JA, Bonilla MM (2005) *Wolbachia*, sex ratio bias and apparent male killing in the harlequin beetle riding pseudoscorpion. Heredity 95:41–49

Zhernakova A, Kurilshikov A, Bonder MJ, Tigchelaar EF, Schirmer M, Vatanen T, Mujagic Z, Vila AV, Falony G, Vieira-Silva S, Wang J, Imhann F, Brandsma E, Jankipersadsing SA, Joossens M, Cenit MC, Deelen P, Swertz MA, Weersma RK, Feskens EJ, Netea MG, Gevers D, Jonkers D, Franke L, Aulchenko YS, Huttenhower C, Raes J, Hofker MH, Xavier RJ, Wijmenga C, Fu J (2016) Population-based metagenomics analysis reveals markers for gut microbiome composition and diversity. Science 352(6285):565–569. https://doi.org/10.1126/science.aad3369

Zhong C, Zhang Y, Chen Y, Jiang Q, Chen Z, Liang J, Pinyopusarerk K, Franche C, Bogusz D (2010) Casuarina research and applications in China. Symbiosis 50:107–114. https://doi.org/10.1007/s13199-009-0039-5

Zhu H, Riely BK, Burns NJ, Ané JM (2006) Tracing nonlegume orthologs of legume genes required for nodulation and arbuscular mycorrhizal symbioses. Genetics 172:2491–2499. https://doi.org/10.1534/genetics.105.051185

Chapter 6
Evolution Underway in Prokaryotes

Sébastien Wielgoss, Pierre Leblond, Catherine Masson-Boivin, and Philippe Normand

Abstract Evolution is a phenomenon that escapes immediate attention because changes occur at a very slow pace and are often considered at odds with a religious vision of the world. Using bacteria that replicate so much faster than eukaryotes has permitted to quantify and discern tendencies. Such laboratory evolution implies growth rate, ability to use this or that substrate, but also synthesis and resistance to antibiotics and the ability to interact with eukaryotic hosts.

Keywords Antibiotic · Artificial selection · Cellular network · Deletion · Directed evolution · DNA repair · Endosymbiosis · Epistasis · Evolutionary constraint · Fitness · Fixation · Fixism · Historical contingency · Hitchhiking · Hypermutagenesis · Infection · Inhibitor · Mobile genetic element · Mutagenesis · Mutation · Mutator · Natural selection · Nitrogen fixation · Nodulation · Phenotype · Punctuated equilibrium · Regulator · Resistance · Saprophyte · Secondary metabolite · Selection · Soil · Stress · Virulence

S. Wielgoss
Institute of Integrative Biology, ETH Zürich, Zürich, Switzerland

P. Leblond
Université de Lorraine UMR INRA 1128 Dynamique des Génomes et Adaptation Microbienne (DynAMic) Faculté des Sciences et Technologies, Vandoeuvre Les Nancy Cedex, France

C. Masson-Boivin
LIPM, Université de Toulouse, INRA, CNRS, Castanet-Tolosan, France

P. Normand (✉)
Laboratoire d'Ecologie Microbienne, UMR 5557, Université Claude Bernard Lyon 1, Villeurbanne, France
e-mail: philippe.normand@univ-lyon1.fr

J.-C. Bertrand et al. (eds.), *Prokaryotes and Evolution*,
https://doi.org/10.1007/978-3-319-99784-1_6

6.1 Replaying in the Lab What Occurs over Eons in the Real World

Darwin waited several years upon completion of his book before publishing it (Darwin 1859), one reason for the delay being the strong opposition he expected from many and his lack of enthusiasm for heated exchanges. He was not deceived on that point as ferocious debates ensued in the following years including the famous one organized by the British Association for the Advancement of Science in 1860 at the Oxford University Museum that included an exchange between his friend T.H. Huxley and English bishop S. Wilberforce. It is uncertain what was actually said, but one account states that at one point Wilberforce asked Huxley if it was on his mother's side or on his father's that the lineage originated from apes. Huxley is said to have answered that he would rather be descended from an ape than from a man who misused his great talents to stifle opponents. Several such heated debates have happened over the years up to this day among several groups, prominent among which are religious fundamentalists who still reject the theory of evolution on the ground it implies there is no more place for God in the creation process. The angles of attack range from day-age literal interpretation of the prophets' scriptures to intelligent design. If most commentators adhere to the general idea that organisms give rise to offsprings that are slightly different from them, many adhere to the fixism hypothesis and use the fossil record as evidence for this as it is often under attack for its "holes."

Fixism is a theory compatible with a literal interpretation of scriptures that is often linked to Carl von Linné who wrote "There exists as many different species as the infinite Being has created different forms at the beginning" (von Linné 1737). Fixism is thus a Godless, factual form of creationism, opposed to evolution. It advocates that species do not evolve into others but remain "forever" as they are and presupposes that they were created. The lacunar fossil record is often invoked as support of the fixism theory.

Eldredge and Gould (1973) have proposed an intermediary "punctuated equilibrium" theory that postulates rapid transitions between stable phenotypes, compatible with major genomic events such as deletions, rearrangements, and chromosome fusions, particularly in small isolated communities. New species are known to be abundant in small islands where empty ecological niches together with founder's effect result in a wide range of morphologies leading, for example, to the Galapagos finches that struck Darwin in 1836 for their markedly different beaks adapted to the different food sources present on the islands.

Man has directed the evolution of all organisms around him as far back as we can decipher. The domestication of wolf, for example, has led to the present-day dog breeds such as the Chihuahua and the Great Dane that are indeed very different morphologically yet that share sufficient common genetic material to be genetically if not sexually compatible (Fig. 6.1). Besides the well-known domestication of cattle in Western Asia, wheat in the Fertile Crescent and grapevine in Greece occurred the concomitant domestication of *Lactobacillus* to yield yogurt, of various fungi to

Fig. 6.1 The Great Dane (left) and the Chihuahua (right) are two breeds of dog that belong to species *Canis lupus familiaris* and hence are sexually compatible. They were selected over the centuries either for hunting or as lapdogs with dramatically contrasted results. (http://en.wikipedia.org/wiki/File:Great_Dane_and_Chihuahua_Skeletons.jpg)

make cheeses and of yeasts to make bread and wine. Yeast varieties have been selected over the millennia for tolerance to alcohol and to sulfur oxides, for vigorous growth, or for production of glycerol, butanediol, acetaldehyde, esters, ketones, lactones, phenols, and acetals to confer particular tastes to wines. A similar process occurred with milk to produce the amazing palette of yogurts and cheeses seen today.

Natural selection is the idea that within a species, there is variation and that the environment favors some phenotypes. This is contrasted with artificial selection where nature is replaced as the agent of change by man. Whatever the agent, the notion that there is variation within a species is completely accepted by all microbiologists, by those who have looked at diarrhea-causing or not *Escherichia coli*, at attenuated or not *Mycobacterium tuberculosis* to make vaccines against tuberculosis, and at good or bad wine-making yeasts. Even within a microbial colony growing on a Petri dish, one can see sectors derived from mutations, yielding different colors, different shines, and different yields (Fig. 6.2). Man uses this natural source of variation to obtain lineages with distinct characteristics. This procedure (cycles of mutation followed by screening) was followed on *Penicillium chrysogenum* to increase a thousand-fold the yield of benzylpenicillin (Macdonald et al. 1964) and to increase ethanol resistance and range of substrates of yeast on various microbes used for the production of an ever-increasing range of compounds in various programs that represent enormous investments.

It is generally accepted nowadays that all organisms carry DNA and that this DNA codes for proteins that are responsible for the biochemical operations that sustain life. It is also commonly accepted that this DNA sequence changes over time resulting in changes in the sequence and thus in the properties of proteins, causing

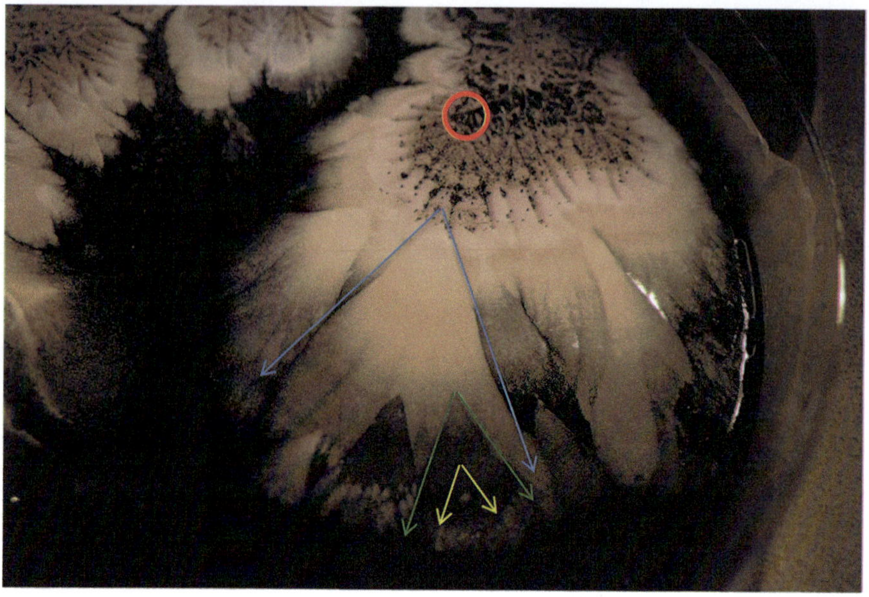

Fig. 6.2 A colony of *Streptomyces lividans* grown on solid medium showing sectors arising from mutant cells (Courtesy of Petar Pujic, CNRS, Lyon). The colony originated in the red circle at the top, a first mutant originated where the two blue arrows start, then a second one from the green arrows, and a third one from the yellow arrows

in turn changes in the physiological properties of the organisms. These changes are commonly known to be associated with devastating and crippling diseases or with antibiotic-resistant microbes.

Evolution is so much part of the life of microbiologists that, for instance, it is used today to modify enzymes to suit particular purposes. For instance, treatment of lipid-rich wastewater depends on the use of lipases; however, the process often breaks down because enzymes are unstable. A thermostable derivative was obtained through error-prone PCR that yielded numerous mutated genes that could then be screened in the *Yarrowia lipolytica* yeast (Bordes et al. 2011). Similarly, wood biomass conversion yield to ethanol is low; it was thus sought to improve it through PCR-based optimization of β-glucosidase and the cellodextrin transporter (Eriksen et al. 2013). Such directed evolution approaches are increasingly adopted to modify key features of enzymes.

Another field where bacterial evolution has played a part has been synthetic chemistry and the environmental fate of its products. Major progress has been accomplished in the industrial synthesis of pigments, of explosives, of pesticides, of hormones, and of thousands of other chemicals over the last century. These molecules have ended in part in the environment where they are often dumped and where they constitute a large source of carbon and nitrogen for starved microbes. There is now a long list of studies that have shown bacteria and fungi able to catabolize the herbicide triazine (Kaufman and Kearney 1970), the explosive trinitrotoluene (Kaplan and Kaplan 1982), the plastic polyethylene (Peixoto et al. 2017), or synthetic contraceptive estrogens (Cajthaml et al. 2009), all molecules that did not exist before the twentieth century and for which microbes had not had time to synthesize adapted enzymes.

We will see in this chapter how microorganisms evolve under our eyes, how their genomes are modified following various treatments, and how cells acquire strikingly different phenotypic characteristics in an emerging discipline called directed evolution.

6.2 Directed Evolution

Any of an organism's phenotypic traits has ultimately been inherited through "descent with modification" and often been shaped by means of natural selection over evolutionary timescales (Darwin 1859). Being able to link hereditary changes with their respective impacts on both phenotype and fitness in the context of the whole population is the holy grail of evolutionary biology, because this allows us to predict evolution (Hindré et al. 2012). Directed or experimental evolution (EE) studies are designed to unravel all processes that govern evolutionary patterns in populations by propagating them for many generations under experimentally controlled conditions (Kawecki et al. 2012). Such an experimental approach (Fig. 6.3) is in stark contrast, and fully complementary, to the classical way evolutionary biologists infer processes from patterns shaped under ecological conditions and long-term evolutionary timescales unfamiliar to us. This is often necessary owing to inherent limitations of the system, e.g., when tracing back the evolution of body plans from the fossil record or when assessing patterns of genetic change and phylogenetic relationships from DNA samples. However, since EE studies allow to observe evolution as it happens, they open up the possibility to directly infer the dynamic processes that ultimately drive evolutionary changes. In recent years, the power of EE has been dramatically increased by the dramatic cost reductions of DNA sequencing and the rise of systems biology focusing on both biological and digital organisms (Hindré et al. 2012). Systems biology aims at connecting population and single cell levels under highly controlled conditions and mainly relies on both miniaturization and high-throughput quantitative data acquisition. These combined approaches also allow to better understand the dynamics of cellular networks and their involvement in evolutionary processes. Here, we will focus mostly on bacterial model systems including their viral parasites. We will see how pattern and process of Darwinian evolution can be observed and understood, respectively, from EE results, which principles will lay the groundwork to better understand the emergence of antibiotics biosynthesis and resistance (Sect. 6.3) and endosymbiosis (Sect. 6.4).

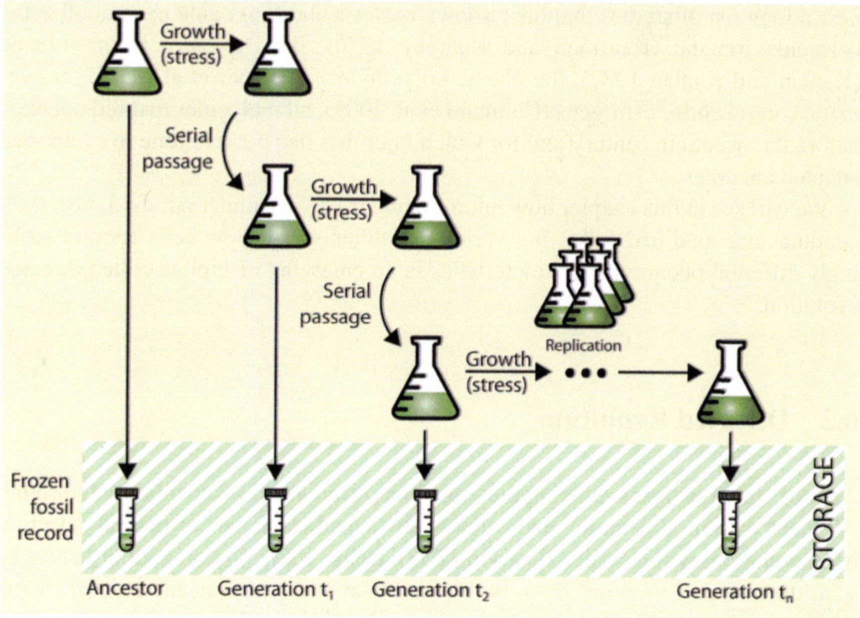

Fig. 6.3 Replicated evolution experiment and frozen fossil record. Twelve replicate lineages of *Escherichia coli* were founded from a common ancestor in 1988 by Richard Lenski and are maintained in shaken glucose-limited culture conditions at 37 °C for 24 h. Each day a one percent fraction of the population is serially transferred to the same, but fresh medium. In parallel, an aliquot of the population is collected at regular intervals from each of the 12 populations and stored in glycerol (a cryoprotectant) in a non-evolving state at −80 °C.

6.2.1 Experimental Models

The majority of EE studies have been conducted under controlled laboratory conditions. These are typically performed in continuous or serially transferred cultures, i.e., in chemostats or batch cultures, respectively; however, alternative experimental settings are also frequently applied that aim at offsetting natural selection by allowing for random evolution via genetic drift, which design is called mutation accumulation (MA) experiments (Barrick and Lenski 2013). MA is utilized to assess the rates and fitness effects of *de novo* mutations.

Escherichia coli is a member of the Gammaproteobacteria that inhabit the intestinal track of warm-blooded animals. Most are harmless but they occasionally colonize other organs causing infections. They are shed in feces regularly and are thus rapidly killed because they are not competitive outside their niche, in the soil or in water. Their only tenuous hope to survive is to infect the intestinal track of a newborn sterile infant. The animal gut is a very stable biotope with a stable temperature, a stable pH, and a steady input of nutrients, but it is also a niche that contains billions of microbial cells belonging to thousands of taxa, and competition is fierce to divide more rapidly than intestinal peristalsis and survive.

Evolution experiments with microorganisms are both attractive and powerful models. Since a flask of rich nutrient broth typically contains in excess of several million *E. coli* cells, it is not surprising that bacterial populations are also fast to evolve both phenotypically and genetically, and this on very short timescales. Ultimately, this is based on a two-step process comprising (1) the chance that a cell retains a spontaneous mutation that has arisen during erroneous replication of its genome and (2) selective pressures that drive advantageous changes to fixation in a population due to the action of natural selection. The first parameter, the mutation rate, is typically low, on the order of 4×10^{-4} per genome per generation in *E. coli* (i.e., after each cell division; Wielgoss et al. 2011). Most *de novo* mutations are thought to be either neutral or nearly neutral, and there are more deleterious mutations than beneficial ones in most organisms (Barrick and Lenski 2013). Thus, in a culture as large as an average *E. coli* population, for example, from the long-term evolution experiment (LTEE; Elena and Lenski 2003), with an effective population size of ~3×10^7 cells and ~6.7 cell divisions until stationary phase, a total of 80,400 mutations will have occurred in the entire culture after less than 24 h! Thus, there is more than enough genetic variation for selection to act upon and optimize the fit of the population to its environment over time, a process called adaptation.

Traditionally, selection experiments have been applied in vaccine development in order to direct evolution toward reduced virulence as is the case for the BCG strain of *Mycobacterium bovis* used as a live vaccine for tuberculosis. However, a more general utilization of such experiments, including that of MA, has been surging as of late in various fields of biological research, in order to tackle a large number of concepts. Among them are themes of general and long-standing interest in evolutionary biology (see Kawecki et al. 2012 for a comprehensive list). For example: how repeatable is evolution if one were able to "rewind the tape of life" as occurs in the movie *It's a Wonderful Life* in which the hero is granted the chance to see an alternate version of his life? To what extent do mutation rates evolve and what is the fate of mutator lineages during adaptation? In what way is genetic change coupled with fitness evolution? Can parasites drive host evolution and *vice versa*? To what extent can cooperation (re-)evolve and be maintained in the face of social cheaters? How do major and complex phenotype transitions evolve *de novo* under simple environmental challenges?

There are three main drivers for the huge success of EE: First, such studies typically involve laboratory organisms which offer huge advantages to the experimentalist. Organisms such as the bacteria *E. coli* (Elena and Lenski 2003) and *Pseudomonas fluorescens* (Rainey and Rainey 2003) are easy to handle and to genetically transform, cheap to propagate, and fast to reproduce while only requiring a modest amount of laboratory space to grow to large population sizes. Very importantly, a large number of these model systems can be maintained in a nonevolving, i.e., cryo-conserved state from which they can be revived at any time to conduct rigorous competition (fitness) assays between an ancestor and its descendant and other crucial experiments to assess phenotypic and genetic changes over evolutionary time. Moreover, the choice of the type of organism is essential to be able to tackle the different abovementioned concepts (Hindré et al. 2012). Second, a range of seminal proof-of-principle studies have demonstrated that EE are a powerful tool that can provide groundbreaking insights into the processes underlying adaptation and biological diversity, including the following study systems: *E. coli* (Lenski 1991) and *P. fluorescens* (Rainey and Travisano 1998), social myxobacteria, such as *Myxococcus xanthus* (Velicer et al. 1998), bacteriophage viruses (Bull et al. 1997), and digital organisms (Adami 2006). Third, it became recently feasible to conduct evolve and resequence (E&R) experiments for a large number of organisms, which utilize the elegance of EE to adapt populations to a novel set of environmental conditions and combine them with the accuracy of emerging next-generation sequencing (NGS) techniques (Long et al. 2015). This allowed pioneering studies to monitor molecular changes at the whole genome level with unprecedented detail and in real time and has resulted in a considerable boost to the field in the past 10 years (Barrick and Lenski 2013; Long et al. 2015).

6.2.2 Phenotypic Evolution

Phenotypic evolution represents the single most studied feature utilizing EE, as it is most amenable to thorough quantitative analysis (Bull et al. 1997; Rainey and Travisano 1998; Velicer et al. 1998; Marchetti et al. 2010; Wiser et al. 2013). One of the hallmark traits studied in EE is the quantitative estimate of fitness change that illustrates the pace of adaptive evolution. Except for cases where natural selection is offset by means of single-cell bottlenecks during MA experiments (e.g., Kibota and Lynch 1996), relative fitness unequivocally increases during adaptation over time. In a hallmark experiment, the long-term evolution experiment (LTEE) with *E. coli* (Fig. 6.3), mean fitness was measured across a total set of 12 populations that have been propagated for more than 64,000 generations as of now (Wiser et al. 2013).

This study indicated a ~70% increase of overall fitness relative to the ancestor after 50,000 generations (Fig. 6.4). The increase was more pronounced in lineages with an ~100-fold elevation of mutation rates early in the experiment, i.e., before generation 10,000 (Fig. 6.4). Furthermore, the trajectories of adaptive evolution are best explained assuming a power-law function, so that the log fitness increases linearly with log time passed. This basically means that while adaptive rates strongly decelerated over the course of the LTEE, the populations will continue to adapt to this rather simple batch culture environment. Finally, the particular power-law relationship itself is most likely generated by the impact of both clonal interference and epistasis through diminishing returns, the two most important population parameters influencing fitness evolution dynamics in the LTEE (Wiser et al. 2013).

In general, the degree of fitness change is dependent on shared, derived phenotypic changes that evolve in response to selection, and several phenotypic traits have been observed to evolve in EE, including features such as cell size, nutrient uptake, stress resistance, global gene expression, production of biofilms or similar structures, (re)emergence of cooperative behavior, or *de novo* evolution of multicellularity (Hindré et al. 2012). From these studies, we can infer a consensus set of three

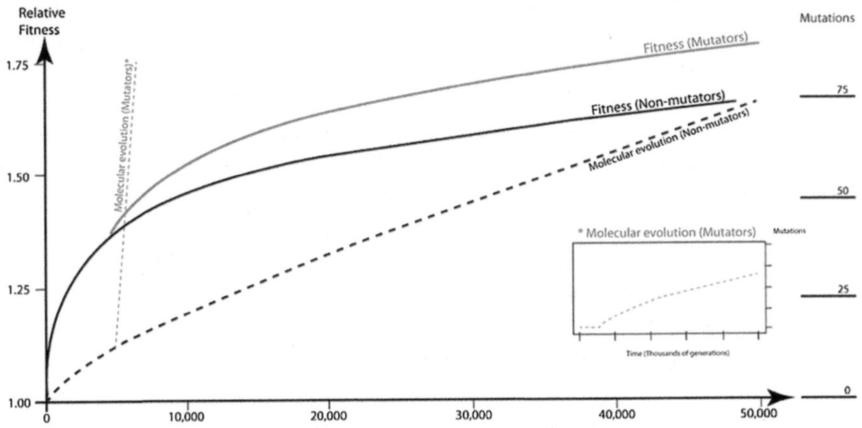

Fig. 6.4 Correlation between fitness and molecular evolution in the long-term evolution lineages with *E. coli*. The best-fit average rates of adaptive evolution for both non-mutator (black solid line) and mutator lineages (green solid line), plotted together with the rate of molecular evolution for both non-mutator (black dashed line) and mutator lineages (Inset: Molecular evolution of a representative early mutator line, Ara-4; fitness trajectories adapted from Wiser et al. 2013; molecular evolution curves adapted from Tenaillon et al. 2016)

global trends: First, phenotypic innovations readily emerge under different conditions if the ancestral population is not optimally adapted to its environment. This is typically achieved very early on in any EE setting, except when historical contingency plays a major role to reach innovation (Blount et al. 2008), but see Van Hofwegen et al. (2016) for an alternative outcome. The causes of that rapid evolution are largely based on parasite-host coevolution, the occupation of new ecological niches which were not accessible by the ancestral population, the recruitment of novel or unused cellular pathways, enhanced metabolic activities, and elevated stress resistance (Hindré et al. 2012). Second, a large number of experimental evolution studies have reported high levels of phenotypic parallelism, which is often accommodated by a similar degree of genetic convergence (Ferenci 2008; Tenaillon et al. 2012; Tenaillon et al. 2016). Interestingly, this outcome of evolution can be idiosyncratic owing to strong constraints on the accessible evolutionary pathways. This deterministic behavior can be offset by genetically removing focal targets of evolution in one metabolic pathway to allow for alternative evolutionary outcomes (Lind et al. 2015). While the trend for parallelism is a major theme, there are exceptions, as evident for the evolution of citrate usage under the oxygenic batch culture conditions of the LTEE (Blount et al. 2008). Citrate is part of the universal TCA cycle, but since there is no dedicated citrate transporter, it usually stays outside the cell where it can chelate divalent cations and fragilize the wall. Citrate is also used as a chelator of iron and is added to bacterial growth medium. The LTEE permitted to identify a lineage with an increased fitness due to a mutation in a gene for a transporter optimized for dicarboxylates (*dctA*) that had acquired the ability to internalize citrate, thus making available another carbon source.

Moreover, not all mutations affect the same position in a given genetic target, and, thus, this allelic divergence may lead to phenotypic divergence, due to different relative effects on fitness in the same environments (Crozat et al. 2010; Tenaillon et al. 2012; Plucain et al. 2014). Third, populations can evolve complex population structures, owing to the emergence and maintenance of stable polymorphisms (Rainey and Travisano 1998; Le Gac et al. 2012). This can be quite unexpected in some cases, given the homogenous environment of well-mixed batch cultures in which many evolution experiments have been conducted and the presence of a single resource (Rozen and Lenski 2000). A well-studied example of the evolution of population structure in a homogeneous environment with a single carbon source was previously reported for 1 of the 12 populations of the seasonal environment of the LTEE (Rozen and Lenski 2000). In this experimental population, Ara-2, two monophyletic sub-lineages emerged after 6500 generations, which were phenotypically differentiated into visibly different colony morphs, i.e., small (S) and large (L) types. These sub-lineages had coevolved and stably coexisted for more than 50,000 generations in a negative-frequency-dependent manner, in which L had an advantage over S ecotypes during exponential growth on glucose, and S clones had an advantage during stationary phase. Importantly, the ecological and evolutionary dynamics of these coexisting lineages is slightly more complex than initially anticipated (Le Gac et al. 2012). In particular, while the latter study confirmed that L and S clones had stable frequency dependence at any given time point throughout the entire course of evolution, L clones were able to invade S clones both from earlier and later generations in reciprocal invasion experiments, while S clones could only invade L clones from earlier time points. This asymmetric interaction is compatible with a model in which the L ecotypes repeatedly encroach on the niches of the S sub-lineages over evolutionary time, thereby continually forcing the S type into ever new positions in niche space. The molecular basis of this complex population structure has been disentangled by whole genome sequencing and gene expression data and showed that only three mutations, all in regulatory genes, allowed for both niche invasion and stable coexistence by strongly altering the central metabolic pathways (Plucain et al. 2014).

6.2.3 Genome Evolution

Phenotype evolution can happen either gradually (optimization regime), or by sudden leaps (innovation regime, saltation), which ultimately depends on the tempo and mode of the underlying genotypic changes. But how are phenotype and genome evolution linked? The advent of novel sequencing techniques and sophisticated computer tools coupled with well-designed evolution experiments has provided unprecedented and rigorous data to tackle this question (Barrick and Lenski 2013). According to JGI's genome online database (GOLD), as of October 2018 there-are 107,218 genome sequencing projects registered (URL: https://gold.jgi.doe.gov/statistics), out of which 88.1% are already completed.

The first genomes that were ever sequenced were those of bacteriophages which coincided with the rise of novel DNA sequencing techniques in the late 1970s. In 1978, the first DNA-based bacteriophage genome from the PhiX174 phage, was sequenced by Fred Sanger, and was found to have a size of 5386 nucleotides that code a mere 11 genes (Sanger et al. 1977). It took almost 20 more years until Craig Venter's lab at Johns Hopkins University in Baltimore reported the first complete genome sequence of a bacterium, *Haemophilus influenzae*, with its relatively small size of ~1.8 M base pairs encoding 1749 genes (Fleischmann et al. 1995). In the field of EE studies, evolve and resequence (E&R) experiments were performed first in bacteriophage-based model systems, before the same approach was feasible for even a single or a handful of bacterial clones (Velicer et al. 2006; Barrick and Lenski 2009). Thus, in the latter cases, authors had to typically rely on one of three approaches to assess the genetic basis of phenotype diversity: targeted and/or random single gene Sanger sequencing (Woods et al. 2006), as well as DNA hybridization to identify mutations related to insertion sequence (IS) elements (Schneider et al. 2000). Nowadays, the utilization of E&R experiments is a standard approach in the field of EE studies, including viruses (Bull et al. 1997) and asexually evolving bacteria (Barrick and Lenski 2009). One of the most important questions that has been tackled and answered with E&R was how changes in fitness are related to the rate and nature of genotypic changes in the abovementioned microbial systems. As for viruses, in a seminal paper, Bull et al. (1997) sequenced the genomes of nine PhiX174 enterobacteriophage plaques that evolved on two different hosts, its original one, *E. coli*, and a novel related one, *Salmonella typhimurium*. This experiment was performed in chemostat environments supplemented with LB broth under a high-temperature regime for a total of 11 days, i.e., approximately 1000 temporal generations with 5 independent lineages. That initial phase was followed by 22 additional days with four lineages at high constant temperatures and in two cases for each temperature subjected to a host switch. Genomes (~5.4kbp each) displayed a substantial degree of convergent evolution: a total of 119 independent substitutions at 68 nucleotide sites were detected after 11 days, 25–50% of which were either parallel substitutions of the same nucleotide at the same site in two or more lineages or reversions from the evolved one to the ancestral nucleotide state. In essence, viral nucleotide substitution rates changed in parallel with fitness improvement. Indeed, they increased more rapidly early on and slowed down significantly in the extended lineages after that, except when subjected to a host switch to *S. typhimurium*. In the latter case, the speed of adaptation was similarly fast as in the early adaptive phase on either host. A follow-up study (Wichman et al. 1999) also revealed the molecular basis of adaptation and its dynamics in a similar setting for two independent evolution lines only adapting to the novel host, *S. typhimurium*, for about 1000 temporal generations. As before, a dozen amino acid substitutions appeared in both lineages during fitness evolution, of which >95% were adaptive and >50% were present in both lineages. Interestingly, however, the order of their appearance was quite different, and parallel substitutions did not reflect the changes with the largest beneficial effects or indicated a common trajectory of adaptation.

How does this pattern of synchronous coupling of fitness and genome evolution during adaptation hold for biological organisms more complex than viruses? The first seminal study to answer this question derives from the long-term evolution experiment (LTEE) with *E. coli* (Barrick and Lenski 2009). The study considered the whole genome sequences of 6 individual isolates sampled after 40,000 generations from one focal population (Ara-1). In that particular case, the genome evolved with a surprisingly constant, clock-like regularity until halfway through the experiment, while the rates of adaptation decelerated strongly. Despite this linear pattern, most of the 45 mutations detected in the 20,000 generation isolate were considered to be beneficial on average and would have hinted at a much more complex coupling of both rates in bacteria. This pattern was in stark contrast to the former studies focusing on bacteriophages. To clarify if that pattern was an outlier or a more general theme across lineages in the LTEE, a recent study analyzed the tempo and mode of evolution for the entire set of 12 populations sampled at each of 11 time points over 50,000 generations of evolution (Tenaillon et al. 2012). As expected, nearly neutral (synonymous) mutations fixed at a constant rate, but the beneficial mutation rates decreased substantially over time. As expected, most mutations were likely to be beneficial and accumulated at a rate of 17.1- and ~3.4-fold the neutral rate over the first 500 generations and the entire length of the LTEE, respectively. Thus, the speed and pattern of molecular evolution qualitatively matches the fitness trajectories over the same period of time (Wiser et al. 2013) and reconciles the earlier studies on phage evolution (Fig. 6.4). Notably, the rise and fixation of mutations that led to ~100-fold mutation rate increases had a slight but clearly positive impact on fitness evolution relative to those lineages that retained the ancestral non-mutator status (Fig. 6.4).

6.2.4 *Evolution of Cellular Networks*

Adaptive evolution happens when environments change, i.e., in response to stress. Importantly, while some stress factors have very limited evolutionary consequences, others have a very broad impact on many cellular reactions at once, such as when experimentally evolved populations are grown under significantly increased temperature regimes, starvation, or suboptimal nutrient supply. In particular, very abrupt environmental changes might call for drastic cellular changes in order for populations to adapt. It is thus not surprising that many bacterial populations in the abovementioned settings have converged to a very common evolutionary solution: instead of relying on hundreds of independent genomic changes, a small number of mutations in so-called transcriptional master regulators (network hubs) allow changes in gene expression on a global scale involving parallel transcriptional changes in hundreds or thousands of genes at once, a phenomenon called pleiotropy. Evolution experiments with both living and digital organisms have played a pivotal role in better understanding the rules that govern the evolution of such complex cellular networks. In particular, the following genes and pathways have been implicated as targets of selection as revealed by genomic, transcriptomic, or proteomic analyses: *rpoS*, encoding alternative RNA polymerase RNAP (sigma-S), which initiates general stress response and transitioning to stationary phase; *rpoB* and *rpoC*, two core subunits B and C of RNA polymerase; *hfq*, encoding the RNA chaperone Hfq, a posttranscriptional regulator of small RNAs, which also regulates sigma-S; *spoT*, encoding (p)ppGpp synthase hydrolase, a sigma-S-mediated stringent response regulator; *topA*, encoding DNA topoisomerase I, regulating the degree of DNA supercoiling; and the *fis-dusB*-operon, encoding the DNA-binding protein Fis and its transcriptional regulator DusB, respectively, which both have a range of broad effects including influencing DNA supercoiling and the onset of transcription and translation (Hindré et al. 2012).

Importantly, master regulators do not have to mutate themselves, but instead cellular networks can be evolutionarily rewired by mutations in the target genes to increase hub gene dependence resulting in pervasive epistatic interactions. This was evident in two focal populations of the LTEE, Ara-1 and Ara+1, respectively. Here, the deletion of master regulator *crp*, which codes for the catabolite repression protein (Crp), leads to drastic changes both in global transcription and in growth rates in two later-generation isolates from generation 20,000, while the effect was relatively mild in their common ancestors (Cooper et al. 2003, 2008). Since there was no mutation in the *crp*-gene of either clones, parallel Crp-regulon expansion must be the result from far-reaching epistatic interactions between the master regulator and mutations in other regulatory genes and effectively lead to the restructured connections. Epistasis, defined as the interaction between genes, is measured as the difference in fitness additivity between the different mutants and the fitness of the combined mutant. In an unpublished follow-up study, Wielgoss et al. (in preparation) subsequently investigated the changes in lineage Ara-1. By deleting the *crp* gene in various genetic backgrounds, including constructed mutants carrying different early mutations in regulatory genes that were eventually fixed in the population, the authors demonstrated that the massively parallel rewiring was initially caused by a single base-pair mutation in the *topA* gene, which leads to global parallel changes in the expression profiles of hundreds of genes at once, many of which were not dependent on *crp* in the ancestor. This demonstrates that interactions between different transcriptional regulators may change dynamically over time, and this based on a small number of mutational changes.

Especially in light of the high complexity of cellular networks, one of the major advantages of applying EE approaches is the frequent discovery of novel mechanisms hitherto unknown using classical genetics. Based on the power of natural selection, rare beneficial mutations are rapidly sorted from the many neutral or minor changes that arise at any given time point in the population, analogous to finding a needle in a haystack, and their frequencies are subsequently boosted during competition for limited resources. For example, a novel regulatory gene, *dusB*, which encodes the tRNA-dihydrouridine synthase B could be identified by applying phylogenetic and experimental analyses to population samples of the LTEE. It was shown that DusB regulates the expression of the global regulator Fis, both genes being part of the same operon. Mutations in both genes were beneficial under the conditions of the LTEE and could be demonstrated to result in similar phenotypic changes in DNA superhelicity (Crozat et al. 2010).

Finally, the influence of challenging environments during EE can result in evolutionary innovations that ultimately help to further our understanding of how cellular network architecture evolves. This is illustrated by a seminal study that focused on cooperative bacteria. In an attempt to better understand evolutionary transitions between cooperative and selfish (cheating) behaviors, the social bacterium *Myxococcus xanthus* was studied, which cells engage in multicellular development and a coordinated group movement in order to survive and reproduce. Cell populations of *M. xanthus* were experimentally evolved in spatially homogeneous asocial environments for ~1000 bacterial generations, and the repeated rise of socially defective phenotypes that dominated the microcosms due to growth advantages was observed (Velicer et al. 1998). Subsequent developmental assays revealed the presence of several potent obligate cheaters (OC) that benefitted from the presence of both their socially proficient ancestors and other wild-type strains when in the minority (Velicer et al. 2000). Strikingly, when allowed to compete with the socially competent ancestors for six sequential cycles of development in response to starvation followed by growth in asocial liquid medium, one OC phenotype evolutionarily restored its social autonomy and also evolved into a superior cooperator following an intermittent, strong population crash (Fiegna et al. 2006). The surviving, dominant phenotype, dubbed PX, in reference to the rise of the phoenix from the ashes in Greek mythology, carried a single mutation in a hitherto unknown small RNA regulatory gene essential for controlling entry in development and thus spore formation (Yu et al. 2010).

6.2.5 General Rules and Patterns

6.2.5.1 Mutation Rates

The intrinsic rates and fitness effects of *de novo* mutations can be assessed using evolution experiments, typically using MA settings since natural selection is offset by applying single-cell bottlenecks. The general conclusion is that the rate of spontaneous base-pair substitutions in both bacteria and eukaryotic microorganisms is quite low, i.e., on the order of 10^{-10} to 10^{-9} per base pair per generation (Barrick and Lenski 2013). Given that the genome of the former consists of 1–10 million base pairs, a single point mutation will emerge *de novo* at rates close to 0.001 to 0.0001 per genome per generation. The rates of insertions and deletions (indels) of one or few consecutive base pairs are typically much lower, around one-tenth to one-hundredth the rate for substitutions (Barrick and Lenski 2013). One common observation derived from experimentally adapting populations is the frequent rise of hypermutator lineages, carrying mutations in genes that affect DNA proofreading or error correction (see Raynes and Sniegowski 2014 for a review). Since many genes can affect the mutation rate, and these mutations *per se* only cause subtle negative effects, hypermutators could reach frequencies of up to 10^{-4} in any *E. coli* population. Hypermutators can produce beneficial mutations at a higher rate than an isogenic non-mutator (Denamur and Matic 2006). Under this scenario, the linked mutator gene can then hitchhike to fixation along with selected beneficial mutations. Finally, in the long term, elevated mutation rates also increase the genetic load, as the number of highly beneficial mutations precipitously decreases in constant environments, while more and more (slightly) deleterious mutations have been acquired along the way. As a result, the mutational defect in DNA repair is expected to be compensated for as time goes by, so that mutation rates evolve dynamically during adaptive evolution (Raynes and Sniegowski 2014). These anticipated rate dynamics were demonstrated in one of the focal lineages of the LTEE, Ara-1 (Wielgoss et al. 2013). The latter evolved a MutT-mutator phenotype that fixed before 30,000 generations in the experiment and had an ~150x elevating effect on the base-pair point mutation rate. Later the same population was invaded independently by two *mutY* mutations that decreased the mutation rate by ~40–60%, while the population was still adapting. This led to a corresponding decrease in the genetic load. In essence, this demonstrated that the mutation rate declined in response to reduced further potential of acquiring highly beneficial mutations.

6.2.5.2 Evolutionary Trade-offs

There is strong empirical evidence that evolution cannot optimize all phenotypes at once in a population and that most evolved genotypes carry strong fitness trade-offs when facing different environmental conditions. A "perfectly" adapted population does not exist because a population adapts to a set of environmental parameters at the "price" of a reduced adaptability to a different set of parameters (Schluter 2000). There is a range of different mechanisms that can result in the evolution of such trade-offs, including two that critically rely on the relative strengths of natural selection and genetic drift: first, antagonistic pleiotropy (AP) which is the simplest form of an evolutionary trade-off, where a beneficial mutation during adaptation to one environment turns out to be harmful when the environment changes and, second, mutation accumulation (MA), in which random genetic drift drives the emergence of mutations that are mainly neutral in one condition, but which later turn out to be strongly deleterious in alternative environments. The 12 populations of the LTEE evolved resource specialization to glucose as single carbon source (Cooper and Lenski 2000). In contrast to increased fitness when consuming glucose, the populations showed a strong tendency for reduced catabolic function when facing a plethora of different carbon sources other than glucose, and the rate of decay was strongest for early time point samples, which trend paralleled the fitness increase during adaptive evolution on glucose as carbon source. Importantly, the decay rate in catabolic function was quite similar for the several mutator lineages in the LTEE even though mutation rates were elevated by a factor of ~100, further strengthening the importance of AP in large populations. However, as indicated above, AP is not always the cause for the evolution of trade-offs. By offsetting the evolutionary force of natural selection by frequently funneling the population through single-cell bottlenecks, MA will be the dominant force ultimately leading to the a gradual decay in phenotypic function and/or fitness (Kibota and Lynch 1996).

6.2.5.3 Epistasis and Evolutionary Constraints

Mutational effects largely depend on the genetic background in which they appear. In its simplest form, the interaction between two (or more) mutations is studied by assessing if the grand total of all phenotypic and fitness effects of the single mutants deviates from their compound effect in an isogenic individual carrying all mutations. That deviation is dubbed the epistatic effect and has been studied in detail only for very few cases such as antibiotic resistance (Long et al. 2015). For example, there are a total of five beneficial mutations in a specific beta-lactamase allele (TEM*) that increase resistance of their bacterial carrier to a third-generation cephalosporin by five orders of magnitude relative to the reference strain harboring the wild-type allele TEM^{wt}. By assessing all 120 possible mutational pathways between these 5 mutations, Weinreich et al. (2006) showed that only very few of them (18) were actually accessible due to the existence of major constraints to protein evolution. One interpretation of this study is that these constraints might help to partly explain the large degree of genetic and phenotypic parallelism in selection experiments in general. Importantly, such constraints on protein evolution might also be influenced by second-order selection for evolvability. In a focal lineage of the LTEE, Ara-1, two beneficial alleles in the same gene, *topA*, which affects DNA superhelicity, emerged early and temporally coexisted in the population. However, only the allele with the slightly lower relative fitness eventually fixed in the population. It could be shown that carriers of that slightly inferior allele were more evolvable, as their fitness strongly increased when adding the next known substitution in the population, namely in *spoT*, coding for a transcriptional regulator of the stringent response, while that same mutation had no positive fitness effect in carriers of the superior *topA* allele (Woods et al. 2011). Recently, it was proposed that by jointly analyzing the trajectories for fitness and accumulated beneficial mutations in evolution experiments, it should be possible to infer the nature of epistasis among the beneficial mutations (Kryazhimskiy et al. 2009). In order to test this prediction, the first 5 mutations in focal lineage Ara-1 of the LTEE were identified, all of their possible 32 combinations constructed, and their fitness effects assessed by performing competition assays (Khan et al. 2011). Based on these data, it could be shown that the decelerating fitness trajectory was due to widespread epistatic interactions among the beneficial mutations that fixed early. Thus, successive beneficial mutations tended to produce diminishing returns with genotype fitness, while the interactions comprising particular mutations had the exact opposite effect. Interestingly, the very strong constraints visible in the abovementioned case study on a beta-lactamase could not be confirmed, and it was suggested that a rather simple function of epistasis could capture most of the dynamics of fitness evolution. Ultimately, however, this demonstrates that our knowledge of how epistatic interactions work on a global scale is still quite rudimentary and that further research is needed (Long et al. 2015).

6.2.5.4 Historical Contingency

Many experiments with bacteria are carried out under conditions in which the organisms are readily capable of exploiting all available nutrients (Hindré et al. 2012). However, sometimes, even if the nutrient is available at high concentrations, it might remain unused until a so-called "actualizing" mutation is evolved that critically hinges on the presence of earlier "potentiating" substitutions, those depicting a case of historical contingency (Barrick and Lenski 2013). One of the most spectacular such examples was found after 30,000 generations of experimental evolution in 1 of the 12 lineages of the LTEE, Ara-3. The latter population has evolved to become a citrate user (Cit+) under aerobic batch culture conditions. This observation represents a critical key innovation compared to the reference strains of *E. coli* that are Cit+ only in the absence of oxygen, a phenotype that is routinely used to taxonomically classify strains as *Escherichia* bacteria in general. The key actualizing mutation was a duplication event that placed a new transcriptional promoter in front of a previously unused citrate transporter gene, *citT*. The final step to boost the Cit+ lineage was a refinement mutation in *dctA* which helps reimport succinate previously lost due to the antiporter activity of CitT (Blount et al. 2012). It could be shown that this particular mutation can repeatedly evolve, i.e., that it is not an extremely rare event, and that its impact must be contingent on the presence of earlier potentiating substitutions (Blount et al. 2008, 2012). First, even though a total of around 1 billion mutations had been tested by natural selection across all populations by 30,000 generations, and though the final phenotype is highly beneficial, Cit+ only emerged once, and this one late in 1 of the 12 populations. Second, by replaying evolution from various time points of the population's history, the Cit+ phenotype did never appear from starting points prior to 15,000 generations, but arose from 20,000 generation samples or later (Blount et al. 2008). Third, by resequencing the genomes of those re-evolved Cit+ phenotypes and comparing them to the actual winning genotype, it is clear that in all cases an IS3 insertion sequence mediated the duplication event repeatedly and actualized the phenotype and that historical contingency was driven by epistatic interactions (Blount et al. 2012).

6.2.5.5 Directed Evolution and the Pitfalls of Evolutionary Misconceptions

It is important to highlight that aerobic citrate usage in *E. coli* has been observed in other settings, too. Most recently, Van Hofwegen et al. (2016) specifically tested how rapidly citrate usage can emerge in *E. coli* and found that it could evolve relatively fast, within 3000 generations. Adaptation was based on principally the same mutational events involving the actualizing *citT* promoter capture and the refinement mutation in *dctA*. Thus, the latter study is a premier example for an initiated directed evolution experiment. Based on the fundamental insights derived from the LTEE, Van Hofwegen et al. (2016) identified those environmental conditions under which an *E. coli* strains would likely and rapidly evolve an intended phenotype, *Cit+* under aerobic conditions, and utilized the combined power ER experiments to achieve this goal. In particular, compared to the LTEE, the authors changed the environment by increasing the time the cell population spends in stationary phase before being transferred to fresh medium from less than 1 to 7 days, i.e., by a full order of magnitude. Thus, both experiments provided strong evidence that gain of function mutations can repeatedly evolve, here, by linking a foreign promoter with a coding sequence not previously expressed under current environmental conditions (Blount et al. 2012; Van Hofwegen et al. 2016). Because any gene consists of at least the coding sequence and all affiliated regulatory elements, promoter capture by definition gives rise to novel gene functions.

We have laid out both important observations and inferred general rules derived from evolution experiments including the longest running evolution experiment with *E. coli* initiated by Richard Lenski in 1988 (the LTEE). It is our hope that the design and conduction of directed evolution experiments with microorganisms will have the same impact that advanced breeding designs with plants had during the green revolution in agriculture as spearheaded by Nobel laureate Norman Borlaug and others. It will hopefully help humankind find truly novel or vastly refined microbial functions that can help improve prosperity and overcome poverty, pollution, and conflict over limited or unequally distributed resources.

6.3 Biosynthesis and Resistance to Antibiotics

Antibiotics are small molecules (molecular weight around 2.000 daltons) either produced by living organisms (mostly bacteria and fungi) or partly synthesized chemically. They act by interaction with specific cellular targets or processes and consequently inhibit cell growth or kill living cells.

Although the first antibiotic to be discovered was penicillin by Alexander Fleming in 1929, the term "antibiotic" was only later introduced by Selman Waksman (awarded the Nobel Prize in 1952), the co-discoverer of streptomycin, which constituted the first efficient drug therapy against tuberculosis that caused in the following years the progressive demise of specialized hospitals called sanatoriums the world over. Antibiotics synthetized by living organisms belong to a broad group called secondary metabolites or natural products (NP). This property gave them their misleading qualification of "secondary," not related to a presumed low importance for the producing organism, but because their production is usually restricted to post-exponential growth phase. Yet, it is important to distinguish the activity for which they are screened and used as drugs in human or veterinary medicines (antibiotics including antifungal, antitumor, immunosuppressant, antiviral activities) from their biological activity and ecological roles in nature which remain for the most part unknown. Bérdy (2005) evaluated at 25,000–30,000 the number of molecules from all sources (microbes, plants, and animals) shown to have antibiotic properties. The main groups of antibiotics used in hospital were discovered in the 1940s and 1950s (tetracyclines, cephalosporins, aminoglycosides, macrolides). Microbial genomes are constantly modifying genes, cutting and pasting domains resulting in new metabolites that interfere positively or negatively with the metabolism of the producing cell or that of neighboring cells.

All classes of antibiotics in use today were first discovered prior to the mid-1980s. The newly discovered antibiotics triggered the decreases in many infectious diseases in Europa and North America. Thanks to the combination of these medical advances and improved practices (food production and distribution, access to safe water for drinking, sanitation) death rates, especially infant mortality, were strongly reduced promoting the rapid expansion of human population size since the end of Second World War. After this golden age, it was expected that the infectious diseases could be definitely defeated and even eradicated as has been the case for smallpox. However, this constituted a new evolutionary challenge for microbes, and infectious diseases still constitute a very serious threat, and the Global Burden of Disease Study (GBD DaHC 2015) estimated yearly deaths due to infectious diseases to reach 7.4 million of deaths in 2013 of 54.8 million in the world.

A worsening factor is the emergence of antibiotic resistance that started soon after the first use of newly identified compounds. This phenomenon took its threatening dimension when associated to the discovery that bacteria were able to acquire resistance genes from their neighbors. This phenomenon was revealed during outbreaks of dysentery in Japan in the 1950s: the emergence of *Shigella dysenteriae* strains resistant to several antibiotics used in therapy suggested that the resistance could be spread in bacterial populations (Watanabe 1963). Horizontal gene transfer is now recognized as the key mechanism for resistance dissemination. The intensive use of antibiotics in human and veterinary medicine but also for other unrelated applications (e.g., growth promotion in animal) by boosting the selection pressure for resistance largely favored the emergence and spreading of pathogenic bacteria resistant to many antibiotics, the so-called multidrug-resistant strains or MDR. Nowadays, the world population is facing the re-emergence of old infectious diseases resulting from the development and spread of antibiotic resistance mechanisms. The main drug-resistant pathogens have emerged from hospital-acquired diseases (*M. tuberculosis*, the enteric Gram-negative pathogens *Escherichia coli*, *Salmonella enterica* and *Klebsiella pneumoniae*, *Pseudomonas aeruginosa*, *Acinetobacter baumannii*, *Staphylococcus aureus*, *Clostridium difficile*) (Davies and Davies 2010). Each year more than 37,000 deaths in Europe and more than 99,000 in the USA are assigned to drug-resistant nosocomial infections (source European Centre for Disease Prevention and Control, http://ecdc.europa.eu/).

In this chapter, we will try to describe the mechanisms underpinning the synthesis of antibiotics, the mechanisms underlying resistance to antibiotics, the role of antibiotics in nature, and how evolution drives the emergence of new bioactive molecules and the concomitant resistance to these new molecules.

6.3.1 Synthesis of Antibiotics

Biosynthesis of secondary metabolites proceeds from precursors originating from primary metabolites. These precursors are diverse and include amino acids, carboxylic acids, or nucleotides and must be activated to be used as substrate by synthases. The resulting scaffolds are then modified by hydroxylation, glycosylation, methylation, or acetylation, i.e., the so-called tailoring enzymes. Two main biosynthetic machineries, the polyketide synthases (PKS) and the non-ribosomal peptide synthases (NRPS), are responsible for the synthesis of a wide diversity of compounds.

Polyketides do provide famous medicines like antibiotics tetracycline, erythromycin, and spiramycin but also anticancer treatments (e.g., daunorubicin), immunosuppressants (rapamycin), or antiparasitics (avermectin). More than one third of the natural products approved as drugs in the 2005–2007 period were polyketides. Polyketide synthases belong to three distinct types, I, II, and III (Staunton and Weissman 2001).

Type I PKSs are large multifunctional enzymes organized into modules, each of them including a set of condensation active domains. The minimal PKS is defined by the three domains acyl carrier protein (ACP), acyltransferase (AT), and keto synthase (KS). A non-iterative condensation procedure is achieved where each module is used once per elongation cycle to incorporate a precursor unit (acyl-CoA or malonyl-CoA). The length of the chain corresponding to the number units incorporated depends of the number of domains of the synthase. Consequently, type I PKS genes may be as large as several tens of kilobases. Additional domains involved in modification of the extending chain are keto reductase (KR), dehydratase (DH), and enoyl reductase (ER). Type II PKSs responsible for the synthesis of the aromatic polyketides are multienzymatic complexes working in an iterative manner; each module is used several times in the extension step. Type III PKSs also called chalcone synthase-like PKSs are homodimeric enzymes possessing a single KS domain and direct binding of free activated precursor to the KS domain (i.e., do not require an ACP).

The different classes of PKSs are closely related to the fatty acid synthases (FAS) whose products are essential in most living organisms for synthesis of the membrane constituents (Jenke-Kodama and Dittmann 2005). Archaea would have acquired this ability by the mean of gene transfer from eubacteria (Dibrova et al. 2014).

The second main machinery is the non-ribosomal peptide synthases (NRPS). In contrast to classical peptides where an mRNA is translated by the ribosome, the sequence of the amino acid and the specificity of the precursors (more than 300 different amino acids and analogs including D forms) are scheduled by the order of catalytic sites within the synthase. The synthases are like in PKS organized in modules containing domains, the minimal one being a domain for amino acid specificity and activation (adenylation, A), a carrier protein (peptidyl carrier protein, PCP), and a condensation domain called C. The amino acid sequence depends on the specificity domains along the genes. The native scaffold is generally circularized, modified in a post-translation step (methylation, formylation, epimerization, oxidation, reduction), and includes diverse moieties (sugars, fatty acids, etc.). A large diversity of end products exhibiting various bioactivities is reported (Schoenafinger and Marahiel 2012) with famous examples being penicillins, cephalosporins, and vancomycin considered a last resort antibiotic against multiresistant organisms.

Since PKSs and NRPSs show similar organizational and enzymatic processes, it is not rare to find PKS and NRPS domains fused in a single enzymatic entity providing hybrid NRP-PK final products.

Polyketides and non-ribosomal peptides are typical products of the filamentous soil bacteria *Streptomyces* and other actinomycetes. The *Streptomyces* genus provides by itself more than two thirds of the active compounds isolated. In more recent periods, fungi and rare actinomycetes, the phylum to which *Streptomyces* and *Micromonospora* belong, became also major providers of NPs. Biosynthetic genes are usually clustered at the same chromosomal or plasmid-borne locus together with regulatory and resistance genes. The size of the cluster varies from 15 kb to more than 140 kb: rapamycin from *Streptomyces hygroscopicus* and stambomycin from *Streptomyces ambofaciens* gene clusters are among the largest ones ever described (Laureti et al. 2011).

Beyond the compounds described above, the wealth of new genomes and metabolic studies have pointed to the wealth of uncharacterized genes and molecules, many of which could play a role in interactions with other organisms.

6.3.2 Mechanisms of Antibiotic Resistance

To induce their biological effect, antibiotics must interact with specific cellular processes and targets. Penicillins, cephalosporins, glycopeptides, carbapenems, and monobactams impair cell wall synthesis. Quinolones inhibit DNA gyrase and block DNA synthesis. Oxazolidinones prevent the formation of the ribosomal initiation complex. Macrolides, chloramphenicol, clindamycin, aminoglycosides, and tetracyclines block protein synthesis by interaction with the ribosomal RNA subunits. Rifampicin blocks RNA polymerase activity and thus prevents mRNA synthesis; sulfonamides and trimethoprim inhibit folic acid metabolism and consequently DNA levels and the activity of various methyltransferases (Lewis 2013).

Resistance to antibiotics can be intrinsic or acquired. The absence of the specific cellular target is an obvious cause of resistance; an example is the absence of cell wall in mycoplasma (involved in lung, urinary, and genital tract infections) and the resistance to β-lactamins that interact with peptidoglycan biosynthesis. The membrane permeability properties can also trigger intrinsic resistance as exemplified with vancomycin resistance in Gram-negative bacteria. Resistance can also be acquired through mutations or by acquisition of resistance genes by horizontal gene transfer. A nice illustration is the acquisition of vancomycin resistance where a 5-gene cluster is responsible for the setting up of a vancomycin insensitive peptidoglycan biosynthetic pathway (Palmer et al. 2010).

The major resistance mechanisms consist either in reduction of the concentration of antibiotics at the intracellular level using efflux pumps or in inactivation of the drug. A variety of enzymes (e.g., acetyltransferases, adenylase, phosphorylases, hydrolases) are able to degrade or modify the antibiotic to render them inert. Another way to resist is to prevent the interaction between the drug and its specific target. For example, methylation of the large ribosomal subunit (23S) prevents the interaction of macrolides with the ribosome. Several resistance mechanisms can coexist to confer resistance to the same antibiotic.

The emergence of antibiotic-resistant bacteria always follows the same scenario: the first report of pathogens resistant to an antibiotic occurred in the years that immediately followed its release on the market and use in therapy (e.g., clinical introduction of penicillin in 1941 followed in 1944 by the report of the first penicillinase associated to resistant staphylococci isolates (Kirby 1944)). More recently, multidrug-resistant bacteria have emerged and constitute nowadays an acute health threat with nosocomial infections (e.g., methicillin-resistant *S. aureus*, multidrug-resistant Gram-negative germs). The massive but inconstant use of antibiotics, by posing an evolutionary challenge to microbial pathogens, was recognized as a major cause for emergence, selection, and rapid spread of antibiotic resistances from environmental to pathogenic bacteria. This phenomenon triggers a serious concern about efficiency of treatments against infectious diseases such as tuberculosis (Davies and Davies 2010).

6.3.3 What Are the Roles of Antibiotics in Nature?

Whatever their ecological roles, there can be no doubt about the benefit provided to the bacteria by antibiotics; the gene set devoted to their biosynthesis (including regulation and resistance genes) can reach 10% of the total genome size and thus represents a very significant genetic investment. Further, synthesizing the enzymatic machinery (e.g., PKS or NRPS complexes) and the final compounds is costly for the producer. It thus means evolutionary pressures to maintain such a comprehensive arsenal are strong and constant over long periods of time.

The classical view of antibiotics used as weapons in competitive biotic interactions has been challenged by the fact that antibiotics may not be found in nature at inhibitory concentrations. Indeed, their expression is highly regulated and would be encountered at inhibitory concentrations only very locally (i.e., in environmental micro-niches) or in anthropogenic niches (e.g., clinical environments). Further, antibiotics have a wide panel of effects in a concentration-dependent manner; this phenomenon is called hormesis (Davies et al. 2006). Low concentrations, between 0.1 and 0.01 relative to MIC, efficiently modulate transcription in the target cells where up to 5–10% of the transcripts are affected. Transcription alteration is associated to changes in the bacterial phenotypes including exoprotein synthesis, biofilm formation, virulence, and motility (Yim et al. 2007). These facts suggest that compounds previously characterized as antibiotics could be considered as signaling molecules. Reciprocally, some signaling molecules such as quorum sensing actors show antibiotic activity at high concentrations (Fajardo et al. 2008).

In this way, it was demonstrated that subinhibitory antibiotic concentrations altered the nutrient use and minimized nutrient overlap in the soil ecosystem between *Streptomyces* isolates (Vaz Jauri et al. 2013). Antibiotic-induced nutrient shift may indeed mediate niche differentiation within soil microbial populations and would help alleviate competitive interactions.

On the other hand, at inhibitory concentrations, the role as weapon and shield for antibiotics and resistance to antibiotics, respectively, is supported. Kinkel and co-workers (2014) identified a correlation between the origin of the *Streptomyces* isolates and the shift of inhibitor phenotype: antibiotic production was more significantly altered in response to sympatric (from the same niche) than allopatric isolates (from distinct niches). In addition, the inhibition phenotype was more intense between closely related strains. These data are consistent with the fact that isolates from the same niche probably share genetic relatedness and common nutrient needs, triggering competitive interactions and an evolutionary force toward synthesis of compounds able to help a given strain dominate. In the same way, Abrudan et al. (2015) showed that competitive interactions of *Streptomyces* strains induced (offensive response) or suppressed (defensive response) inhibitory capacities supporting the involvement of antibiotics in competition for niche occupation. Although in the latter study there was no association between genetic relatedness and the inhibitory phenotype, this social behavior may maintain the capacity to respond to threats by keeping together cells with different antibiotic arsenal. A similar picture was also reported in populations of *Vibrionaceae* isolates originating from the ocean (Cordero et al. 2012). Antibiotics production is restricted to a few individuals called "super-killers" possessing a unique antibiotic gene cluster (i.e., a hybrid PKS/NRPS). Antibiotics could be considered as public goods and would benefit to populations rather than to the sole individuals producing the compound. This strategy may confer a strong benefit by reducing the cost of interpopulation competitive interactions; only a few individuals are paying the high metabolic cost of synthesizing antibiotics, while the rest of the population are paying the much lower cost associated to the resistance. As in *Streptomyces* populations in soil (Cheng et al. 2015), the populations of *Vibrionaceae* are highly recombinogenic and are thus prone to frequent gene acquisitions and losses by horizontal gene transfer. Their potential to gain new antibiotics gene clusters and resistance remains active.

Antibiotics may also influence co-living eukaryotic cells and may contribute to major ecological issues such as plant health and growth by modulation of fungal development (e.g., pathogenic or symbiotic (Kurth et al. 2015)) or in animal physiology by modulating their immune system. In *Pseudomonas aeruginosa*, a major opportunistic pathogens in hospitalized and cystic fibrosis patients, it was shown that antibiotics belonging to different chemical families (aminoglycoside, tetracycline, and norfloxacin/quinolone) induced physiological answers related to biofilm formation, iron uptake, oxidative stress response, motility, and cytotoxicity (Linares et al. 2006). For example, tetracycline at subinhibitory concentrations increases the expression of type III secretion system genes in *Pseudomonas* nearly fourfold, correlated with an increased toxicity on macrophages.

Metabolites produced by certain endophytes, i.e., bacteria associated to plant cells, are analogs of plant hormones and modulate development and stress responses in plant (Brader et al. 2014). They can also provide protective compounds such as antibacterial, antifungal, or insecticidal activities (e.g., aminocyclitols produced by *Candidatus* Burkholderia kirkii). Interestingly, pseudomonads associated to eukaryotic cells as endophytes/epiphytes (*Pseudomonas fluorescens*) possess a higher content of metabolite gene clusters than their free-living counterparts (*P. putida*). It is tempting to interpret that these secondary metabolites serve to adapt to this complex niche.

Bacteria also produce a variety of volatile molecules (more than 1000 reported so far) which remain unknown and unexplored to a large extent (Audrain et al. 2015). These may exert antagonistic effects on other microorganisms in the rhizosphere as well as in the animal digestive tracts. Bacterial volatile compounds were also shown to interfere with antibiotic resistance, biofilm formation, and virulence and to be modulators of host cell physiology. Finally, understanding the roles of the large diversity of small molecules synthetized by the human microbiota (peptides, non-ribosomal peptides, polyketides, etc.) is a challenge to understand the impacts of microbe-host interactions maintenance on human health and development of diseases (Donia and Fischbach 2015).

Questioning the roles of secondary metabolites in the environment is of absolute relevance for the future development of our arsenal of antibiotics. Since most antibiotic gene clusters are silent under laboratory culture conditions, understanding antibiotic regulation and identifying factors (biotic or abiotic) eliciting their synthesis can readily help to better exploit the secondary metabolism of environmental microorganisms (bacteria as well as fungi). For that purpose, several methods have been applied to activate the silent gene clusters including genetic manipulation of pathway-specific activator or global transcriptional regulators, mutation in the RNA polymerase subunits, ribosome engineering, co-cultures, and fermentation in different growth conditions or heterologous expression in a host engineered for expression (superhost) (Liu et al. 2013, Aigle et al. 2014). In *S. ambofaciens* ATCC23877, we successfully activated two biosynthetic gene clusters, the stambomycin and the kinamycin gene clusters. In the first case, the overexpression of a LAL family regulator led to the discovery of a new family of glycosylated macrolides with promising antitumoral properties (Laureti et al. 2011) (Fig. 6.5). In the case of kinamycin, the depletion of *alpW*, encoding a member of the TetR regulator family, caused the constitutive synthesis of kinamycins. AlpW was shown to be a late repressor of the cellular control of kinamycin biosynthesis (Bunet et al. 2011).

On the other hand, these elicitors have highly specific partners or targets within the cell (bacteria, fungi or other living organisms of the ecological niche); then, identifying their cellular targets and deciphering molecular interactions can provide new hints/leads to develop new bioactive compounds. The extreme diversity of the

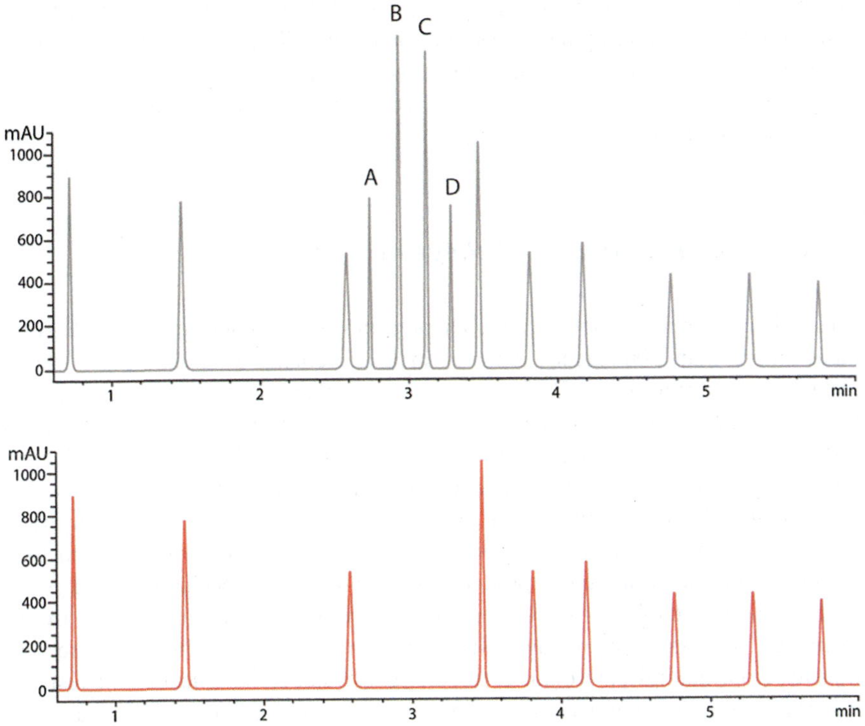

Fig. 6.5 HPLC chromatogram of two strains, in red the chromatogram of the wild type and in black above, that of the mutant where overexpression of a LAL family regulator causes the synthesis of stambomycins, a new family of glycosylated macrolides with antitumoral but no marked antibiotic activities. (Drawn from Laureti et al. 2011)

chemical structures of the secondary metabolites and the enormous number of cognate cellular targets (at every level of the cell) open large perspectives to identify new bacterial inhibitors.

An approach found to be effective is to manipulate the regulators identified within the gene cluster, to overexpress positive ones or reciprocally to delete negative ones when the regulation mode can be assessed from genome mining. In *S. ambofaciens* ATCC23877, this strategy was successfully applied to activate two biosynthetic gene clusters, the stambomycine and the kinamycin gene clusters (Bunet et al. 2011; Laureti et al. 2011). In the first case, the overexpression of a LAL family regulatory gene led to the discovery of a new family of glycosylated macrolides with promising antitumoral but no marked antibiotic activities (Fig. 6.5).

In the case of kinamycin, a deletion mutant of *alpW*, encoding a member of the TetR regulator family, was shown to constitutively produce kinamycins. AlpW was shown to be a key late repressor of the cellular control of kinamycin biosynthesis.

On the other hand, these elicitors have highly specific partners or targets within the cell (bacteria, fungi or other living organisms of the ecological niche); identifying their cellular targets and deciphering molecular interactions can thus provide new hints/leads to develop new bioactive compounds. The extreme diversity of the chemical structures of the secondary metabolites and the enormous number of cognate cellular targets (at every level of the cell) open large perspectives to identify new bacterial inhibitors.

6.3.4 From Resistome to Pathogenic Bacteria

The resistome includes not only the already known antibiotic resistance determinants (more than 3000 genes reported in Comprehensive Antibiotic Resistance Database (McArthur et al. 2013)) but also genes that could generate a resistance phenotype, i.e., proto-resistance genes or r genes according to Davies and Davies (2010). The advent of metagenomic approaches considerably increased the knowledge on the resistomes of natural environments and anthropogenic habitats (Nesme et al. 2014). The soil, habitat of the main prolific antibiotic producers, constitutes the main reservoir of resistance genes. Hence, to protect themselves from their harmful products, the antibiotic producers evolve resistance mechanisms and genes.

The emergence of a resistance gene from a proto-resistance gene results from a three-step process: first, mutagenesis (mutation and/or recombination); second, transfer from natural environment to pathogens; and third, selection of resistant bacteria by the pressure exerted by the massive use of antibiotics in human practices. Direct transfer of a proto-gene can confer resistance in a new host owing to awakening or stimulating its expression. This scenario is considered for the emergence of drug-specific and MDR efflux pumps in human pathogens. While they confer a strong advantage to pathogens under the antibiotic pressure, they participate to a diversity of biological functions (detoxification, virulence, trafficking, etc.) other than resistance in environmental microorganisms. It is suggested that the acquisition of efflux systems associated to their relaxed expression in the new host may have been strongly advantageous under conditions of high antibiotic pressure and favored further dissemination in pathogenic bacteria (Martinez et al. 2009). Mutations can directly convert a proto-resistance gene into a resistant determinant (e.g., resistance to macrolides) or can change the original substrate specificity of an enzyme and confer on it the capacity to degrade or inactivate an antibiotic. This is the evolutionary scenario envisaged for the emergence of β-lactamases, and their capacity to degrade of the antibiotic β-lactams, from original activities devoted to peptidoglycan development or reshaping (Meroueh et al. 2003). This mechanism is likely to occur through duplication of a gene (e.g., penicillin-binding protein, PBP) and further accumulation of point mutations within one of the two copies. Recombination is a one-step mechanism leading to emergence of new resistance genes. Examples are the *erm*(33) gene, conferring resistance to macrolides such as lincosamide, streptogramin, ketolide, and oxazolidinone (MLSKO) antibiotics normally active on *Staphylococcus* that likely resulted from recombination between the two preexisting resistant genes *erm*(A) and *erm*(C) genes (Schwarz et al. 2002). The

New Delhi metallo-β-lactamase (NDM-1) present in *Shigella boydii* and *Vibrio cholera* which confers resistance to all classes of β-lactam antibiotics (penicillins, cephalosporins, and carbapenems) is another striking example. The in-frame fusion of a preexisting metallo-β-lactamase gene with the aminoglycoside resistance gene *aphA6* (Fig. 6.6) alters the properties of the NDM-1 protein, i.e., NDM-1 differs from other metallo-β-lactamases by its ability to associate with the bacterial outer membrane. In addition, the recombination event put the newly shaped $bla_{NDM\text{-}1}$ gene under the influence with a strong promoter (Fig. 6.6). This structure is frequently encountered on plasmids favoring its successful dissemination in *Acinetobacter* spp. notably (Jones et al. 2015).

The selection for new emerging resistance genes and mechanisms is effective even at subinhibitory concentrations. Antibiotics may be encountered in natural environments at low concentrations because of the production by endogenous microorganisms or result from contamination of the environment following their massive use by humans (agriculture, hospital, wastewater treatment, etc. (Davies

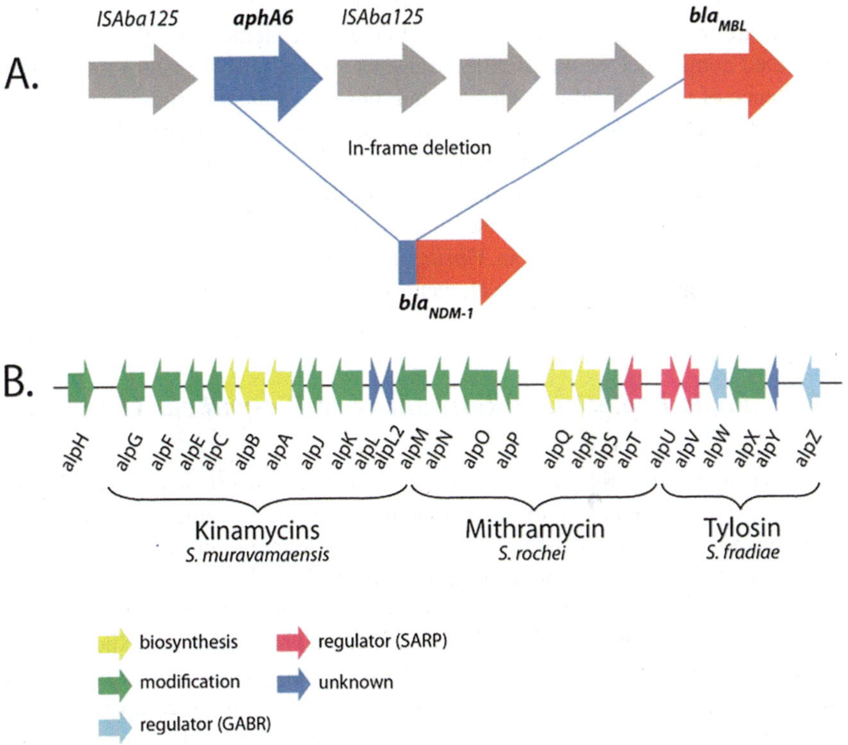

Fig. 6.6 A possible recombination event leading to the chimeric structure of resistance and metabolite gene clusters. (**a**) The New Delhi metallo-β-lactamase gene *blaNDM-1* from *Acinetobacter baumannii*. (Adapted from Toleman et al. 2012). (**b**) Modular organization of the kinamycin gene cluster from *Streptomyces ambofaciens*. (Adapted from Pang et al. 2004)

and Davies 2010)). Low antibiotic concentrations were shown to have different impacts favoring the (1) selection for preexisting resistance or de novo resistance, (2) emergence for mutator lineages producing among other resistant mutants at higher frequencies, (3) induction of the SOS response (e.g., ciprofloxacin in *Pseudomonas aeruginosa* (Torres-Barcelo et al. 2015)), and (4) stimulation of horizontal gene transfer, through the formation of biofilm (Andersson and Hughes 2014). Hence, while high antibiotic concentrations select only for preexisting resistant lineages, advantageous mutations are selected from concentrations 10 to 100 lower than the minimal inhibitory concentration (defined as the minimal selective concentration). This weak antibiotic pressure leads to the emergence of variants harboring a broad range of mutations conferring a low resistance from which secondary mutations conferring high resistance will be selected under antibiotic pressure. The induction of the general stress regulon (depending of RpoS in *E. coli*) and of the SOS response (DNA damaging induced response) under subinhibitory antibiotic conditions is a well-known phenomenon (for review (Baharoglu and Mazel 2014). These stress answers include error-prone DNA repair mechanisms causing an increase of mutation frequencies.

Antibiotic resistance genes and other virulence genes or restriction/modification systems are frequently associated with mobile genetic elements (plasmids, phages, integrated conjugative elements/ICEs, transposons, etc.). These genes may ensure persistence of the element in their host and thus constitute addiction systems. The induction of the DNA damaging response following the exposure to antibiotics stimulates indirectly the mobilization of a wide range of mobile elements through alleviation of the repression of their regulatory circuit (Bellanger et al. 2014). This phenomenon participates to the dissemination of antibiotic resistances. A spectacular example is the ICE SXT from *Vibrio cholerae* where resistance gene cassettes (i.e., an integron) are found on the element (Poulin-Laprade and Burrus 2015).

6.3.5 Rapid Evolution of Resistance and Biosynthetic Genes

Once transmitted to pathogenic bacteria, resistant genes are efficiently spread under antibiotic pressure, thanks to their mobility and the advantages they confer. Consequently, the diversity of resistance genes is strongly reduced, and the expansion within bacterial communities is almost clonal. It could be expected that in the absence of the antibiotic pressure (after cessation of antibiotic use), the frequency of resistant bacteria would progressively decrease. This optimistic expectation is founded on the high cost of resistance under antibiotic pressure: growth, survival, and competitive performance of resistant bacteria are diminished compared to their sensitive counterparts. Unfortunately, this expectation is not verified due the occurrence of compensatory mutations. A famous example was documented in *M. tuberculosis* (Sherman et al. 1996). To escape the oxidative stress in macrophages and survive during infection, isoniazid-resistance mutations in *katG* (encoding catalase) are compensated by suppressive mutations, resulting in hyperexpression of a hydroxyperoxidase (AhpC). In the same way in *E. coli*, the growth impairment associated to defect of the major porins OmpCF conferring the resistance to different classes of antibiotics could be compensated by mutations triggering a high-level expression of an alternative porin (ChiP) while maintaining resistance (Knopp and Andersson 2015).

In natural environments such as soils, the antibiotic pressure may mostly result from endogenous production by soil microorganisms. While resistance confers an obvious advantage in the immediate environment of the antibiotic producer, the wide distribution and the accumulation of resistance genes may result from the low fitness cost of their maintenance. The shield provided by accumulated antibiotic resistance genes may help face the changing antibiotic threat encountered in the ecological niche with constant confrontations with new competing isolates. These forces may drive the evolution of the antibiotic production arsenal and the dynamics of the soil bacterial populations.

What are the molecular mechanisms diversifying the antibiotic arsenal of the bacteria? The close observation of the structure of the secondary gene clusters, of the genomes hosting them, and of their distribution across phylogenetic reconstruction gives some clues about their evolution.

The *Streptomyces* genus indeed constitutes a paradigm to study the evolution of the antibiotic biosynthetic genes. The genetic organization of the *Streptomyces* is remarkable; they possess among the largest bacterial genomes (from about 6 to 12 Mb) with for the most part a linear configuration. While the origin of replication region is roughly conserved across the species, the replication terminus regions (over several hundreds of kilobases) are variable and include "contingency" genes (Choulet et al. 2006). This might reflect the gene flux, net balance between DNA acquisitions, and losses accumulating in the terminal regions, driving the divergence between isolates within an environmental niche or a bacterial population. Thanks to the advent of new generation sequencing technologies, an unexpected and untapped reservoir of antibiotic biosynthetic genes was revealed from the 2000s in the *Streptomyces* genomes. Hence, while each producer species was known and exploited for synthesizing only a single or a few compounds, 20–30 antibiotic genes or gene clusters were found within each sequenced genome (Aigle et al. 2014). A "core" secondary metabolism is shared by almost all the species and may be essential in the ecosystem (e.g., siderophores, volatile terpenoids such as geosmin, carotenoids, etc.). Reciprocally, gene clusters, for the most encoding antibiotic activities, are unique to strains or even to isolates and constitute the "contingent secondary metabolism." These gene clusters are either located in the variable regions or constitute genomic islands in the conserved central region of the chromosome. Horizontal gene transfer and recombination are the two key phenomena involved in biosynthetic gene cluster evolution. Transfer is facilitated by the clustering of the biosynthetic genes together with regulators and resistance genes. In addition to this organization suitable for single event transfer, these clusters are frequently located on large linear plasmids (reviewed by Kinashi 2011). The conjugative transfer of linear plasmids is well-established in streptomycetes (Hopwood and Kieser 1993) and may provide a very efficient way to mobilize large DNA fragments of the same order of magnitude as the largest gene clusters identified. In addition, the terminal regions are enriched in secondary metabolite gene clusters with on average half of the clusters in a quarter of the total genome size. The terminal location may itself favor conjugational transfer. Hence, Lee et al. (2011) reported that linear plasmids mobilize themselves together with the linear chromosomes from its telomeres by the mean of terminally associated proteins. This hypothesis called the "end-first model" is supported by the early report by Stonesifer et al. (1986) of the high frequency of conjugational transfer of the tylosin biosynthetic gene cluster together with an amplifiable unit of DNA in *Streptomyces fradiae*.

Finally, recombination is a hallmark of the terminal region of the *Streptomyces* chromosome and can favor both transfer and gene cluster evolution. Terminal recombination events are frequent under laboratory conditions as revealed by the large deletions (up 2 Mb) and intense DNA amplifications observed in spontaneous mutants (Thibessard et al. 2015). This may result from intrinsic properties of the subtelomeres, e.g., proximity of telomeres, presence of mobile elements, conformation and partition of the chromosome, etc. Consequently, recombinational exchanges between linear replicons either promote the acquisition of antibiotic gene clusters within the subtelomeric regions or reciprocally promote their transfer to a linear plasmid as in the case of the oxytetracycline gene cluster in *Streptomyces rimosus* (Pandza et al. 1998). Another spectacular example of reciprocal recombination event exchanged the termini between the chromosome (8.6 Mb) and the SCP1 linear plasmid (356 kb) in *Streptomyces coelicolor* (Yamasaki and Kinashi 2004) resulting in a 7.2 Mb chromosome plus a 1.85 Mb linear plasmid, SCP1.

The frequent transfer of antibiotic gene clusters in *Streptomyces* was revealed by inconsistent phylogenetic reconstructions when comparing trees based on 16S rDNA sequences (considered as retracing the "true" evolutionary history of the species) and on the biosynthetic gene sequences (Jenke-Kodama et al. 2005). In the same way, it was shown that in the marine actinomycete *Salinispora*, more than half of the identified gene clusters (124 biosynthetic pathways in 75 strains sequenced) showed a sporadic distribution (i.e., present in only one or two of the isolates) (Ziemert et al. 2014). Interestingly, it was observed that different biosynthetic gene clusters can occupy the same locus in closely related strains. The existence of these exchange hotspots reported in *Salinispora* as well as in *Streptomyces* species supports the concept that acquisition/loss may occur through homologous recombination between flanking conserved regions.

The structure of the antibiotic gene clusters may also hold some traces of their evolutionary past (Fischbach et al. 2008). Type I PKSs exhibit a multimodular organization providing a favorable substrate for homologous recombination events within a PKS gene (or evenly between PKS genes). Deletions as well as duplications/amplifications of DNA are likely to occur at high frequencies. In-frame deletion of modules may consequently reduce the number of precursors incorporated in the scaffold. Reciprocally duplications/amplifications may increase the number of modules and increase the chain length of the product. Hence, assessment of the duplication event number within *Streptomyces* phylogenetic trees suggested that type I PKS had mostly evolved by this mean and that HGT had a lesser impact (Jenke-Kodama et al. 2005). In *S. ambofaciens*, PKS encoding loci constitute hotspots for DNA rearrangements. Two main amplifiable loci are described in *S. ambofaciens*, one of which is the stambomycin gene cluster (Thibessard et al. 2015). Deletion and amplification were shown to affect the PKS genes either spontaneously (Aigle et al. 1996) or after induction of double-strand breaks in the chromosomal DNA (Hoff et al. in preparation) and to alter antibiotic production (Schauner et al. 1999). Amplifications involve overlapping stretches of the locus and lead to tandem duplication of all or any part of the stambomycin gene cluster.

Recombination between gene clusters (preexisting or newly transferred) can also lead to the formation of hybrid clusters. This situation is strongly suggested for the gene cluster responsible for simocyclinone D8 biosynthesis in *Streptomyces antibioticus* Tü6040 which appears as a chimera of four different clusters (Zotchev 2014). Another example is the structure of the kanamycin gene cluster in *S. ambofaciens* (Pang et al. 2004) which is strongly suspected to derive from horizontal gene transfer (Fig. 6.6b). Two blocks of genes related to the biosynthesis of kanamycin (angucyclic antibiotics) and of mithramycin (aureolic acid group) are fused. In addition, a third block including five regulatory genes (*Streptomyces* antibiotic regulatory proteins, SARPs, and to the γ-butyrolactone receptor proteins, GABRs) showed a strong identity and synteny with the regulatory gene subset of tylosin. This situation is striking since tylosin (macrolide) is produced by a type I PKS gene cluster (Cundliffe et al. 2001), while kanamycin biosynthesis involves a type II PKS gene cluster, further convincing of the hybrid structure of the gene cluster.

A correlation was established between total genome size and the number of PKS genes; a trend to duplicate and diversify PKS genes in large bacterial genomes is thus evident (Jenke-Kodama et al. 2005). Having a large genome is a prerogative of free-living microorganisms, particularly those living in soils that permit slow-living microbes to survive. There, they evolve under biotic and abiotic changing conditions that cause them to adapt and to cope with competitors through the production of signaling molecules and weapons (Giovannoni et al. 2014). The intrinsic structure of the antibiotic gene clusters as well as their location in recombinogenic chromosomal regions could confer a high capacity to evolve (i.e., evolvability). Thanks to horizontal gene transfer, *Streptomyces* strains can acquire new gene clusters at a low cost. This newly acquired genetic information may confer an immediate selective advantage, but more probably confer no advantage to the owner or only to a fraction of the population. Due to their low fitness contribution, these biosynthetic genes are favorable targets for the recombination and mutational processes. Finally, a wide diversity of bioactive compounds can be produced and can confer a selective advantage in certain environmental conditions (Fig. 6.7). Antibiotic synthesis and antibiotic resistance are thus the two facets of a general Red Queen never-ending race leading to the survival of the fittest.

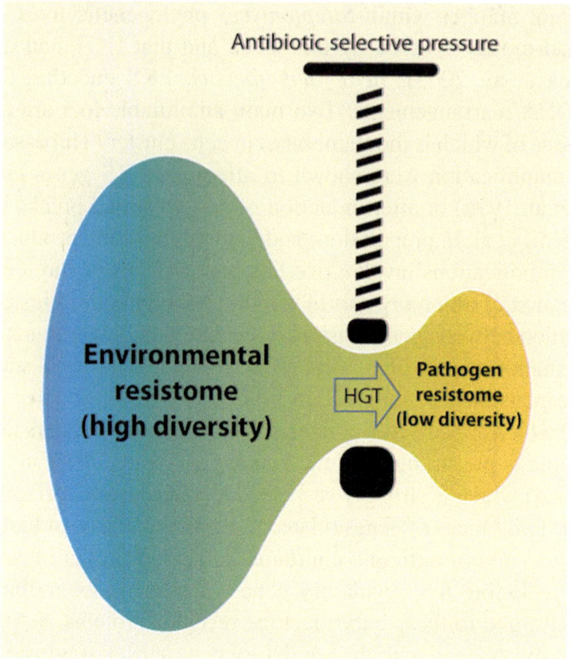

Fig. 6.7 Antibiotic resistance gene flux from environmental reservoirs to pathogen bacteria. Natural environments are huge reservoirs for highly diverse resistance (or proto-) genes. A few are carried by mobile genetic elements and are efficiently mobilized and spread within the commensal and/or pathogenic flora

6.4 Emergence of Endosymbiosis

6.4.1 Rhizobia, Nitrogen-Fixing Symbionts of Legumes

Soil is a biological creation, the common work of plants that fix carbon, die, and fall on the ground and of microbes that recycle this carbon and amalgamate it with the mineral substratum. Plants ancestors have invaded the continents about 400 MY ago and injected enormous amounts of organic matter into the mineral terrain that has slowly become the soils. This organic matter comprises easily catabolized compounds such as saccharides, organic acids, proteins, and hard to catabolize compounds such as aromatics and polymers; these will tend to accumulate and form the organic matter of soils. Soils are thus seas of low concentrations of complex organic matter that house the highest density and diversity of microbes on earth (Torsvik et al. 2002) together with plants that are rich islands of available carbon, thus constituting a tremendous driver of bacterial evolution.

Microbes represent the majority of soil microorganisms and act as important regulators of the plant diversity and productivity through direct (mutualism, pathogenesis) or indirect (commensalism) effects on growth, health, and composition of plant populations. Elucidating how microbes adapt to new plant environments is crucial for the long-term management of agro- and ecosystems.

Vascular plants are thought to have colonized the continents about 400 MY ago at the same time as the endomycorhizal fungi (Simon et al. 1993) that permitted plants to explore the soil and gain access to water and nutrients. Plants thus evolved a complex signaling machinery to let these beneficial fungi penetrate roots but not the millions of saprophytes and pathogens that are also present, machinery that consist in kinases with *lysM* motifs that recognize compounds such as peptidoglycan and chitin as well as the Nod factor (Stracke et al. 2002). Legumes are thought to have emerged about 60 MY ago from the rosid lineage (Bell et al. 2010) and to have recruited parts of this complex signaling pathway and to have yielded the immensely successful family Fabaceae with 700 genera and over thousands of species, making it one the most abundant plant families.

Soil bacteria termed rhizobia are a remarkable example of bacteria with a formidable impact on plant nutrition and growth. Rhizobia elicit on legume roots the formation of specialized organs called nodules that work as miniature nitrogen-fixing factories in which nitrogen is fixed or transformed into ammonium to the benefit of the plant (Masson-Boivin et al. 2009). This mutualistic symbiosis of major ecological and agricultural importance makes legumes an exception within the plant kingdom. Most cultivated plants indeed depend on pollution-causing nitrogen fertilizers for growth, nitrogen being the second limiting plant growth factor after water in most agricultural areas. N-fertilizer synthesis, transport, and application on fields consume large amounts of fossil energy, and a significant part of applied N-fertilizers is leached into the soil, contributing to water pollution and eutrophication in lands of intensive agriculture. The rhizobium-legume symbiosis accounts for 50 million tons of nitrogen injected into agriculture every year, as compared to 90 million tons of N-fertilizers.

Rhizobia are facultative endosymbionts that alternate saprophytic and symbiotic phases (Fig. 6.8). They are primarily soil bacteria that can establish a mutualistic symbiosis when they encounter a compatible legume host. The formation of nodules occurs through a complex symbiotic process that involves three main successive and overlapping steps, the induction of the formation of nodules (nodulation), the infection of root tissues and nodule cells, and nitrogen fixation that occurs within nodule cells. The capacity of rhizobia to fix nitrogen within nodules can however vary widely from no fixation to levels equivalent to nitrogen-fed controls, depending of environmental conditions and legume partners, meaning that the relation may be mutualistic or not. On the bacterial side, bacteria benefit from a specific ecological niche, the nodule, where bacteria are protected from many environmental stresses and grow without competitors. Usually a single bacterium forms a nodule and grows up to 10^{10} bacteria which are released to the soil following nodule senescence.

The last decades have seen tremendous progress in our understanding of the molecular mechanisms that control nodulation and early infection (Oldroyd et al. 2011). This endosymbiosis is controlled by a large number of genes in both partners including a set of essential nodulation (*nod*) and nitrogen fixation (*nif*) bacterial genes.

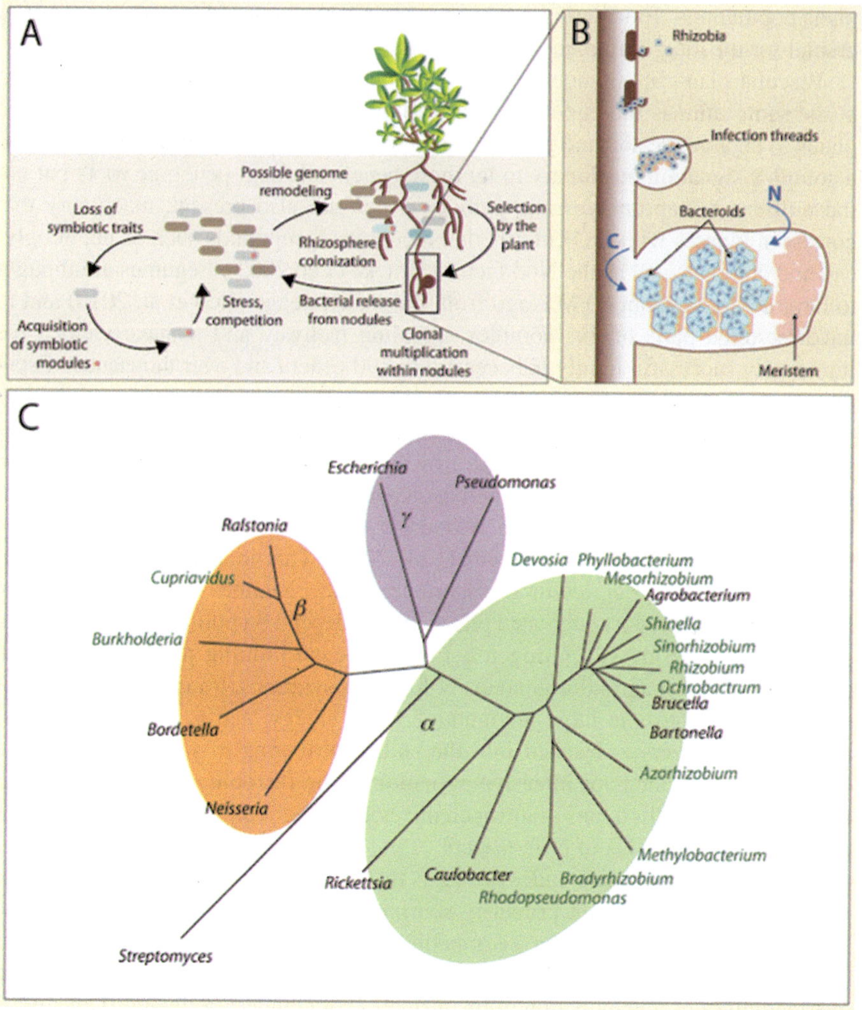

Fig. 6.8 (**a**) In the soil, rhizobia are in competition with other bacteria for survival and rhizosphere colonization. Competitive strains form on the root of legume hosts nodules within which they multiply. After a period of active N_2 fixation, the nodule senesces and bacteria are released to the soil, where they return to a saprophytic life. They may lose their symbiotic modules or transfer them to other soil bacteria. (**b**) Rhizobia (r) invade roots via infection threads (it) that propagate into emerging nodules delivering thousands of bacteria into the plant cells. C. Unrooted phylogenetic tree of 16S rDNA sequences from selected alpha-, beta-, and gamma-proteobacteria. Rhizobia are in green. (Adapted from Remigi et al. 2016 (**a**) and Masson-Boivin et al. 2009(**c**))

The *nod* genes play a central role in nodulation and infection of plant hosts. In response to plant compounds, mainly flavonoids, the *nod* genes determine the synthesis and export of small lipo-chitooligosaccharidic signal molecules called Nod factors that trigger the plant developmental program leading to nodule organogenesis and infection, although other genes such as polysaccharide synthesis genes also are required for infection. At the end of this process, bacteria fix nitrogen, thanks to the expression of *nif* genes involved in the synthesis, processing, and assembly of the nitrogenase enzymatic complex.

Rhizobia have in all likelihood emerged from a saprophytic nitrogen-fixing soil microbe; the most likely candidate to be considered the first nodulating rhizobia are the slow-growing *Bradyrhizobium* (Hungria et al. 2015); these thus gained a protected niche and were multiplied billions of time. The gained determinants permitting to synthesize the Nod factors (*nod* genes) and trigger the plant signaling machinery initially developed for mycorhizal fungi. These genetic determinants were then spread laterally together with nitrogen fixation genes to distantly related soil bacteria, initiating their conversion into rhizobia. Indeed *nod* are generally clustered with *nif* genes into mobile genetic elements such as symbiotic plasmids or genomic islands. This explains why extant rhizobia do not form a homogenous taxonomical group. They are currently distributed in a dozen or so of genera and hundreds of species of alpha- and beta-proteobacteria that also contain saprophytes and/or pathogens (Remigi et al. 2016). The spread of symbiotic proficiency in many different bacteria, exhibiting different metabolic properties and adapted to different soil and plant environments, has likely been key in the evolutionary success of symbiosis which has become established in 70% of the legume species (Masson-Boivin and Sachs 2017). Depending on the symbiotic partners, there are variations in the nodulation, infection, and nitrogen fixation processes (Masson-Boivin et al. 2009).

6.4.2 *Replaying* Rhizobium *Evolution in the Laboratory*

The natural evolution of rhizobia has been partly elucidated. Genetic, genomic, phylogenetic, and experimental data have established that rhizobia arose through horizontal transfer of essential symbiotic genes (*nod* and *nif* genes), converting various likely nitrogen-fixing soil bacteria and plant pathogens into mutualistic symbionts of legumes (Sullivan and Ronson 1998). Although compared phylogenies of rhizobia and nodulation genes predict that symbiotic genes have been transferred over large phylogenetic distances (Moulin et al. 2001), transfer of symbiotic proficiency is quite hard to achieve in the lab between distantly related species (Marchetti et al. 2010). This suggests that the recipient bacteria need to undergo further adaptations, beyond the acquisition of essential symbiotic genes to achieve mutualistic symbiosis. Each plant indeed represents a complex ecosystem with specific requirements, e.g., in terms of immunity and metabolism, to which the bacterium must adapt. There must be selective pressures favoring the expression of the acquired symbiosis traits and adaptive mechanisms to deal with these pressures. Experimental evolution coupled with WGS (whole genome sequencing) is a powerful tool to elucidate post-transfer mechanisms allowing bacteria to go to the symbiotic route.

To replay the evolution of rhizobia in the laboratory under fast-forward mode, the natural evolutionary history of rhizobia was imitated. The logic was i) to introduce a rhizobial symbiotic plasmid into a non-rhizobial soil bacterium, mimicking the natural transfer of symbiotic modules, and ii) to further evolve this chimeric strain under legume selection pressure (Fig. 6.9). In short, the symbiotic plasmid of the *Mimosa* rhizobial symbiont *Cupriavidus taiwanensis*, which contains the *nif* and *nod* genes, coding for nitrogenase and Nod factor synthesis, was transferred into the root pathogen *Ralstonia solanacearum*. This chimera was then led to evolve using serial *ex planta-in planta* (*Mimosa pudica)* cycles (Marchetti et al. 2014). Nine parallel lineages of 21-day cycles and 9 parallel lineages of 42-day cycles, representing 288 nodule bacterial populations, were derived from *Ralstonia solanacearum* (Fig. 6.9). This experiment lasted 1 (short cycles) or 2 (long cycles) years. All these evolved populations were kept at −80 °C as fossil records. One clone was isolated from each population for further phenotypic analysis and genome resequencing.

The most suitable biological material was used as starting material to experimentally replay rhizobium evolution. *C. taiwanensis*, a symbiont of *M. pudica*, was chosen as donor of symbiotic genes for *C. taiwanensis* is very likely a recently evolved rhizobium, as well as a minimal rhizobium that possesses the lowest number of nodulation genes described so far (Amadou et al. 2008) (Gyaneshwar et al. 2011). *R. solanacearum* was chosen as *nod-nif* genes receiver because (1) it belongs to a genus that contains no rhizobia so far and thus is suitable to evolve into a new rhizobial lineage; (2) its phylogenetic distance to *C. taiwanensis* is ideal for laboratory evolution, not too close to allow space for evolution and not too far to allow evolution in the laboratory timescale; and (3) it is a plant pathogen, allowing to evaluate symbiosis-pathogenesis relationships, e.g., whether virulence functions are inactivated, modulated, or recruited for symbiosis. Both genomes were completely sequenced. *M. pudica*, the legume host of the rhizobium donor, was chosen as selection pressure, as it possesses the appropriate receptors for *C. taiwanensis* Nod factors.

Experimental conditions were optimized to increase the chances of success. Series of *ex planta-in planta* passages were used to tentatively recapitulate the alternation of saprophytic-symbiotic lives that may have shaped the evolution of rhizobia *in natura*. Indeed in the soil, following symbiosis gene transfer, a combination of molecular events, e. g., spontaneous mutations and genetic rearrangements of the recipient genome, may lead to increased compatibility to the plant. Indeed plants have developed control mechanisms that select for compatible rhizobia and defend against strains that provide little or no benefit. Since each nodule generally hosts a monoclonal symbiont population (Gage 2002), iterations of clonal amplification of the bacteria within the nodule niche, followed by the release of the bacterial population when the nodule senesces, ensure the fixation of the beneficial mutations and progressive adaptation to a legume host through a ratchet-like phenomenon. In the first cycle (selection cycle), *M. pudica* was used to trap rare spontaneous nodulating mutants of the ancestral chimera. In subsequent cycles, each of the three nodulating

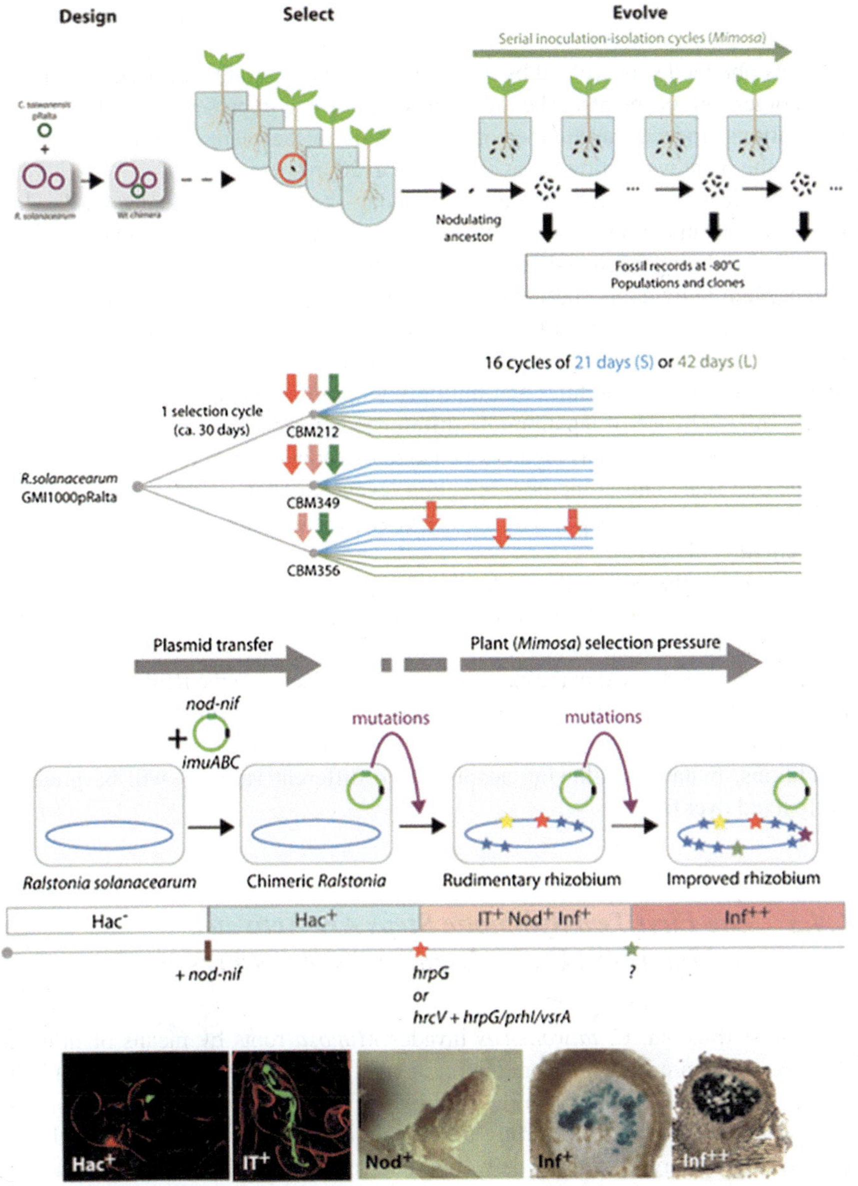

Fig. 6.9 (**a**) The symbiotic plasmid of *C. taiwanensis* was introduced into *R. solanacearum* (Design). Three nodulating derivatives of this chimera were selected using *M. pudica* as a trap (Select) and further left to evolve using serial plant-bacteria co-cultures (Evolve). (**b**) Eighteen independent lineages have been derived from the three nodulating ancestors, CBM212, CBM349, and CBM356. (**c**) *R. solanacearum* evolved in three main steps. (1) Potentialization: acquisition of *nod* genes allowed *R. solanacearum* to induce root hair curling (Hac+) on *M. pudica*. (2) Activation: root entry (IT+), nodulation (Nod+), and intracellular infection (Inf+) were activated via mutations affecting the *R. solanacearum* virulence pathway. (3) Optimization: intracellular infection and associated properties were improved (Inf++), via still unknown adaptive mutations. The *C. taiwanensis* symbiosis plasmid harbors *imuABC* genes encoding stress-responsive error-prone DNA polymerases that transiently elevated the mutation rate in the recipient genome, increasing the number of symbiotically improved variants. P, phenotype; G, genotype. (Adapted from Remigi et al. 2014)

mutants obtained was evolved by inoculating the whole nodule bacterial population to another set of plants. The 17 cycles performed represent ca 400 bacterial generations. Indeed the number of generations in the plant culture medium and within the plant was estimated at ca. 5 and ca. 20, respectively.

Two selection regimes were used: a 21-day-short (S) series and a 42-day-long (L) series that may impact differently on bacterial evolution. S cycles may enrich for bacteria exhibiting improved nodulation and infectiveness. Indeed following inoculation, plant individuals continuously form nodules from 5 to 7 days after inoculation and until the number of nitrogen-fixing nodules is sufficient to provide the nitrogen required. Short cycles may thus select bacteria that are the most competitive, i.e., which form nodules and multiply within nodules the most rapidly. Long cycles may increase occupation rate difference between Fix+ and Fix- nodules, if nitrogen fixation occurs as legumes penalize rhizobia that fail to fix nitrogen at the nodule cell scale and at the post-infection level (Kiers et al. 2003; Daubech et al. 2017). We indeed observed that the selection regime affects symbiotic evolution (Marchetti et al. 2017).

To increase the bacterial population recovered in each cycle, we inoculated 30 plantlets in each cycle. A mean of 300 nodules in each cycle was recovered, i. e., potentially up to 3×10^9 bacteria.

M. pudica seeds used in the experiment were collected in the field, they are likely a mix of different varieties. This makes these evolution experiment closer to natural conditions where bacteria encounter different varieties of *M. pudica*. Under these conditions, mutations allowing adaptation to different varieties will be preferentially fixed over time.

6.4.3 The First Two Symbiotic Steps Are Activated and Improved Under Legume Selection Pressure

Like most rhizobia, *C. taiwanensis* invades *Mimosa* roots by means of infection threads, which are initiated from microcolonies entrapped within curled root hairs, whose deformations have been induced by the bacteria (Fig. 6.8). Concomitantly plant cells divisions occur, leading to the formation of nodules. Infection threads elongate into emerging nodules delivering bacteria into the plant cells. Each infected cell houses thousands of internalized bacteria (called bacteroids) that persist without eliciting plant defense reactions and ultimately fix nitrogen to the benefit of the plant (Fig. 6.8). Mature *Mimosa* nodules induced by *C. taiwanensis* have the typical histology of indeterminate nodules, i. e., a single distal persistent meristem, and bacteroids are not terminally differentiated, i. e., they are able to resume growth when resuspended in culture medium (Marchetti et al. 2010).

R. solanacearum interacts with plants in a drastically different way. *R. solanacearum* is a soil-borne plant pathogen internationally recognized as one of the leading models in the study of pathogenicity determinants toward plants (Genin and

Denny 2012); it invades plants intercellularly and heavily colonizes their root systems, where excessive production of extracellular polysaccharides blocks water traffic, causing wilting. Its analysis has already led to the identification of a wide range of pathogenicity determinants including plant cell wall degradative enzymes, attachment factors, a type III secretion pathway encoded by *hrp-hrc* genes, and type III-dependent effectors (Genin and Denny 2012). *M. pudica is* not a host plant for *R. solanacearum.*

The first questions that arose were the following: which symbiotic properties (nodulation, infection, and nitrogen fixation) could be activated and improved at the lab scale? How fast is lab evolution? Would pathogenesis and symbiosis traits coexist? To answer these questions, we analyzed the symbiotic and pathogenesis traits of evolved clones along the evolution experiment.

In the evolution experiment, bacterial populations were evolved via serial *ex planta-in planta* passages. However to get information on the extent of the evolutionary changes, individual clones isolated from intermediate and final populations were analyzed with respect to different symbiotic traits and compared to the reference rhizobium *C. taiwanensis* and/or the ancestral strains.

The different symbiotic functions (nodulation, infection, nitrogen fixation) are phenotypically clear-cut and their acquisition easily detectable. Nodulation is assessed by counting the number of genuine nodules over time. Infection is evaluated by counting the number of bacteria recovered from nodules after nodule surface sterilization, crushing and plating onto solid medium. Nitrogen fixation is assessed from the green color of plants and by an acetylene reduction assay.

Other traits such as infection thread formation, infection of nodule cells (intracellular infection), and persistence of intracellular bacteria (bacteroids) require technically advanced methods. Intracellular infection is evaluated by analyzing sections of nodules formed by labelled bacterial mutant clones and measuring the relative surface of infected cells (Marchetti et al. 2014). Bacteroid persistence was analyzed by electron microscopy which permitted to visualize premature degradation of bacterial or plant cells or cell sorting by flow cytometry of nodule cells treated with a propidium stain (Marchetti et al. 2014). Finally two strains (e.g., an evolved clone and its ancestor) can be compared for their symbiotic capacities by co-inoculating them to *Mimosa* plants and counting the respective number of nodules formed by each strain, to assess nodulation competitiveness or the respective number of bacteria present in the nodules, to assess relative *in planta* fitness.

Through the introduction of the *C. taiwanensis* symbiotic plasmid, *R. solanacearum* acquired nodulation genes required for the synthesis of lipochitooligosaccharide Nod factors that when secreted in the rhizosphere trigger the plant developmental program of nodule organogenesis (Oldroyd et al. 2011). Although these genes were fully turned on by the aromatic luteolin, known to induce *nod* genes in *C. taiwanensis*, the chimeric *Ralstonia* was unable to nodulate the *C. taiwanensis* host *M. pudica*. It promoted root hair proliferation and deformations typical of those induced by Nod factors but showed a clear defect in infection thread initiation and elongation. Thus the designed chimeric strain had a symbiotic potential that was however not expressed. This symbiotic potential was progressively activated through various intermediary stages, yet incomplete after 17 cycles of plant-bacteria co-culture.

The first symbiotic property, root entry and nodulation, was obtained after one selection cycle, where the ancestral chimeric *Ralstonia* was inoculated to thousands of *M. pudica* seedlings grown under nitrogen-free conditions, until nodules were obtained. During this first cycle, *M. pudica* was thus capable to trap nodulating mutants and let them multiply. Three nodules, which appeared 3–4 weeks after inoculation, were recovered from three independent experiments. The three clones isolated from the three nodules nodulated *Mimosa* plants with different kinetics, and their nodulation ability was reduced relative to *C. taiwanensis*, indicating that the three nodulating ancestors were different and thus derived from three independent events.

The second symbiotic property, penetration and multiplication within nodule cortical cells, was acquired in all lineages but three, at different times. The capacity to intracellularly infect nodules was acquired concomitantly to nodulation in two of the nodulating ancestors, whereas this property was not activated in the third nodulating clone, CBM356. It was acquired later on during laboratory evolution in the three CBM356-derived lineages, at cycles 5, 9, and 11. However, when just acquired intracellular infection was rudimentary compared with the *M. pudica* symbiont *C. taiwanensis*. Fewer bacteria were routinely isolated per nodule, the intracellular zone defined by infected cells was limited, bacteroids did not persist within cells, and bacteria were found in intercellular spaces of the nodules (Marchetti et al. 2010) (Guan et al. 2013). Moreover bacteria induced plant defense reactions as evidenced by necrosis zones within nodules, which are never observed in nodules formed by *C. taiwanensis*.

Infection was further improved in most of the lineages, both quantitatively and qualitatively. Intracellular infection and bacteroid persistence increased, while defense reactions decreased, suggesting that the expression of symbiotic competence requires the capacity to limit plant immunity. In some clones the size of the intracellular infection and necrosis zones was not statistically different from those of the reference rhizobium *C. taiwanensis*, demonstrating that the legume, i.e., a plant genetically programmed for symbiosis, is a highly selective environment for selecting intracellular symbionts. However, the best evolved clones were still approximately 100 times less competitive, thus less fit than the reference rhizobium, as assessed by co-inoculation experiments.

Interestingly, all evolved clones, including the first nodulating ancestors, were not pathogenic anymore on *Arabidopsis*, indicating a trade-off between symbiosis and pathogenesis.

Nitrogen fixation is the ultimate step that turns a parasitic rhizobium-legume association into a mutualistic one with full trophic exchange of ammonium against organic acids. Although first steps in the evolutionary process of reciprocal cooperation have been reached, mutualism has not been achieved in any lineage in this evolutionary short experiment. Indeed none of the clones fixed nitrogen after 17 evolution cycles. Although their persistence was significantly improved as compared to their respective ancestors (Marchetti et al. 2014), bacteroids persisted very poorly within nodules as compared to a *nifH* mutant of *C. taiwanensis*. This could stem from a defect in metabolic integration or in avoidance of plant defenses. Evolved symbiotic properties are thus likely insufficient to sustain effective nitrogen fixation, which requires a massive and persistent infection of root nodules. Pursuing this laboratory evolution should allow further improvement of persistence and, possibly, nitrogen fixation.

6.4.4 Highly Adaptive Mutations Target the Virulence Pathway of R. solanacearum

Comparative analysis of alpha- and beta-rhizobia that have been conducted for years indicated that the genetic innate skills of the genome having acquired a symbiotic region have been exploited for legume symbiosis. Indeed a significant part of the symbiotic genes are located outside the symbiotic island/plasmid and could even be specific to the recipient genome lineage. Genomic modifications, involving point mutations and genetic rearrangements, may have modulated the symbiotic capacity of emerging rhizobia along evolution, via rewiring of regulatory circuits and inactivation, modulation, or recruitment of indigenous functions. Thanks to advances in DNA technology and affordability, our evolution experiment now provides a tractable system to monitor and analyze post-HGT genomic changes.

To sort out genomic changes responsible for the symbiotic evolution in our laboratory evolution, there were two possibilities. The first strategy consisted in resequencing final clones; identifying genomic changes relative to the common ancestor; searching for convergent modifications, e. g., identical mutations or genes mutated in several parallel lineages; evaluating at which cycle these modifications arose; and finally assessing their adaptive nature by genetic reconstruction of the genomic change in an ancestral strain. The second strategy consisted in searching for phenotypic shifts, i. e., clear changes in the phenotype along lineages, resequencing and comparing the genome of clones framing each phenotypic shift, and evaluating the adaptive nature of all identified changes by genetic reconstruction in the clone preceding the shift.

While the first strategy has most often been chosen in evolution experiments, the second one was chosen because i) resequencing of final clones revealed a huge number of mutations and very few genetic convergences and ii) clear and sharp phenotypic shifts have been identified along some lineages, i. e., activation of nodulation, acquisition of intracellular infection, and drastic improvement of intracellular infection.

Using this strategy, it was found that mutations responsible for the pathogenesis-symbiosis transition targeted the genes of the virulence pathway of *R. solanacearum*.

In *R. solanacearum*, the virulence network is complex and partly understood (Genin and Denny 2012). It is orchestrated by the central regulator HrpG that controls a type III secretion machinery (T3SS) required for pathogenesis and associated effectors via the intermediate regulator HrpB, as well as many genes involved in virulence and host adaptation via a still unknown circuitry. HrpG is activated by plant cell contact through the PrhARIJ cascade. Beside HrpG, the sensor histidine kinase VsrA controls many other virulence functions. In addition the global metabolic repressor EfpR also acts as a regulator of virulence (Perrier et al. 2016).

In one of the nodulating ancestors (CBM356), root entry (IT+) and nodulation (Nod+) were activated through inactivation of the *hrcV* gene (Marchetti et al. 2010), which encodes an essential structural component of the type III secretion system thus inactivating it. Later on, intracellular infection was activated via mutations affecting the virulence regulators *prhI*, *vsrA*, or *hrpG* (Guan et al. 2013). In the two other nodulating ancestors (CBM212 and CBM349), root entry (IT+), nodulation (Nod+), and intracellular infection (Inf+) were obtained via inactivation of the virulence regulator *hrpG*. Since both the genes *hrpG* and *hrcV* are required for pathogenesis, their inactivation accounts for the observed trade-off between symbiosis and pathogenesis. In several lines, further improvement of intracellular infection capacity (Inf++) was gained via mutations in the *efpR* dual regulator of metabolism and virulence (Capela et al. 2017).

How these mutations allow nodulation and infection activation/improvement is still unknown. Very likely one or several secreted effectors, whose delivery in plant cells is impaired in *hrcV* or *hrpG* mutants, block nodulation, either by triggering the host plant immunity responses or by specifically interfering with the Nod factor signaling pathway. Other virulence functions controlled by HrpG, EfpR, and/or VsrA may interfere with intracellular infection.

Interestingly, it was observed that for all the strongly adaptive mutations, any mutation that positively impacts infection also improves nodulation and vice versa, suggesting a genetic link between nodulation or infection thread formation and nodule cell infection. Such coupling sounds biologically meaningful with respect to dissemination of rhizobia in various host plant taxa. Since the legume selects very few bacteria to form nodules, a rhizospheric bacterium has nearly no chance to enter the root if it is not highly competitive for nodulation/initial infection. The coupling of bacterial capacities to nodulate and intracellularly infect nodules may thus have favored the spread and maintenance of intracellular infection capacity in nodulating bacteria.

Altogether these findings supports evidence that transfer of symbiotic proficiency between phylogenetically distant bacteria requires subsequent genome remodeling, allowing the newly acquired functions to adjust to both the recipient cell and the new plant environment.

6.4.5 Transient Hypermutagenesis Accelerates the Adaptation Process

Evolution of microorganisms in response to changing environments relies on the natural selection of genetic variants harboring beneficial phenotypic traits. Organisms have evolved mechanisms enabling them to increase genetic variability, such as constitutive (heritable) and stress-associated (nonheritable) mutagenesis (Foster 2007). Constitutive hypermutators display high mutation rates due to alteration in genetic systems that control fidelity of DNA replication and repair. Selection of constitutive mutator lineages has been observed during in vitro and in vivo evolution experiments (Barrick et al. 2009; Yang et al. 2011) and is quite common in nature. However, while initially capable of fast adaptation through generation of adaptive mutations, mutator bacteria rapidly accumulate numerous deleterious mutations. Mutation rates in bacteria can also be increased by stress-induced reversible activation of some gene functions, which result in a transient mutator phenotype, the SOS response being a paradigm of such a process. Under a variety of stressful conditions, global responses are induced and thus permit to adapt to and survive these stresses that result in sweeping changes in gene expression and cellular metabolism but also in activating error-prone polymerases, downregulating error-correcting enzymes and favoring movement of mobile genetic elements, and thus inducing genetic changes. Transient mutability, also called stress-associated mutability, which generates genetic variability only under stress conditions, is thought to be important for adaptive evolution (Bjedov et al. 2003), although its adaptive role is difficult to demonstrate. In addition, transient mutability has not been evidenced in evolution experiments so far.

Our lab evolution let us uncover a mechanism involving transient hypermutability that accelerates the symbiotic evolution of a soil bacterium under legume selection pressure: key symbiotic genes are co-transferred with genes encoding stress-responsive ImuABC error-prone DNA polymerases that transiently elevate the mutation rate in the recipient genome, creating a burst in genetic diversity favorable to the emergence of beneficial variants (Remigi et al. 2014).

This discovery comes from the observation that each of the cycle 16-evolved clones carries a higher number of mutations than expected, scattered all along the genome. Since no mutation was detected in the DNA repair system and the mutation rate of evolved *Ralstonia* did not differ from that of the wild-type *R. solanacearum* strain in rich medium, we hypothesized that bacteria underwent transient hypermutability during the lab experiment and thus increased their genetic diversity at some stage of the laboratory experiment. Increase of genetic diversity could have occurred either *ex planta* or *in planta* or both. Indeed in each cycle, *Mimosa* plantlets in tubes were inoculated with bacteria which first diffused in the nitrogen-free and carbon-free plant culture medium before entering the roots and colonizing the newly induced nodules.

To analyze this phenomenon, i. e., which environmental and genetic factors induce hypermutability, two methods were used aiming at evaluating the genomic heterogeneity of bacterial populations grown under different conditions: (1) whole genome sequencing (Illumina) of pools of individuals randomly isolated from a population and (2) classical mutability tests which determine the frequency of appearance of antibiotic resistance mutants in the population (see (Remigi et al. 2014)). This last method, which can be carried out routinely, allows estimating the mutation rate from the rate of appearance of specific adaptive mutations. Sequencing of bacterial pools, more costly and technically heavier, is very powerful, allowing the identification and analysis of all point mutations, which are either adaptive or neutral, which have arisen in the investigated subpopulation.

Characterizing pools of bacteria isolated from nodules or from the plant culture medium 21 days after *Mimosa* inoculation showed that mutations accumulated outside the plant but probably not in the nodules (Remigi et al. 2014). An excess of synonymous mutations over nonsynonymous mutations and very few convergent mutations indicated a signature of purifying selection, suggesting that most mutations are not adaptive and have been fixed by hitchhiking with adaptive ones. In a second step, mutability tests showed that this increase in genetic diversity was controlled by a locus (*imuA2B2C2*) located on the alien *C. taiwanensis* symbiotic plasmid that has been introduced to *R. solanacearum* as a first step of the laboratory experiment (Remigi et al. 2014). This locus contains an error-prone polymerase designated ImuC2. Error-prone DNA polymerases are polymerases that can bypass DNA lesions and achieve replication but that make errors and thus generate mutations. This cassette, either complete or without the *imuA* gene, is widespread in bacteria and has been shown to mediate stress-induced mutagenesis as part of the SOS response (Erill et al. 2006; Ippoliti et al. 2012).

The environmental factors that induce *C. taiwanensis imuA2B2C2*-based hypermutability are still unknown. We speculate nevertheless that nutrient starvation, a condition frequently encountered in the soil and endured by bacteria in the Jensen plant culture medium, could be involved. Plant factors seem to be also involved since hypermutability is higher in the presence of *Mimosa* plants than without.

The fact that *imuABC* mutagenesis cassette accelerates evolution in the lab and *in natura* was supported both experimentally and by in silico analysis. Chimeric *Ralstonia* that differed only by the presence or absence of the *imuABC* cassette were evolved independently through serial *ex planta-in planta* passages. After only five cycles of laboratory evolution, bacteria with the intact cassette were found to be more symbiotically efficient than their cassette-less counterparts in competition experiments. It is proposed that the mutagenesis cassette increases the evolvability (capacity to evolve) of the recipient *Ralstonia* genome *ex planta* and that the plant then selects, among spontaneously arisen mutants, the most beneficial. Rounds of *ex planta* phenotypic diversification/plant selection/*in planta* clonal amplification may have driven the adaptation process in the laboratory and *in natura*. Overrepresentation of plasmid *imuBC* genes in rhizobial lineage plasmids provides further evidence of the positive role of *imuABC*-dependent hypermutagenesis in the acquisition of complex rhizobial traits.

Altogether these results indicate that the successful colonization of new bacterial hosts by nodulation genes is assisted by error-prone polymerases encoded in the transferred genetic element, which are more active in the presence of the plant.

Recent advances in DNA sequencing technologies have revolutionized the field of microbial experimental evolution by permitting to connect phenotypic responses to genetic changes. However, experimental evolution coupled to rapid and cheap whole genome sequencing has been predominantly used to study genetic adaptation to simple and well-controlled conditions such as defined change in culture media (Tenaillon et al. 2012; Plucain et al. 2014). Although these selection experiments provide essential lessons on general evolutionary processes, they cannot mirror the complexity of natural environments, where many selective environmental forces act simultaneously. Significant progress has already been made in the field of virus-bacteria coevolution (Scanlan et al. 2011) and adaptation to pathogen to human environments (Yang et al. 2011). Our result highlights the potential of this approach to tackle challenging complex questions such as the emergence and evolution of biotic interactions, which imposes considerably different selective pressures in terms of immunity and metabolism.

Phylogenetic analyses together with field experiments have previously established the role of horizontal transfer in rhizobium evolution. The restriction of legume symbiosis to a dozen or so bacterial genera, as compared to the abundance of bacterial species within some genera, however predicted that successful symbiotic transfers across wide phylogenetic distances were rare. Our work demonstrated that phylogenetically distant recipient bacteria require further adaptations to go down to the symbiotic route. Some adaptive mutations identified resulted in the inactivation of bacterial functions interfering with symbiosis. Many other adaptive genomic changes, possibly involving integration of incoming functions into preexisting regulatory circuitries, and recruitment and modulation of local functions, are likely to be identified through further analysis of the various evolution experiments underway. Moreover, this work revealed that symbiosis plasmids not only encode a genetic toolkit for fundamental symbiotic functions such as Nod factor signaling and nitrogen fixation but also the wherewithal to accelerate the evolution of the host bacterial genome and thereby repurpose it more rapidly for symbiosis in the laboratory and likely *in natura*. Altogether these approaches have shed light on the molecular and evolutionary mechanisms that allowed effective dissemination of complex symbiotic traits over large phylogenetic distances and open the way to exploit mutagenesis cassettes for manipulating the evolvability of bacteria in experimental evolution and synthetic biology.

The phylogenetic proximity of bacteria with very diverse lifestyles such as pathogenicity, mutualism, and saprophytism suggests that ecological transitions have been frequent over the course of evolution (Ochman and Moran 2001). Plants are rich in available carbon, whereas it is comparatively so scarce is soils that this represents an immense evolutionary pressure for soil microbes. Pathogenic and mutualistic associations between plants and microbes are the extreme outcomes of a continuum of possible inter-organism interactions under environmental constraints. The fact that very few mutations allows to shift from extracellular pathogenesis to endosymbiosis supports the view that host-microbe interactions are evolutionary labile.

6.5 General Conclusion

Directed evolution is a subdiscipline of evolutionary biology that is in rapid progression with a high proportion of papers published in high-impact journals, due in part to the resonance its findings have in the general population. It has shown that in a few years, bacteria could be made to evolve dramatically, spanning the ecological distances bacteria have taken millions of years to cover when left alone. Directed evolution is also a discipline that can have applications in the development of enzymes tailored for specific purposes, of antibiotics targeting specific pathogens, and of microbes with optimized properties. It also depends on rapidly developing technical techniques such as high-throughput sequencing, highly resolutive mass spectrometry, and highly specific mutation techniques. Directed evolution will thus in all likelihood continue to move forward at a fast rate in the coming years.

References

Abrudan MI, Smakman F, Grimbergen AJ, Westhoff S, Miller EL, van Wezel GP, Rozen DE (2015) Socially mediated induction and suppression of antibiosis during bacterial coexistence. Proc Natl Acad Sci U S A 112:11054–11059

Adami C (2006) Digital genetics: unravelling the genetic basis of evolution. Nat Rev Genet 7:109–118

Aigle B, Schneider D, Morilhat C, Vandewiele D, Dary A, Holl AC, Simonet JM, Decaris B (1996) An amplifiable and deletable locus of *Streptomyces ambofaciens* RP181110 contains a very large gene homologous to polyketide synthase genes. Microbiology 142(Pt 10):2815–2824

Aigle B, Lautru S, Spiteller D, Dickschat JS, Challis GL, Leblond P, Pernodet JL (2014) Genome mining of *Streptomyces ambofaciens*. J Ind Microbiol Biotechnol 41:251–263

Amadou C, Pascal G, Mangenot S, Glew M, Bontemps C, Capela D, Carrere S, Cruveiller S, Dossat C, Lajus A et al (2008) Genome sequence of the beta-rhizobium *Cupriavidus taiwanensis* and comparative genomics of rhizobia. Genome Res 18:1472–1483

Andersson DI, Hughes D (2014) Microbiological effects of sublethal levels of antibiotics. Nat Rev Microbiol 12:465–478

Audrain B, Farag M, Ryu C, Ghigo J (2015) Role of bacterial volatile compounds in bacterial biology. FEMS Microbiol Rev 39:222–233

Baharoglu Z, Mazel D (2014) SOS, the formidable strategy of bacteria against aggressions. FEMS Microbiol Rev 38:1126–1145

Barrick JE, Lenski RE (2009) Genome-wide mutational diversity in an evolving population of *Escherichia coli*. Cold Spring Harb Symp Quant Biol 74:119–129

Barrick JE, Lenski RE (2013) Genome dynamics during experimental evolution. Nat Rev Genet 14:827–839
Barrick JE, Yu DS, Yoon SH, Jeong H, Oh TK, Schneider D, Lenski RE, Kim JF (2009) Genome evolution and adaptation in a long-term experiment with *Escherichia coli*. Nature 461:1243–1247
Bell CD, Soltis DE, Soltis PS (2010) The age and diversification of the Angiosperms re-revisited. Am J Bot 97:1–8
Bellanger X, Payot S, Leblond-Bourget N, Guedon G (2014) Conjugative and mobilizable genomic islands in bacteria: evolution and diversity. FEMS Microbiol Rev 38:720–760
Berdy J (2005) Bioactive microbial metabolites. J Antibiot (Tokyo) 58:1–26
Bjedov I, Tenaillon O, Gerard B, Souza V, Denamur E, Radman M, Taddei F, Matic I (2003) Stress-induced mutagenesis in bacteria. Science 300:1404–1409
Blount ZD, Borland CZ, Lenski RE (2008) Historical contingency and the evolution of a key innovation in an experimental population of *Escherichia coli*. Proc Natl Acad Sci U S A 105:7899–7906
Blount ZD, Barrick JE, Davidson CJ, Lenski RE (2012) Genomic analysis of a key innovation in an experimental Escherichia coli population. Nature 489:513–518
Bordes F, Tarquis L, Nicaud J, Marty A (2011) Isolation of a thermostable variant of Lip2 lipase from *Yarrowia lipolytica* by directed evolution and deeper insight into the denaturation mechanisms involved. J Biotechnol 156:117–124
Brader G, Compant S, Mitter B, Trognitz F, Sessitsch A (2014) Metabolic potential of endophytic bacteria. Curr Opin Biotechnol 27:30–37
Bull JJ, Badgett MR, Wichman HA, Huelsenbeck JP, Hillis DM, Gulati A, Ho C, Molineux IJ (1997) Exceptional convergent evolution in a virus. Genetics 147:1497–1507
Bunet R, Song L, Mendes MV, Corre C, Hotel L, Rouhier N, Framboisier X, Leblond P, Challis GL, Aigle B (2011) Characterization and manipulation of the pathway-specific late regulator AlpW reveals *Streptomyces ambofaciens* as a new producer of Kinamycins. J Bacteriol 193:1142–1153
Cajthaml T, Kresinova Z, Svobodova K, Sigler K, Rezanka T (2009) Microbial transformation of synthetic estrogen 17alpha-ethinylestradiol. Environ Pollut 157:3325–3335
Capela D, Marchetti M, Clérissi C, Perrier A, Guetta D, Gris C, Valls M, Jauneau A, Cruveiller S, Rocha EPC et al (2017) Recruitment of a lineage-specific virulence regulatory pathway promotes intracellular infection by a plant pathogen experimentally evolved into a legume symbiont. Mol Biol Evol 34:2503–2521
Cheng K, Rong X, Pinto-Tomas AA, Fernandez-Villalobos M, Murillo-Cruz C, Huang Y (2015) Population genetic analysis of Streptomyces albidoflavus reveals habitat barriers to homologous recombination in the diversification of streptomycetes. Appl Environ Microbiol 81:966–975
Choulet F, Aigle B, Gallois A, Mangenot S, Gerbaud C, Truong C, Francou FX, Fourrier C, Guerineau M, Decaris B et al (2006) Evolution of the terminal regions of the streptomyces linear chromosome. Mol Biol Evol 23:2361–2369
Cooper VS, Lenski RE (2000) The population genetics of ecological specialization in evolving *Escherichia coli* populations. Nature 407:736–739
Cooper TF, Rozen DE, Lenski RE (2003) Parallel changes in gene expression after 20,000 generations of evolution in Escherichiacoli. Proc Natl Acad Sci U S A 100:1072–1077
Cooper TF, Remold SK, Lenski RE, Schneider D (2008) Expression profiles reveal parallel evolution of epistatic interactions involving the CRP regulon in Escherichia coli. PLoS Genet 4:e35
Cordero OX, Wildschutte H, Kirkup B, Proehl S, Ngo L, Hussain F, Le Roux F, Mincer T, Polz MF (2012) Ecological populations of bacteria act as socially cohesive units of antibiotic production and resistance. Science 337:1228–1231
Crozat E, Winkworth C, Gaffe J, Hallin PF, Riley MA, Lenski RE, Schneider D (2010) Parallel genetic and phenotypic evolution of DNA superhelicity in experimental populations of *Escherichia coli*. Mol Biol Evol 27:2113–2128
Cundliffe E, Bate N, Butler A, Fish S, Gandecha A, Merson-Davies L (2001) The tylosin-biosynthetic genes of *Streptomyces fradiae*. Antonie Van Leeuwenhoek 79:229–234

Darwin C (1859) On the origin of species by means of natural selection, or the preservation of favoured races in the struggle for life. John Murray, Londres
Daubech B, Remigi P, Doin de Moura G, Marchetti M, Pouzet C, Auriac M, Gokhale C, Masson-Boivin C, Capela D (2017) Spatio-temporal control of mutualism in legumes helps spread symbiotic nitrogen fixation. elife 6:e28683
Davies J, Davies D (2010) Origins and evolution of antibiotic resistance. Microbiol Mol Biol Rev 74:417–433
Davies J, Spiegelman GB, Yim G (2006) The world of subinhibitory antibiotic concentrations. Curr Opin Microbiol 9:445–453
Denamur E, Matic I (2006) Evolution of mutation rates in bacteria. Mol Microbiol 60:820–827
Dibrova DV, Galperin MY, Mulkidjanian AY (2014) Phylogenomic reconstruction of archaeal fatty acid metabolism. Environ Microbiol 16:907–918
Donia MS, Fischbach MA (2015) Human Microbiota. Small molecules from the human microbiota. Science 349:1254766
Eldredge N, Gould S (1973) Punctuated equilibria: an alternative to phyletic gradualism. In: Schopf TJM (ed) Models in paleobiology. Freeman, Cooper & Co, San Francisco
Elena SF, Lenski RE (2003) Evolution experiments with microorganisms: the dynamics and genetic bases of adaptation. Nat Rev Genet 4:457–469
Eriksen DT, Hsieh PC, Lynn P, Zhao H (2013) Directed evolution of a cellobiose utilization pathway in *Saccharomyces cerevisiae* by simultaneously engineering multiple proteins. Microb Cell Factories 12:61
Erill I, Campoy S, Mazon G, Barbe J (2006) Dispersal and regulation of an adaptive mutagenesis cassette in the bacteria domain. Nucleic Acids Res 34:66–77
Fajardo A, Martinez-Martin N, Mercadillo M, Galan JC, Ghysels B, Matthijs S, Cornelis P, Wiehlmann L, Tummler B, Baquero F et al (2008) The neglected intrinsic resistome of bacterial pathogens. PLoS One 3:e1619
Ferenci T (2008) Bacterial physiology, regulation and mutational adaptation in a chemostat environment. Adv Microb Physiol 53:169–229
Fiegna F, Yu YT, Kadam SV, Velicer GJ (2006) Evolution of an obligate social cheater to a superior cooperator. Nature 441:310–314
Fischbach MA, Walsh CT, Clardy J (2008) The evolution of gene collectives: how natural selection drives chemical innovation. Proc Natl Acad Sci U S A 105:4601–4608
Fleischmann RD, Adams MD, White O, Clayton RA, Kirkness EF, Kerlavage AR, Bult CJ, Tomb JF, Dougherty BA, Merrick JM et al (1995) Whole-genome random sequencing and assembly of *Haemophilus influenzae* Rd. Science 269:496–512
Foster PL (2007) Stress-induced mutagenesis in bacteria. Crit Rev Biochem Mol Biol 42:373–397
Gage DJ (2002) Analysis of infection thread development using Gfp- and DsRed-expressing *Sinorhizobium meliloti*. J Bacteriol 184:7042–7046
GBD DaHC (2015) Global, regional, and national disability-adjusted life years(DALYs) for 306 diseases and injuries and healthy life expectancy (HALE) for 188 countries, 1990-2013: quantifying the epidemiological transition. Lancet 386:2145–2191
Genin S, Denny TP (2012) Pathogenomics of the *Ralstonia solanacearum* species complex. Annu Rev Phytopathol 50:67–89
Giovannoni SJ, Cameron Thrash J, Temperton B (2014) Implications of streamlining theory for microbial ecology. ISME J 8:1553–1565
Guan SH, Gris C, Cruveiller S, Pouzet C, Tasse L, Leru A, Maillard A, Medigue C, Batut J, Masson-Boivin C et al (2013) Experimental evolution of nodule intracellular infection in legume symbionts. ISME J 7:1367–1377
Gyaneshwar P, Hirsch AM, Moulin L, Chen WM, Elliott GN, Bontemps C, Estrada-de Los Santos P, Gross E, Dos Reis FB, Sprent JI et al (2011) Legume-nodulating betaproteobacteria: diversity, host range, and future prospects. Mol Plant-Microbe Interact 24:1276–1288
Hindré T, Knibbe C, Beslon G, Schneider D (2012) New insights into bacterial adaptation through in vivo and in silico experimental evolution. Nat Rev Microbiol 10:352–365

Hopwood D, Kieser T (1993) Conjugative plasmids of *Streptomyces*. In: Clewell D (ed) Bacterial conjugation. Plenum Press, New York, pp 293–311

Hungria M, Menna P, Delamuta JRM (2015) Bradyrhizobium, the ancestor of all rhizobia: Phylogeny of housekeeping and nitrogen-fixation genes. In: de Bruijn FJ (ed) Biological nitrogen fixation. Wiley, Hoboken. https://doi.org/10.1002/9781119053095.ch18

Ippoliti PJ, Delateur NA, Jones KM, Beuning PJ (2012) Multiple strategies for translesion synthesis in bacteria. Cell 1:799–831

Jenke-Kodama H, Dittmann E (2005) Combinatorial polyketide biosynthesis at higher stage. Mol Syst Biol 1:2005 0025

Jenke-Kodama H, Sandmann A, Muller R, Dittmann E (2005) Evolutionary implications of bacterial polyketide synthases. Mol Biol Evol 22:2027–2039

Jones CL, Clancy M, Honnold C, Singh S, Snesrud E, Onmus-Leone F, McGann P, Ong AC, Kwak Y, Waterman P et al (2015) Fatal outbreak of an emerging clone of extensively drug-resistant Acinetobacter baumannii with enhanced virulence. Clin Infect Dis 61:145–154

Kaplan DL, Kaplan AM (1982) Thermophilic biotransformations of 2,4,6-trinitrotoluene under simulated composting conditions. Appl Environ Microbiol 44:757–760

Kaufman DD, Kearney PC (1970) Microbial degradation of s-triazine herbicides. Residue Rev 32:235–265

Kawecki TJ, Lenski RE, Ebert D, Hollis B, Olivieri I, Whitlock MC (2012) Experimental evolution. Trends Ecol Evol 27:547–560

Khan AI, Dinh DM, Schneider D, Lenski RE, Cooper TF (2011) Negative epistasis between beneficial mutations in an evolving bacterial population. Science 332:1193–1196

Kibota TT, Lynch M (1996) Estimate of the genomic mutation rate deleterious to overall fitness in *E. coli*. Nature 381:694–696

Kiers ET, Rousseau RA, West SA, Denison RF (2003) Host sanctions and the legume-rhizobium mutualism. Nature 425:78–81

Kinashi H (2011) Giant linear plasmids in *Streptomyces*: a treasure trove of antibiotic biosynthetic clusters. J Antibiot (Tokyo) 64:19–25

Kinkel LL, Schlatter DC, Xiao K, Baines AD (2014) Sympatric inhibition and niche differentiation suggest alternative coevolutionary trajectories among *Streptomycetes*. ISME J 8:249–256

Kirby WM (1944) Extraction of a highly potent penicillin inactivator from penicillin resistant *Staphylococci*. Science 99:452–453

Knopp M, Andersson DI (2015) Amelioration of the fitness costs of antibiotic resistance due to reduced outer membrane permeability by upregulation of alternative porins. Mol Biol Evol 32:3252–3263

Kryazhimskiy S, Tkacik G, Plotkin J (2009) The dynamics of adaptation on correlated fitness landscapes. Proc Natl Acad Sci U S A 106:18638–18643

Kurth D, Belfiore C, Gorriti MF, Cortez N, Farias ME, Albarracin VH (2015) Genomic and proteomic evidences unravel the UV-resistome of the poly-extremophile Acinetobacter sp. Ver3. Front Microbiol 6:328

Laureti L, Song L, Huang S, Corre C, Leblond P, Challis GL, Aigle B (2011) Identification of a bioactive 51-membered macrolide complex by activation of a silent polyketide synthase in *Streptomyces ambofaciens*. Proc Natl Acad Sci U S A 108:6258–6263

Le Gac M, Plucain J, Hindre T, Lenski RE, Schneider D (2012) Ecological and evolutionary dynamics of coexisting lineages during a long-term experiment with *Escherichia coli*. Proc Natl Acad Sci U S A 109:9487–9492

Lee HH, Hsu CC, Lin YL, Chen CW (2011) Linear plasmids mobilize linear but not circular chromosomes in *Streptomyces*: support for the 'end first' model of conjugal transfer. Microbiology 157:2556–2568

Lenski RE (1991) Quantifying fitness and gene stability in microorganisms. Biotechnology 15:173–192

Lewis K (2013) Platforms for antibiotic discovery. Nat Rev Drug Discov 12:371–387

Linares JF, Gustafsson I, Baquero F, Martinez JL (2006) Antibiotics as intermicrobial signaling agents instead of weapons. Proc Natl Acad Sci U S A 103:19484–19489

Lind P, Farr A, Rainey P (2015) Experimental evolution reveals hidden diversity in evolutionary pathways. elife 4:e07074

Liu Z, Zhao X, Bai F (2013) Production of xylanase by an alkaline-tolerant marine-derived Streptomyces viridochromogenes strain and improvement by ribosome engineering. Appl Microbiol Biotechnol 97:4361–4368

Long A, Liti G, Luptak A, Tenaillon O (2015) Elucidating the molecular architecture of adaptation via evolve and resequence experiments. Nat Rev Genet 16:567–582

Macdonald K, Hutchinson J, Gillett W (1964) Properties of heterozygous diploids between strains of *Penicillium chrysogenum* selected for high penicillin yield. Antonie Van Leeuwenhoek 30:209–224

Marchetti M, Capela D, Glew M, Cruveiller S, Chane-Woon-Ming B, Gris C, Timmers T, Poinsot V, Gilbert LB, Heeb P et al (2010) Experimental evolution of a plant pathogen into a legume symbiont. PLoS Biol 8:e1000280

Marchetti M, Jauneau A, Capela D, Remigi P, Gris C, Batut J, Masson-Boivin C (2014) Shaping bacterial symbiosis with legumes by experimental evolution. Mol Plant-Microbe Interact 27:956–964

Marchetti M, Clerissi C, Yousfi Y, Gris C, Bouchez O, Rocha E, Cruveiller S, Jauneau A, Capela D, Masson-Boivin C (2017) Experimental evolution of rhizobia may lead to either extra- or intracellular symbiotic adaptation depending on the selection regime. Mol Ecol 26:1818–1831

Martinez JL, Fajardo A, Garmendia L, Hernandez A, Linares JF, Martinez-Solano L, Sanchez MB (2009) A global view of antibiotic resistance. FEMS Microbiol Rev 33:44–65

Masson-Boivin C, Sachs JL (2017) Symbiotic nitrogen fixation by rhizobia-the roots of a success story. Curr Opin Plant Biol 44:7–15

Masson-Boivin C, Giraud E, Perret X, Batut J (2009) Establishing nitrogen-fixing symbiosis with legumes: how many rhizobium recipes? Trends Microbiol 17:458–466

McArthur AG, Waglechner N, Nizam F, Yan A, Azad MA, Baylay AJ, Bhullar K, Canova MJ, De Pascale G, Ejim L et al (2013) The comprehensive antibiotic resistance database. Antimicrob Agents Chemother 57:3348–3357

Merouch SO, Minasov G, Lee W, Shoichet BK, Mobashery S (2003) Structural aspects for evolution of beta-lactamases from penicillin-binding proteins. J Am Chem Soc 125:9612–9618

Moulin L, Munive A, Dreyfus B, Boivin-Masson C (2001) Nodulation of legumes by members of the beta-subclass of Proteobacteria. Nature 411:948–950

Nesme J, Cecillon S, Delmont TO, Monier JM, Vogel TM, Simonet P (2014) Large-scale metagenomic-based study of antibiotic resistance in the environment. Curr Biol 24:1096–1100

Ochman H, Moran NA (2001) Genes lost and genes found: evolution of bacterial pathogenesis and symbiosis. Science 292:1096–1099

Oldroyd GE, Murray JD, Poole PS, Downie JA (2011) The rules of engagement in the legume-rhizobial symbiosis. Annu Rev Genet 45:119–144

Palmer KL, Kos VN, Gilmore MS (2010) Horizontal gene transfer and the genomics of enterococcal antibiotic resistance. Curr Opin Microbiol 13:632–639

Pandza S, Biukovic G, Paravic A, Dadbin A, Cullum J, Hranueli D (1998) Recombination between the linear plasmid pPZG101 and the linear chromosome of *Streptomyces rimosus* can lead to exchange of ends. Mol Microbiol 28:1165–1176

Pang X, Aigle B, Girardet JM, Mangenot S, Pernodet JL, Decaris B, Leblond P (2004) Functional angucycline-like antibiotic gene cluster in the terminal inverted repeats of the *Streptomyces ambofaciens* linear chromosome. Antimicrob Agents Chemother 48:575–588

Peixoto J, Silva LP, Kruger RH (2017) Brazilian Cerrado soil reveals an untapped microbial potential for unpretreated polyethylene biodegradation. J Hazard Mater 324:634–644

Perrier A, Peyraud R, Rengel D, Barlet X, Lucasson E, Gouzy J, Peeters N, Genin S, Guidot A (2016) Enhanced in planta fitness through adaptive mutations in EfpR, a dual regulator of virulence and metabolic functions in the plant pathogen Ralstonia solanacearum. PLoS Pathog 12:e1006044

Plucain J, Hindre T, Le Gac M, Tenaillon O, Cruveiller S, Medigue C, Leiby N, Harcombe WR, Marx CJ, Lenski RE et al (2014) Epistasis and allele specificity in the emergence of a stable polymorphism in *Escherichia coli*. Science 343:1366–1369

Poulin-Laprade D, Burrus V (2015) A lambda cro-like repressor is essential for the induction of conjugative transfer of SXT/R391 elements in response to DNA damage. J Bacteriol 197:3822–3833

Rainey PB, Rainey K (2003) Evolution of cooperation and conflict in experimental bacterial populations. Nature 425:72–74

Rainey PB, Travisano M (1998) Adaptive radiation in a heterogeneous environment. Nature 394:69–72

Raynes Y, Sniegowski PD (2014) Experimental evolution and the dynamics of genomic mutation rate modifiers. Heredity (Edinb) 113:375–380

Remigi P, Capela D, Clerissi C, Tasse L, Torchet R, Bouchez O, Batut J, Cruveiller S, Rocha EPC, Masson-Boivin C (2014) Transient hypermutagenesis accelerates the evolution of legume endosymbionts following horizontal gene transfer. PLoS Biol 12:e1001942

Remigi P, Zhu J, Young JPW, Masson-Boivin C (2016) Symbiosis within symbiosis: evolving nitrogen-fixing legume symbionts. Trends Microbiol 24:63–75

Rozen DE, Lenski RE (2000) Long-term experimental evolution in *Escherichia coli*. VIII. Dynamics of a balanced polymorphism. Am Nat 155:24–35

Sanger F, Air GM, Barrell BG, Brown NL, Coulson AR, Fiddes CA, Hutchison CA, Slocombe PM, Smith M (1977) Nucleotide sequence of bacteriophage phi X174 DNA. Nature 265:687–695

Scanlan PD, Hall AR, Lopez-Pascua LD, Buckling A (2011) Genetic basis of infectivity evolution in a bacteriophage. Mol Ecol 20:981–989

Schauner C, Dary A, Lebrihi A, Leblond P, Decaris B, Germain P (1999) Modulation of lipid metabolism and spiramycin biosynthesis in *Streptomyces ambofaciens* unstable mutants. Appl Environ Microbiol 65:2730–2737

Schluter D (2000) The ecology of adaptive radiation. Oxford University Press, New York

Schneider D, Duperchy E, Coursange E, Lenski RE, Blot M (2000) Long-term experimental evolution in *Escherichia coli*. IX. Characterization of insertion sequence-mediated mutations and rearrangements. Genetics 156:477–488

Schoenafinger G, Marahiel M (2012) Nonribosomal peptides: biosynthesis. In: Natural products in chemical biology. https://doi.org/10.1002/9780470048672.wecb398 (ed. I John Wiley and Sons)

Schwarz S, Kehrenberg C, Ojo KK (2002) *Staphylococcus sciuri* gene *erm*(33), encoding inducible resistance to macrolides, lincosamides, and streptogramin B antibiotics, is a product of recombination between *erm*(C) and *erm*(A). Antimicrob Agents Chemother 46:3621–3623

Sherman DR, Mdluli K, Hickey MJ, Arain TM, Morris SL, Barry CE 3rd, Stover CK (1996) Compensatory *ahpC* gene expression in isoniazid-resistant *Mycobacterium tuberculosis*. Science 272:1641–1643

Simon L, Bousquet J, Levesque RC, Lalonde M (1993) Origin and diversification of endomycorrhizal fungi and coincidence with vascular land plant. Nature 363:67–69

Staunton J, Weissman KJ (2001) Polyketide biosynthesis: a millennium review. Nat Prod Rep 18:380–416

Stonesifer J, Matsushima P, Baltz RH (1986) High frequency conjugal transfer of tylosin genes and amplifiable DNA in *Streptomyces fradiae*. Mol Gen Genet 202:348–355

Stracke S, Kistner C, Yoshida S, Mulder L, Sato S, Kaneko T, Tabata S, Sandal N, Stougaard J, Szczyglowski K et al (2002) A plant receptor-like kinase required for both bacterial and fungal symbiosis. Nature 417:959–962

Sullivan J, Ronson C (1998) Evolution of rhizobia by acquisition of a 500-kb symbiosis island that integrates into a phe-tRNA gene. Proc Natl Acad Sci U S A 95:5145–5149

Tenaillon O, Rodríguez-Verdugo A, Gaut RL, McDonald P, Bennett AF, Long AD, Gaut BS (2012) The molecular diversity of adaptive convergence. Science 335:457–461

Tenaillon O, Barrick J, Ribeck N, Deatherage D, Blanchard J, Dasgupta A, Wu G, Wielgoss S, Cruveiller S, Medigue C et al (2016) Tempo and mode of genome evolution in a 50,000-generation experiment. Nature 536:165–170

Thibessard A, Haas D, Gerbaud C, Aigle B, Lautru S, Pernodet JL, Leblond P (2015) Complete genome sequence of *Streptomyces ambofaciens* ATCC 23877, the spiramycin producer. J Biotechnol 214:117–118

Toleman M, Spencer J, Jones L, Walsh T (2012) blaNDM-1 is a chimera likely constructed in Acinetobacter baumannii. Antimicrob Agents Chemother 56:2773–2776

Torres-Barcelo C, Kojadinovic M, Moxon R, MacLean RC (2015) The SOS response increases bacterial fitness, but not evolvability, under a sublethal dose of antibiotic. Proc Biol Sci 282:20150885

Torsvik V, Ovreas L, Thingstad TF (2002) Prokaryotic diversity—magnitude, dynamics, and controlling factors. Science 296:1064–1066

Van Hofwegen DJ, Hovde CJ, Minnich SA (2016) Rapid evolution of citrate utilization by *Escherichia coli* by direct selection requires *citT* and *dctA*. J Bacteriol 198:1022–1034

Vaz Jauri P, Bakker MG, Salomon CE, Kinkel LL (2013) Subinhibitory antibiotic concentrations mediate nutrient use and competition among soil streptomyces. PLoS One 8:e81064

Velicer GJ, Kroos L, Lenski RE (1998) Loss of social behaviors by *Myxococcus xanthus* during evolution in an unstructured habitat. Proc Natl Acad Sci U S A 95:12376–12380

Velicer G, Kroos L, Lenski R (2000) Developmental cheating in the social bacterium *Myxococcus xanthus*. Nature 404:598–601

Velicer GJ, Raddatz G, Keller H, Deiss S, Lanz C, Dinkelacker I, Schuster SC (2006) Comprehensive mutation identification in an evolved bacterial cooperator and its cheating ancestor. Proc Natl Acad Sci U S A 103:8107–8112

von Linné C (1737) Genera plantarum: eorumque characteres naturales secundum numerum, figuram, situm, et proportionem omnium fructificationis partium. Lugduni Batavorum & Apud Conradum Wishoff, Leiden

Watanabe T (1963) Infective heredity of multiple drug resistance in bacteria. Bacteriol Rev 27:87–115

Weinreich DM, Delaney NF, Depristo MA, Hartl DL (2006) Darwinian evolution can follow only very few mutational paths to fitter proteins. Science 312:111–114

Wichman HA, Badgett MR, Scott LA, Boulianne CM, Bull JJ (1999) Different trajectories of parallel evolution during viral adaptation. Science 285:422–424

Wielgoss S, Barrick JE, Tenaillon O, Cruveiller S, Chane-Woon-Ming B, Medigue C, Lenski RE, Schneider D (2011) Mutation rate inferred from synonymous substitutions in a long-term evolution experiment with *Escherichia coli*. G3 genes. Genomes Genet (Bethesda) 1:183–186

Wielgoss S, Barrick JE, Tenaillon O, Wiser MJ, Dittmar WJ, Cruveiller S, Chane-Woon-Ming B, Medigue C, Lenski RE, Schneider D (2013) Mutation rate dynamics in a bacterial population reflect tension between adaptation and genetic load. Proc Natl Acad Sci U S A 110:222–227

Wiser MJ, Ribeck N, Lenski RE (2013) Long-term dynamics of adaptation in asexual populations. Science 342:1364–1367

Woods R, Schneider D, Winkworth CL, Riley MA, Lenski RE (2006) Tests of parallel molecular evolution in a long-term experiment with *Escherichia coli*. Proc Natl Acad Sci U S A 103:9107–9112

Woods RJ, Barrick JE, Cooper TF, Shrestha U, Kauth MR, Lenski RE (2011) Second-order selection for evolvability in a large Escherichia coli population. Science 331:1433–1436

Yamasaki M, Kinashi H (2004) Two chimeric chromosomes of *Streptomyces coelicolor* A3(2) generated by single crossover of the wild-type chromosome and linear plasmid scp1. J Bacteriol 186:6553–6559

Yang L, Jelsbak L, Marvig RL, Damkiaer S, Workman CT, Rau MH, Hansen SK, Folkesson A, Johansen HK, Ciofu O et al (2011) Evolutionary dynamics of bacteria in a human host environment. Proc Natl Acad Sci U S A 108:7481–7486

Yim G, Wang HH, Davies J (2007) Antibiotics as signalling molecules. Philos Trans R Soc Lond Ser B Biol Sci 362:1195–1200

Yu YT, Yuan X, Velicer GJ (2010) Adaptive evolution of an sRNA that controls *Myxococcus* development. Science 328:993

Ziemert N, Lechner A, Wietz M, Millan-Aguinaga N, Chavarria KL, Jensen PR (2014) Diversity and evolution of secondary metabolism in the marine actinomycete genus *Salinispora*. Proc Natl Acad Sci U S A 111:E1130–E1139

Zotchev S (2014) Genomics-based insights into the evolution of secondary metabolite biosynthesis in Actinomycete Bacteria. In: Pontarotti P (ed) Evolutionary biology: genome evolution, speciation, coevolution and origin of life. Springer, Cham, pp 35–45

Index

J.-C. Bertrand et al. (eds.), *Prokaryotes and Evolution*,
https://doi.org/10.1007/978-3-319-99784-1

9783319997827